Probability and Mathematical Statistics (Continued)

WILLIAMS • Diffusions, Markov Processes, and Marting[illegible] Foundations
ZACKS • Theory of Statistical Inference

Applied Probability and Statistics

ANDERSON, AUQUIER, HAUCK, OAKES, VANDAELE, and WEISBERG • Statistical Methods for Comparative Studies
ARTHANARI and DODGE • Mathematical Programming in Statistics
BAILEY • THE Elements of Stochastic Processes with Applications to the Natural Sciences
BAILEY • Mathematics, Statistics and Systems for Health
BARNETT • Interpreting Multivariate Data
BARNETT and LEWIS • Outliers in Statistical Data
BARTHOLOMEW • Stochastic Models for Social Processes, *Third Edition*
BARTHOLOMEW and FORBES • Statistical Techniques for Manpower Planning
BECK and ARNOLD • Parameter Estimation in Engineering and Science
BELSLEY, KUH, and WELSCH • Regression Diagnostics: Identifying Influential Data and Sources of Collinearity
BENNETT and FRANKLIN • Statistical Analysis in Chemistry and the Chemical Industry
BHAT • Elements of Applied Stochastic Processes
BLOOMFIELD • Fourier Analysis of Time Series: An Introduction
BOX • R. A. Fisher, The Life of a Scientist
BOX and DRAPER • Evolutionary Operation: A Statistical Method for Process Improvement
BOX, HUNTER, and HUNTER • Statistics for Experimenters: An Introduction to Design, Data Analysis, and Model Building
BROWN and HOLLANDER • Statistics: A Biomedical Introduction
BROWNLEE • Statistical Theory and Methodology in Science and Engineering, *Second Edition*
BURY • Statistical Models in Applied Science
CHAMBERS • Computational Methods for Data Analysis
CHATTERJEE and PRICE • Regression Analysis by Example
CHOW • Analysis and Control of Dynamic Economic Systems
CHOW • Econometric Analysis by Control Methods
CLELLAND, BROWN, and deCANI • Basic Statistics with Business Applications, *Second Edition*
COCHRAN • Sampling Techniques, *Third Edition*
COCHRAN and COX • Experimental Designs, *Second Edition*
CONOVER • Practical Nonparametric Statistics, *Second Edition*
CONOVER and IMAN • Introduction to Modern Business Statistics
CORNELL • Experiments with Mixtures: Designs, Models and The Analysis of Mixture Data
COX • Planning of Experiments
DANIEL • Biostatistics: A Foundation for Analysis in the Health Sciences, *Third Edition*
DANIEL • Applications of Statistics to Industrial Experimentation
DANIEL and WOOD • Fitting Equations to Data: Computer Analysis of Multifactor Data, *Second Edition*
DAVID • Order Statistics, *Second Edition*
DAVIDSON • Multidimensional Scaling
DEMING • Sample Design in Business Research
DODGE and ROMIG • Sampling Inspection Tables, *Second Edition*
DOWDY and WEARDEN • Statistics for Research
DRAPER and SMITH • Applied Regression Analysis, *Second Edition*
DUNN • Basic Statistics: A Primer for the Biomedical Sciences, *Second Edition*
DUNN and CLARK • Applied Statistics: Analysis of Variance and Regression
ELANDT-JOHNSON • Probability Models and Statistics Methods in Genetics
ELANDT-JOHNSON and JOHNSON • Survival Models and Data Analysis

continued on back

University of Birming
BU 13194321

BU 13194321

BIRMINGHAM UNIVERSITY LIBRARY
WITHDRAWN

PER ARDUA
AD ALTA

TO JIM AND BETTY

Preface

This textbook is designed for the population of students we have encountered while teaching a two-semester introductory statistical methods course for graduate students. These students come from a variety of research disciplines in the natural and social sciences. Most of the students have no prior background in statistical methods but will need to use some, or all, of the procedures discussed in this book before they complete their studies. Therefore, we attempt to provide not only an understanding of the concepts of statistical inference but also the methodology for the most commonly used analytical procedures.

Experience has taught us that students ought to receive their instruction in statistics early in their graduate program, or perhaps, even in their senior year as undergraduates. This ensures that they will be familiar with statistical terminology when they begin critical reading of research papers in their respective disciplines and with statistical procedures before they begin their research. We frequently find, however, that graduate students are poor with respect to mathematical skills; it has been several years since they completed their undergraduate mathematics and they have not used these skills in the subsequent years. Consequently, we have found it helpful to give details of mathematical techniques as they are employed, and we do so in this text.

We should like our students to be aware that statistical procedures are based on sound mathematical theory. But we have learned from our students, and from those with whom we consult, that research workers do not share the mathematically oriented scientists' enthusiasm for elegant proofs of theorems. So we deliberately avoid not only theoretical proofs but even too much of a mathematical tone. When statistics was in its infancy, W. S. Gosset replied to an explanation of the sampling distribution of the partial correlation coefficient by R. A. Fisher:†

† From letter No. 6, May 5, 1922, in *Letters From W. S. Gosset to R. A. Fisher 1915–1936*, Arthur Guinness Sons and Company, Ltd., Dublin. Issued for private circulation.

> . . . I fear that I can't conscientiously claim to understand it, but I take it for granted that you know what you are talking about and thankfully use the results!
>
> It's not so much the mathematics, I can often say "Well, of course, that's beyond me, but we'll take it as correct" but when I come to 'Evidently' I know that means two hours hard work at least before I can see why.

Considering that the original "Student" of statistics was concerned about whether he could understand the mathematical underpinnings of the discipline, it is reasonable that today's students have similar misgivings. Lest this concern keep our students from appreciating the importance of statistics in research, we consciously avoid theoretical mathematical discussions.

We want to show the importance of statistics in research, and we have taken two specific measures to accomplish this goal. First, to explain that statistics is an integral part of research, we show from the very first chapter of the text how it is used. We have found that our students are impatient with textbooks that require eight weeks of preparatory work before any actual application of statistics to relevant problems. Thus, we have eschewed the traditional introductory discussion of probability and descriptive statistics; these topics are covered only as they are needed. Second, we try to present a practical example of each topic as soon as possible, often with considerable detail about the research problem. This is particularly helpful to those who enroll in the statistical methods course before the research methods course in their particular discipline. Many of the examples and exercises are based on actual research situations that we have encountered in consulting with research workers. We attempt to provide data that are reasonable but that are simplified for ease of computation. We realize that in an actual research project a statistical package on a computer will probably be used for the computations, and we considered including printouts of computer analyses. But the multiplicity of the currently available packages, and the rapidity with which they are improved and revised, makes this infeasible.

It is probable that every course has an optimum pace at which it should be taught; we are convinced that such is the case with statistical methods. Because our students come to us unfamiliar with inductive reasoning, we start slowly and try to explain inference in considerable detail. The pace quickens, however, as soon as the students seem familiar with the concepts. Then when new concepts, such as bivariate distributions, are introduced, it is necessary to pause and reestablish the gradual acceleration. Testing helps to maintain the pace, and we find that our students benefit from frequent testing. The exercises at the end of each section are often taken directly from these tests.

A textbook can never replace a reference book. But, many people, because they are familiar with the text they used when they studied statistical methods, often refer to that book for information during later professional activities. We have kept this in mind while designing the text and have included some features that should be helpful: Summaries of procedures are clearly set off, references to articles and books that further develop the topics discussed are given at the end of each chapter, and explanations on reading the statistical tables are given in the table section.

We thank Professor Donald Butcher, Chairman of the Department of Statistics and Computer Science at West Virginia University, for his encouragement of this project. We are also grateful for the assistance of Professor George Trapp and computer science graduate students Barry Miller and Benito Herrera in the production of the statistical tables. Finally, for their helpful suggestions, we thank all of our students who studied statistical methods with us during the preliminary version of the text.

SHIRLEY DOWDY
STANLEY WEARDEN

Morgantown, West Virginia
December 1982

Contents

STATISTICS FOR RESEARCH

1

The Role of Statistics

In this chapter we informally discuss how statistics is used to answer questions raised in research. Since this is an overview, we make no attempt to give precise definitions. The formal development will follow in later chapters.

1.1. THE BASIC STATISTICAL PROCEDURE

Scientists sometimes use statistics to describe the results of an experiment or an investigation. This process is referred to as *data analysis* or *descriptive statistics*. Scientists also use statistics another way; if the entire population of interest is not accessible to them for some reason, they often observe only a portion of the population (a sample) and use statistics to answer questions about the whole population. This process is called *inferential statistics*. Statistical inference is the main focus of this book (Figure 1.1).

The basic process in inferential statistics is to assign probabilities so that we can reach conclusions. The inferences we make are either decisions or estimates about the population. The tool for making inferences is probability (Figure 1.2).

We can illustrate this process by the following simple example.

Example 1.1. Assigning Probabilities in Order to Reach a Conclusion

A behavioral biologist needs a shipment of hamsters within two weeks so that his experiment can proceed on schedule. There are three places from which he can order the animals. The quality and price are the same at all three. However, in the past he has noticed that the time required to fill an

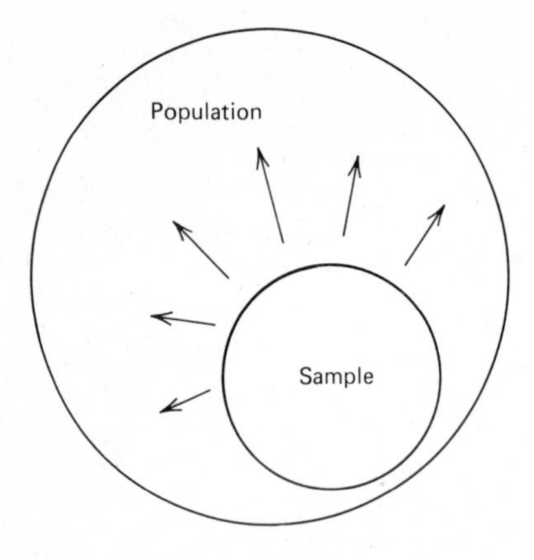

FIGURE 1.1. Statistical inference.

order varies from one supply company to another. Looking over his records, he finds that the proportions of orders that arrive within two weeks from the companies are as follows:

Supply Company:	*A*	*B*	*C*
Proportions of Orders Arriving Within Two Weeks:	0.60	0.80	0.35

These proportions are estimates of the probability of receiving an order within two weeks. Looking at these estimated probabilities, the biologist concludes that he should order the hamsters from Company *B*.

When we reach a conclusion through the use of probability, we feel comfortable about our decision but we do not have absolute certitude. For

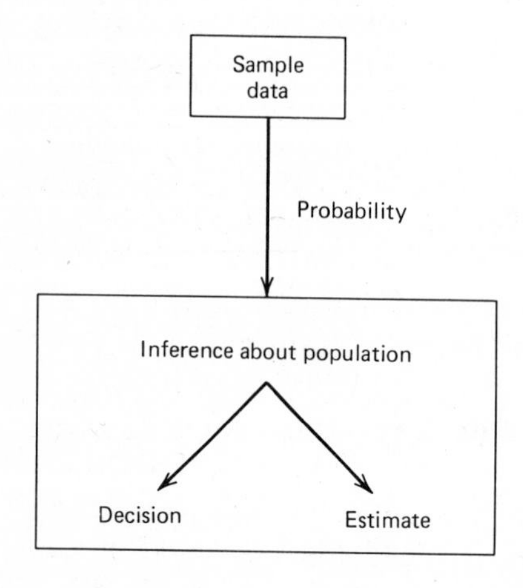

FIGURE 1.2. Statistics in research.

example, the biologist might not receive the animals in time even though he ordered them from Company *B*. It is because of the uncertainty of the outcome when this particular order is placed that probability, a measure of certainty for the total collection of orders, is used to reach a conclusion.

In the following example, we again see probability used to reach a conclusion.

Example 1.2. Using Probabilities to Make a Decision
A sociologist has two large sets of file cards, set *A* and set *B*, on which she has been recording data for her research. The sets each contain about 10,000 cards. Set *A* concerns a group of people half of whom are women. In set *B*, 80% of the cards are for women. The two files look alike. Unfortunately, the sociologist loses track of which is *A* and which is *B*. She does not want to sort and count the cards, so she decides to use probability to identify the sets. The sociologist selects a set. She draws a card at random from the selected set, notes whether or not it concerns a woman, replaces the card, and repeats this procedure 10 times. She finds that all 10 cards contain data about women. She must now decide between two possible conclusions:

1. This is set *B*.
2. This is set *A*, but an unlikely sample of cards has been chosen.

In order to decide in favor of one of these conclusions, she computes the probabilities of obtaining ten cards all for females. Using the symbol $P(E)$ to stand for the probability that the event E occurs,

$$P(\text{10 females}) = P(\text{first is female}) \times P(\text{second is female}) \times \cdots \times P(\text{tenth is female})$$

because each choice is independent of the others. For the set *A*, the probability of selecting 10 cards for females is $(0.50)^{10} = 0.00098$ (rounded to two significant digits). For set *B*, the probability of 10 cards for females is $(0.80)^{10} = 0.11$ (again rounded to two significant digits). Since the probability of all 10 of the cards being for women if the set is *B* is about 100 times the probability if the set is *A*, she decides that the set is *B*, that is, she decides in favor of the conclusion with the higher probability.

When we use a strategy based on probability, we are not guaranteed success every time. However, if we repeat the strategy, we will be correct more often than mistaken. In the above example, the sociologist could make the wrong decision because 10 cards chosen at random from set *A* could all be cards for women. In fact, in repeated experiments using set *A*, 10 cards for females will appear 0.098% of the time, that is, almost once in every thousand 10-card samples.

The examples of the hamsters and the files are artificial and oversimplified. In real life, we use statistical methods to reach conclusions about some significant aspect of research in the natural, physical, or social sciences. Statistical procedures do not furnish us with proofs, as do many mathematical techniques. Rather, statistical procedures establish probability bases on which we can accept or reject certain hypotheses.

Example 1.3. Using Probability to Reach a Conclusion in Science
A real example of the use of statistics in science is the analysis of the effectiveness of Salk's polio vaccine.

A great deal of work had to be done prior to the actual experiment and the statistical analysis. Salk first had to gather enough preliminary information and experience in his field to know which of the three polio viruses to use. He had to solve the problem of how to culture that virus. He also had to determine how long to treat the virus with formaldehyde so that it would die but retain its protein shell in the same form as the live virus; the shell could then act as an antigen to stimulate the human body to develop antibodies. At this point, Salk could conjecture that the dead virus might be used as a vaccine to give patients immunity to paralytic polio.

Finally, Salk had to decide on the type of experiment that would adequately test his conjecture. He decided on a double blind experiment in which neither patient nor doctor knew whether the patient received the vaccine or a saline solution. The patients receiving the saline solution would form the *control group*, the standard for comparison. Only after all of these preliminary steps could the experiment be carried out.

When Salk speculated that patients inoculated with the dead virus would be immune to paralytic polio, he was formulating the *experimental hypothesis*, the expected outcome if the experimenter's speculation is true. Salk wanted to use statistics to make a decision about this experimental hypothesis. The decision was to be made solely on the basis of probability. He made the decision in an indirect way; instead of considering the experimental hypothesis itself, he considered a statistical hypothesis called the *null hypothesis*—the expected outcome if the vaccine is ineffective and only chance differences are observed between the two sample groups, the inoculated group and the control group. The null hypothesis is often called the hypothesis of no difference, and it is symbolized H_0. In Salk's experiment, the null hypothesis is that the incidence of paralytic polio in the general population will be the same whether it receives the proposed vaccine or the saline solution. In symbols†

$$H_0: \quad \pi_I = \pi_C$$

† This use of the symbol π has nothing to do with the geometry of circles or the irrational number 3.1416. . . .

in which π_I is the proportion of cases of paralytic polio in the general population if it were inoculated with the vaccine, and π_C is the proportion of cases if it received the saline solution. If the null hypothesis is true, then the two sample groups in the experiment should be alike except for chance differences of exposure and contraction of the disease.

The experimental results were as follows:

	Proportion with Paralytic Polio	Number in Study
Inoculated Group	0.00016	200,745
Control Group	0.00057	201,229

The incidence of paralytic polio in the control group was almost four times higher than in the inoculated group.

Salk then found the probability that these experimental results, or more extreme ones, could have happened with a true null hypothesis. The probability that $\pi_I = \pi_C$ and the difference between the two experimental groups was caused by chance was less than 1 in 10,000,000, so Salk rejected the null hypothesis and decided that he had found an effective vaccine for the general public.†

Usually when we experiment, the results are not as conclusive as the results obtained by Salk. The probabilities will always fall between 0 and 1, and we have to establish a level below which we reject the null hypothesis and above which we accept the null hypothesis. If the probability associated with the null hypothesis is small, we reject the null hypothesis and accept an alternative hypothesis (usually the experimental hypothesis). If the probability associated with the null hypothesis is large, we accept the null hypothesis. This is one of the basic procedures of statistical methods, to ask: What is the probability that we would get these experimental results (or more extreme ones) with a true null hypothesis?

Since the experiment has already taken place, it may seem after the fact to ask for the probability that only chance caused the difference between the observed results and the null hypothesis. Actually, when we calculate the probability associated with the null hypothesis, we are asking: If this experiment were performed over and over, what is the probability that chance will produce experimental results as different as are these results from what is expected on the basis of the null hypothesis?

We should also note that Salk was interested not only in the samples of 401,974 people who took part in the study; he was interested in *all* people, then and in the future, who could receive the vaccine. He wanted to make inference to the entire population from the portion of the population that he was able to observe.

† This probability is found using a chi-square test (see Section 5.3).

Sometimes in science the inference we should like to make is not in the form of a decision about a hypothesis; rather, it consists in an estimate. For example, perhaps we want to estimate the proportion of adult Americans who approve of the way in which the President is handling the economy, and we want to include some statement about the amount of error possibly related to this estimate. Estimation of this type is another kind of inference, and it also depends on probability. For simplicity, we focus on tests of hypotheses in this introductory chapter. The first example of inference in the form of estimation is discussed in Chapter 3.

EXERCISES

1.1.1. A trial mailing is made to advertise a new science dictionary. The trial mailing list is made up of random samples of current mailing lists of several popular magazines. The number of advertisements mailed and the number of people who ordered the dictionary are as follows:

	Magazine				
	A	*B*	*C*	*D*	*E*
Mailed	900	810	1100	890	950
Ordered	18	15	10	30	45

Estimate the probability that a subscriber to each of the magazines will buy the dictionary. Make a decision about the mailing list that will probably produce the highest percentage of sales if the entire list is used.

1.1.2. If 60% of the population of the United States needs corrective lenses for proper vision, we say that the probability that an individual chosen at random from the population needs lenses is $P(L) = 0.60$.

a. Find the probability that an individual chosen at random does not need lenses. *Hint:* Use the fact that the population can be divided into two nonoverlapping groups, those who need corrective lenses and those who do not, and that the sum of the probabilities of being in these groups must equal 1.

b. If three people are chosen at random from the population, what is the probability that all three need lenses, $P(LLL)$? *Hint:* Use the multiplication law of probability, which states that the probability of an event that is the simultaneous occurrence of two or more independent elementary events is the product of the probabilities of the elementary events.

c. If three people are chosen at random from the population, what is the probability that the second person does not need lenses, but the first and the third do, $P(LNL)$?

d. If three people are chosen at random from the population, what is the probability that one out of the three needs lenses, $P(LNN$ or NLN or $NNL)$? *Hint:* Use the addition law of probability, which states that the probability of a specified outcome is the sum of the probabilities of the mutually exclusive events making up that outcome.

e. What is the probability that John Smith, a particular person, needs lenses?

1.2. THE SCIENTIFIC METHOD

The natural, physical, and social scientists who use statistical methods to reach conclusions all approach their problems by the same general procedure, the *scientific method.* The steps involved in the scientific method are:

1. State the problem.
2. Formulate the hypothesis.
3. Design the experiment or survey.
4. Make observations.
5. Interpret the data.
6. Draw conclusions.

We use statistics mainly in Step 5, "interpret the data." In an indirect way we also use statistics in Steps 2 and 3, since the formulation of the hypothesis and the design of the experiment or survey must take into consideration the type of statistical procedure to be used in analyzing the data.

The main purpose of this book is to examine Step 5. We frequently discuss the other steps, however, because an understanding of the total procedure is important. A statistical analysis may be flawless but it is not valid if data are gathered incorrectly. A statistical analysis may not even be possible if a question is formulated in such a way that a statistical hypothesis cannot be tested. Considering all of the steps also helps those who study statistical methods before they have had much practical experience in using the scientific method. A full discussion of the scientific method is outside the scope of this book, but in this section we make some comments on the five steps.

STEP 1. STATE THE PROBLEM. Sometimes when we read reports of research, we get the impression that research is a very orderly analytic process. Nothing could be further from the truth. A great deal of hidden work and also a tremendous amount of intuition are involved before a solvable problem can even be stated. Technical information and experience are indispensable before anyone can hope to formulate a reasonable problem, but they are not sufficient. The mediocre scientist and the outstanding scientist may be equally familiar with their field; the difference between them is the intuitive insight and skill that the outstanding scientist has in identifying relevant problems that he or she can reasonably hope to solve.

One simple technique for getting a problem in focus is to formulate a clear and explicit statement of the problem and put the statement in writing. This may seem like an unnecessary instruction for a research scientist; however, it is frequently not followed. The consequence is a vagueness and lack of focus that makes it almost impossible to proceed. It leads to the collection of unnecessary information or the failure to collect essential information. Sometimes the original question is even lost as the researcher gets involved in the details of the experiment.

STEP 2. FORMULATE THE HYPOTHESIS. The "hypothesis" in this step is the experimental hypothesis, the expected outcome if the experimenter's speculations are true. The experimental hypothesis must be stated in a precise way so that an experiment can be carried out that will lead to a decision about the hypothesis. A good experimental hypothesis is comprehensive enough to explain a phenomenon and predict unknown facts and yet is stated in a simple way. Classic examples of good experimental hypotheses are Mendel's laws, which can be used to explain hereditary characteristics (such as the color of flowers) and to predict what form the characteristics will take in the future.

Although the null hypothesis is not used in a formal way until the data are being interpreted, it is appropriate to formulate the null hypothesis at this time in order to verify that the experimental hypothesis is stated in such a way that it can be tested by statistical techniques.

Several experimental hypotheses may be connected with a single problem. Once these hypotheses are formulated in a satisfactory way, the investigator should do a literature search to see whether the problem has already been solved, whether or not there is hope of solving it, and whether or not the answer will make a worthwhile contribution to the field.

STEP 3. DESIGN THE EXPERIMENT OR SURVEY. Included in this step are several decisions. What *treatments* or conditions should be placed on the objects or subjects of the investigation in order to test the hypothesis? What are the *variables* of interest, that is, what variables should be measured? How will this be done? With how much precision? Each of these decisions is complex and requires experience and insight into the particular area of investigation.

Another group of decisions involves the choice of the *sample*, that portion of the population of interest that will be used in the study. The investigator tries to utilize samples that are:

(a) Random.
(b) Representative.

(c) Sufficiently large.

(d) Controlled for extraneous variables.

We discuss each of these conditions in turn by giving examples of how researchers can be misled.

In order to make a decision based on probability, it is necessary that the sample be *random*. Random samples make it possible to determine the probabilities associated with the study. A sample is random if it is just as likely that it will be picked from the population of interest as any other sample of that size. Strictly speaking, statistical inference is not possible unless random samples are used. (Specific methods for achieving random samples are discussed in Section 2.1.)

Random, however, does not mean haphazard. Haphazard processes often have hidden factors that influence the outcome. For example, one scientist using guinea pigs thought that time could be saved in choosing a treatment group and a control group by drawing the treatment group of animals from a box without looking. The scientist drew out half of the guinea pigs for testing and reserved the rest for the control group. It was noticed, however, that most of the animals in the treatment group were larger than those in the control group. Repetition of the selection process produced the same results. For some reason, perhaps because they were larger, or slower, the heavier guinea pigs were drawn first. Instead of this haphazard selection, the experimenter could have recorded the animals' ear-tattoo numbers on plastic disks and drawn the disks at random from a box.

We might think that if we use naturally occurring phenomena in the order in which they occur, they will be random. But the more we learn about naturally occurring phenomena, the more orderly and less random they seem to be. An accountant for a bank thought that the records of mortgages filed in the order in which they were issued would be random with respect to the amount of the mortgage. In fact, since the records spanned several years, the amounts were, in general, increasing because of rising real estate values.

Other examples of phenomena that occur naturally and are not random are: the sex of people waiting to enter a movie theater (many people attend movies as couples, so *MFMFMF* . . . sequences are common); the grades of examinations relative to the order in which the examination papers are given to the instructor (many good students finish early); the amount of overpayment or underpayment of income tax relative to the order in which the tax return is filed (people getting a rebate return their forms earlier than those who must pay an additional amount).

Unfortunately, in many fields of investigation random sampling is not possible; for example, meteorology, some medical research, and certain areas of economics. Random samples are the ideal, but sometimes only nonrandom data are available. In these cases the investigator may decide to proceed with statistical inference, realizing, of course, that it is somewhat risky. Any final report of such a study should include a statement of the author's awareness that the requirement of randomness for inference has not been met.

The second condition that an investigator often seeks in a sample is that it be *representative*. Usually we do not know how to find truly representative samples. Even when we think we can find them, we are often governed by a subconscious bias.

A classic example of a subconscious bias occurred at a midwestern agricultural station in the early days of statistics. Agronomists were trying to predict the yield of a certain crop in a field. To make their prediction, they chose several 6-ft by 6-ft sections of the field which they felt were representative of the crop. They harvested those sections, calculated the arithmetic average of the yields, then multiplied this average by the number of 36-ft^2 sections in the field to estimate the total yield. A statistician assigned to the station suggested that instead they should have picked random sections. After harvesting several random sections, a second average was calculated and used to predict the total yield. At harvest time, the actual yield of the field was closer to the yield predicted by the statistician. The agronomists had predicted a much larger yield, probably because they chose sections that looked like an ideal crop. An entire field, or course, is not ideal. The unconscious bias of the agronomists prevented them from picking a representative sample. Such unconscious bias cannot occur when experimental units are chosen at random.

Although representativeness is an intuitively desirable property, in practice it is usually an impossible one to meet. How can a sample of 30 possibly contain all the properties of a population of 2000 individuals? The 2000 certainly have more characteristics than can possibly be proportionately reflected in 30 individuals. So although representativeness seems necessary for proper reasoning from the sample to the population, statisticians do not rely on representative samples—rather, they rely on random samples. (Large random samples will very likely be representative.) If we do manage to deliberately construct a sample that is representative but is not random, we will be unable to compute probabilities related to the sample and, strictly speaking, we will be unable to do statistical inference.

It is also necessary that samples be *sufficiently large*. No one would question the necessity of repetition in an experiment or survey. We all know the danger of generalizing from a single observation. Sufficiently large, however, does not mean massive repetition. When we use statistics, we are trying to get information from relatively small samples. This is possible if we carry out the study wisely. To determine whether a factory was producing unbiased scales, we would be using unwise repetition if we weighed a standard on one of its scales 1000 times. Although this is a great deal of repetition, it yields information about only one scale produced by the factory. If we tested 1000 different scales one time each and computed the average, we would also be unwise; this procedure would fail to detect faulty scales if the bias were not the same from scale to scale, that is, if some scales overweighed and others underweighed. Instead, we should use several scales and several trials on each scale.

Determining a reasonable sample size for an investigation is often difficult. The size depends upon the magnitude of the difference we are trying to detect, the variability of the variable of interest, the type of statistical procedure we are using, the seriousness of the errors we might make, and the cost involved in sampling. (We make further remarks on sample size as we discuss various procedures throughout this text.)

We can satisfy the last condition mentioned for a good sample, that it be *controlled for extraneous variables*, by one of several methods. One way is to work with a homogeneous population. This is why white mice are often chosen for a preliminary experiment—they are close to uniform. If the results are satisfactory, then the experiment may be tried on the actual population of concern. The uniformity of white mice, however, also mitigates their usefulness as experimental animals. They are such a restricted population that any phenomenon found to occur in white mice cannot be inferred to occur in other populations as well. A cure for cancer in white mice may not be a cure for cancer in any other population or species. Thus a homogeneous population is primarily useful in gaining preliminary information. As in the Salk experiment, the final study must be performed on a sample from the population of interest to which the inference will be applied.

If more than one treatment group is involved, there is a second way of controlling nuisance variables. This is by balancing them across the treatment groups. We do this under the assumption that if such effects are equally distributed in all groups, they will not interfere with the detection of the effects of primary concern. An example in which this balance was *not* achieved is a test carried out by a commercial pork producer. He was asked by an agricultural pharmaceutical company to test a drug that would prevent blood-sucking insects from attacking pigs. The drug could possibly kill the pigs, or it might end up in their flesh. The pork producer valued his barrows (castrated male pigs) more than his females because the barrows were the main pigs marketed for pork, so he used mostly barrows for the control group and gave the drug mostly to females. The experiment was useless because there was no way to tell if any difference in insect attacks was due to the drug or to the different physiologies of the barrows and females. Ideally, there should have been the same ratio of barrows to female pigs in both groups. This second plan would even have allowed the experimenter to determine whether the drug was suitable or effective for one of the groups but not for the other. Of course, in order to compute probabilities, it is necessary to choose the subjects of the subgroups (female/drug, female/control, male/drug, male/control) randomly.

A third approach is to compensate mathematically for the effect of another variable. To do this, we must know the precise effect of the other variable. For example, the IQ scores of students are related to their classroom performance. If experimental classes are used to test new teaching techniques, balanced groups of students with various IQ scores can be used, or unbalanced groups can be used and a mathematical compensation made for the IQ scores. Similar compensations could be made for an experiment involving weights of men when each sample is not representative of all heights, or for the effect of age on blood pressure when the groups have not been balanced for age. This statistical compensation is accomplished by a technique called *analysis of covariance* (the subject of Chapter 13).

If it is impossible to use a homogeneous population or balance the secondary variables, or if the effects (or even the existence and number) of these other variables are not precisely known, then the experimenter uses random samples of the entire population of interest in an effort to achieve balance and thus control for other variables that might influence the outcome of the experiment. This latter design may be less powerful than a balanced design, that is, when real differences exist, they may be difficult to detect.

One further condition for an experiment or survey is not essential but very helpful: equal-sized sample groups. In most cases of multiple treatments or multiple populations, the statistical procedure is simplified if equal numbers of cases are observed from each category. This design is also usually more powerful than one with unequal sample sizes, that is, it will be more likely to detect real differences.

STEP 4. MAKE OBSERVATIONS. Once the procedure for the investigation has been decided upon, the researcher must see that it is carried out in a rigorous manner. The study should be free from all errors except random measurement errors, that is, slight variations that are due to the limitations of the measuring instrument.

Care should be taken to avoid bias. For example, bias could occur from an instrument out of calibration, an interviewer who influences the answers of a respondent, or a judge who sees the scores given by other judges. Equipment should not be changed in the middle of an experiment, nor should judges be changed halfway through an evaluation.

The data should be examined for unusual values, *outliers*, which do not seem to be consistent with the rest of the observations. Each outlier should be checked to see whether or not it is due to a recording error. If it is an error, it should be corrected. If it cannot be corrected, it should be discarded. If an outlier is not an error, it should be given special attention when the data are analyzed. For further discussion, see Barnett and Lewis (1978).

Finally, the investigator should keep a complete, legible record of the results of the investigation. All original data should be kept until the analysis is completed and the final report written. Summaries of the data are often not sufficient for a proper statistical analysis.

STEP 5. INTERPRET THE DATA. The general statistical procedure was illustrated in Example 1.3, in which the Salk vaccine experiment was discussed. To interpret the data, we set up the null hypothesis and then decide whether the experimental results are a rare outcome if the null hypothesis is true.

That is, we decide whether the difference between the experimental outcome and the null hypothesis is due to more than chance; if so, this indicates that the null hypothesis should be rejected.

If the results of the experiment are unlikely when the null hypothesis is true, we reject the null hypothesis; if they are expected, we accept the null hypothesis. We must remember, however, that statistics does not *prove* anything. Even Salk's result, with a probability of 1 in 10,000,000 that chance was causing the difference between the experimental outcome and the null hypothesis, does not prove that the null hypothesis is false. An extremely small probability, however, does make the scientist believe that the difference is not due to chance alone and that some additional mechanism is operating.

Two slightly different approaches are used to evaluate the null hypothesis. In practice, they are often intermingled. Some researchers compute the probability that the experimental results, or more extreme values, could occur if the null hypothesis is true; then they use that probability to make a judgment about the null hypothesis. In research articles this is often reported as the *observed significance level*, or the *significance level*, or the *P value*. If the P value is large, they conclude that the data are consistent with the null hypothesis. If the P value is small, then either the null hypothesis is false or the null hypothesis is true and a rare event has occurred.

Other researchers prefer a second, more decisive approach. Before the experiment they decide on a *rejection level*, the probability of an unlikely event (sometimes this is also called the *significance level*). An experimental outcome, or a more extreme one, that has a probability below this level is considered to be evidence that the null hypothesis is false. Some research articles are written with this approach. It has the advantage that only a limited number of probability tables are necessary. Without a computer, it is often difficult to determine the exact P value needed for the first approach. For this reason the second approach is emphasized in this text.

The sequence in this second procedure is:

(a) Assume H_0 is true and determine the probability P that the experimental outcome, or a more extreme one, would occur.

(b) Compare that probability to a preset rejection level symbolized by α (the Greek letter alpha).

(c) If $P \leq \alpha$, reject H_0. If $P > \alpha$, accept H_0.

If $P > \alpha$ we say, "Accept the null hypothesis." Some statisticians prefer not to use that expression, since in the absence of evidence to reject the null hypothesis they choose to simply withhold judgment about it. This group would say, "The null hypothesis may be true," or, "There is no evidence that the null hypothesis is false."

If the probability associated with the null hypothesis is very close to α, more extensive testing may be desired. Notice that this is a blend of the two approaches.

An example of the total procedure follows.

Example 1.4. Using a Statistical Procedure to Interpret Data
A manufacturer of baby food gives samples of two types of baby cereal, Type A and Type B, to a random sample of four mothers. Type A is the manufacturer's brand, Type B a competitor's. The mothers are asked to report which type they prefer. The manufacturer wants to detect any preference for their cereal if it exists.

The null hypothesis, or the hypothesis of no difference, is H_0: $\pi_A = \frac{1}{2}$, in which π_A is the proportion of mothers in the general population who prefer Type A. The experimental hypothesis, which often corresponds to a second statistical hypothesis called the *alternative hypothesis*, is that there is a preference for cereal A, H_a: $\pi_A > \frac{1}{2}$.

Suppose that four mothers are asked to choose between the two cereals. If there is no preference, the following 16 outcomes are possible with equal probability:

AAAA	*AAAB*	*ABBA*	*BBAB*
BAAA	*BBAA*	*ABAB*	*BABB*
ABAA	*BABA*	*AABB*	*ABBB*
AABA	*BAAB*	*BBBA*	*BBBB*

The manufacturer feels that only one of these 16 cases, *AAAA*, is very different from what would be expected to occur under random sampling, when the null hypothesis of no preference is true. Since the unusual case would appear only $\frac{1}{16}$ of the time when the null hypothesis is true, α (the rejection level) is set equal to $\frac{1}{16} = 0.0625$.

If the outcome of the experiment is in fact four choices of Type A, then $P = P(AAAA) = \frac{1}{16}$, and the manufacturer can say that the results are in the region of rejection, or the results are *significant*, and the null hypothesis is rejected. If the outcome is three choices of Type A, however, then $P = P(3 \text{ or more } A\text{'s}) = P(AAAB \text{ or } AABA \text{ or } ABAA \text{ or } BAAA \text{ or } AAAA) = \frac{5}{16} > \frac{1}{16}$, and he does not reject the null hypothesis. (Notice that P is the probability of this type of outcome, or a more extreme one, in the direction of the alternative hypothesis, so *AAAA* must be included.)

The way in which we set the rejection level α depends on the field of research, on the seriousness of an error, on cost, and to a great degree on tradition. In the example above, the sample size is four so an α smaller than $\frac{1}{16}$ is impossible. Later (in Section 3.2), we discuss using the seriousness of errors to determine a reasonable α. If the possible errors are not serious and cost is not a consideration, traditional values are often used.

Experimental statistics began about 1920 and was not used much until 1940, but it is already tradition bound. In the early part of this century Karl Pearson had his students at University College, London, compute tables of probabilities for reasonably rare events. Now computers are programmed to produce these tables, but the traditional levels used by Pearson persist for the most part. Tables are usually calculated for α equal to 0.10, 0.05, and 0.01. Many times there is no justification for the use of one of these values except tradition and the availability of tables. If an α close to but less than or equal to 0.05 were desired in the example above, a sample size of at least five would be necessary, then $\alpha = 1/32 = 0.03125$ if the only extreme case is *AAAAA*.

STEP 6. DRAW CONCLUSIONS. If the procedure just outlined is followed, then our decisions will be based solely on probability and will be consistent with the data from the experiment. If our experimental results are not unusual for the null hypothesis, $P > \alpha$, then the null hypothesis seems to be right and we should not reject it. If they are unusual, $P \leq \alpha$, then the null hypothesis seems to be wrong and we should reject it. We repeat that our decision could be incorrect, since there is a small probability, α, that we will reject a null hypothesis when in fact that null hypothesis is true; there is also a possibility that a false null hypothesis will be accepted. (These possible errors are discussed in Section 3.2.)

In some instances, the conclusion of the study and the statistical decision about the null hypothesis are the same. The conclusion merely states the statistical decision in specific terms. In many situations, the conclusion goes further than the statistical decision. For example, suppose that an orthodontist makes a study of malocclusion due to crowding of the adult lower front teeth. The orthodontist hypothesizes that the incidence is as common in males as in females, H_0: $\pi_M = \pi_F$. (Note that in this example the experimental hypothesis coincides with the null hypothesis.) In the data gathered, however, there is a preponderance of males and P is less than or equal to α. The statistical decision is to reject the null hypothesis; but this is not the final statement. Having rejected the null hypothesis, the orthodontist *concludes* the report by stating that this condition occurs more frequently in males than in females and advises family dentists of the need to watch more closely for tendencies of this condition in boys than in girls.

EXERCISES

1.2.1. Put the example of the cereals in the framework of the scientific method, elaborating on each of the six steps.

1.2.2. State a null and alternative hypothesis for the example of the file cards in Section 1.1., Example 1.2.

1.2.3. In the Salk experiment described in Example 1.3 of Section 1.1:
a. Why should Salk not be content just to reject the null hypothesis?
b. What conclusion could be drawn from the experiment?

1.2.4. Describe an experiment in your major field of study, placing it in the framework of the scientific method. Elaborate on:
a. Randomization.
b. Representativeness.
c. Repetition.
d. Other variables that might obscure the experimental results, and how these variables might be controlled.

1.2.5. Two college roommates decide to perform an experiment in extrasensory perception (ESP). Each produces a snapshot of his hometown girl friend, and one snapshot is placed in each of two identical brown envelopes. One of the roommates leaves the room and the other places the two envelopes side by side on the desk. The first roommate returns to the room and tries to pick the envelope that contains his girl friend's picture. The experiment is repeated 10 times. If the one who places the envelopes on the desk tosses a coin to decide which picture will go to the left and which to the right, the probabilities for correct decisions are listed below.

Number of Correct Decisions	Probability	Number of Correct Decisions	Probability
0	1/1024	6	210/1024
1	10/1024	7	120/1024
2	45/1024	8	45/1024
3	120/1024	9	10/1024
4	210/1024	10	1/1024
5	252/1024		

a. State the null hypothesis based on chance as the determining factor in a correct decision. (Make the statement in words and in symbols.)
b. State an alternative hypothesis based on the power of love.
c. If α is set as near 0.05 as possible, what is the region of rejection, that is, what numbers of correct decisions would provide evidence for ESP?

d. What is the *region of acceptance*, that is, those numbers of correct decisions that would not provide evidence of ESP?

e. Suppose the first roommate is able to pick the envelope containing his girl friend's picture 10 times out of 10; which of the following statements are true?

   i. The null hypothesis should be rejected.
   ii. He has demonstrated ESP.
   iii. Chance is not likely to produce such a result.
   iv. Love is more powerful than chance.
   v. There is sufficient evidence to suspect that something other than chance was guiding his selections.
   vi. With his luck he should raise some money and go to Las Vegas.

1.2.6. The mortality rate of a certain disease is 50% during the first year after diagnosis. The chance probabilities for the number of deaths within a year from a group of six persons with the disease are:

Number of Deaths:	0	1	2	3	4	5	6
Probability:	1/64	6/64	15/64	20/64	15/64	6/64	1/64

A new drug has been found that is helpful in cases of this disease, and it is hoped that it will lower the death rate. The drug is given to six persons who have been diagnosed as having the disease. After a year, a statistical test is performed on the outcome in order to make a decision about the effectiveness of the drug.

a. What is the null hypothesis, in words and symbols?

b. What is the alternative hypothesis, based on the prior evidence that the drug is of some help?

c. What is the region of rejection if α is set as close to 0.10 as possible?

d. What is the region of acceptance?

e. Suppose that four of the six persons die within one year. What decision should be made about the drug?

1.2.7. A company produces a new kind of instant coffee which is thought to have a superior taste when compared with the three currently most popular brands. In a preliminary random sample, 20 consumers are presented with all four kinds of coffee (in unmarked containers and in random order), and they are asked to report which one tastes best. If all four taste equally good, there is a one-in-four chance that a consumer will report that the new product tastes best. If there is no difference, the probabilities for various numbers of consumers indicating by chance that the new product is best are:

Number Picking New Product:	0	1	2	3	4
Probability:	0.003	0.021	0.067	0.134	0.190

Number Picking New Product:	5	6	7	8	9
Probability:	0.202	0.169	0.112	0.061	0.027

Number Picking New Product:	10	11	12	13–20
Probability:	0.010	0.003	0.001	less than 0.001

a. State the null and alternative hypotheses, in words and symbols.
b. If α is set as near 0.05 as possible, what is the region of rejection? What is the region of acceptance?
c. Suppose that six of the 20 consumers indicate that they prefer the new product. Which of the following statements is correct?
   i. The null hypothesis should be rejected.
   ii. The new product has a superior taste.
   iii. The new product is probably inferior because fewer than half of the people selected it.
   iv. There is insufficient evidence to support the claim that the new product has a superior taste.

1.3. EXPERIMENTS VS. SURVEYS

An *experiment* involves the collection of measurements or observations about populations that are treated or controlled by the experimenter. A *survey*, in contrast to an experiment, is an examination of a system in operation in which the investigator does not have an opportunity to assign different conditions to the objects of the study.

For example, we might use a survey to compare two countries with different types of economic systems. If there is a significant difference in some economic measure, such as per capita income, it does not mean that the economic system of one country is superior to the other. The survey takes conditions as they are and cannot control other variables that may effect the economic measure, such as comparative richness of natural resources, population health, or level of literacy. All that can be concluded is that at this particular time a significant difference exists in the economic measure. Unfortunately, surveys of this type are frequently misinterpreted.

A similar mistake could have been made in a recent survey of the life expectancy of men and women. The life expectancy of men was found to be 69 and of women, 77. Without control for risk factors—smoking, drinking, physical inactivity, stressful occupation, overweight, poor sleeping patterns, and poor life satisfaction—these results would be of little value. Fortunately, the investigators gathered information on these factors and found that women have more high-risk characteristics than men but still live longer. Because this was a carefully planned survey, the investigators were able to conclude that women biologically have greater longevity.

Surveys in general do not give answers that are as clear-cut as those of experiments. If an experiment is possible, it is preferred. For example, in order to determine which of two methods of teaching reading is more effective, we might conduct a survey of two schools that are each using a different one of the methods. But the results would be more reliable if we could conduct an experiment and set up two balanced groups within one school, teaching each group by a different method.

From this brief discussion it should not be inferred that surveys are not trustworthy. Most of the data presented as evidence for an association between heavy smoking and lung cancer come from surveys. Surveys of voter preference cause certain people to seek the presidency and others to decide not to enter the campaign. Quantitative research in many areas of social, biological, and behavioral science would be impossible without surveys. However, in surveys we must be alert to the possibility that our measurements may be affected by variables that are not of primary concern. Since we do not have as much control over these variables as we have in an experiment, we should record all concomitant information of pertinence for each observation. We can then study the effects of these other variables on the variable of interest, and possibly adjust for their effects.

EXERCISES

1.3.1. Explain how a researcher might obtain misleading results in each of the following surveys.

  a. An estimation of per capita wealth for a city is made from a random sample of people listed in the city's telephone directory.
  b. A political preference survey is made by an interviewer taking a random sample of Monday morning bank customers.
  c. The average length of fish in a lake is estimated by:
     i. The average length of fish caught, reported by anglers.
     ii. The average length of dead fish found floating in the water.
  d. The average number of words in the working vocabulary of first-grade children in a given county is estimated by a vocabulary test given to a random sample of first-grade children in the largest school in the county.

e. The frequency of people who can distinguish between two similar tones is estimated on the basis of a test given to a random sample of university students in a music appreciation class.

1.3.2. In each of the research situations described below, determine whether the researcher is conducting an experiment or a survey.

a. Traps are set out in a grain field to determine whether rabbits or raccoons are the more frequently found pests.

b. A graduate student in English literature uses random 500-word passages from the writings of Shakespeare and Marlowe to determine which author uses the conditional tense more frequently.

c. A random sample of hens is divided into two groups at random. The first group is given minute quantities of an insecticide containing an organic phosphorus compound; the second group acts as a control group. The average difference in eggshell thickness between the two groups is then determined.

d. A political scientist selects a city at random from among those that have had a recent racial incident of serious proportions. For a control group, he selects a random city of similar size that has been relatively free of racial tension. He then gives to random samples of police officers from each city a test that measures the relative racial tolerance of the police in these two cities.

e. To determine whether honeybees have a color preference in flowers, an apiarist mixes a sugar-and-water solution and puts equal amounts in two equal-sized sets of vials of different colors. Bees are introduced into a cage containing the vials, and the frequency with which bees visit vials of each color is recorded.

1.3.3. In 1971, *Time* magazine reported that El Paso's water was heavily laced with lithium, a tranquilizing chemical, whereas Dallas had a low lithium level. *Time* also reported that in 1971, FBI statistics showed that El Paso had 2,889 known crimes per 100,000 population and Dallas had 5,970 known crimes per 100,000 population. The article reported that a University of Texas biochemist felt that the reason for the lower crime rate in El Paso lay in El Paso's water. Comment on the biochemist's conjecture.

REVIEW EXERCISES

Decide whether each of the following statements is true or false. If a statement is false, explain why.

1.1. Determining a suitable sample size is never a problem when statistical techniques are being used.

1.2. To say that the null hypothesis is rejected does not necessarily mean it is false.

1.3. In a practical situation, the null hypothesis, alternative hypothesis, and the level of rejection should be specified before the experimentation.

1.4. The probability of choosing a random sample of three persons in which the first two say "yes" and the last person says "no" from a population in which $P(\text{"yes"}) = 0.7$ is (0.7)(0.7)(0.3).

1.5. If a particular couple is normal, then the probability that their next child will be a girl is approximately 0.5.

1.6. If the experimental hypothesis is true, chance does not enter into the outcome of the experiment.

1.7. The alternative hypothesis is often the experimental hypothesis.

1.8. A decision made on the basis of a statistical procedure will always be correct.

1.9. The probability of choosing a random sample of three persons in which exactly two say "yes" from a population with $P(\text{"yes"}) = 0.6$ is (0.6)(0.6)(0.4).

1.10. In the total process of investigating a question, the very first thing a scientist does is state the problem.

1.11. A scientist completes an experiment and then forms a hypothesis on the basis of the results of the experiment.

1.12. Scientific decisions cannot be made on the basis of a single observation.

1.13. In an experiment, the scientist should always collect as large an amount of data as is humanly possible.

1.14. Even a specialist in a field may not be capable of picking a sample that is truly representative, so it is better to choose a random sample.

1.15. "Random" in statistics means "haphazard."

1.16. One of the main reasons for using random sampling is to find the probability that an experiment will yield a particular outcome by chance if the null hypothesis is true.

1.17. The α level in a statistical procedure depends on the field of investigation, the cost, and the seriousness of error; however, traditional levels are often used.

1.18. A conclusion reached on the basis of a correctly applied statistical procedure is based solely on probability.

1.19. The null hypothesis may be the same as the experimental hypothesis.

1.20. The "α level" and the "region of rejection" are two expressions for the same thing.

1.21. If a correct statistical procedure is used, it is possible to reject a true null hypothesis.

1.22. The probability of rolling two sixes on two dice is 1/6 + 1/6 = 1/3.
1.23. In general, a survey is just as good as an experiment.
1.24. A weakness of many surveys is that there is little control of secondary variables.

SELECTED READINGS

Anscombe, F. J. (1960). Rejection of outliers, *Technometrics,* **2,** 123–147.

Barnard, G. A. (1947). The meaning of a significance level, *Biometrika,* **34,** 179–182.

Barnett, V., and T. Lewis (1978). *Outliers in Statistical Data*, Wiley, New York.

Berkson, J. (1942). Tests of significance considered as evidence, *Journal of the American Statistical Association,* **37,** 325–335.

Box, G. E. P. (1976). Science and statistics, *Journal of the American Statistical Association,* **71,** 791–799.

Cox, D. R. (1958). *Planning of Experiments*, Wiley, New York.

Duggan, T. J., and C. W. Dean (1968). Common misinterpretation of significance levels in sociology journals, *The American Sociologist,* **3,** 45–46.

Edgington, E. S. (1966). Statistical inference and nonrandom samples, *The Psychological Bulletin,* **66,** 485–487.

Edwards, W. (1965). Tactical note on the relation between scientific and statistical hypotheses, *The Psychological Bulletin,* **63,** 400–402.

Gibbons, J. D., and J. W. Pratt (1975). *P*-values: Interpretation and methodology, *The American Statistician,* **29,** 20–25.

Gold, D. (1969). Statistical tests and substantive significance, *The American Sociologist,* **4,** 42–46.

Greenberg, B. G. (1951). Why randomize?, *Biometrics,* **7,** 309–322.

Labovitz, S. (1968). Criteria for selecting a significance level: A note on the sacredness of .05, *The American Sociologist,* **3,** 220–222.

McGinnis, R. (1958). Randomization and inference in sociological research, *American Sociological Review,* **23,** 408–414.

Plutchik, R. (1974). *Foundations of Experimental Research*, 2nd ed., Harper & Row, New York.

Rosenberg, M. (1968). *The Logic of Survey Analysis*, Basic Books, New York.

Selvin, H. C. (1957). A critique of tests of significance in survey research, *American Sociological Review,* **22,** 519–527.

2

Populations, Samples, and Probability Distributions

In Chapter 1 we showed that statistics often plays a role in the scientific method; it is used to make inference about some characteristic of a population that is of interest. In this chapter we define some terms that are needed to explain more formally how inference is carried out in various situations.

2.1. RANDOM SAMPLING

We use the term *population* rather broadly in research. A population is commonly understood to be a natural, geographical, or political collection of people, animals, plants, or objects. Some statisticians use the word in the more restricted sense of the set of measurements of some attribute of such a collection; thus they might speak of "the population of heights of male college students." Or they might use the word to designate a set of categories of some attribute of a collection; for example, "the population of religious affiliations of U.S. government employees."

In statistical discussions, we often refer to the physical collection of interest as well as to the collection of measurements or categories derived from the physical collection. In order to clarify which type of collection is being discussed, in this book we use the term *population* as it is used by the research scientist: The population is the physical collection. The derived set of measurements or categories is called the set of *values of the variable of interest*. Thus, in the first example above, we speak of "the set of all values of the variable *height* for the population of male college students."

This distinction may seem overly precise but it is important because in a given research situation, more than one variable may be of interest in relation to the population under consideration. For example, an economist might wish to learn about the economic condition of Appalachian farmers. He first defines the population. Involved in this is specifying the geographical area "Appalachia" and deciding whether a "farmer" is the person who owns land suitable for farming, the person who works on it, or the person who makes managerial decisions about how the land is to be used. The economist's decision depends on the group in which he is interested. After he has specified the population, he must decide on the variable or variables, that characteristic or set of characteristics of these people, that will give him information about their economic condition. These characteristics might be money in savings accounts, indebtedness in mortgages or farm loans, income derived from the sale of livestock, or any of a number of other economic variables. The choice of variables will depend on the objectives of his study, the specific questions he is trying to answer. The problem of choosing characteristics that are relevant to an issue is not trivial and requires a great deal of insight and experience in any given field.

Once the population and the related variable or variables are specified, we must be careful to restrict our conclusions to this population and these variables. For example, if the above study reveals that Appalachian farm managers are heavily in debt, it cannot be inferred that owners of Kansas wheat farms are carrying heavy mortgages. Nor if Appalachian farm workers are underpaid can it be inferred that they are suffering from malnutrition, poor health, or any other condition that was not directly measured in the study.

After we have defined the population and the appropriate variable, we usually find it impractical if not impossible to observe all the values of the variable. For example, all the values of the variable *miles per gallon in city driving for this year's model of a certain type of car* could not be obtained since some of the cars probably are yet to be produced. Even if they did exist, the task of obtaining a measurement from each car is not feasible. In another example, the values of the variable *condition of all flashbulbs—good or defective—manufactured on a particular day by a certain factory* could be obtained, but this is not desirable since the product would be destroyed in the process of testing. Instead we consider a *sample* (a portion of the population), obtain measurements or observations from this sample (the *sample data*), and then use statistics to make an inference about the entire set of values. In order to carry out this inference, the sample must be random. We discussed the need for randomness in Chapter 1; here we outline the mechanics.

Most statistics departments have entire courses in which different sampling techniques and their efficiencies are studied; only a brief description of sampling can be given here. If we have a population of N items from which a sample of n is to be drawn, and we choose the n items in such a way that every combination of n items has an equally likely chance of being chosen, then this is a *completely random sample*, also called a *simple random sample*.

In an attempt to ensure that all combinations are equally likely, we often use a lottery or other gambling technique in drawing a sample. Thus, if we have five pairs of human twins in whom we wish to compare two methods of teaching speed reading, we may toss a coin to decide which twin is assigned to a particular method. Or a physiologist may have 35 frogs and want a sample of 10 for use in testing an antispasmodic drug. In one technique, he paints with vegetable dye the numerals 1 through 35 on the backs of the frogs and numbers 35 index cards with the same numerals. He then shuffles the cards and draws 10 cards. The 10 numbers determine which frogs will be in the treatment group.

Such methods are only as reliable as the gambling or lottery device used. In the 1970 military draft, each date of the year was placed in a capsule, but the capsules were separated by month to ensure that every day of every month was included. The first month's capsules were checked and placed in a container. The second month's capsules were checked and added to the container, and both groups were mixed together. Then the third month was checked, added, and mixed. This process was repeated for each of the succeeding months. Thus January was mixed 11 times, February 10 times, March 9 times, and so on. Finally, the capsules were poured into a different container and the lottery began. Young men of draft age were to be called into service in the order in which their birth dates were drawn. However, later analysis of the order indicated that those born in certain months were much more likely to be called into service than those born in other months. The Selective Service System was criticized and was unable to defend the randomness of its procedure. In 1971 the procedure was modified; it made use of two containers, one holding a capsule for every date of the year and the other the numbers from 1 to 365. Two capsules were picked at each draw, one from each container, and the number drawn indicated the order of call-up for the date drawn. This order was acceptably random.

Instead of a gambling device, the use of *random numbers* is usually advisable. If we have access to a computer, it probably has a random number generator. From this, we can obtain a random listing of n of the available N numbered items. Some hand-held calculators produce random numbers. If a computer or a random number generator is not available, many tables of random numbers are in existence. Table A.1 in the Appendix of Useful Tables at the back of this book is an example of a small table of random numbers. There are various ways to use a table of random numbers; the example that follows illustrates one method.

Example 2.1. Using a Table of Random Numbers to Choose a Random Sample

The physiologist who wants a random sample of 10 of his 35 frogs might use Table A.1 in the following fashion.

1. He points blindly to a five-digit group, say 39140 (in row 32). This number is used to determine which portion of the table will be read and to guarantee that in repeated samplings different portions of the table are used.
2. He takes the first two digits, 39, to indicate the row at which he should begin. If the first two digits had been between 51 and 00 (00 counts as 100), he would subtract 50.
3. The second two digits, 14, indicate the column at which he should begin. (Columns are composed of single digits; the five-digit groups are to help when counting columns. If the second two digits are greater than 50, he subtracts 50 to determine the column.)
4. The fifth digit is an even number, 0, so he reads the table horizontally. Had it been an odd number, he would read the table vertically.
5. He reads the table as pairs of digits because the largest number for a frog (35) requires a two-digit number. To save time, he may want to use not only 01 through 35, but also 36 through 70. In order to use this latter group, he subtracts 35 from each of its members, and the difference indicates the number of the frog to be included in the sample. He does not use values between 71 and 00 (100) because this group does not have 35 members. If he used them similarly to 36 through 70, there would then be three ways in which frogs 1 through 30 could be in the sample but only two ways that frogs 31 through 35 could be included, and the probability of selecting 1 through 30 would be higher than the probability of selecting 31 through 35.
6. The pairs of digits as he finds them in Table A.1 are as follows, with parentheses around the pair that cannot be used:

   04, (85), 50, 62, 67, (62), 24, (84), 14, (72), 26, 34, (74), 69, 03, 02

   The frogs to be included in the sample are:

   $$04,\ 50 - 35 = 15,\ 62 - 35 = 27,\ 67 - 35 = 32,\ 24$$
   $$14,\ 26,\ 34,\ (69 - 35 = 34),\ 03,\ 02$$

Table A.1 in the Appendix is suitable for most small or moderate-sized samples. Should a very large sample be required, however, one would need a list of random digits generated by a computer program or would need to refer to a published listing such as *A Million Random Digits With 100,000 Normal Deviates* by the Rand Corporation.

Sometimes it is not possible to sample from the entire population of interest because part of the population is not available for sampling. A geologist may be interested in the heavy minerals in a certain layer of sandstone in a sequence of shale, but the layer of sandstone is only available at a few exposed ledges. The rest is buried and hidden from view. Similarly, a sociologist may be interested in a characteristic of all of the families in a certain city, but the only feasible list of families for sampling purposes is a current commercially published city directory. Some families have moved into the city since the directory was compiled, and some have left. Using the directory makes it impossible to include any of the new families in the sampling process. In situations such as these, the researcher often modifies the description of the population so that it coincides with the population available for sampling. Statistical inference from the sample is made only to the available population, then a judgment is made from within the specialized area whether or not the conclusion can be applied to the entire population of interest.

There are other methods of sampling besides simple random sampling. One is *stratified random sampling*. This consists in dividing the population into groups, or *strata*, and then taking a simple random sample from each stratum. This is done to improve the accuracy of estimates, to reduce cost, or to make it possible to compare strata. The sampling is often proportional so that the sizes of the samples from the strata are proportional to the sizes of the strata.

Thus if the number of male students at a university is $N_m = 9000$ and the number of female students is $N_f = 6000$, and a survey of $n = 100$ students is desired, a proportional sample would contain $n_m = 60$ males and $n_f = 40$ females. It would be a stratified random sample if all combinations of $n_m = 60$ males are equally likely to be chosen and if all combinations of $n_f = 40$ females are equally likely. Irrespective of sampling design, the key to randomness lies in all combinations being equally likely within a component of the sample.

In this book we assume that all random samples are completely random samples. If a sampling design other than completely random sampling is employed, then adjustments of the techniques we describe are usually necessary.

EXERCISES

2.1.1. Use Table A.1 and the rules given in this section to enter the table in each of the situations below.

a. Select three of eight items if the starting point is indicated by 35201.

b. Give the first two random digits if the starting point is indicated by 78807.

c. Five of 45 items are to be selected. What are they if the starting point is indicated by 13923?
d. Select four of 25 items when the starting point is indicated by 02150.

2.1.2. Use Table A.1 to pick a random sample of 15 people out of a group of 100.

2.1.3. Use Table A.1 to pick a random sample of five mice out of a collection of 25 mice.

2.1.4. Use Table A.1 to pick a random sample of 10 out of 40 plants.

2.1.5. Use Table A.1 to pick a random sample of 25 out of 2135 people.

2.1.6. A random sample of 100 people who have telephones listed in the current Cleveland phone book is desired for an opinion survey of the telephone company's service. Describe a method of obtaining this random sample.

2.1.7. Describe a method for obtaining a random sample of 1% of all elementary school children in a given county.

2.1.8. Heights (in Inches) of 50 Male Students

Student Number (Tens)	(Units)									
	00	01	02	03	04	05	06	07	08	09
00	—	64	65	65	66	66	67	67	67	68
10	68	68	69	69	69	69	69	69	69	69
20	70	70	70	70	70	70	70	70	70	70
30	71	71	71	71	71	71	71	72	72	72
40	72	72	72	72	73	73	73	74	74	74
50	75									

a. The accompanying table represents the values of the variable *height* for a population of 50 male students. Use the table of random digits to draw a random sample of 10 men from this population, and record the corresponding sample data.
b. Compute the arithmetic average of your sample data and compare it to 70, which is the mean of the variable height for the entire population.

2.2. RECORDING VALUES OF VARIABLES

When we make observations about a sample from some population of interest, we are collecting the sample data. These data may consist in lists of measurements, tallies of particular categories, answers to questions, and so on. The attribute we are observing will take on different values, or will vary, from observation to observation, so we have been calling these attributes *variables*. Thus, collecting sample data consists in recording the various values the variables assume for each member of the sample.

We often have a choice of levels when recording the values of the variable. For example, after a class test an instructor could record various types of data depending upon how precisely he wants to perceive the students' performances. He might simply record pass or fail for each student—a rough categorization involving a low level of perception. He might choose a higher level of perception and rank the test papers from best to worst and record these ranks. If the test is a 20-point true–false test, the instructor could record the number of correct answers, a higher type of perception than ranks. If the test is an essay examination, he could use a still higher level of perception; he could think of the grades as forming a continuum, give a percentage grade to each test paper, and record these grades. A different level of perception is used in each of these cases. These levels are called the nominal scale, the ordinal scale, the discrete numerical scale, and the continuous numerical scale, respectively.

Levels of Perception	Example
Numerical scales	
Continuous	Grade in percentage
Discrete	Number correct out of 20
Ordinal scale	Rank in class
Nominal scale	Pass, fail

We are using the *nominal scale* when we put observations into categories that have no natural numerical relationship to each other. Examples are: sex; occupation; color of eyes; state of residence. When choosing categories for a nominal scale, it is necessary that there be a class for each observation and that no observation belong to more than one class.

The *ordinal scale* is a higher level of perception than the nominal scale. We are using the ordinal scale if we rank the observations. For example, we could rank the pelts of 10 foxes from the lightest color to the darkest. When the ordinal scale is used the ranks give some numerical information about the categories, but the underlying classification need not be numerical, as in this case of the color of the pelts. If the underlying categories are numerical, the difference between any two consecutive ranks need not be constant. For example, if we rank the weights of five research animals, the difference between the first and second weight might be three ounces, while the difference between the second and third weight might be only one ounce. In this example there is more precise underlying information, but we choose not to record it. If the only information available is on the ordinal scale, then it is not possible to specify the underlying difference between any two ranks.

We are using the *discrete numerical scale* when the observations are naturally numerical, the scale is uniform, and there is a built-in limit to how precisely the measurements can be taken. If data are on a discrete numerical scale, there are only a finite number of values possible, or possibly a countable infinity—as many as the counting numbers.† Examples are: the number of offspring in a litter; the number of rooms in a house; the number of quarts of milk ordered by a supermarket (the count here could be in $\frac{1}{4}$-quarts, but no more precise measurement is usually possible); the values of various coins; shoe sizes (for a fixed width); the number of wells drilled until oil is found.

The *continuous numerical scale* is the highest level of perception. A variable is continuous when it is a measurement, the scale is uniform, and it can be observed as precisely as we choose. Continuous variables theoretically can assume as many values as there are real numbers. In practice, we measure in whole numbers or to a few decimal places so the data are collected on the discrete numerical scale, but theoretically there is a more precise underlying scale of measurement. Examples are: weight; blood pressure; age; length; temperature.

If we have collected data using either numerical scale, it is possible to decrease the level of perception to the ordinal scale. For example, if the measurements are the heights in inches of five men, these measurements can be reduced to ranks. The scale could even be reduced to a nominal scale by classifying the men as tall or short.

Although we can reduce the scale from a higher to a lower level of perception, it is impossible for us to move the other way. If it is known that a certain number of men are tall and another number short, there is no way of determining how many men are 69 in. tall. It is important to be aware of this during the planning of an experiment. We must be sure to make our observations at a level high enough to give us pertinent information. If data are collected at too low a level of perception, it is impossible to recover more precise information. On the other hand, no one should go to extreme efforts to obtain a very fine measurement if this information is not necessary, or if it is distracting. For example, it is sufficient to know that an insecticide kills termites within a 24-hour period. There is no advantage to knowing whether it attains 100% mortality in 17 hours 13 minutes and 49 seconds compared with another insecticide that attains 100% mortality in 18 hours 31 minutes and 11 seconds.

Knowledge of the different levels of perception has another use in addition to enabling us to make decisions about the desired level of precision. The scale on which the data are recorded determines which statistical procedures are appropriate for analyzing the data. One set of procedures applies only to the nominal scale, another set to the ordinal scale, and still others are applicable to the discrete or the continuous numerical scale. Unless we can recognize the level of perception being used, we will be unable to choose

† The nominal and ordinal scales are also discrete.

an appropriate analysis. Chapters 3 through 5 deal with procedures for analyzing data collected on a nominal scale. The rest of the chapters deal with numerical data. Ordinal data are not treated. The reader is referred to one of the texts on nonparametric statistics in the Selected Readings for the analysis of ordinal data: Conover (1980), Daniel (1978), Hollander and Wolfe (1973).

EXERCISES

2.2.1. Which is the highest level of perception possible for each of the following variables?

a. The daily high temperature for a given year in Chicago.
b. The marital status of the applicants for a particular job.
c. Class standings at a university (freshman, sophomore, and so on).
d. Colors of roses.
e. Weights of all American-made cars.
f. Number in attendance per day at a particular high school.
g. Birthdays of people in a certain group.

2.2.2. Which of the following sets of categories are suitable for a nominal scale?

a. Female, only child, under 66 in. tall.
b. Only child, has only brothers, has only sisters, has both brothers and sisters.
c. Less than three children in a family, more than three children in a family.
d. Left-handed, right-handed.
e. Blue-eyed, female, blond.

2.2.3. Correct each of the unsuitable sets in Exercise 2.2.2.

2.2.4. In Exercise 2.1.8:

a. The level of perception used to record height for this population is the numerical scale. Is it discrete or continuous?
b. Could a higher level of perception have been employed to record the data?
c. Could height have been measured more accurately?

2.3. RANDOM VARIABLES AND PROBABILITY DISTRIBUTIONS

In Example 1.4, a test of hypothesis is carried out to determine if there is a preference for Type A baby cereal over Type B. The sample is a randomly chosen group of four mothers and the variable is recorded on the nominal scale (A or B). The test of hypothesis amounts to comparing the empirical results of sampling and recording outcomes in the real world with a theoretical *model* of what happens if the null hypothesis is true. The theoretical model is called a *probability distribution*. In this section we discuss the nature of probability distributions and how they act as models for studies that involve random sampling.

In order to develop the theoretical model for the test in Example 1.4, the possible outcomes of the study are associated with numbers, the number of mothers out of the four in the sample who prefer cereal A. The outcomes of this study are associated with 0, 1, 2, 3, or 4 (Figure 2.1). Numbers of this type, that is, those that are associated with the possible outcomes of an experiment or survey, are called the *values of the random variable y*. The random variable is the process of association. The random variable in this example is a *discrete random variable* because it has a countable number of values: 0, 1, 2, 3, 4.

To build the model, we assume that the null hypothesis is true and we determine the probability of each of the values of the random variable. Since the null hypothesis in this example is that the mothers have no preference between A and B (that is, a randomly chosen mother will prefer A with probability 1/2 and B with probability 1/2), the 16 outcomes in Figure 2.1 are equally likely. The value of the random variable is 0 if no mothers prefer A, thus the probability of 0 is 1/16 since there is only one outcome of this type ($BBBB$) among the 16 equally likely outcomes. We write $p(0) = 1/16$ to indicate that the probability that the value of the random variable will be 0 is 1/16.

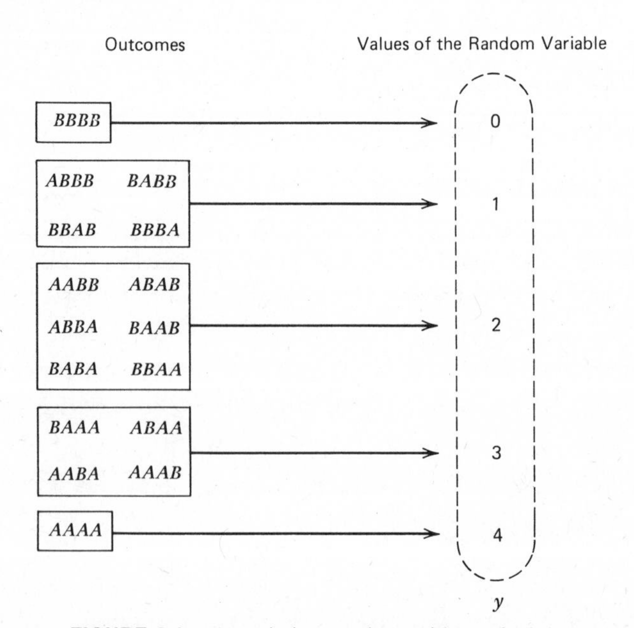

FIGURE 2.1. Associating numbers with nominal data.

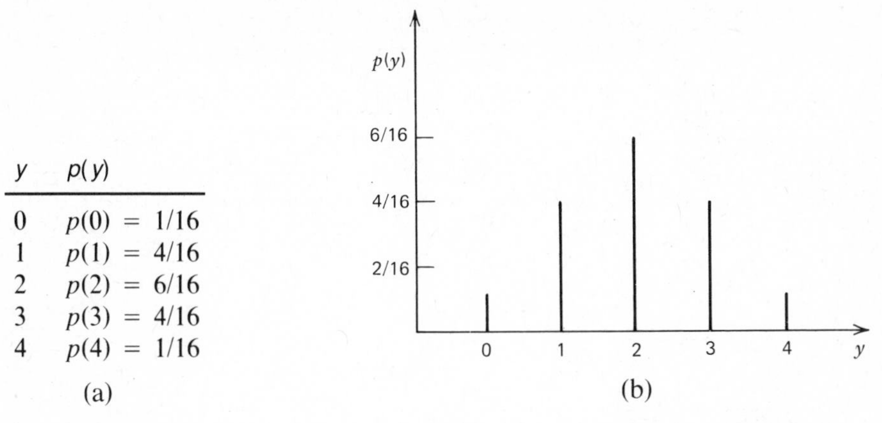

y	$p(y)$
0	$p(0) = 1/16$
1	$p(1) = 4/16$
2	$p(2) = 6/16$
3	$p(3) = 4/16$
4	$p(4) = 1/16$

(a) (b)

FIGURE 2.2. A discrete probability distribution. (a) Tabular form. (b) Graph.

To find $p(1)$, we note that there are four cases in which exactly one mother out of four prefers A, thus $p(1) = 4/16$. The general rule for calculating the *probability of an event* when all outcomes are equally likely is:

$$\text{probability of an event} = \frac{\text{number of outcomes giving the event}}{\text{total number of outcomes}}$$

All of the probabilities are summarized in the table of Figure 2.2a and in the graph of Figure 2.2b.

The values of a discrete random variable y together with their associated probabilities are called a *probability distribution*, and $p(y)$ is called the *probability function*. In order for $p(y)$ to be a probability function, two conditions are necessary:

1. $0 \le p(y) \le 1$ for all values of y.
2. $\sum_y p(y) = 1$, that is, the sum of $p(y)$ over all values of y is 1.

Note that in the baby cereal example these two conditions are satisfied.

There are many functions that satisfy these two conditions. For example, in Table 2.1, columns A through D represent discrete probability distributions. In column D the random variable has a countable infinity of values, and $p(y)$ can be given by the formula $p(y) = (1/2)^y$. In many cases it is possible to represent the probability function by a formula.

It is not difficult to find functions with the two properties required for a probability function. However, a probability distribution will only be of value statistically if it represents—models—a real life situation. Some examples of probability distributions used as models occur in Exercises 1.2.5 through 1.2.7. The method for determining the probabilities in these examples is explained in Chapter 3. An example of a test of hypothesis that uses a different type of discrete probability distribution follows.

TABLE 2.1 Four Discrete Probability Distributions

A		B		C		D	
y	$p(y)$	y	$p(y)$	y	$p(y)$	y	$p(y)$
0	1/4	5	1/5	0.5	0.125	1	1/2
1	1/2	6	1/5	1.0	0.125	2	1/4
2	1/4	7	1/5	1.5	0.125	3	1/8
		8	1/5	2.0	0.625	4	1/16
		9	1/5			5	1/32
						⋮	⋮

Example 2.2. Testing a Hypothesis Using a Discrete Probability Distribution

A new salesperson for a company is told that the probability of making a sale on a single call is 1/4. The salesperson calls on seven people and makes no sales. Finally, on the eighth attempt, a sale is completed. The salesperson wonders if there is any evidence (at the 0.05 level of significance) that the probability of 1/4 for a sale is too high.

The null hypothesis is H_0: $\theta = 1/4$, that is, the probability of a sale is 1/4 on a single attempt.† The alternative is H_a: $\theta < 1/4$ because the salesperson is looking for evidence that the figure is too high.

If the probability of a sale is 1/4, then the probability of no sale on a single trial is 3/4. Using these values, the probability model can be found. The probability of a sale on the first call is

$$p(1) = 1/4$$

and the probability that the first sale occurs on the second call is

$$p(2) = (3/4)(1/4) = 3/16$$

since there is no sale on the first call and there is a sale on the second call. The probabilities are multiplied because the calls are assumed to be *independent* of each other, that is, we assume the customers are randomly chosen, do not influence each other, and that the salesperson behaves the same way on each call.

Similarly,

$$p(3) = (3/4)(3/4)(1/4) = 9/64$$

and

$$p(y) = (3/4)^{y-1}(1/4)$$

is the general formula for the probability that the first sale occurs on the yth call.

† The Greek letter θ is read "theta."

The beginning of the probability distribution that is the model of this study can be summarized as follows.

y	$p(y)$		
1	1/4	= 0.2500	0.8665 (sum of rows 1–7)
2	3/16	= 0.1875	
3	9/64	= 0.1406	
4	27/256	= 0.1055	
5	81/1,024	= 0.0791	
6	243/4,096	= 0.0593	
7	729/16,384	= 0.0445	
8	2,187/65,536	= 0.0335	
⋮		⋮	

If $\theta < 1/4$, a larger number of calls will be necessary before the first sale than if $\theta = 1/4$. Thus the P value associated with this study is

$$\begin{aligned} P &= P(\text{8 or more calls needed for the first sale}) \\ &= 1 - P(\text{1 through 7 calls needed for the first sale}) \\ &= 1 - 0.8665 \\ &= 0.1335 \end{aligned}$$

Since $P = 0.1335 > \alpha = 0.05$, the null hypothesis is accepted. There is no evidence that the figure given to the salesperson is too high.

If the data are recorded on a continuous scale, the variable of interest corresponds to a *continuous random variable*. In this type of model it is not possible to represent the related probabilities by a table or a line graph; instead, a smooth curve is used to indicate the *continuous probability distribution* which is the model for the study.

Example 2.3. A Continuous Probability Distribution

On an unbiased spinner it is equally likely that the pointer will land at any point on the circumference. One fixed point on the circumference is designated 0 and any other point is named by measuring in a clockwise direction the angle formed by 0, the center, and the point being named (Figure 2.3). A graph of the probability distribution of the random variable y, representing the point at which the spinner stops, appears in Figure 2.4. The curve is a horizontal line which indicates that the spinner is equally likely to stop at any point on the circumference. The line $f(y) = 1/360$ is called the *density function* of the random variable y. The curve (the horizontal line) is placed at 1/360 so that the area of the rectangle under the line is 1. The probability of landing between 90 and 180 is represented by the area between 90 and

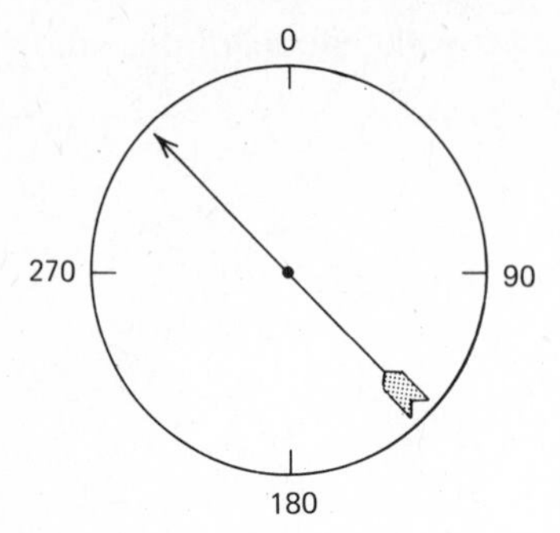

FIGURE 2.3. An unbiased spinner.

180 and under the curve; the probability is 1/4 (Figure 2.4). The probability of landing between 0 and 95 is given by the area under the curve and to the left of 95. The probability is $95/360 = 19/72$ (Figure 2.5).

Notice that the density function, unlike a probability function for a discrete random variable, does not indicate a probability directly, rather the density function is used to find an area that corresponds to the probability. Because areas correspond to probabilities, the probability that the spinner will stop at a particular point, say $y = 95$, is 0. This becomes clear by noticing that rather than a region, there is only a vertical line segment at 95, and that a line segment has no area. It follows that $P(y \leq 95) = P(y < 95)$ in a continuous probability distribution, but this is not true in a discrete distribution.

In many models for continuous random variables, the continuous probability distribution is given by a curve that is not a straight line nor a figure formed from straight lines. In these cases, areas are difficult to determine and calculus must be used. Fortunately, tables are available for most of the commonly encountered distributions, and thus even those who are not familiar with calculus are able to use continuous probability distributions that are represented by curves. The first distribution of this type is discussed in Chapter 5.

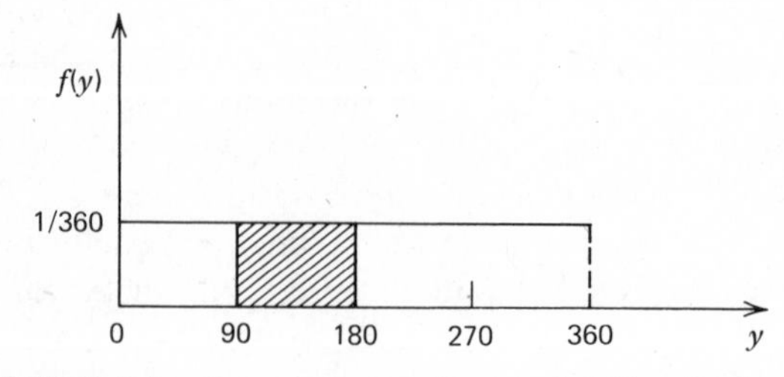

FIGURE 2.4. A continuous uniform probability distribution.

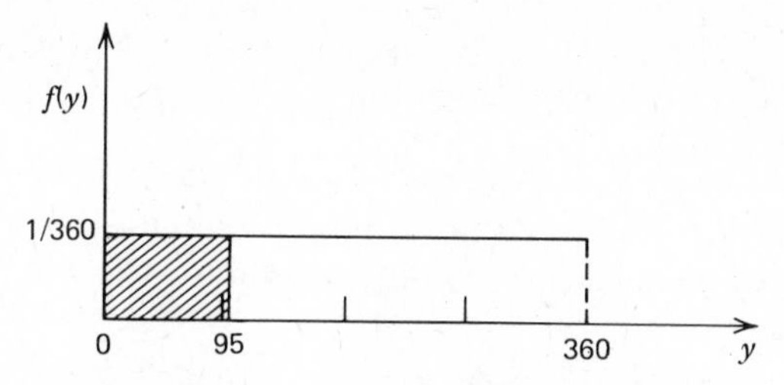

FIGURE 2.5. Shaded area indicates $P(0 \leq y \leq 95)$.

EXERCISES

2.3.1.

y:	2	4	6	8	10
$p(y)$:	1/6	2/6	1/6	—	1/6

a. If the table above represents a probability distribution, what is the value of $p(8)$?

b. Graph the probability distribution.

c. Find: $P(y \leq 6)$, $P(y < 6)$, $P(y = 6)$, $P(y > 6)$, $P(y \geq 6)$.

2.3.2. If $p(y) = 1/5$ for $y = 1, 2, 3, 4$, and 5:

a. Show that this is a probability distribution.

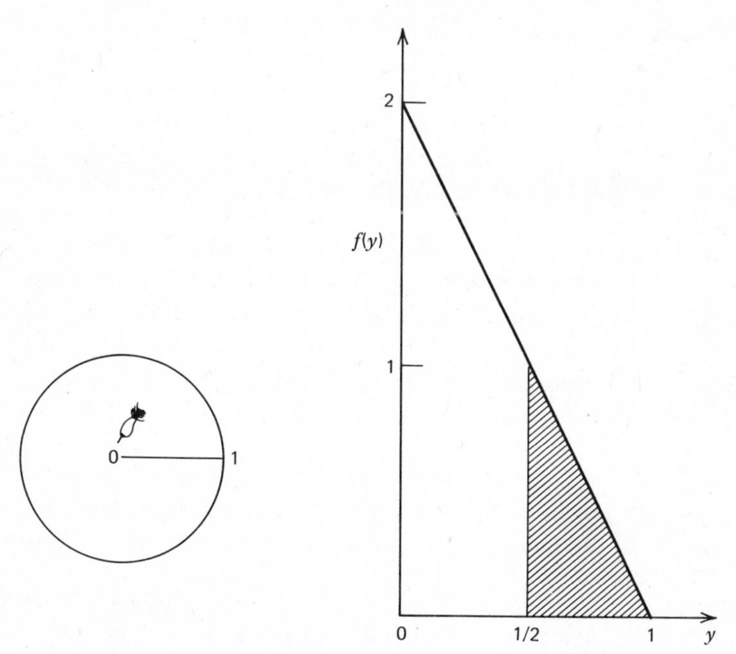

FIGURE 2.6. A continuous triangular probability distribution.

b. Draw the graph.

c. Find: $P(y > 3)$, $P(y = 3)$, $P(y \leq 3)$, and $P(y < 3)$.

2.3.3. Given the continuous probability distribution in Figure 2.6, imagine that the distribution represents the probability that a certain expert dart thrower will hit a 1-ft. target within a certain distance y from the center 0.

a. What is the total area within the triangle?

b. What is the area of the shaded portion of the distribution?

c. What is the probability that the dart will hit at a point that is from 6 in. to 1 ft. from the center of the target?

d. What is the area of the unshaded portion of the distribution?

e. What is the probability that the dart will hit at a point that is less than 6 in. from the center of the target?

2.3.4. An oil company believes that the probability of striking oil on a single random drilling in a certain field is 1/3. They drill and hit oil on the sixth attempt. Is there any evidence that the probability of a strike is less than 1/3?

2.4. EXPECTED VALUE AND VARIANCE OF A PROBABILITY DISTRIBUTION

Since probability distributions are the key to statistical inference, it is helpful to study some of their characteristics. Two useful characteristics of a probability distribution are its *expected value* and its *variance*. Expected value is a measure of the location of the distribution, while variance is a measure of its spread.

In order to introduce the idea of *expected value*, let us consider a certain electronic game that involves hitting a random target. To make the game sufficiently challenging to hand–eye coordination, it has been programmed so that the position of the target, the moment that the target appears, and the number of targets that appear during the period of play all vary. The number of targets to appear can be 11, 12, 13, 14, 15, or 16. They occur randomly and with equal frequency over a large number of periods of play. A player of the game is unable to predict the number of targets that will appear during any one playing period, but the player can determine the expected number of targets, that is, the average number per playing session if the game is played many times.

The number of targets can be modeled by a discrete uniform probability distribution in which the values of the random variable y are 11, 12, 13, 14, 15, and 16, and the probability function $p(y)$ is 1/6 for each of the values because they occur with equal frequency.

y	$p(y)$
11	1/6
12	1/6
13	1/6
14	1/6
15	1/6
16	1/6

The expected number of targets, $E(y)$, per playing period is

$$E(y) = \frac{11 + 12 + 13 + 14 + 15 + 16}{6} = \frac{81}{6} = 13.5$$

that is, the arithmetic average of the six equally frequent numbers. If many games are played, on the average 13.5 targets will appear per session. Note that the expected value need not be one of the possible values of the random variable; 13.5 targets never appear in a playing session.

Another way to compute the expected value is to use the formula

$$E(y) = \sum yp(y)$$

that is, the expected value of y is the sum of the products of the values of y times their corresponding probabilities. The following table illustrates how this formula is used.

y	$p(y)$	$yp(y)$
11	1/6	11/6
12	1/6	12/6
13	1/6	13/6
14	1/6	14/6
15	1/6	15/6
16	1/6	16/6

$$E(y) = \sum yp(y) = 81/6 = 13.5$$

A third column is computed from the probability distribution. This third column is obtained by finding the product of the corresponding elements in the first two columns. The expected value of y is the sum of the products in the third column. The advantage of this second approach is that it can be used to find an expected value even if the probabilities are not all the same. The following example illustrates this general type of problem.

Example 2.4. The Expected Value of a Discrete Probability Distribution

A teacher gives frequent short quizzes that consist of two multiple-choice questions. Each question is followed by four answers, and only one is correct. Because these quizzes are so short, the teacher wonders if they are useful for determining which students have learned the material. The teacher decides to find out how many questions a student can be expected to answer correctly if the student has no knowledge of the material and is choosing answers in a random fashion.

On a single question, the probability of a correct guess is 1/4 because each answer is equally likely to be chosen and only one answer is correct. For two questions, the number of correct responses y can be 0, 1, or 2, and the probability distribution, which is a model of the number of correct responses under guessing, is

y	$p(y)$
0	9/16
1	6/16
2	1/16

The probabilities in this distribution are obtained by computing $p(0) = P(\text{two incorrect}) = (3/4)(3/4) = 9/16$ and $p(2) = P(\text{two correct}) = (1/4)(1/4) = 1/16$; then $p(1)$ must equal 6/16 so that the sum of the probabilities is equal to 1.

If a large number of quizzes of this type are given, then the expected number of correct answers per quiz is

$$E(y) = \sum yp(y)$$

In tabular form:

y	$p(y)$	$yp(y)$
0	9/16	0
1	6/16	6/16
2	1/16	2/16

$$E(y) = \sum yp(y) = 8/16 = 0.5$$

On the average, the student will guess correctly only 0.5 of an answer. Although it is impossible to get 0.5 of an answer correct on a single quiz, the expected value is meaningful for a large number of quizzes.

The teacher decides that the quizzes are useful for distinguishing those who are guessing from those who have knowledge of the material. For example, if 40 such two-question quizzes are given, then the student who is guessing is expected to answer correctly about 20 out of the 80 questions asked. A student who answers many more correctly, for example 60 out of the 80 questions, demonstrates some knowledge of the material.

The expected value can be thought of as the location, or center, of the probability distribution. This seems reasonable if we visualize a uniform

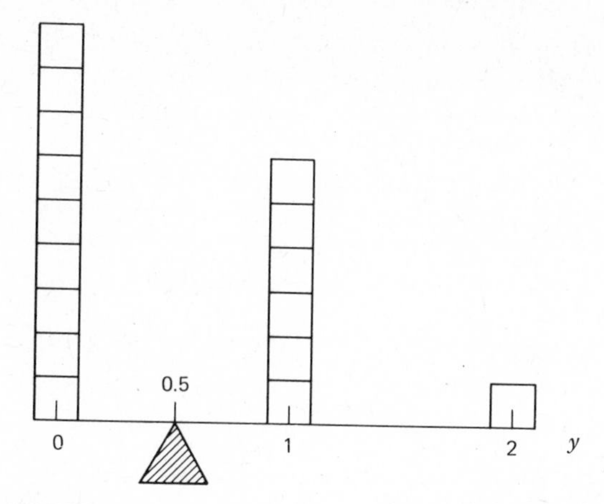

FIGURE 2.7. The expected value as the balancing point.

calibrated bar on which we place weights (all of equal magnitude): nine at 0, six at 1, and one at 2 (Figure 2.7). The bar will balance at 0.5, the expected value.

Another useful characteristic of a probability distribution is its *variance*. Variance is a measure of the spread of a distribution relative to its expected value. In the electronic game example, the random variable y had values 11, 12, 13, 14, 15, 16 with equal frequency. The deviations of these values from the expected value of 13.5 are

y	$y - E(y)$
11	$11 - 13.5 = -2.5$
12	$12 - 13.5 = -1.5$
13	$13 - 13.5 = -0.5$
14	$14 - 13.5 = 0.5$
15	$15 - 13.5 = 1.5$
16	$16 - 13.5 = 2.5$

The deviations are shown graphically in Figure 2.8.

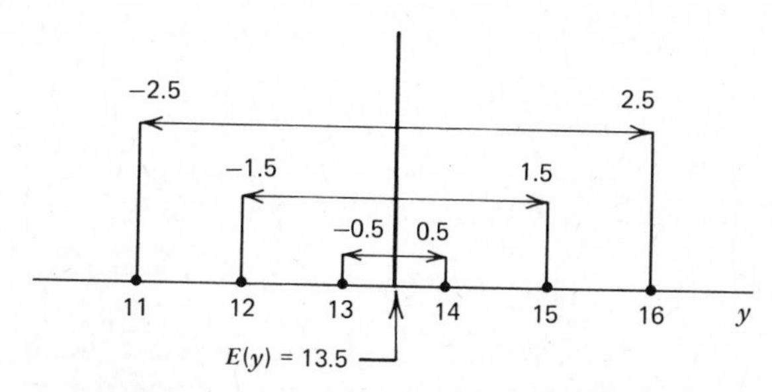

FIGURE 2.8. Deviations from the expected value.

We might expect to measure spread by averaging these deviations. However, since the sum of the deviations from the expected value is always 0, this is not a useful measure. To obtain a meaningful average, we use the squares of the deviations. The variance of a probability distribution is the average squared deviation from its expected value. Using the probabilities, the formula for the variance of y is

$$V(y) = \sum[y - E(y)]^2 p(y)$$

In tabular form (using fractions to avoid rounding error), the computations are

y	$p(y)$	$y - E(y)$	$[y - E(y)]^2$	$[y - E(y)]^2 p(y)$
11	1/6	$-2.5 = -5/2$	25/4	25/24
12	1/6	$-1.5 = -3/2$	9/4	9/24
13	1/6	$-0.5 = -1/2$	1/4	1/24
14	1/6	$0.5 = 1/2$	1/4	1/24
15	1/6	$1.5 = 3/2$	9/4	9/24
16	1/6	$2.5 = 5/2$	25/4	25/24
			$V(y) =$	70/24

This formula is used even if the probabilities are not all equal.

Variance measures the spread of a distribution. The larger the variance the larger the spread. If we take the positive square root of the variance we obtain the *standard deviation* of the random variable, sd(y). In this example

$$\text{sd}(y) = \sqrt{V(y)} = \sqrt{70/24} = 1.71$$

If we are told only the expected value and standard deviation of a probability distribution, we know a surprising amount about the nature of the distribution. Values of the random variable that are more than two or three standard deviations from the mean have very low probabilities associated with them. For example, in the case of the electronic game

$$E(y) = 13.50$$

$$\text{sd}(y) = 1.71$$

and

$$2[\text{sd}(y)] = 3.42$$

Two standard deviations below the expected value is

$$E(y) - 2[\text{sd}(y)] = 13.50 - 3.42 = 10.08$$

and the probability of 10 or fewer targets in a single playing period is very low; in fact, it is zero. Two standard deviations above the expected value is

$$E(y) + 2[\text{sd}(y)] = 13.50 + 3.42 = 16.92$$

and the probability of 17 or more targets is zero.

In practice, the computation of the variance from the formula

$$V(y) = \sum[y - E(y)]^2 p(y)$$

is sometimes tedious because of the subtractions and squaring. A mathematically equivalent formula may be used:

$$V(y) = \sum y^2 p(y) - [E(y)]^2$$

We illustrate this formula for the probability distribution of the two-question multiple-choice quizzes.

Example 2.5. The Variance of a Probability Distribution

For the short quizzes, a fourth column $y^2p(y)$ is computed after the computation of the expected value. The fourth column is obtained by multiplying the elements in the first column times the corresponding elements in the third column.

y	$p(y)$	$yp(y)$	$y^2p(y)$
0	9/16	0	0
1	6/16	6/16	6/16
2	1/16	2/16	4/16
		$\sum y^2p(y) =$	10/16

Then

$$\begin{aligned} V(y) &= \sum y^2 p(y) - [E(y)]^2 \\ &= 10/16 \quad - (1/2)^2 \\ &= 6/16 \end{aligned}$$

Note that in this example

$$\begin{aligned} E(y) &= 0.5 \\ \text{sd}(y) &= \sqrt{6/16} = 0.61 \end{aligned}$$

and two standard deviations below and above the expected value are

$$\begin{aligned} E(y) - 2[\text{sd}(y)] &= 0.5 - 2(0.61) = -0.72 \\ E(y) + 2[\text{sd}(y)] &= 0.5 + 2(0.61) = 1.72 \end{aligned}$$

There is zero probability that the value of the random variable is below -0.72, and 1/16 probability that the random variable will have a value above

1.72. Using only these facts, if a student frequently answered both questions correctly, the teacher decides that the model based on guessing does not fit this student and the student probably has knowledge of the material.

The main use of the variance (or standard deviation) is for purposes of inference. This application is developed more fully in later chapters. The discussion in this section is restricted to discrete random variables. It is also possible to consider the expected value and variance of a continuous random variable; in such cases, calculus is usually needed to find the values.

EXERCISES

2.4.1. Find the mean and the variance of the probability distributions in Table 2.1.

2.4.2. In Mendel's experiments on pea plants, he found that the trait of being tall is dominant over being short. His theory indicates that if pure-line tall and pure-line short plants are cross-pollinated and then the next generation of hybrids are cross-pollinated, in the resulting population approximately 3/4 of the plants will appear tall and 1/4 will appear short. If four plants are chosen at random from such a population, the best model for the number of tall plants in four is:

y:	0	1	2	3	4
$p(y)$:	1/256	12/256	54/256	108/256	81/256

a. Find the expected value of this probability distribution.

b. Find the variance of the probability distribution.

c. What is the probability that the value of the random variable will be more than two standard deviations below the expected value?

d. What is the probability that the value of the random variable will be more than two standard deviations above the expected value?

2.4.3. A gambling game is played in which there is a group of 100 cards with one \$25 winning card, two \$10 winning cards, and three \$5 winning cards. After paying a certain fee, a player selects one card at random. If it is one of the winning cards, the player receives the designated amount. If it is one of the other cards, the player wins nothing. The card is returned to the deck, the cards shuffled, and they are ready for the next play.

a. Find the probability distribution for y, the number of dollars won (use the rule for equally likely events).

b. If a large number of plays are purchased, what are the expected winnings per play, or in statistical terms, what is the expected value of y?

c. Would it be reasonable to pay $1 to play this game?

d. Find the variance of this probability distribution.

e. What proportion of the time will the winnings be within two standard deviations of the expected value?

2.4.4.

y:	1	2	3	4	5
$p(y)$:	1/5	1/5	1/5	1/5	1/5

a. Find the expected value of y.

b. Find $V(y)$.

c. Compare your answers with those found in Exercise 2.4.1 for Table 2.1, column B. Explain why there is a difference in the expected values but the variances are the same.

2.4.5.

y:	1	2	3	4
$p(y)$:	1/4	1/4	1/4	1/4

a. Find $E(y)$.

b. Compare this result with that of Exercise 2.4.4; find a simple general formula for the expected value of a discrete uniform distribution of successive integers from a to b.

REVIEW EXERCISES

Decide whether each of the following statements is true or false. If a statement is false, explain why.

2.1. The objective of statistics is to make inference about a population based on information contained in a sample from that population.

2.2. A single population may have several variables of interest to the investigator.

2.3. If an investigator works hard enough, there is always a way to find the values of the variable of interest for every member of the population.

2.4. Completely random samples and simple random samples are two different types of samples.

2.5. A lottery device may be an acceptable way to obtain a completely random sample.

2.6. When using a random number table to select a sample, always begin at the beginning of the table.

2.7. The choice of sampling design has no effect on the choice of the procedure used for statistical analysis.

2.8. When choosing categories for the nominal scale, the only condition is that there is a category for each piece of data.

2.9. Data on the numerical scale can be easily changed to the nominal scale.

2.10. The ordinal scale is sometimes used even though more precise numerical information is available.

2.11. Data on an ordinal scale can be easily changed to the numerical scale.

2.12. Barometric pressure is usually recorded on the ordinal scale.

2.13. Yearly wages to the nearest dollar are recorded on the discrete numerical scale.

2.14. Age is actually a continuous random variable, but it is recorded using a discrete numerical scale.

2.15. In a continuous probability distribution, the total area between the curve representing the distribution and the horizontal axis is one.

2.16. In a continuous probability distribution, the probability of any particular value is the vertical distance at the value between the horizontal axis and the curve representing the distribution.

2.17. In a discrete probability distribution, the length of a vertical line at a certain value can be interpreted as the probability that such a value will result from random sampling.

2.18. Continuous variables can theoretically be measured as precisely as desired, but they are usually rounded to a convenient unit.

2.19. In a probability distribution, the probability that the random variable will assume a particular value is 0.

2.20. If a population is infinite in size, the variable of interest is continuous.

2.21. Random variables always have numerical values.

2.22. Nominal variables cannot be directly modeled by a probability distribution.

2.23. The expected value of a probability distribution can be thought of as the center of balance.

2.24. The variance of a probability distribution is a measure of location, and the expected value indicates the spread.

2.25. If two probability distributions have equal variances, then their expected values are equal also.

2.26. The variance of a probability distribution can be defined symbolically as $E[y - E(y)]^2$.

SELECTED READINGS

Anderson, N. H. (1961). Scales and statistics: Parametric and nonparametric, *The Psychological Bulletin*, **58**, 305–316.

Cochran, W. G. (1963). *Sampling Techniques*, 2nd ed., Wiley, New York.

Conover, W. J. (1980). *Practical Nonparametric Statistics*, 2nd ed., Wiley, New York.

Daniel, W. W. (1978). *Applied Nonparametric Statistics*, Houghton Mifflin.

Hollander, M., and D. S. Wolfe (1973). *Nonparametric Statistical Methods*, Wiley, New York.

Kish, L. (1965). *Survey Sampling*, Wiley, New York.

Rand Corporation (1955). *A Million Random Digits with 100,000 Normal Deviates*, Free Press, Glencoe, Illinois.

Scheaffer, R. L., W. Mendenhall, and L. Ott (1979). *Elementary Survey Sampling*, 2nd ed., Wadsworth, Belmont, California.

3

Binomial Distributions

In many experiments and surveys in which the variable of interest is being recorded at the nominal level, there are only two possible values or outcomes for the variable. For example, a salesman either makes a sale or does not make a sale, a newborn child is either a girl or a boy, an insecticide may kill an insect or fail to kill it. Under certain conditions, samples involving dichotomous variables of this type can be represented by a theoretical probability distribution called a *binomial distribution*, binomial because of the two possible outcomes. In this chapter we look at the statistical interpretation of experimental results that can be modeled by binomial distributions.

3.1. THE NATURE OF BINOMIAL DISTRIBUTIONS

The population of human beings can be classified as "having type-O blood" or "not having type-O blood." There is no way that we can get exact information about the entire population, since this group is so large. It has been estimated that the proportion of people with type-O blood is 0.40. Assume that the estimate is correct. If we observe a single person selected at random, the probability that the person will have type-O blood is 0.40 and the probability that the person will not have type-O blood is 0.60.

Now let us imagine that a large metropolitan hospital has a list of several thousand people willing to donate blood. If four people are chosen at random from the list, how likely is it that none have type-O blood? One has type-O? Two? Three? Four?

We first list the different possible outcomes for a sample of four people. Let O mean that a person has type-O blood, and let N mean that the person does not have type-O blood. The sequence of symbols indicates the results in the order in which they occur in the experiment, so *NNON* is a different outcome from O*NNN*.

Number with Type-O Blood	Possible Outcomes
0	*NNNN*
1	O*NNN* *N*O*NN* *NN*O*N* *NNN*O
2	OO*NN* O*N*O*N* O*NN*O *N*OO*N* *N*O*N*O *NN*OO
3	*N*OOO O*N*OO OO*N*O OOO*N*
4	OOOO

When we ask a question like "How likely is it that two persons out of four have type-O blood?" we have shifted our focus from the underlying variable of blood type (O or not-O) on the nominal scale to a count that is on the discrete numerical scale. Since it is numerical, the count can be thought of as a random variable, and we are looking for the probability distribution of this discrete random variable. We have already seen an example like this in the baby-cereal preference study (Example 1.4 and Section 2.3), except in that case the probabilities were all equal.

Since not all of the 16 outcomes in this example are equally likely, in order to find the probabilities associated with 0, 1, 2, 3, and 4 we must use the laws of probability.

Laws of Probability

1. If p is a probability, $0 \le p \le 1$.

2. If A and $\bar{A}$ are two mutually exclusive events that include all possible outcomes, then $P(A) + P(\bar{A}) = 1$. (Two events A and B are *mutually exclusive* if they are nonoverlapping, that is, if $P(A \text{ and } B) = 0$.)

3. *Addition Law.* The probability of a specified outcome is the sum of the probabilities of the mutually exclusive events making up that outcome.

4. *Multiplication Law.* The probability of an event that is the simultaneous occurrence of two or more independent elementary events is the product of the probabilities of the elementary events. (Two events A and B are *independent* if the occurrence or nonoccurrence of A has no effect on the probability of B and vice versa.)

We already used the second law when we stated that $P(N) = 0.60$. We reasoned that $P(N) = 1 - P(\text{O}) = 1 - 0.40 = 0.60$. Now we find that the probability of zero out of four having type-O blood is

$$p(0) = P(NNNN) = [P(N)]^4 = (0.60)^4 = 0.1296$$

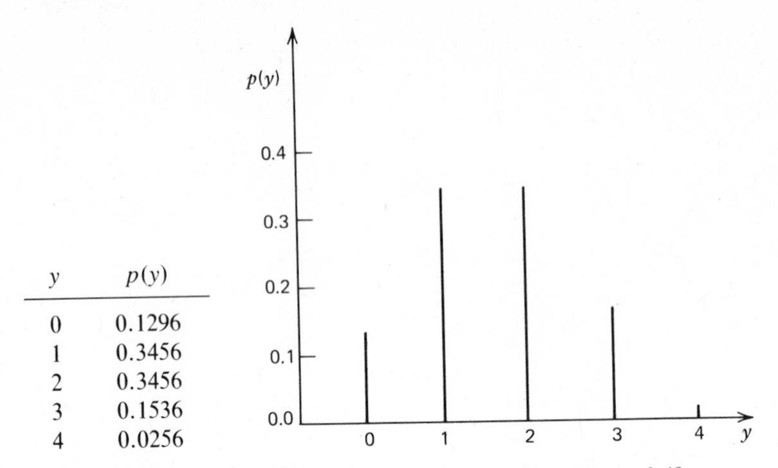

y	$p(y)$
0	0.1296
1	0.3456
2	0.3456
3	0.1536
4	0.0256

FIGURE 3.1. The binomial distribution with $n = 4$, $\pi = 0.40$.

and the probability that one out of four will have type-O blood is

$$\begin{aligned} p(1) &= P(ONNN \text{ or } NONN \text{ or } NNON \text{ or } NNNO) \\ &= P(ONNN) + P(NONN) + P(NNON) + P(NNNO) \\ &= (0.40)(0.60)^3 + (0.60)(0.40)(0.60)^2 \\ &\quad + (0.60)^2(0.40)(0.60) + (0.60)^3(0.40) \\ &= 4(0.40)(0.60)^3 \\ &= 0.3456 \end{aligned}$$

In a similar way, we find that

$$\begin{aligned} p(2) &= 6(0.40)^2(0.60)^2 = 0.3456 \\ p(3) &= 4(0.40)^3(0.60) = 0.1536 \\ p(4) &= (0.40)^4 = 0.0256 \end{aligned}$$

In summary, for this example the probability distribution is as appears in Figure 3.1. The discrete random variable with values 0, 1, 2, 3, 4 represents the number of people with type-O blood in a random sample of four people, and $p(y)$ is the probability function of y. This probability distribution is called a *binomial probability distribution*. Note that a binomial probability distribution is a model of an experiment with only two possible outcomes. We concentrate on one of the outcomes, type-O blood, and count the number of occurrences (successes) in the sample. The probability of type-O blood does not change from observation to observation,† and the observations are independent of each other. We call such a survey or experiment a *binomial experiment*.

† Each time we remove a person from the population the probability of type-O blood does in fact change slightly. However, since we are selecting only four people from several thousand, the changes are negligible.

A *binomial experiment* is an experiment in which

1. There are only two possible outcomes, success S or failure F, with $P(S) = \pi$ and $P(F) = 1 - \pi$.
2. The experiment is repeated n times, that is, there are n trials.
3. $P(S) = \pi$ is constant from trial to trial.
4. The trials are independent of each other.
5. We are interested in y, the number of successes, with $y = 0, 1, 2, \ldots, n$.

The probability of success π is called the *binomial parameter*. A parameter is a numerical characteristic of a population. In the blood-type example, $\pi = 0.40$ is the proportion of the population with type-O blood. The parameter π also specifies the theoretical model for the experiment, the binomial distribution with $n = 4$ trials and $P(S) = \pi = 0.40$.

In the seventeenth century, members of the Bernoulli family found a formula to calculate the binomial distribution for any number of trials and any probability of success. Before examining their formula it may be best to explain the notation that occurs in it.

The symbol π^y means $(\pi)(\pi)\cdots(\pi)$, that is the product when π is used as a factor y times. For example,

$$(3/4)^5 = (3/4)(3/4)(3/4)(3/4)(3/4) = 243/1024$$

Similarly,

$$(1 - \pi)^{n-y} = \underbrace{(1 - \pi)(1 - \pi) \cdots (1 - \pi)}_{n - y \text{ times}}$$

so that

$$(1 - 3/4)^{7-5} = (1/4)(1/4) = 1/16$$

The symbol $\binom{n}{y}$ is read "the number of combinations of n things taken y at a time." For example, if there are four slips of paper marked A, B, C, and D in a box, and two slips are drawn at random, the possible combinations are

$$AB, AC, AD, BC, BD, CD$$

In this case $\binom{4}{2} = 6$. We are not interested in which letter is drawn first, so AB and BA are the same combination.

The symbol $\binom{n}{y}$ can also be applied to the blood-type example. Here $\binom{4}{2}$ means the number of different places that two O's can appear in a sequence of four symbols, that is, we are picking two positions out of the four possible positions. If first, second, third, and fourth are the positions, O can occur

1st and 2nd	1st and 3rd	1st and 4th
2nd and 3rd	2nd and 4th	3rd and 4th

or

$$\begin{matrix} OONN & ONON & ONNO \\ NOON & NONO & NNOO \end{matrix}$$

In general,

$$\binom{n}{y} = \frac{n!}{y!\,(n-y)!}$$

where $n! = n(n-1)(n-2)\cdots(2)(1)$, and $n!$ is read "*n factorial.*" Some examples are:

$$\binom{4}{2} = \frac{4!}{2!(4-2)!} = \frac{4\cdot3\cdot2\cdot1}{(2\cdot1)(2\cdot1)} = 6$$

and

$$\binom{4}{0} = \frac{4!}{0!(4-0)!} = \frac{4\cdot3\cdot2\cdot1}{1(4\cdot3\cdot2\cdot1)} = 1$$

because $0! = 1$ by definition.

Table A.2 in the Appendix of Useful Tables is a table for $n!$, and Table A.3 is a table for $\binom{n}{y}$, the *binomial coefficients*. It should be noted that $\binom{n}{y} = \binom{n}{n-y}$, since this will often shorten calculations.

The Bernoulli formula for calculating binomial probabilities will now be understandable. To find $b(y;n,\pi)$, the probability in the binomial distribution of y successes when the number of trials is n and the probability of success on a single trial is π, we use the following formula:

$$b(y;n,\pi) = \binom{n}{y}\pi^y(1-\pi)^{n-y}$$

Thus the mathematical model in the blood-type example is the random variable y having values 0, 1, 2, 3, 4 and probability function $b(y;4,0.40)$. The probabilities are computed in Table 3.1. This is the same result we previously computed by listing all possible experimental outcomes.

Since the Bernoulli formula can be used for any sample size and any probability of success, there is no need to go back to the list of all possible outcomes. If the number of trials is 20 and π is 0.30, then the probability

TABLE 3.1. Computing Binomial Probabilities

Y	$b(y;\,4,\,0.4)$
0	$\binom{4}{0}(0.4)^0(1-0.4)^{4-0} = (1)(0.4)^0(0.6)^4 = 0.1296$
1	$\binom{4}{1}(0.4)^1(1-0.4)^{4-1} = (4)(0.4)^1(0.6)^3 = 0.3456$
2	$\binom{4}{2}(0.4)^2(1-0.4)^{4-2} = (6)(0.4)^2(0.6)^2 = 0.3456$
3	$\binom{4}{3}(0.4)^3(1-0.4)^{4-3} = (4)(0.4)^3(0.6)^1 = 0.1536$
4	$\binom{4}{4}(0.4)^4(1-0.4)^{4-4} = (1)(0.4)^4(0.6)^0 = 0.0256$

TABLE 3.2. Four Binomial Distributions

y	$b(y; 20, 0.30)$	$b(y; 20, 0.50)$	$b(y; 20, 0.70)$	$b(y; 20, 0.75)$	y
0	0.001	0.000	0.000	0.000	0
1	0.007	0.000	0.000	0.000	1
2	0.028	0.000	0.000	0.000	2
3	0.072	0.001	0.000	0.000	3
4	0.130	0.005	0.000	0.000	4
5	0.179	0.015	0.000	0.000	5
6	0.192	0.037	0.000	0.000	6
7	0.164	0.074	0.001	0.000	7
8	0.114	0.120	0.004	0.001	8
9	0.065	0.160	0.012	0.003	9
10	0.031	0.176	0.031	0.010	10
11	0.012	0.160	0.065	0.027	11
12	0.004	0.120	0.114	0.061	12
13	0.001	0.074	0.164	0.112	13
14	0.000	0.037	0.192	0.169	14
15	0.000	0.015	0.179	0.202	15
16	0.000	0.005	0.130	0.190	16
17	0.000	0.001	0.072	0.134	17
18	0.000	0.000	0.028	0.067	18
19	0.000	0.000	0.007	0.021	19
20	0.000	0.000	0.001	0.003	20

of 7 successes out of 20 trials is

$$b(7;20,0.30) = \binom{20}{7}(0.30)^7(1 - 0.30)^{20-7}$$

$$= 77520(0.30)^7(0.70)^{13}$$

$$= 0.16$$

Most of the time it is not even necessary to use this formula, since tables are available for many sample sizes and probabilities. Computers can easily be programmed to produce other tables of binomial distributions. It is useful, however, to know the formula so that the tables are meaningful.

Table 3.2 is an example of a table for four binomial distributions. The value of $b(7;20,0.30)$, which was calculated earlier in this section, can be found in the eighth row of the first column.

Note that there are entries of 0.000 in some positions, for example, $b(1;20,0.50)$. This does *not* mean that there is zero probability of getting one successful outcome in a sample of 20 when π is 0.50, but rather that the probability of one successful outcome is smaller than 1/1000.

The most likely outcome(s) for each value of π can be read from this table. If $\pi = 0.30$, the most likely outcome is 6. If $\pi = 0.50$, the most likely outcome is 10. For $\pi = 0.70$ it is 14, and for $\pi = 0.75$ it is 15.

Since a binomial distribution is a probability distribution, we can find its expected value, $E(y)$, and variance, $V(y)$, by using the formulas introduced in Section 2.4. However, because of the special nature of the binomial distributions, shorter formulas exist. For a binomial distribution

$$E(y) = n\pi$$

$$V(y) = n\pi(1 - \pi)$$

Thus, for $b(y;20,0.50)$:

$$E(y) = 20(0.5) = 10$$

$$V(y) = 20(0.5)(0.5) = 5$$

$$\text{sd}(y) = \sqrt{5} = 2.24$$

If we consider an interval from two standard deviations below the expected value to two standard deviations above the expected value, that is,

$$10 \pm 2(2.24)$$

or

$$5.52 \quad \text{to} \quad 14.48$$

we find there is a probability of 0.958 that a value of the random variable will be within this interval and only a 0.042 probability that the value will be outside this interval.

In the next two sections we see how binomial distributions can help interpret the results of experiments.

EXERCISES

3.1.1. In a certain large college course, past records show that grades of *A*, *B*, *C*, *D*, and *F* are equally likely. If one student is chosen at random, find the following probabilities:

a. $P(C)$.

b. $P(A \text{ or } B)$.

c. $P(\text{a grade higher than } D)$.

d. $P(A, B, C, D, \text{ or } F)$.

e. $P(B \text{ and } D)$.

f. $P(E)$.

g. $P(\text{not-}A)$.

h. $P(\text{not-}A \text{ and not-}F)$.

3.1.2. If two people who are not friends take the course described in Exercise 3.1.1, find:

a. P(two A's).
b. P(same grade).
c. P(different grades).
d. P(both higher than D).
e. P(both fail).
f. P(one passes and one fails).

3.1.3. In a certain city, a fourth of the families take their children to the doctor for regular checkups. Five families are chosen at random.
a. What is the probability that exactly three families out of the five take their children to the doctor for regular checkups?
b. What is the probability that, at most, two families out of the five take their children for regular checkups?
c. What is the probability that more than one family out of the five take their children?

3.1.4. Assume a standard deck of 52 cards is used in the following problems.
a. Find the probability of drawing a heart or a picture card when selecting one card at random. Explain why P(heart or picture card) $\neq P$(heart) $+ P$(picture card).
b. Find the probability of drawing two cards of the same color if the first card is randomly selected and kept out of the deck, and the second card is then selected at random. Explain why P(two red cards) $\neq (1/2)(1/2)$.

3.1.5. In the game of Yahtzee, five ordinary dice are tossed.
a. How likely is it that a player will get exactly four two's on a random roll of the dice?
b. In this game, 30 points are awarded if all five dice show the same number. How likely is this to happen on a random toss?

3.1.6. Find:
a. $4!$.
b. $0!$.
c. $5!$.
d. $1!3!$.
e. $2!(6 - 2)!$.
f. $(10 - 2)!$.

3.1.7. Compute:
a. $\binom{4}{4}$.
b. $\binom{3}{2}$.
c. $\binom{5}{0}$.
d. $\binom{5}{3}$.

e. $\binom{5}{1}$.
f. $\binom{4}{3}$.

3.1.8. Use Exercise 3.1.7 to find the following without doing any further computations:
a. $\binom{5}{5}$.
b. $\binom{3}{1}$.
c. $\binom{5}{2}$.
d. $\binom{5}{4}$.
e. $\binom{4}{1}$.
f. $\binom{4}{0}$.

3.1.9. Compute:
a. $\binom{7}{3}(0.20)^3(0.80)^4$.
b. $\binom{8}{0}(0.70)^0(0.30)^8$.
c. $\binom{10}{8}(0.10)^8(0.90)^2$.

3.1.10. Compute the following binomial probabilities:
a. $b(y;3,0.25)$ for $y = 0, 1, 2, 3$.
b. $b(y;4,0.30)$ for $y = 0, 1, 2, 3, 4$.
c. $b(y;5,0.10)$ for $y = 0, 1, 2, 3, 4, 5$.

3.1.11. Use Exercise 3.1.10b to find the binomial distribution $b(y;4,0.70)$ without doing any further computations.

3.1.12. Find the expected value and variance for the blood-type example by using the formulas given in Section 2.4.

3.1.13. Repeat Exercise 3.1.12 using the special formulas for the expected value and variance of a binomial distribution that are given in this section.

3.1.14. An experimental psychologist has 20 volunteers for a sensory perception experiment and wishes to draw a random sample of 10 of these volunteers. Suppose that he decides to write all combinations of 10 names on index cards and then draw one of the cards at random. How many combinations will there be?

3.1.15. A geneticist studying dairy cattle has four bulls and eight cows that can be used in an experiment. How many different matings are possible?

3.1.16. There are six teams in a baseball conference.
a. How many games are necessary before each team plays every other team once?
b. If there are no ties in standings, how many ways can the teams be ranked on the basis of number of games won?

3.1.17. Twelve school photographs (all the same size) are placed in random order face down on a table. Two of them are of identical twin boys. One of the twins is brought into the room and asked to select a photograph.

a. What is the probability that he will select his own by chance?

b. What is the probability that he will select his own or his brother's?

c. If he is asked to select two photographs, what is the probability that he will select his own and his brother's?

3.1.18. There is evidence that among lower forms of animal life, behavioral characteristics can be transferred from one individual to another along with the transfer of the chemical substance known as RNA. In an experimental study of this transfer behavior, eight salamanders are divided at random into two equal-sized groups of four. One group will be the experimental group and the other the control group.

a. Show that there are 70 different ways the two groups can be formed.

b. What is the probability that the four fastest swimmers are all in the same group?

c. What is the probability that three of the four fastest swimmers are in the same group?

d. All of the salamanders in one group (called the experimental group) receive RNA from a salamander that has been trained to swim fast. The other group (called the control group) receives RNA from an untrained salamander. Before one could believe that behavior is transferred with RNA, what should the number of fastest swimmers in the experimental group be? Explain.

3.1.19. Many candy manufacturers who have used artificial chocolate claim that their customers cannot tell it from real chocolate. Suppose five customers are selected at random and each is allowed to taste a candy bar made with real chocolate and the same kind of bar made with artificial chocolate. They are not told which contains real chocolate and they are asked which one it is.

a. If the manufacturer is correct about their inability to tell real from artificial chocolate, find the probability that a taster will correctly choose the one that is the real chocolate.

b. What is the probability that all five tasters will choose correctly?

3.1.20. A certain basketball player has a success record of one in three for making attempted field goals. Suppose he attempts seven field goals in a game.

a. What conditions must be true in order to use the binomial distribution to produce reliable probability statements?

b. Assuming the necessary conditions are met, compute the probability that the player will make exactly four field goals.

3.1.21. A night watchman must check in at nine stations in a warehouse during each round of inspection. He decides to try all possible sequences of the nine stations and use the shortest of these as his routine round of inspection. There are 9! possible different sequences of the stations.

a. Why are there 9! different sequences?

b. How many sequences must he try?

c. If he walks four rounds of inspection each night, how many nights will he require to try all possible sequences?

3.1.22. A sociologist examines six northern cities that have the same percentage of racial minorities. He is able to rank the cities according to employment opportunities for high-school graduates from the minority groups. He then orders the cities on the basis of truancy among minority high-school students.

a. How many ways is it possible to order six cities on the basis of truancy among minority students?

b. If ordering by truancy and by job opportunities are unrelated, how likely is it that truancy will have a perfect reverse ordering to job opportunities?

c. If the truancy ordering is the exact reverse ordering of that for job opportunities, should the sociologist decide that this happened by chance and that there is no relationship between the two?

3.1.23. A person claims the extrasensory ability of looking at a photograph and telling whether the subject of the photograph is still living or has died. In an experiment to test his claimed ability, he is shown 10 photographs of people unknown to him. (To improve the experiment the subjects should be of the same age and the photographs taken at the same time; a high-school yearbook would meet both conditions.) He is asked to point out the five subjects who are now dead.

a. How many ways can he select five of the 10 photographs?

b. How many ways can he select the photographs of the five dead subjects?

c. What is the probability of selecting the correct five photographs by guessing rather than by extrasensory ability?

d. Why should this be a double-blind experiment?

3.1.24. The grading of laboratory reports is tedious, so a laboratory instructor decides that he will grade only a randomly chosen two of the five reports that each student has submitted. If both are acceptable the student will be given an *A* as his laboratory grade; if one is acceptable he will receive a *B*; a *C* will be given if neither is acceptable.

a. How likely is a student to receive an *A* when he has submitted five acceptable reports? Four? Three? Two? One? Zero?

b. How likely is a student to receive a *C* when he has submitted five acceptable reports? Four? Three? Two? One? Zero?

3.2. TESTING HYPOTHESES

We return to the basic statistical problem of using probability to make decisions about populations that are not totally accessible. The following example shows how the probabilities in a theoretical binomial distribution can help to interpret the results of an experiment. (We have already seen an example in the baby-cereal preference study, Example 1.4 and Section 2.3.)

Example 3.1. Using a Binomial Distribution to Test a Hypothesis
Dairy farmers need many more cows than bulls, so it would be advantageous to them if a method were found to control the sex ratio of calves. Left to chance, the ratio of female to male calves is approximately one to one. Many biological experiments have attempted to alter sex ratio. Some experimenters have tried to separate the sperm cells that produce male offspring when an egg cell is fertilized from those that produce females. Others have attempted selective inactivation of one of the different kinds of sperm cell by raising or lowering the acidity (pH) of the fertilization environment.

A reproductive physiologist claims that by treating the semen of the bull with a mild acid and using artificial insemination, he can change the sex ratio of calves. (This is the scientific hypothesis.) He decides to perform an experiment and observe 20 calves that have been produced by this method. He is going to use statistics in order to generalize the result from these 20 calves to the entire population of calves that could be produced by this method. Thus, the statistical procedure begins at this point, prior to the actual experiment.

The steps in the statistical procedure are:

1. State the null hypothesis.
2. State the alternative hypothesis.
3. Establish α, the level of rejection, and the region of rejection.
4. Perform the experiment and observe the outcome.
5. Draw conclusions.

STEP 1. STATE THE NULL HYPOTHESIS. In this experiment, H_0: $\pi = 0.5$, that is, under chance alone, the probability of a newborn calf being female is 0.5. In other words, the treatment has no effect on the sex ratio. The theoretical probability distribution if the null hypothesis is true is $b(y;20,0.50)$.

This experiment can be done in such a way that it satisfies the five conditions of a binomial experiment: There are only two possible outcomes, a male calf or a female calf. There will be a repeated number of trials, 20. If the null hypothesis is true, $P(\text{female calf}) = 0.5$ for each trial. The 20 cows can be selected at random, and the semen can also be selected at random from different bulls, ensuring independence from trial to trial. The physiologist is interested in the statistic y, in this experiment the number of female calves born.

STEP 2. STATE THE ALTERNATIVE HYPOTHESIS. In this experiment, the alternative hypothesis is H_a: $\pi \neq 0.5$. Since the physiologist does not know ahead of time what effect the mild acid will have on the sex of newborn calves, this is a two-sided, or a *two-tailed test*. He will reject the null hypothesis if the outcome is an extreme case in either tail of the binomial distribution.

STEP 3. ESTABLISH α, THE LEVEL OF REJECTION, AND THE REGION OF REJECTION. Looking at the binomial distribution $b(y;20,0.50)$, he wants to set a rejection level as close to 0.05 as possible (because this is a traditional level used). Since this is a two-tailed test, he wants to reject the null hypothesis if he obtains an outcome with a probability of less than 0.025 at either side of the distribution. He notes from Table 3.2 that

$$\begin{aligned} &P(0 \text{ or } 1 \text{ or } 2 \text{ or } 3 \text{ or } 4 \text{ or } 5) \\ &= P(0) + P(1) + P(2) + P(3) + P(4) + P(5) \\ &= 0.000 + 0.000 + 0.000 + 0.001 + 0.005 + 0.015 \\ &= 0.021 \end{aligned}$$

and that

$$\begin{aligned} &P(15 \text{ or } 16 \text{ or } 17 \text{ or } 18 \text{ or } 19 \text{ or } 20) \\ &= P(15) + P(16) + P(17) + P(18) + P(19) + P(20) \\ &= 0.015 + 0.005 + 0.001 + 0.000 + 0.000 + 0.000 \\ &= 0.021 \end{aligned}$$

so he sets $\alpha = 0.042$. The region of rejection is all y such that $0 \leq y \leq 5$ and $15 \leq y \leq 20$, and y is called the *test statistic*. Including any more values in the region of rejection would have made α further from 0.05. The symbol y here stands for the number of female calves born (alternatively, y could stand for the number of male calves born).

STEP 4. PERFORM THE EXPERIMENT AND OBSERVE THE OUTCOME. The experiment is now performed. For the sake of this example, let us assume that two males and 18 females are born. Even without formal statistical procedures, the experimenter would be encouraged by such results. A nine to one ratio, in either direction, does not seem likely to occur by chance.

STEP 5. DRAW CONCLUSIONS. The α level and the region of rejection merely specify, prior to the experiment, those outcomes that can be considered plausible and those that would be unusual when the null hypothesis is true. In this experiment, outcomes of less than six or more than 14 occur only 0.042 of the time if the null hypothesis is true. Since $y = 18$, the physiologist decides to reject the null hypothesis, that is, it seems that the change in acidity has an effect on the sex ratio. Looking at the direction of the result, 18 females to two males, he further concludes that the mild acid treatment increases the proportion of females above 0.5.

The statistical procedure was two-sided. The experimenter would have been encouraged by a significant deviation in either direction. It would indicate the validity of the association between sex ratio and the *p*H of the fertilization environment. In further experiments he could adjust the acidity in the manner that should produce more females.

Once again, let us remember that it is not known for sure that the experimenter in the above example has found an effective method of changing the sex ratio of calves. The result of the experiment could have been a rare chance outcome. The only basis the experimenter had for his decision was that the outcome was not probable under the null hypothesis and chance alone.

This process of setting up the null hypothesis may still seem rather roundabout, since the null hypothesis is usually the opposite of the decision the scientist is hoping to make. However, since there is no information about the probability associated with the experimental hypothesis, the null hypothesis must be set up so that known probabilities can be used.

Not all tests of hypotheses are two-tailed. Sometimes the experimenter is looking for evidence in a particular direction. The following example will illustrate a *one-tailed test* of hypothesis.

Example 3.2. Testing a Hypothesis Using a Binomial Distribution

The staff of a reading clinic is interested in determining the sex ratio of children who have a certain reading problem. The children reverse the letter sequences in words; for example, they read "saw" for "was." Someone has claimed that more than 70% of the children with this disorder are boys. The staff decides to look at a random sample of 20 children who have this reading problem. The null hypothesis is H_0: $\pi = 0.7$ and H_a: $\pi > 0.7$ because they are looking for evidence to substantiate the claim. Assuming the null hypothesis is true, they use the binomial distribution $b(y;20,0.70)$ as the theoretical model. The number of boys in the random sample of children with this disorder is represented by y.

The level of rejection in this survey is chosen to be as close to 0.05 as possible. Looking at Table 3.2 in Section 3.1, the actual α is seen to be 0.036 and the region of rejection is 18, 19, 20.

Assume the survey reveals that 18 out of the 20 afflicted children are boys. Since the test statistic 18 is in the region of rejection, the null hypothesis is rejected and it is concluded that there is evidence that more than 70% of the children with this disorder are boys.

We have noted that with this type of test there is no way to be certain whether the null hypothesis is true or false. Although the null hypothesis was rejected in the example above, it is of course possible that it is actually true and a very unlikely outcome just happened to occur. To reject a true null hypothesis is called a *Type I error*. The probability of committing a Type I error in the survey above is 0.036 because $\alpha = 0.036$, that is, there is a 3.6% chance that the null hypothesis is true and sample results lead to rejection of it. The probability of a Type I error is always α, the level of rejection, and is chosen by the experimenter.

If the results had been different, the null hypothesis might not have been rejected. For example, the survey might have shown that 15 out of 20 children displaying reading reversals were boys. Since 15 is not in the region of rejection, the null hypothesis would not have been rejected, and it could be concluded that among the children with reading reversals 70% or fewer may be boys. In this case, it is possible that the null hypothesis is false, but it has not been rejected. To fail to reject a null hypothesis when it is false is called a *Type II error*.

It is more difficult to determine the probability of a Type II error than a Type I error. The probability of a Type I error, rejecting a true null hypothesis, is α. The probability of a Type II error is, in this case, the probability that y is not in the region of rejection of the null hypothesis if π is not 0.70. This cannot be determined in this form because there is no specific value for π; $\pi \neq 0.70$ is an infinity of values.

In order to determine the probability of a Type II error:

1. Choose a reasonable *specific* alternative value of the parameter, $\pi = \pi_a$, that is of clinical importance.
2. Find β, the cumulative frequency in $b(y;n,\pi_a)$ for y in the acceptance region of H_0; that is, $\beta = P(y$ is in the region of acceptance of H_0 if $\pi = \pi_a)$.

The probability β is the probability of failing to reject the null hypothesis when it is false by a specific amount. In more positive terms, the *power* of the experiment or survey, that is, the probability of detecting the specific alternative hypothesis, is $1 - \beta$.

In the example above, in which 15 out of 20 children with reading reversals were boys, the null hypothesis was not rejected. What is the probability that a false null hypothesis may have been accepted? From knowledge of reading problems, the staff might agree that a reasonable alternative value is $\pi = 0.75$. Power depends on the "degree of falseness" of the null hypothesis, so they specify the smallest degree of falseness of practical interest. This means that if in fact 75% of the cases of reading reversals occur in boys, the clinic would examine boys very carefully for this problem, but if fewer than 75% were boys they would not examine boys more closely than girls. Referring to the table in the previous section under $\pi = 0.75$, we find that the probability that $0 \leq y \leq 17$ is $\beta = 0.909$. This means that there is a 90.9% chance of *failing* to reject the null hypothesis if in fact 75% of the children with reading reversals are boys! The chance of detecting the difference is only $1 - 0.909 = 0.091$; the power of this survey is very low.

A *powerful* experiment means a power of 0.70 or greater, so the survey above is very poor. This illustrates the need to design an experiment in such a way that there is a reasonable chance of detecting a clinically important difference if it exists. In order to increase the power in this survey a much larger sample size is necessary. Another way to increase the power (decrease β) is to increase α.

In practice, many times we do not have enough information to choose a reasonable specific alternative and thus we are not able to compute β. Fortunately, the power of an experiment usually increases with the size of the sample, so we work with samples that seem large enough to make the experiment powerful. If we can specify the alternative value of the parameter, it is possible to use a repetitive process (possibly with the aid of a computer) to determine how large the sample size must be in order to have a specified power. In the reading-reversal example, it is necessary to use a sample size of $n = 501$ in order to achieve a power of 0.80 in detecting $\pi_a = 0.75$ when the null hypothesis is H_0: $\pi = 0.70$ and $\alpha = 0.05$ (Buckalew 1974, p. 61). This large size is required because a relatively small difference is specified.

We usually try to achieve a balance between the α level and the power. We want a moderately low α level (as 0.05) and try to get the power as high as possible, usually by taking relatively large samples.

Which type of error is worse depends on the situation. For example, imagine that an experimenter is testing a new antibiotic for effectiveness against a particular bacterium. Currently used antibiotics are known to have a cure rate of $\pi = 0.75$. The two types of error could occur under the following circumstances:

Type I. The experimenter is testing H_0: $\pi = 0.75$ against H_a: $\pi > 0.75$. The new antibiotic actually has a cure rate of 0.75, but the results of the experiment lead her to conclude that it is better than the antibiotics currently used. If the new one is equal to the others in all other respects, such as price and side effects, then this Type I error is not serious. If, however, the price is higher or the side effects are more serious, then the Type I error is serious.

Type II. The experimenter is again testing H_0: $\pi = 0.75$ against H_a: $\pi > 0.75$. Now, however, let us assume that the new antibiotic is actually better but she fails to detect this from the results of the experiment. The

Type II error here means that a more effective medication will not be used. The seriousness of the error depends on the seriousness of the illness and how much better the new medicine would be. If π is actually 0.78, this would not be much of an improvement so the error is not as serious as if π were 0.98 and a very effective medication were not being used.

The diagram in Figure 3.2 summarizes the various possibilities that occur when testing hypotheses. The specific probabilities listed refer to the reading-reversal study (Example 3.2) used in this section.

Note that the probabilities in the columns of this diagram sum to one. Also, once the decision is made only one type of error is possible. If the null hypothesis is rejected, there is then no possibility of a Type II error. Similarly, if we fail to reject the null hypothesis, we no longer need to worry about a Type I error.

In the discussion of hypothesis testing and errors in this section, we have used only examples that fit the small table of binomial distributions given in Section 1 of this chapter. Two similar but larger tables are found in the Appendix: Table A.4a, for samples of size $n = 20$, and Table A.4b, for samples of size $n = 25$. These tables are used in the same manner as the smaller table in Section 3.1.

If $\alpha = 0.10$ and the test is two-tailed, the horizontal lines indicate the regions of rejection and acceptance. If $\alpha = 0.05$ and the test is one-tailed, the line in the appropriate tail may be used to indicate the region of rejection. Other α levels can be used, but then the regions must be determined by the user of the table. The probability of a Type II error can also be found from these larger tables; the method is the one just described in this section.

Many other tables are readily available in statistics books and in reference books. If the particular table needed is not available, it can be computed using the Bernoulli formula. Approximation methods are also possible; these are discussed in Chapter 7.

A brief summary of this section follows.

Decision About H_0 Based on Test of Significance		H_0 is really TRUE	H_0 is really FALSE
	ACCEPT	*No error*: $1 - \alpha = 0.964$	*Type II error*: P (Type II error) $= \beta$; $\beta = 0.909$
	REJECT	*Type I error*: P (Type I error) $= \alpha$; $\alpha = 0.036$	*No error*: Power $= 1 - \beta = 0.091$

FIGURE 3.2. Type I and Type II errors.

■ ■ ■

Procedure. Test of Hypotheses for a Binomial Parameter π.

H_0: $\pi = \pi_0$

H_a: $\pi \neq \pi_0$ or $\pi > \pi_0$ or $\pi < \pi_0$

Significance level: α

Test statistic:

y, the number of successes out of n trials

Using a table for the binomial distribution with probability function $b(y;n,\pi_0)$, determine the region of rejection.

For H_a: $\pi \neq \pi_0$, the region of rejection is $0 \leq y \leq c_L$ and $c_U \leq y \leq n$ such that

$$\sum_{0}^{c_L} b(y;n,\pi_0) \quad \text{and} \quad \sum_{c_U}^{n} b(y;n,\pi_0)$$

are each as close as possible to $\alpha/2$.

For H_a: $\pi > \pi_0$, the region of rejection is $c_U \leq y \leq n$ such that

$$\sum_{c_U}^{n} b(y;n,\pi_0)$$

is as close as possible to α.

For H_a: $\pi < \pi_0$, the region of rejection is $0 \leq y \leq c_L$ such that

$$\sum_{0}^{c_L} b(y;n,\pi_0)$$

is as close as possible to α.

Reject H_0 if y is in the region of rejection.

P(Type I error) $= \alpha$.

P(Type II error if $\pi = \pi_a$) $= P$(y is in the region of acceptance of H_0 if $\pi = \pi_a$).

■ ■ ■

EXERCISES

3.2.1. Use Tables A.4a and A.4b in the Appendix to find the following:

a. $P(4 < y < 8)$ when $n = 20$, $\pi = 0.8$.

b. $P(y \leq 2)$ when $n = 25$, $\pi = 0.6$.

c. $P(y \geq 4)$ when $n = 25$, $\pi = 0.25$.

d. $P(y > 15)$ when $n = 20$, $\pi = 0.70$.

e. $P(y < 19)$ when $n = 20$, $\pi = 0.55$.

f. $P(6 \le y \le 9)$ when $n = 25$, $\pi = 0.35$.

3.2.2. A teacher gives a student a make-up test consisting of 20 true–false questions. The intent of the test is to determine whether the student answers the questions correctly through knowledge of the material or merely by making lucky guesses. Assume the correct answers are a random sequence of "true" and "false," and that the student's guesses are also random.

a. State a null hypothesis based on the probability of guessing the correct answer to a question.

b. State a one-tailed alternative hypothesis based on the probability of arriving at the correct answer through knowledge.

c. Find the region of rejection when α is set as close to 0.05 as possible. (Remember that the null hypothesis will be rejected only if an extreme value occurs on one side of the distribution.)

d. What should the teacher conclude if the student correctly answers 16 of the 20 questions?

3.2.3. A carnival operator wants a game that can be won about 30% of the time. If the game is won more frequently, it will not be economical for the operator; if winning is less frequent, potential players will be reluctant to risk their money. He devises a dart-tossing game that he thinks will suit his criterion and tests it on 20 random players.

a. State a null hypothesis based on his criterion.

b. State a two-tailed alternative hypothesis.

c. If the region of rejection is set at $0 \le y \le 2$ and $11 \le y \le 20$, what is the α level?

d. What conclusion should the operator draw about the game if there are nine winners among the first 20 players? What must be assumed about the players in order to accept this conclusion?

3.2.4. A campus parking lot contains 20 spaces, all reserved for faculty members. The administration decides that students may park their cars in the lot after 4:00 PM if faculty usage then drops to less than 70%. A random weekday afternoon is chosen to sample the faculty usage after 4:00 PM.

a. State the null hypothesis.

b. State a one-tailed alternative hypothesis that would lead to student usage of the lot.

c. Find the region of rejection for α as close to 0.05 as possible.

d. What decision should be reached about student parking if there are 18 faculty cars in the lot at the time of the survey?

e. Do you see any difficulties in the design of this survey? Suggest a better design.

3.2.5. In the experiment concerning the altering of the sex ratio in newborn calves (Example 3.1), the null hypothesis is H_0: $\pi = 0.5$ and H_a: $\pi \neq 0.5$. There are 20 trials and the region of rejection is $0 \leq y \leq 5$ and $15 \leq y \leq 20$.

a. The physiologist would consider the experiment a success if the proportion of female calves is 0.70. How likely is it that a change of this magnitude will be detected by the statistical procedure?

b. What would you suggest to the physiologist if he does not think that this experimental design is powerful enough to detect this useful change?

3.2.6. In an effort to control mosquitoes without having to use dangerous insecticides, entomologists have taken advantage of two factors in the biological nature of mosquitoes: male mosquitoes are not bloodsuckers, and nearly all female mosquitoes mate but once. Thus the entomologists release massive numbers of sterilized male mosquitoes to reduce the probability of a female mating with a fertile male and consequently producing more mosquitoes. After such a release, the entomologists hypothesize that the probability of a female mating with a fertile male is H_0: $\pi = 0.30$. If 20 females are captured and examined for fertile eggs:

a. Find the region of rejection if the alternative hypothesis is H_a: $\pi > 0.30$.

b. What is the power of the experiment if $\pi_a = 0.50$?

c. What is the power if $\pi_a = 0.70$?

3.2.7. A large corporation is going to purchase 150 company cars for its salesmen and executives. The corporation has already eliminated many makes and models and now must choose between two specific types of cars, *A* and *B*, which are comparable in size, purchase price, and maintenance cost. The corporation will base its final decision on the gasoline mileage of these two types. It is known that 70% of the cars of type *A* average more than 20 miles per gallon, and it is strongly believed that car *B* has a better record. If *B* is proved better they will buy *B*, otherwise they will buy *A*.

a. State the two outcomes that should be considered for a random sample of cars of type *B*.

b. State the null hypothesis in terms of cars of type *B*.

c. State the one-tailed alternative hypothesis for car *B*.

d. Which type of error should be kept to a minimum in this experiment? How can this be accomplished?

3.2.8. A behavioral scientist feels that right-handed people have a tendency to make right-hand turns when they have no other basis for choosing the direction in which they should turn. To conduct a statistical test, he draws a random sample of 20 right-handed individuals from a large group of volunteers. To keep the subjects unaware of the nature of the experiment, he pretends to be conducting a survey of family dietary habits. He has the subjects brought into his office one at a time, questions them about the eating habits of their families, and then directs them out by a different way from the one by which they entered. They are told to go down a hall and out either door at the end. The experimenter watches each subject leave and records whether the subject chooses the door to the right or left as he or she exits.

a. State a null hypothesis which specifies that only chance leads to the choice of the door to the right.

b. For a two-tailed alternative hypothesis, the region of rejection could be $0 \le y \le 5$ and $15 \le y \le 20$. What is the α level?

c. For a one-tailed alternative hypothesis, the region of rejection could be $14 \le y \le 20$. What is the α level?

d. For the specific alternative $\pi_a = 0.70$, which is more powerful, the one-tailed or the two-tailed test?

e. Comment on the deception involved in this experiment.

3.2.9. For a binomial experiment in which $n = 20$ and H_0: $\pi = 0.30$:

a. Find the region of rejection with an α as near 0.05 as possible when H_a: $\pi \neq 0.30$.

b. Find the region of rejection with an α as near 0.05 as possible when H_a: $\pi > 0.30$.

c. For the specific alternative $\pi_a = 0.50$, how much more powerful is the one-tailed test than the two-tailed test?

d. Which of the following statements is true?

i. The one-tailed test is more powerful because it has a greater α level.

ii. The one-tailed test is more powerful because it has a greater β.

iii. The one-tailed test is more powerful because there are more possible y values in its region of rejection.

iv. The one-tailed test is more powerful because the sum of the probabilities associated with the region of rejection is greater for the specified alternative $b(y;20,0.50)$.

3.2.10. After a flood or storm, insurance companies buy damaged goods from stores that carry their policies. To recover some of the loss, they sell the damaged goods to salvage companies. Suppose 30,000 flood-damaged flashbulbs are offered for sale by an insurance company with the claim that 25% of them are too damaged to flash.

a. State a null hypothesis that would test the insurance company's claim.

b. State the alternative hypothesis of greatest concern to the insurance company.

c. State the alternative hypothesis of greatest concern to a salvage company.

d. Suppose the insurance company's statement about the 30,000 bulbs is correct. How likely is it that a random sample of 20 bulbs will have:

   i. Exactly 10 bulbs that fail to flash?

   ii. At least 10 (that is, $10 \leq y \leq 20$) that fail to flash?

e. Suppose the insurance company's statement is incorrect, and actually 40% are too damaged to flash.

   i. What is the probability that exactly 10 will fail to flash?

   ii. What is the probability that at least 10 will fail to flash?

f. Suppose H_0: $\pi = 0.25$ is being tested, what is the power of the test when α is as near as 0.05 as possible and π is really 0.40?

3.2.11. With respect to Exercise 3.2.10, suppose a demonstration is given for salvage company representatives. A random sample of 20 flashbulbs is taken, the bulbs are tested in a camera, and just 10 of them flash. What decision should be made about H_0 if $\alpha = 0.05$ and:

a. The insurance company's alternative hypothesis is used?

b. The salvage company's alternative hypothesis is used?

3.2.12. A consumer advocate claims that 35% of the cars of a certain make are being equipped with faulty mufflers. A concerned local dealer assumes that 25 cars he has in stock constitute a random sample, so he has his mechanics detach the mufflers and test them for the stated defect.

a. What is the most logical null hypothesis?

b. Which alternative would be most favorable to the automobile dealer?

c. Find the region of rejection suitable for this alternative hypothesis with α as near 0.05 as possible.

3.2.13. Describe how a Type I or Type II error could occur in the following situations, and give some of the factors that would determine the seriousness of the errors.

a. A bookstore is trying to determine what proportion of the students buying a certain textbook will also buy an optional student guide. In the past, 40% of the students buying the text have also bought the guide. The bookstore wants to test H_0: $\pi = 0.40$ against H_a: $\pi > 0.40$.

b. A seed company wants to claim on a certain seed package that at least 90% of the seeds will germinate. The company decides to check this before the packages are printed and test H_0: $\pi = 0.90$ against H_a: $\pi < 0.90$.

c. A recreation specialist is planning campsite facilities for a state forest and wants to include several rustic tent-only campsites that will be inaccessible to campers on wheels. He thinks that only 20% of the people camping in the area would desire such facilities. He tests H_0: $\pi = 0.20$ against H_a: $\pi \neq 0.20$.

3.2.14. Archaeologists use pelvic bones to determine whether a skeleton is that of a man or woman. Primitive cultures often buried their outstanding members (rulers, warriors, athletes, and so on) with greater ceremony than ordinary members. Using this fact, much can be learned about the status of women in an early culture by observing the frequency of skeletons of females in ceremonial graves. Suppose that an archaeologist discovers 20 graves that can be assumed to be a random sample of the ceremonial graves of a Stone Age culture in Wiltshire, England.

a. What is the most logical statistical hypothesis to be tested?

b. Suppose the region of rejection is: The number of skeletons of females is less than eight. What is the value of α?

c. Suppose $\pi_a = 0.30$; what is the numerical value of β?

d. What assumption is necessary to use this test procedure?

3.2.15. A certain dental condition, which can be corrected if detected early enough, occurs in the population with a frequency of $\pi = 0.20$. An orthodontist believes this condition occurs more frequently in children who were born with cleft palates, and that parents of such children should be warned to watch for early evidence of the dental condition. To test his hypothesis, he follows the dental development of a random sample of 25 children born with cleft palates.

a. What is the most logical null hypothesis for the orthodontist to check? What alternative hypothesis should he use?

b. Suppose he wants α to be as close to 0.05 as possible, what region of rejection should he set for y, the number of children in the sample who develop this dental condition?

c. Suppose eight of the children in his sample develop the condition. Should he reject the null hypothesis? Why, or why not? What conclusion should he draw?

3.2.16. Sickle-cell anemia is a potentially lethal genetic disease in the Black race. It is estimated that 30% of Blacks in a certain Gulf Coast region have the disease or carry the trait for it. This figure seems too large to a physician in the region, so he takes a random sample of 25 of his Black patients and examines blood smears.

a. State the physician's most logical null and alternative hypotheses.

b. What region of rejection would you suggest he use? What is the α level for this region?

c. If the percentage in question is really 15, what is the power of his test?

d. Which type of error is more serious in his study, Type I or Type II? Why?

e. Suppose 12 patients of his sample have the disease or seem to be genetic carriers, should he reject his null hypothesis or not? Why? What conclusion should he draw about the proportion of sickle-cell anemia in the Black population?

3.2.17. In a very large population of females and males, 20% are color-blind. A geneticist wishes to determine whether the type of color blindness found in this population is more prevalent in males than in females. He examines a random sample of 100 members of the population.

a. What is the most logical null hypothesis for the geneticist to test about the sex ratio?

b. What is the most logical null hypothesis for him to test about color blindness?

c. Assuming independence, and that half the population is male, what is the expected number of color-blind females in the sample?

d. Suppose he takes a random subsample of 25 females.

  i. What is the most logical null hypothesis for him to test about color blindness? What is the alternative hypothesis?

  ii. Given α near 0.10, what is the region of rejection?

  iii. Suppose none of the females in the subsample are color-blind; what decision does he make about the null hypothesis? Interpret this decision.

e. Suppose he takes a random subsample of 25 males.

  i. What is the most logical null hypothesis for him to test about color blindness? What is the alternative hypothesis?

  ii. Given α near 0.10, what is the region of rejection?

  iii. Suppose seven of the males in this subsample are color-blind; what decision should he make about the null hypothesis? What conclusion should he draw?

3.2.18. Cryobiologists have been experimenting for many years with methods of freezing human corneas so that when thawed, the membranes can be safely used in "eye transplants." If corneas are suspended in ethylene glycol, 70% of membranes survive freezing and thawing.

Unfortunately the chemical compound is toxic, and therefore a cornea soaked in it is unsafe for transplant. Suppose a cryobiologist finds a nontoxic chemical that has similar protective properties. He wants to compare its effectiveness with ethylene glycol in the freezing–thawing process.

a. State the null and alternative hypotheses.
b. If 20 corneas are to be used in his experiment, give the region of rejection for $\alpha = 0.10$.
c. Suppose $y = 10$ is the number that survive; should the experimenter feel encouraged or discouraged by the results? Give a reason for your answer.

3.2.19. Vegetable farmers try to avoid the use of insecticides because of expense and health hazards. However, if crops become too heavily infested, it becomes necessary to spray them. Suppose a farmer decides that he will spray his cabbages if their infestation with moth larvae is greater than 20%.

a. If the farmer samples the crop to determine the percentage of infested cabbages, what is the null hypothesis?
b. What is the most logical choice for the alternative hypothesis? Why?
c. For $n = 20$ and α as close to 0.05 as possible, choose the region of rejection that is consistent with the alternative hypothesis.

3.2.20. In times of stress, some people hyperventilate to the point of dizziness and fainting. To determine whether this behavior is equally likely in men and women, a researcher takes a random sample of 25 cases from a hospital emergency room's file on those treated for hyperventilation.

a. What hypothesis should be tested about the percentage of males among those treated?
b. What should the region of rejection be if α is to be as near 0.01 as possible?
c. If 16 of the 25 persons in the sample are men, should the researcher conclude that men are more likely to hyperventilate than women? Why or why not?

3.3. ESTIMATION

So far, our discussion of statistical methods has dealt with only one of the general problems of statistics, decisions about hypotheses. Tests of hypotheses are possible only when we have quite a bit of information about the experimental situations. For example, to analyze the results of the ex-

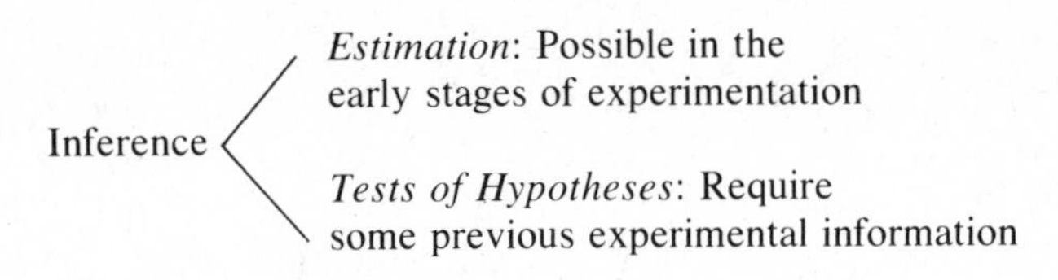

FIGURE 3.3. Types of inference.

periment on the sex ratio of calves, the experimenter had to know the sex ratio of newborn calves in an untreated population. In the early stages of experimentation, when less information is available, the scientist often uses *estimation* (Figure 3.3).

Estimation will answer questions like "What proportion of ex-prisoners who have gone through a certain group therapy program will be arrested again within the first two years after release?" If we consider the entire population of prisoners who have gone through or will have gone through the program during their incarceration, and we use as the variable of interest whether or not they are arrested again within two years after release, what is the appropriate value of π, the proportion arrested again?

Since we cannot observe the entire population, we will instead examine a random sample from it and count the number of subsequent arrests in the sample. Recall that this count, based on the results of sampling, is called a *statistic*. Then, using the binomial distribution as a model for this study, we will use the statistic to make a statement about the unknown parameter π, the true proportion of ex-prisoners who will be arrested again (Figure 3.4).

In trying to estimate the unknown parameter, two types of estimates are possible:

1. A *point estimate*—a statistic based on a sample.
2. An *interval estimate*—an inference based on a statistic.

The natural point estimator of a proportion π is

$$\hat{\pi} = \frac{y}{n}$$

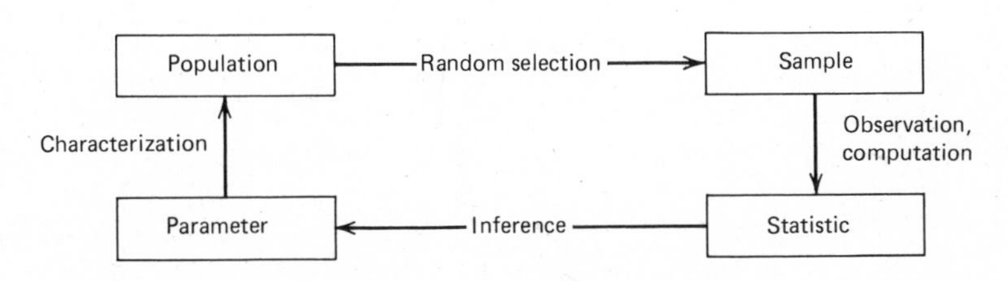

FIGURE 3.4. The inferential process.

in which y is the number of successes in a sample size of n. The estimator is read "π hat." In general, placing a "hat" on a Greek letter indicates an estimator of the parameter.

The estimator $\hat{\pi}$ is not only the natural point estimator but also the best because it has three desirable properties of an estimator:

1. $\hat{\pi}$ is a *maximum likelihood estimator*, that is, the estimate of π that we are most likely to get using this estimator is the true parameter π. We can see this by using Table 3.2. We can see that the value of y with the greatest probability gives the best estimate $\hat{\pi} = y/n$ of the binomial parameter π. In the distribution with probability function $b(y;20,0.30)$, $y = 6$ is the most probable outcome and $6/20 = 0.30$; in $b(y;20,0.50)$, $y = 10$ is the most probable outcome and $10/20 = 0.50$; in $b(y;20,0.70)$, $y = 14$ is the most probable outcome and $14/20 = 0.70$ (see Figure 3.5).
2. $\hat{\pi}$ is *unbiased*, that is, if we were to repeat the estimation process, the average of all possible estimates would be the true parameter π.

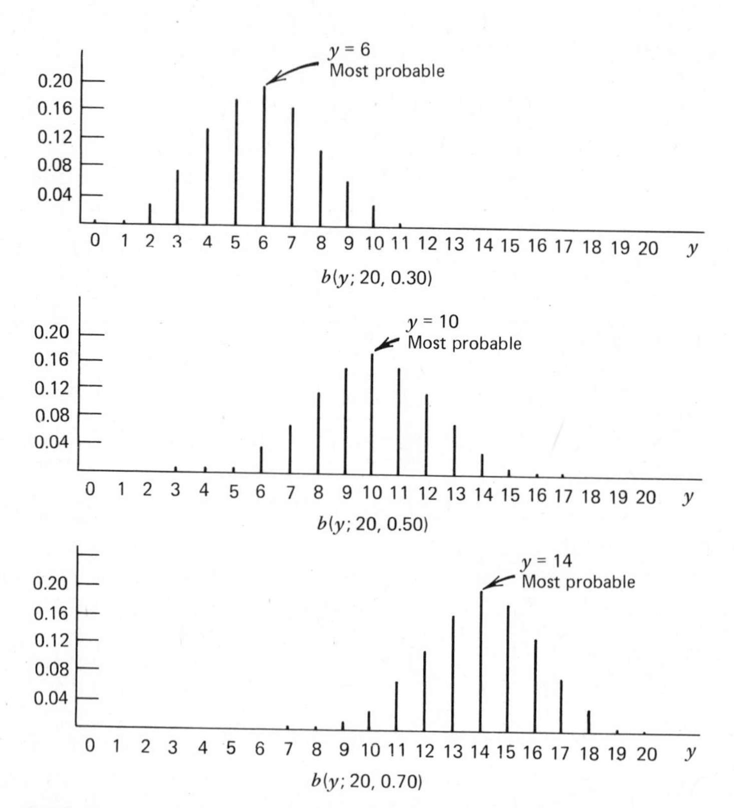

FIGURE 3.5. The most probable outcome in three binomial distributions.

3. $\hat{\pi}$ has a *minimum variance*, that is, the possible estimate closer to π than for any other unbiased estimator.

Thus if we observe a random sample of 20 prisoners who had the therapy program and we find that six of them have been a then the best point estimate of the proportion of subsequent

$$\hat{\pi} = \frac{6}{20} = 0.30$$

Because of the properties of this estimator, we can be confident that this is likely to be close to the true value. Unfortunately, it will usually not be the true value. A repetition of the survey might yield

$$\hat{\pi} = \frac{8}{20} = 0.40$$

Although we know that both of these estimates are close, we also know that probably neither of them is exactly correct.

One way to avoid this difficulty is to use an *interval estimate*, an inference that the parameter is between certain bounds.

We use the following steps to find an interval estimate.

■ ■ ■

Procedure. Central Confidence Intervals for π

1. Specify an α level.
2. Take a sample of size n.
3. Find y, the number of successes.
4. Give the interval of all values of π for which y would fall in the region of acceptance for a two-sided α-level test.†

For example, if $\alpha = 0.10$, $n = 20$, and $y = 8$, we use Table A.4a in the Appendix; 8 is in the region of acceptance for π between 0.25 and 0.55. Since $\alpha = 0.10$, when we use this procedure about 90% of the intervals obtained will include the actual parameter being estimated. The interval is written

$$\text{CI}_{0.90}: \quad 0.25 \leq \pi \leq 0.55$$

and is called the 90% *confidence interval* for π. This method yields a *central* confidence interval since two-sided regions of acceptance are employed. Note that the best point estimate, $\hat{\pi} = 8/20 = 0.40$, is within this interval.

† The authors are indebted to H. C. Fryer for the graphic determination of confidence intervals in this section and in Tables A.4a and A.4b.

For any given sample size, the method we just outlined gives the narrowest $CI_{1-\alpha}$. The confidence interval in this example is quite wide; this is because the sample size $n = 20$ is small. If a larger sample is used (and α remains constant), the same statistic $\hat{\pi} = y/n$ will yield a smaller confidence interval. To see this, Tables A.5a through A.5e in the Appendix can be used. These tables list the confidence intervals for various sample sizes and various α levels. (Instructions for reading these tables precede the group.) In order to see the effect of increased sample size, let $\alpha = 0.10$, $n = 100$, $y = 40$, then $\hat{\pi} = 40/100 = 0.40$ (as in the previous example), and from Table A.5c

$$CI_{0.90}: \quad 0.318 \leq \pi \leq 0.487$$

which is a smaller interval than the one found for $n = 20$.

Tables A.5a and A.4b give slightly different 90% confidence intervals for sample size $n = 25$. This difference occurs because Tables A.5a through A.5e were calculated by a different procedure than Tables A.4a and A.4b. The method for finding confidence intervals used in Tables A.4a and A.4b is very instructive, but lengthy to compute. The alternative shorter method used for Tables A.5a through A.5e will not be explained here; it is an approximate method and is known to produce reliable confidence intervals.

We can find one-sided confidence intervals as well as central confidence intervals. The method is the same except that the region of acceptance for a one-sided α-level test is used in step 4 of the Procedure given above. If Tables A.5a through A.5e are used, we refer to the α column that is twice as large as the desired α level and use only one of the values L or U that are given. (Example 3.3 demonstrates a one-sided procedure.)

Linear interpolation can be used to obtain confidence intervals for sample sizes between those listed in the tables, or it can be used for statistics that fall between values listed in the tables. This method of interpolation of confidence intervals is a conservative estimate because the confidence intervals actually decrease along curves within the straight lines along which interpolation occurs. Since the interpolated values are outside the actual curves, they more than preserve the α level of the tables (Figure 3.6).

As mentioned before, by using an interval estimate we avoid the almost certain error of a point estimate. If an interval estimate includes the true proportion, then it is correct. It is possible for two different interval estimates to be correct. For example, two polls on the proportion of the American population that approves of the President's economic policy could yield

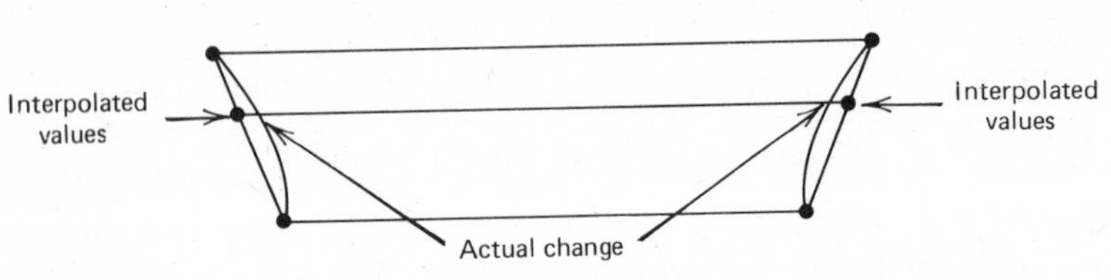

FIGURE 3.6. Linear interpolation yields conservative confidence intervals.

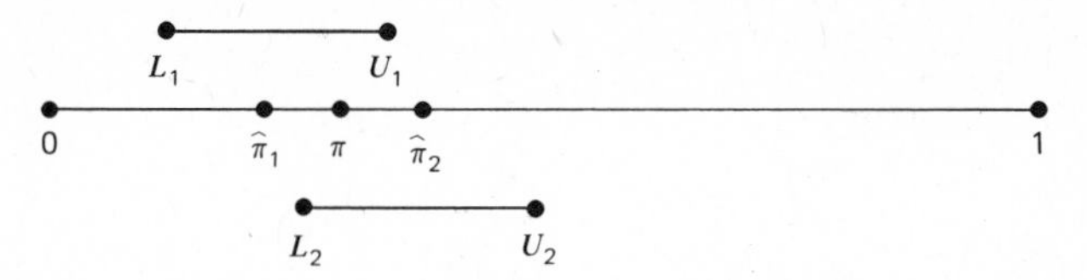

FIGURE 3.7. Confidence intervals for the same parameter obtained from different samples.

point estimates $\hat{\pi}_1$ and $\hat{\pi}_2$ and interval estimates as in Figure 3.7. If π is the true proportion, both point estimates are wrong. However, both interval estimates are correct. In this particular case, neither interval contains both point estimates, but both intervals are still correct.

The question of Type I or Type II errors does not apply to the inference of confidence intervals since no decisions concerning hypotheses are being made. However, the reliability of the estimate made by the confidence interval is expressed in the percentage of confidence. A level of confidence of 95% means that 95% of the intervals that could be determined by this method contain the true population parameter.

Although Tables A.5a through A.5e list confidence intervals, they may also be used to test hypotheses. This is demonstrated in the following example.

Example 3.3. Using Confidence Intervals to Test Hypotheses

It is generally felt that those opposed to the issuance of a new school bond are more likely to go to the polls to vote than those who favor the bond. Thus a local school board feels that a bond issue must be favored by more than 70% of the registered voters to have a chance of being approved in the bond election.

Since the school board is concerned about detecting whether enough people are in favor of the bond issue, it wants to determine a one-sided confidence interval on π that makes a statement about the smallest possible value that π might be.

Suppose a random sample of $n = 250$ registered voters is surveyed by the school board and $y = 190$ favor the bond issue while $n - y = 60$ oppose it. Using Table A.5d and $y/n = 190/250 = 0.76$, the table is entered at $1 - 0.76 = 0.24$ and the lower bound is $1 - 0.289 = 0.711$. The 95% confidence interval that puts a lower bound on π is

$$\text{one-sided CI}_{0.95}: \quad 0.711 \leq \pi \leq 1.00$$

(The 0.10 column is used because only the lower bound is needed.) This interval shows that the school board can schedule an election and feel confident that the bond issue will pass.

If the board preferred to phrase its investigation in terms of a test of hypothesis, it would test

$$H_0: \quad \pi = 0.70 \qquad \text{(bond issue may not pass)}$$

against

$$H_a: \quad \pi > 0.70 \qquad \text{(bond issue will pass)}$$

The board would find the one-sided confidence interval for the lowest value of π and conclude that the null hypothesis should be rejected at the 5% significance level because $\pi = 0.70$ is not in the interval.

Similar approaches can be used for two-sided alternatives and one-sided less-than alternatives. The correspondence between confidence intervals and tests of hypotheses is summarized in the following procedure.

■ ■ ■

Procedure. Testing Hypotheses Using Confidence Intervals

Confidence Interval	Test
Central $CI_{1-\alpha}$: $L \leq \pi \leq U$	H_0: $\pi = \pi_0$ H_a: $\pi \neq \pi_0$ α level of rejection Reject H_0 if π_0 is *not* in the confidence interval, that is: $\pi_0 < L$ or $\pi_0 > U$
Upper bound one-sided $CI_{1-\alpha}$: $0 \leq \pi \leq U$	H_0: $\pi = \pi_0$ H_a: $\pi < \pi_0$ α level of rejection Reject H_0 if π_0 is *not* in the confidence interval, that is: $\pi_0 > U$
Lower bound one-sided $CI_{1-\alpha}$: $L \leq \pi \leq 1.00$	H_0: $\pi = \pi_0$ H_a: $\pi > \pi_0$ α level of rejection Reject H_0 if π_0 is *not* in the confidence interval, that is: $\pi_0 < L$

■ ■ ■

EXERCISES

3.3.1. In a random sample of 250 inmates of federal prisons, 175 are found to have committed nonviolent crimes.

a. What is the best estimate of the proportion of such federal offenders?

b. Place a 95% confidence interval on the proportion of all federal prisoners convicted of nonviolent crimes.

c. Can you deduce from this that the majority of inmates of all federal prisons have been convicted of nonviolent crimes?

3.3.2. A random sample of 25 precocious readers is drawn and their family backgrounds carefully studied. In 40% of the cases, the child's father is at least 15 years older than the mother. Place a 90% confidence interval on the proportion of such age disparities between the parents of precocious readers.

3.3.3. A random sample of 100 persons suffering from mental depression reveals that 75 of them cannot properly evaluate their job skills.

a. Give the maximum likelihood estimate of the binomial parameter.

b. Set up a 95% confidence interval for this parameter.

3.3.4. A survey is taken to evaluate the marketability of detergents containing enzyme presoaks. A random sample of 50 families is taken in a certain region; 60% of the persons interviewed state that they are currently using such detergents to do their laundry.

a. Place a 90% confidence interval on the true proportion of families in that region using detergents containing enzyme presoaks.

b. Using only the information given above, which of the following statements can be properly concluded from the survey?

i. The proportion of families in the region who use detergents containing enzymes may be as large as 75%.

ii. A majority of families in the region may use some other type of detergent.

iii. If the proportion using detergents containing enzymes is about 60% in that region, it is also probably about the same in any other region of the United States.

3.3.5. In a random sample of 50 kindergarten children, there are seven who hold crayons in their left hands while coloring a picture.

a. Give the best point estimate of the proportion of left-handed kindergarten children.

b. Explain what "best" means in this exercise.

3.3.6. A random sample of 125 schoolchildren are given their choice of candy made with either light or dark chocolate, but otherwise the candy is the same. Only 30% of them choose the dark chocolate. If a candymaker wants not more than a 1 in 100 chance of being misled by sampling variability, what is the estimate of the proportion of children who prefer dark chocolate?

3.3.7. A random sample of 250 married couples is given sample ballots containing the names of all candidates for contested offices in the coming election. Husband and wife mark their ballots independently, and their ballots are compared; 130 couples are in perfect agreement in their voting.

a. What is the estimated numerical value of the binomial parameter for the distribution that models this situation?

b. Set up a 95% confidence interval for the binomial parameter.

3.3.8. A random sample of 375 football fans who attended last Saturday's game is questioned about the team's strategy. Their responses are as follows:

Response	Frequency
Team threw ball too often	105
Team threw ball about right number of times	135
Team did not throw ball often enough	120
Too busy watching cheerleaders to notice	15
	375

a. Of those fans who observed the team, estimate the proportion who felt the team threw the ball too often.

b. Place a 99% confidence interval on the above estimate.

3.3.9. A random sample of 200 apples from an orchard that had not been sprayed with insecticide contains 162 apples that bear evidence of insect damage.

a. What is the best estimate of the proportion of damaged fruit in the orchard?

b. In what range would you say the "true" proportion lies, if you want to have only a 1 in 100 chance of being wrong?

3.3.10. If you are told that $0.033 \leq \pi \leq 0.218$ is a $\mathrm{CI}_{0.95}$ based on a random sample, how large is the sample?

3.3.11. In each case below, the sample size n, the statistic y, the level of confidence $1 - \alpha$, the lower confidence limit L, or the upper confidence limit U are given. Use tables for placing a confidence interval on the binomial parameter π to fill in the missing values in each case.

Case	n	y	$1 - \alpha$	L	U
1	50	20	0.99	—	—
2	—	—	0.95	0.300	0.423
3	250	80	0.95	—	—
4	500	430	0.99	—	—
5	50	16	0.99	—	—
6	—	—	0.95	0.102	0.258
7	200	22	0.90	—	—
8	100	—	—	0.216	0.374
9	—	30	—	0.036	0.093
10	20	—	0.90	0.250	—

3.3.12. In a northern county in West Virginia, a random sample of 500 voters is taken; 265 of the voters indicate that they will vote for the Democratic candidate for governor.

a. Set a 99% confidence interval for the proportion of voters in the county who will vote for the Democratic candidate.

b. The Republican candidate claims that he will win the county by 1% of the votes.

i. State a null hypothesis for his claim.

ii. Does the confidence interval in Part a lead to acceptance or rejection of this null hypothesis? Why?

iii. With what α level was the hypothesis tested?

REVIEW EXERCISES

Decide whether each of the following statements is true or false. If the statement is false, explain why.

3.1. In a binomial experiment, the outcomes fall into two mutually exclusive classes.

3.2. In a binomial experiment with n trials, y can take on any of n values.

3.3. Binomial distributions are not symmetrical, except when $\pi = 1 - \pi$.

3.4. Because the binomial is a discrete distribution, the expected value will be an integer value.

3.5. If the binomial parameter π is 0.60, the probability of exactly 60 successes out of 120 trials is greater than the probability of 72 successes out of 120 trials.

3.6. If A and B are mutually exclusive events, then $P(A \text{ or } B) = P(A) \times P(B)$.

3.7. The variance for discrete distributions can be computed by using the formula $V(y) = n\pi(1 - \pi)$.

3.8. The addition rule of probability applies only to mutually exclusive events.

3.9. The binomial distribution is an example of a continuous probability distribution.

3.10. In order to calculate the probabilities in a binomial distribution, the number of trials n and the binomial parameter π must be known.

3.11. The null hypothesis may be H_0: $\pi = 0.05$ and $y/n = 0.05$, but the null hypothesis may still be false.

3.12. A Type I error is defined as "the probability of rejecting the null hypothesis when it is true."

3.13. When the null hypothesis is true, the probability of making a Type I error is equal to α.

3.14. Using good statistical techniques, true null hypotheses are probably rejected about as often as false ones.

3.15. It is impossible to make a Type I error when the null hypothesis is false.

3.16. The symbol β represents the probability of rejecting H_0 when H_0 is false.

3.17. The power of a test of hypothesis is $1 - \alpha$.

3.18. It is impossible to make a Type II error when the null hypothesis is false.

3.19. It is impossible to make a Type II error when the null hypothesis is rejected.

3.20. If large sample sizes are used, there is less likelihood of a Type I error and a Type II error.

3.21. If an experiment is well designed and both α and β are small, it should be a good experiment.

3.22. Even when correct statistical procedure is used, it is possible to accept the null hypothesis when it is false.

3.23. The greater the region of rejection, the more powerful the experiment.

3.24. $P(y$ is in region of rejection$) = \alpha$ whether the null hypothesis is true or false.

3.25. The two primary types of inference are obtaining point estimates and interval estimates of population parameters.

3.26. The best point estimate $\hat{\pi} = y/n$ of the parameter π will lie exactly in the middle of the 95% confidence interval for π.

3.27. If the degree of certainty is increased from 0.95 to 0.99, the confidence interval becomes narrower.

3.28. Two methods of estimation are confidence intervals and tests of hypotheses.

3.29. Confidence intervals that are based on large samples are more likely to include the population parameter than those based on smaller samples.

3.30. Other things remaining the same, the larger the value of $\hat{\pi}$ the wider the confidence interval.

3.31. Other things being equal, the greater the level of confidence desired the wider will be the confidence interval.

3.32. In estimating a population parameter, a confidence interval is more likely to be correct than a point estimate.

3.33. The binomial parameter π must be known in order to set confidence limits.

3.34. Repeated samples of the same size from the same population will always produce 99% confidence intervals of the same width on the binomial parameter π.

3.35. If the confidence interval does not contain some hypothesized value π_0 of the binomial parameter, the hypothesis can be rejected.

SELECTED READINGS

Anderson, T. W., and H. Burstein (1967). Approximating the upper binomial confidence limit, *Journal of the American Statistical Association,* **62,** 857–861.

——— (1968). Approximating the lower binomial confidence limit, *Journal of the American Statistical Association,* **63,** 1413–1415, Correction, **64** (1969), 669.

Buckalew, I., Jr. (1974). A comparison of the efficiency of the normal approximation to the binomial with the binomial distribution, Master's report, West Virginia University.

Burke, C. J. (1954). A brief note on one-tailed tests, *The Psychological Bulletin,* **50,** 384–387.

——— (1954). Further remarks on one-tailed tests, *The Psychological Bulletin,* **51,** 587–590.

Clopper, C. J., and E. S. Pearson (1934). The use of confidence or fiducial limits illustrated in the case of the binomial, *Biometrika,* **26,** 404–413.

Crow, E. L. (1956). Confidence intervals for a proportion, *Biometrika,* **43,** 423–435.

Feinberg, W. E. (1971). Teaching the Type I and Type II errors: The judicial process, *The American Statistician,* **25** (June), 30–32.

Fryer, H. C. (1966). *Concepts and Methods of Experimental Statistics*, Allyn & Bacon, Boston.

Jones, L. V. (1952). Tests of hypotheses: One-sided versus two-sided alternatives, *The Psychological Bulletin,* **49,** 43–46.

——— (1954). A rejoinder on one-tailed tests, *The Psychological Bulletin*, **51,** 585–586.

Natrella, M. G. (1960). The relation between confidence intervals and tests of significance, *The American Statistician,* **14** (February), 20–22, 38.

4

Poisson Distributions

In this chapter we look at a second family of probability distributions, Poisson distributions. Poisson distributions are the appropriate probability model for certain types of experiments. There is an interesting relationship between binomial distributions and Poisson distributions, and this relationship provides a way to approximate some binomial probabilities that are very difficult to compute directly.

4.1. THE NATURE OF POISSON DISTRIBUTIONS

Many scientific experiments involve observing the number of discrete events in a fixed time interval or in a fixed length, area, or volume. For example, a forester might count the number of white-oak trees damaged by deer within sampling quadrants (square areas); an epidemiologist might count the number of cases of hepatitis in a certain county in one month; a quality control manager might count the number of defects in 25-ft lengths of wire; an ecologist might count the number of parasites per host. In each case the event of interest (damaged white oak, incidence of disease, defect, parasite) is counted for a certain interval (a quadrant, a month, 25 feet, per host).

The outcomes in experiments of this type often have the characteristics of a Poisson process. This process is named after Siméon-Denis Poisson (1781–1840), a French mathematician who first studied variables of this type in 1837.

A *Poisson process* consists of discrete events that occur in an interval (as time, length, area, volume, or on an object) and:

1. The probability of a single occurrence of the event is directly proportional to the size of the interval.
2. If the interval is sufficiently small, the probability of two or more occurrences of the event is negligible.
3. The occurrences of the event in nonoverlapping intervals are inde-

pendent, that is, what happens in one interval has no effect on what happens in another nonoverlapping interval.

If an experiment generates a Poisson process, then the appropriate probability model for the *number of occurrences* of the event in the specified interval is a *Poisson distribution*. The Poisson distribution is a discrete probability distribution with probability function

$$p(y;\lambda) = \frac{e^{-\lambda}\lambda^y}{y!}$$

for $y = 0, 1, 2, \ldots$. In this probability function y is the value of the random variable, $y!$ has the usual meaning of y factorial, e is the constant which is the base of the natural logarithms† (equal to 2.7183 if rounded to four decimal places), and λ (the Greek letter "lambda") is the expected number of occurrences in the specified interval. Table A.6 in the Appendix of Useful Tables gives values of $e^{-\lambda}$ for selected values of λ.

Note that this probability distribution is completely determined by the parameter λ. If we know λ we can compute the distribution as in the following example.

Example 4.1. A Poisson Probability Distribution

If the suicides in a particular city are distributed such that the expected number is three per month, then the probability distribution that models this situation is

$$p(y;3) = \frac{e^{-3}(3)^y}{y!}$$

for $y = 0, 1, 2, \ldots$

y	$p(y;3)$		
0	$e^{-3}\cdot 3^0/0!$	$= e^{-3}$	$= 0.0498$
1	$e^{-3}\cdot 3^1/1!$	$= p(0)(3/1)$	$= 0.1494$
2	$e^{-3}\cdot 3^2/2!$	$= p(1)(3/2)$	$= 0.2240$
3	$e^{-3}\cdot 3^3/3!$	$= p(2)(3/3)$	$= 0.2240$
4	$e^{-3}\cdot 3^4/4!$	$= p(3)(3/4)$	$= 0.1680$
5	$e^{-3}\cdot 3^5/5!$	$= p(4)(3/5)$	$= 0.1008$
6	$e^{-3}\cdot 3^6/6!$	$= p(5)(3/6)$	$= 0.0504$
7	$e^{-3}\cdot 3^7/7!$	$= p(6)(3/7)$	$= 0.0216$
8	$e^{-3}\cdot 3^8/8!$	$= p(7)(3/8)$	$= 0.0081$
9	$e^{-3}\cdot 3^9/9!$	$= p(8)(3/9)$	$= 0.0027$
10	$e^{-3}\cdot 3^{10}/10!$	$= p(9)(3/10)$	$= 0.0008$
11	$e^{-3}\cdot 3^{11}/11!$	$= p(10)(3/11)$	$= 0.0002$
12	$e^{-3}\cdot 3^{12}/12!$	$= p(11)(3/12)$	$= 0.0001$
13	$e^{-3}\cdot 3^{13}/13!$	$= p(12)(3/13)$	$= 0.0000$

and $p(y) = 0.0000$ (rounded to four decimal places) for $y > 13$.

---

† The irrational number e can also be defined as the limit of the series $(1 + 1/n)^n$, that is, $(1 + 1/1)^1 = 2.0000$, $(1 + 1/2)^2 = 2.500$, $(1 + 1/3)^3 = 2.3704, \ldots$

Note that the Poisson probabilities are easy to compute since the probability of any value y can be computed from the probability of the previous value $y - 1$,

$$p(y;\lambda) = p(y-1;\lambda)\frac{\lambda}{y}$$

Poisson probability distributions have some interesting properties. The expected value of y is equal to λ and the variance of y is also λ, that is, $E(y) = V(y) = \lambda$. Also, the sum of two Poisson random variables is a Poisson random variable; thus if y_1 and y_2 are Poisson random variables with parameters λ_1 and λ_2 respectively, then $y_1 + y_2$ is a Poisson random variable with expected value $\lambda_1 + \lambda_2$.

EXERCISES

4.1.1. If the expected number of water mites found on a host, the chironomid fly, is 2.5 and this is a Poisson process, what is the probability that exactly one mite will be found on a fly?

4.1.2. If the accident rate at a certain factory is 7.0 per year and this is a Poisson process:

a. Find the probability that fewer than three accidents will occur in a year.

b. Find the probability that three or more accidents will occur in a year.

4.1.3. If the expected number of flaws in 20-ft intervals of insulated wire is 5.0, what is the probability that there are 4 flaws in 10 ft of wire?

4.1.4. In Example 4.1 in this section, involving the number of suicides per month:

a. What is the probability that no suicide will occur in a month?

b. What is the probability that more than six suicides will occur?

c. What percentage of months will have at least one suicide but not more than six suicides?

4.1.5. Additives such as trace minerals, antibiotics, vermifuges, and insecticides are incorporated into animal feeds in parts per million (ppm). For effective mixing, the additives may be compressed into pellets the size of the ground grain in the feed and then colored with vegetable dye for easy identification. Quality control for thoroughness of mixing can be maintained by scooping out a known volume of the mixed feed and counting the number of colored pellets of additives. If properly mixed feed yields a Poisson process with $\lambda = 2.5$ per scoop, find the probability that:

a. A scoop will contain no pellets of additive.
b. A scoop will contain exactly one such pellet.
c. A scoop will contain at least one pellet.
d. Find the outcomes that are most likely to occur approximately 80% of the time.

4.1.6. In the feed-mixing problem described in Exercise 4.1.5, suppose customary quality control procedures require 10 independently drawn scoops from each batch of mixture. In 10 scoops of properly mixed feed, find:

a. The expected total number of colored pellets.
b. The probability that there will be no such pellets.

4.1.7. a. Compute the Poisson distribution for each of the following values of λ: 0.25, 0.50, 1.00, 10.00. Round the probabilities to four decimal places.
b. Graph the Poisson distributions of Part a.
c. Describe the behavior of the graphs of Part b.

4.1.8. a. Use the probabilities in Exercise 4.1.7a for $\lambda = 0.25$ to find the expected value of that Poisson distribution. Why is this value slightly different from $E(y) = \lambda = 0.25$?
b. Use the probabilities computed in Exercise 4.1.7a and $E(y) = 0.25$ to find $V(y)$ for that Poisson distribution. Why is this value slightly different from $V(y) = \lambda = 0.25$?

4.1.9. If y_1 and y_2 are independent Poisson random variables with $\lambda = 0.25$, then $y_1 + y_2$ is a Poisson random variable with $\lambda = 0.50$. Use Exercise 4.1.7 to show that this is true for $y_1 + y_2 = 3$. (*Hint:* Remember that $0 + 3 = 3$, $3 + 0 = 3$, $1 + 2 = 3$, and $2 + 1 = 3$.)

4.2. TESTING HYPOTHESES

Using Appendix Table A.7, which contains the Poisson distributions for selected values of λ, we can test hypotheses with a procedure similar to the one we used for the binomial distribution.

Example 4.2. Test of Hypothesis for a Poisson Parameter

A biologist studying yeast cells believes that after a certain treatment the cells will be present at a rate of 0.55 per square of a hemacytometer (a microscopic plate usually used to count blood cells). He finds 13 yeast cells in 20 squares and wonders if $13/20 = 0.65$ indicates that a rate of 0.55 is incorrect. In order to determine whether 13 cells in the 20 squares are likely to occur if his conjectured rate is correct, he uses the Poisson distribution.

The null and alternative hypotheses are:

$$H_0: \lambda = 0.55$$
$$H_a: \lambda \neq 0.55$$

Since the sum of two Poisson random variables is also a Poisson random variable, if $\lambda = 0.55$ for one square then $\lambda = 20(0.55) = 11$ for 20 squares. Using Table A.7, the biologist finds that for α as close to 0.10 as possible, the region of rejection is

$$y = 0, 1, 2, 3, 4, 5, 17, 18, 19, \ldots$$

if the test statistic is the number of yeast cells per 20 squares. The actual α level is 0.0934. The count is 13 yeast cells in 20 squares after this treatment, and since 13 does not lie in the region of rejection, the biologist concludes that after the treatment the average number of yeast cells per square may be 0.55.

For small values of λ the Poisson distributions have relatively large probabilities in the lower tail, so it may be impossible to designate a small α level for a two-tailed alternative or for a one-tailed less-than alternative hypothesis. The technique of using several units—such as the 20 squares in the above example—helps overcome this difficulty.

Table A.7 lists a limited number of values of λ; and the necessary one may not be there. If λ is not too large, the necessary probability distribution can be calculated. For large λ's approximation methods are available; these are discussed in Chapter 7.

■ ■ ■

Procedure. Test of Hypothesis for a Poisson Parameter

H_0: $\lambda = \lambda_0$ (λ = expected number of occurrences in a specified interval)

H_a: $\lambda \neq \lambda_0$, $\lambda < \lambda_0$, or $\lambda > \lambda_0$

Significance level: α

Test statistic: y, the number of occurrences of the phenomenon of concern in a multiple of k specified intervals

Using a table for the Poisson distribution with probability function $p(y;\lambda_0 k)$, determine the region of rejection.

For H_a: $\lambda \neq \lambda_0$, the region of rejection is $0 \leq y \leq c_L$ and $c_U \leq y \leq \infty$ suc that $\sum_{0}^{c_L} p(y;\lambda_0 k)$ and $\sum_{c_U}^{\infty} p(y;\lambda_0 k)$ are each as close as possible to $\alpha/2$.

For H_a: $\lambda < \lambda_0$, the region of rejection is $0 \leq y \leq c_L$ such that $\sum_{0}^{c_L} p(y;\lambda_0 k)$ is as close as possible to α.

For H_a: $\lambda > \lambda_0$, the region of rejection is $c_U \leq y \leq \infty$ such that $\sum_{c_U}^{\infty} p(y;\lambda_0 k)$ is as close as possible to α.

■ ■ ■

EXERCISES

4.2.1. A physicist wants to verify that the level of radiation does not have a rate of more than four radioactive particles per millisecond. He measures the radioactivity with a Geiger counter and it records 18 particles in three milliseconds. Use a test of hypothesis to determine if the radioactivity level is too high. Let α be as close to 0.05 as possible.

4.2.2. A certain area of the United States has a rate of 4.5 tornadoes per year. A local religious cult claims that their rituals can reduce this rate. They conduct their rituals and that year two tornadoes hit. Use a test of hypothesis with α as close to 0.10 as possible to determine if the rate is less than 4.5 per year. What assumptions are you making as you perform this test?

4.2.3. A certain hospital emergency ward handled victims of auto accidents at the rate of 10 per week when the local highway had a speed limit of 70 miles per hour. After the speed limit was reduced to 55 miles per hour, four highway-accident victims were admitted in a randomly selected week. Does this indicate a reduction in emergency admissions for auto accidents? Could you conclude that lowering the speed limit has reduced highway accidents? Why, or why not?

4.2.4. Grain sorghum is a naturally tall-growing plant, but dwarf varieties have been developed so that the crop can be harvested with conventional farm equipment. However, back mutation occurs frequently and tall offspring reappear in a field with an expected value of 1.5 tall plants per 200 ft^2. With each development of a new grain sorghum hybrid, plant breeders must satisfy the farmer that the amount of back mutation has not increased. A hybrid seed company has many experimental hybrids under consideration at a time, and ides to allot only three 200 ft^2 plots per hybrid. Set up a test pothesis for the amount of back mutation.

ive the null hypothesis for three plots.
ive the alternative hypothesis.
ive the region of rejection for α as close to 0.05 as possible.
d. Suppose that for a particular hybrid, the back mutation doubles to 3.0 per 200 ft^2; what is the power of the test for three plots?
e. What is the power if only one plot is used? Is it advisable to use more than one plot?

4.2.5. The rarest white blood cell is the basophil, which constitutes only 1% of the total white blood cells. Students who are learning to perform white cell counts are inclined to mistake other cells for basophils until they have seen them often enough to recognize them. Thus a student's proficiency in performing differential white blood cell counts can be tested by checking whether too many cells have been recorded as basophils. This can be thought of as a Poisson process in which the interval is a count of 100 white blood cells.

a. State a null hypothesis indicating that the student can accurately identify the different kinds of white blood cells.
b. State an alternative hypothesis indicating that the student mistakes other cells for basophils.
c. The instructor decides that any student who records four or more basophils per 100 cells counted cannot yet distinguish these cells properly. How likely is it that a student will record cells correctly but have an unusual random sample of cells?
d. The frequency of basophils increases after surgery. Suppose the student is counting white blood cells from a blood smear taken under such conditions and $\lambda = 2.4$ per 100 cells. How likely is it that fewer than four basophils are among the 100 cells counted? Should the instructor take precautions that the students are not using blood smears from postoperative patients?

4.3. ESTIMATION

The best point estimate of the Poisson parameter λ is y, the number of occurrences of the event of interest in a randomly selected interval. If several intervals are sampled, the total number of occurrences is the best estimate for the combined intervals. Central and one-sided confidence intervals can be found in a manner similar to finding confidence intervals for the binomial parameter π. Table A.7 in the Appendix is used to find the confidence intervals for the Poisson parameter. Because of the relatively large probabilities for low values of y, the horizontal lines in Table A.7 are drawn so that α is as close to 0.20 as possible, thus these lines correspond to approximate 80% central confidence intervals.

Example 4.3. A Central Confidence Interval for a Poisson Parameter

Foresters are concerned about the number of young trees destroyed by deer. Suppose a forester chooses four quarter-acre quadrants at random and finds that in the four plots, eight young trees have been destroyed by deer. He wants to estimate the damage rate per acre by an approximate 80% confidence interval.

Using Table A.7, he finds that eight is in the region of acceptance for $\lambda = 5.0$ to $\lambda = 12.0$, so the confidence interval is

$$\text{CI}_{0.80}\text{:} \quad 5.0 \le \lambda \le 12.0$$

in which λ is the damage rate per acre.

Note in this example that the use of the interval of one acre avoids the necessity of counting all of the trees in the area to determine the rate.

One-sided confidence intervals can also be determined.

Example 4.4. A One-Sided Confidence Interval for a Poisson Parameter

The architect for a new hospital in a small city needs to know the maximum number of emergency cases that can be expected in a half-hour period in order to plan adequate facilities. He examines the records at the existing city hospital, which is being replaced; a random selection of 10 half-hour periods gives a total of six emergency cases. He can use Table A.7 to find an approximate 90% one-sided confidence interval.

$$\text{One-sided CI}_{0.90}\text{:} \quad \lambda \le 9.0$$

if λ is for a five-hour period because 9.0 is the largest value of λ for which six would be in the region of acceptance. Or he could write

$$\text{One-sided CI}_{0.90}\text{:} \quad \lambda \le 0.90$$

if λ is for a half-hour period.

The one-sided confidence interval indicates that the largest expected value of the Poisson distribution that is likely is 0.90, that is, the largest average number of cases in a 30-minute period is 0.90. Since 0.90 is the average, some of the 30-minute periods will have more cases and others less. Since the number of cases in a 30-minute period will usually be within two standard deviations of the expected value λ, and in a Poisson distribution $\lambda = V(y)$, the architect can prepare for the worst situation.

$$\lambda = V(y) = 0.90$$
$$\mathrm{sd}(y) = \sqrt{\lambda} = 0.95$$

and the largest number of cases is not likely to be more than

$$\lambda + 2\mathrm{sd}(y) = 0.90 + 2(0.95) = 2.80$$

To be safe, he plans to be able to accommodate three cases each half hour.

■ ■ ■

Procedure. Confidence Intervals for λ

Central

1. Specify α.
2. Take a sample of k intervals of the specified unit.
3. Observe y, the number of occurrences of the phenomenon of interest in the k intervals.
4. Give the interval of all values of λ for which y would fall in the region of acceptance for a two-sided α-level test.
5. Divide the confidence limits by k to determine the central confidence interval for the rate λ for intervals of the specified unit.

One-Sided, Upper Confidence Limit. Proceed as for a central confidence interval, but in step 4 use the region of acceptance for a one-tailed less-than test of hypothesis.

One-Sided, Lower Confidence Limit. Proceed as for a central confidence interval, but in step 4 use the region of acceptance for a one-tailed greater-than test of hypothesis.

■ ■ ■

EXERCISES

4.3.1. If three noxious weeds are found in a 0.25-oz random sample of grass seed, use the Poisson probability distribution to find an 80% confidence interval for the expected number of weeds per 0.25 oz of seed. (Note that using the Poisson model here avoids the necessity of counting all the seeds, a tedious task.)

4.3.2. If eight defects are found in a production process during a random five-minute interval, find with 90% of confidence the largest mean number of defects that could be expected to occur in a five-minute period.

4.3.3. It is found that there are six fatal accidents in an underground coal mine for a sample of 20,000,000 employee hours of exposure. Place an approximate 80% confidence interval on the Poisson parameter if the interval is 100,000 employee hours.

4.3.4. In the quality control process described in Exercise 4.1.5, place an approximate 90% confidence interval on the smallest mean number of pellets expected in one scoop if seven pellets are found in four random scoops.

4.4. POISSON DISTRIBUTIONS AND BINOMIAL DISTRIBUTIONS

Besides being useful in its own right, the Poisson distribution is often used as an approximation of the binomial distribution if the number of trials n is large and the probability of success on a single trial π is small. The approximation is possible because it can be shown mathematically that if π becomes very small while n becomes very large and the product $n\pi$ remains constant, then the binomial distribution will be approximately a Poisson distribution with $\lambda = n\pi$ and the Poisson interval the set of n trials.

Example 4.5. Using a Poisson Distribution to Approximate a Binomial Distribution

A geneticist believes that in a certain experiment the mutation rate is 4 in 1,000,000. He would like to find the probability that in a random sample of 25,000 he will observe no more than one mutation. This experimental situation is appropriately modeled by the binomial distribution $b(y;25{,}000,0.000004)$ and he wants to compute

$$\begin{aligned} P(y \leq 1) &= b(0;25{,}000,0.000004) + b(1;25{,}000,0.000004) \\ &= \binom{25{,}000}{0}(0.000004)^0(0.999996)^{25{,}000} \\ &\quad + \binom{25{,}000}{1}(0.000004)^1(0.999996)^{24{,}999} \end{aligned}$$

This computation is not feasible directly, and logarithms would have to be used to compute an approximate answer.

Instead, the geneticist could approximate this probability by using a Poisson distribution. The Poisson parameter would be $\lambda = n\pi = 25{,}000(0.000004) = 0.100000$, that is, the expected number of mutations per 25,000 trials is

0.1. For the Poisson distribution

$$\begin{aligned} P(y \le 1) &= p(0;0.1) + p(1;0.1) \\ &= \frac{e^{-0.1}(0.1)^0}{0!} + \frac{e^{-0.1}(0.1)^1}{1!} \\ &= 0.904837 + 0.904837(0.1) \\ &= 0.995321 \end{aligned}$$

Using this very simple computation, the geneticist can be relatively certain that in a random sample of size 25,000 he will observe no more than one mutation.

This approximation of the binomial distribution by the Poisson distribution is good only for small π and large n. Some statisticians suggest as a rule of thumb that $\lambda = n\pi$ should be less than seven.

■ ■ ■

Procedure. Poisson Approximation of a Binomial Distribution

For $n\pi < 7$, a binomial distribution may be approximated by a Poisson distribution: $b(y;n,\pi)$ is approximated by $p(y;n\pi)$.

■ ■ ■

It is important that we recognize the difference between a Poisson distribution and a binomial distribution so that we use the proper one to model an experiment and so that we know when it is appropriate to approximate a binomial by a Poisson. The following summary may be helpful:

Binomial	Poisson
1. Random variable: y = number of successes in n trials	1. Random variable: y = number of successes in a specified interval
2. Number of trials: n, a finite number	2. Number of trials: infinite, since we count discrete events (successes) in an interval
3. Two parameters: π = probability of success for a single trial; n = number of trials	3. One parameter: λ = average number of successes per interval
4. $E(y) = n\pi$; $V(y) = n\pi(1 - \pi)$	4. $E(y) = V(y) = \lambda$

EXERCISES

4.4.1. If it is known that the probability of having a bad reaction to a certain injection is 0.001, what is the probability that more than one person in 100 will have a bad reaction?

4.4.2. If the rate of accidental drownings per year is 0.000003 (that is, 3 per 1,000,000 population), what is the probability that there will be more than two drownings in a city with a population of 400,000?

4.4.3. A manufacturer of TV sets initiates an inspection system to reduce the number of defective sets leaving the plant. Prior to this system the proportion of defective sets was 1 in 80. After the new system is in effect, in a random sample of 320 sets there are two defective sets. Use a test of hypothesis to decide if the proportion of defects has been reduced.

4.4.4. Suppose routine blood typing for 400 army recruits reveals that six of them have AB-negative blood.

a. What assumptions would you have to make for this to be considered a random sample of army personnel? Of the entire country?

b. Place an approximate 80% confidence interval on the proportion with AB-negative blood among army recruits.

c. Assuming it can be justified, place an approximate 80% confidence interval on the proportion of those with AB-negative blood in the entire country.

4.4.5. Fish and game commissions measure the hunting pressure on large game in their states by taking random samples of hunters and recording their successes during the hunting season. The following data record the number of white-tailed deer taken by a random sample of 50 Texas deer hunters.

Number of Deer Killed	Hunters
0	45
1	4
2	1

Because the fish and game commission wishes to protect against overhunting, place an approximate 90% confidence interval on the largest mean number of deer taken per 50 hunters in the state.

REVIEW EXERCISES

Decide whether each of the following statements is true or false. If the statement is false, explain why.

4.1. In a Poisson distribution, $E(y) = n\pi$ and $V(y) = n\pi(1 - \pi)$.

4.2. Poisson data consist of discrete, countable observations.

4.3. Because $E(y)$ is usually small for a Poisson distribution, a relatively large number of intervals is needed to estimate λ effectively.

4.4. A unique characteristic of Poisson distributions is that for any specified distribution, the expected value will be numerically greater than the variance.

4.5. The Poisson distribution is sometimes called the "distribution of rare events" and hence is seldom encountered in experimentation.

4.6. The shape of a Poisson frequency distribution is symmetrical around its expected value.

4.7. In testing a hypothesis about the Poisson parameter, the alternative hypothesis may be one-tailed or two-tailed.

4.8. Confidence intervals for a Poisson parameter are symmetrical around the point estimate y.

4.9. There is a separate Poisson distribution for every value of λ and n.

4.10. The Poisson distribution can always be used to approximate the probabilities of a binomial distribution.

4.11. Because λ is usually small, small values of y are much more probable than large values when sampling from a Poisson distribution.

4.12. The power of a test of hypothesis for the Poisson parameter is increased as the number of intervals sampled is increased.

SELECTED READINGS

Haight, F. (1967). *Handbook of the Poisson Distribution*, Wiley, New York.

Hoaglin, David C. (1980). A Poissonness plot, *The American Statistician,* **34,** 146–149.

"Student" [William Sealy Gosset] (1906). On the error of counting with a haemacytometer, *Biometrika,* **5,** 351–360.

5

Chi-square Distributions

In this chapter we study some uses of a continuous probability distribution called the *chi-square distribution*. Although this theoretical probability distribution is usually not a direct model of a population distribution, it has many uses when we are trying to answer questions about populations. For example, the chi-square distribution can be used to decide whether or not a set of data fits a specified theoretical probability model—a "goodness-of-fit" test. It can also be used to decide whether or not several samples came from the same population even when the model of the population is unspecified—a chi-square test of homogeneity. It is possible to make these decisions about populations because the chi-square distribution is often a model for the distribution of some statistic obtained by sampling from the population.

5.1. THE NATURE OF CHI-SQUARE DISTRIBUTIONS

In 1876, Frederick R. Helmert did some of the early work on the theoretical chi-square distributions. Their probability density functions are known, and we can get some feeling for the nature of these distributions from their graphs (Figure 5.1). The symbol usually used for the chi-square random variable is the compound symbol χ^2 (this should not be confused with the squaring operation).

If χ^2 is a random variable with a chi-square distribution:

1. χ^2 is a positive real number.
2. The density function $f(\chi^2)$ for χ^2 depends on only one parameter ν (pronounced "nu"), called the *degrees of freedom*.

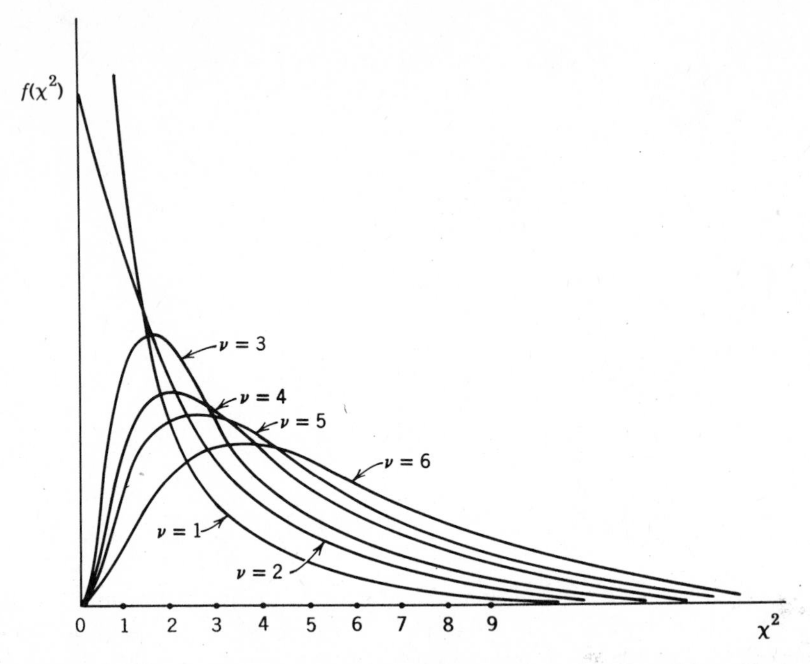

FIGURE 5.1. Chi-square distributions with ν degrees of freedom. (Adapted from *Elementary Statistics*, 4th ed., P. G. Hoel, Wiley, 1976, page 249.)

3. The expected value of χ^2 is equal to the degrees of freedom, that is, $E(\chi^2) = \nu$.
4. The variance of χ^2 is two times the degrees of freedom, that is, $V(\chi^2) = 2\nu$.
5. The maximum value of $f(\chi^2)$ is at $\chi^2 = \nu - 2$ if $\nu > 2$.
6. The graph of $f(\chi^2)$ is not symmetrical but approaches symmetry as the degrees of freedom increase.

Table A.8 in the Appendix of Useful Tables gives selected critical values for some of the chi-square distributions. The degrees of freedom are listed at the left, thus each row is from a different chi-square distribution. The headings at the top of the columns give α, the area to the right of the chi-square values listed in the tables. For example, if χ^2 has a chi-square distribution with four degrees of freedom, then a vertical line at $\chi^2 = 0.484$ divides the chi-square distribution so that $\alpha = 0.975$ of the area under the curve is to the right of 0.484 and $1 - \alpha = 0.025$ of the area is to the left (see Figure 5.2). We write $\chi^2_{0.975,4} = 0.484$. Critical values are used to determine regions of rejection because for continuous random variables, areas correspond to probabilities. The probability that a chi-square random variable with four degrees of freedom has a value greater than 0.484 is equal to 0.975.

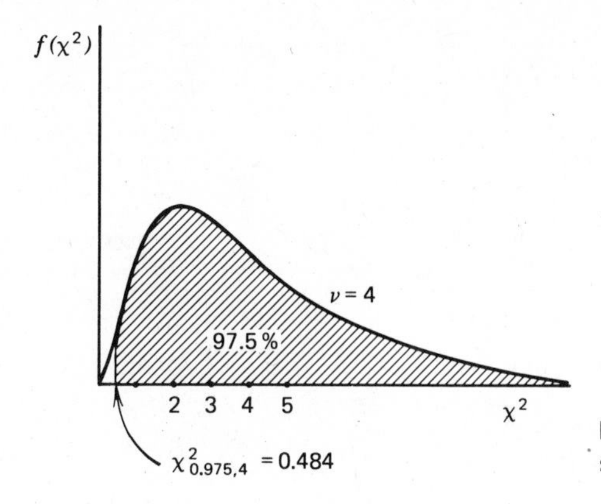

FIGURE 5.2. Meaning of values in the chi-square table.

Another example is given in Figure 5.3. If χ^2 is a chi-square random variable with 15 degrees of freedom, then 5% of the area is to the right of a vertical line at $\chi^2 = 24.996$ and 95% of the area is to the left of this line, or $\chi^2_{0.05,15} = 24.996$. This distribution has a mean of 15, a variance of 30, and the graph has a maximum at 13.

Helmert studied these theoretical distributions with apparently no idea that they could be used for a test of significance. In 1900 Karl Pearson was able to use Helmert's chi-square distributions to test hypotheses about multinomial experiments. A multinomial experiment is a generalization of a binomial experiment.

A *multinomial experiment* is an experiment in which:

1. There are k possible outcomes and the probability of the ith outcome is π_i with $\sum_{i=1}^{k} \pi_i = 1$.
2. The experiment is repeated n times, that is, there are n trials.
3. The π_i's are constant from trial to trial.
4. The trials are independent.
5. We are interested in y_i, the number of times the ith outcome occurs; $\sum_{i=1}^{k} y_i = n$.

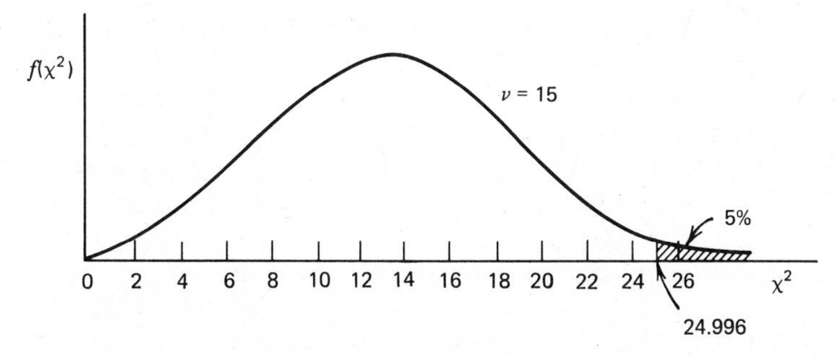

FIGURE 5.3. A chi-square distribution.

Note that a binomial experiment is a multinomial experiment with $\pi_1 = \pi$, $\pi_2 = 1 - \pi$ in which π is the probability of success on a single trial, and $y_1 = y$, $y_2 = n - y$ in which y is the number of successes in n trials. Like the binomial distribution, the expected number of occurrences of the ith outcome is $n\pi_i$.

Example 5.1. A Multinomial Experiment

If palomino horses are bred to other palominos, they produce progeny in the ratio of one dark-colored colt to two palominos to one light-colored colt. An experiment involving a random sample of 96 colts of palominos would be a multinomial experiment.

1. There are $k = 3$ outcomes: dark, palomino, light. $P(\text{dark}) = 1/4 = \pi_1$; $P(\text{palomino}) = 1/2 = \pi_2$; $P(\text{light}) = 1/4 = \pi_3$; $1/4 + 1/2 + 1/4 = 1$.
2. $n = 96$.
3. The π_i's are constant from trial to trial.
4. Since this is a random sample, the trials are independent.
5. We are interested in the number of colts of each color type.

If a geneticist questioned whether the ratios specified above were correct, he could use Pearson's approach to resolve the question. Pearson was looking for a simple statistic, a value that could be easily computed and that would indicate whether the results of an experiment deviated from expected results. He proposed the following statistic:

$$u = \sum_{i=1}^{k} \frac{(y_i - e_i)^2}{e_i}$$

in which $e_i = n\pi_i$, the expected value of y_i. A small value of u would indicate close agreement of the experimental results with the theory and a large value would indicate disagreement with the theory.

Pearson's statistic is a discrete random variable since it is composed of arithmetic operations on the discrete random variables $y_1, y_2, \ldots, y_k$. The probability distribution of u can be shown to be approximately Helmert's chi-square distribution with $k - 1$ degrees of freedom. Since the probabilities have been tabulated for the theoretical chi-square distribution, it is possible to use Pearson's statistic in a more precise way than just as a descriptive statistic; we can do a statistical test of hypothesis. Since Pearson's statistic is approximately a chi-square random variable, many people write

$$\chi^2 = \sum_{i=1}^{k} \frac{(y_i - e_i)^2}{e_i}$$

We also write χ^2 instead of u. It should be remembered, however, that the theoretical chi-square distribution studied by Helmert is a continuous probability distribution, whereas Pearson's statistic, which arises from multinomial experiments, is a discrete random variable. A test of hypothesis to check that specified probabilities in a multinomial experiment are correct is called the *multinomial chi-square test.*

Example 5.2. A Multinomial Chi-Square Test

The geneticist mentioned above found that in the random sample of 96 colts of palominos there are 21 dark-colored colts, 52 palomino colts, aand 23 light-colored colts. He wants to check whether $\pi_1 = 1/4$, $\pi_2 = 1/2$, and $\pi_3 = 1/4$ are correct parameters for a probability model. Thus he decides to test

$$H_0: \quad \pi_1 = 1/4, \quad \pi_2 = 1/2, \quad \pi_3 = 1/4$$

against

$$H_a: \quad \pi_1 \neq 1/4 \quad \text{or} \quad \pi_2 \neq 1/2 \quad \text{or} \quad \pi_3 \neq 1/4$$

that is, at least one inequality. He will reject the null hypothesis if the experimental results are unusual when the null hypothesis is true, that is, if they occur by chance alone less than $\alpha = 0.05$ of the time.

The expected number in each category is:

$$\begin{aligned} e_1 &= n\pi_1 = 96(1/4) = 24 \\ e_2 &= n\pi_2 = 96(1/2) = 48 \\ e_3 &= n\pi_3 = 96(1/4) = 24 \end{aligned}$$

He then uses the following table to organize his computations.

Category	Observed y_i	Expected e_i	$y_i - e_i$	$(y_i - e_i)^2$	$(y_i - e_i)^2/e_i$
Dark	21	24	−3	9	0.375
Palomino	52	48	4	16	0.333
Light	23	24	−1	1	0.042
					$\chi^2 = 0.750$

Since there are $k = 3$ categories, this statistic is distributed approximately as the chi-square random variable with $\nu = 3 - 1 = 2$ degrees of freedom. Referring to Table A.8 and recalling that large deviations from the expected values will give a large chi-square statistic, the geneticist finds that for $\nu = 2$ the theoretical chi-square value of 5.991 divides the lower 95% of the distribution from the upper 5%. He will reject the null hypothesis if the chi-square statistic is greater than or equal to 5.991. Since this is not the case, he concludes that there is no evidence that the theory is incorrect and that the specified ratios may be correct.

Since binomial experiments are a special case of multinomial experiments, the multinomial chi-square test can be used to test the correctness of a binomial parameter. There will be two categories, success and failure, thus one degree of freedom. This procedure has an advantage over the test given in Chapter 3; it is independent of sample size and the specified binomial parameter, so a multitude of binomial tables is unnecessary—Table A.8 is sufficient. If the experimenter had to rely on available binomial tables he might be tempted to tailor the experiment to fit the table. He might pick a sample size that appears in the table even if it is not the best sample size; or he might discard data if he cannot control the sample size (as in many genetics experiments) so that it fits the tables. Needless to say, these are not ideal scientific procedures. The multinomial chi-square test helps to avoid these pitfalls.

There are two disadvantages, however, to using a multinomial chi-square test when testing a binomial parameter. First, because of the nature of the chi-square statistic, one-tailed alternatives are more involved than we can discuss here. Thus if a one-tailed alternative is desired, the exact binomial distribution should be used (in the case of large sample sizes, the approximation procedure that will be explained in Chapter 7 may be used). The second disadvantage is that the approximation of the discrete sampling chi-square distribution by the continuous theoretical chi-square distribution is not very good for one degree of freedom with small sample sizes. For $n \leq 25$, a continuity correction should be made in the chi-square statistic:

$$\text{corrected } \chi^2 = \sum_{i=1}^{k} \frac{(|y_i - e_i| - 0.5)^2}{e_i}$$

For degrees of freedom other than one, there is no appropriate continuity correction. However, except for very small samples, the chi-square approximation is good. In general, the approximation of the discrete chi-square distribution by the continuous one is good. Some statisticians recommend that all expected values should be at least five in order to have an acceptable approximation. Others feel this is too conservative and indicate that no expected value should be less than one, and not more than 20% of the expected values should be less than five. We suggest these latter guidelines. If these conditions are not met, it is sometimes possible to combine categories to raise the expected value. Care should be taken, however, that the experimental question can still be answered when the categories are combined.

Besides being convenient, the chi-square test has another property to recommend it. In many situations the chi-square test is the most powerful one available—that is, it is the test that is most likely to detect a deviation from the null hypothesis if one exists.

■ ■ ■

Procedure. Multinomial Chi-Square Test

H_0: $\pi_1 = \pi_{1_0}, \pi_2 = \pi_{2_0}, \ldots, \pi_k = \pi_{k_0}$

H_a: *At least one inequality*

Significance level: α

Test statistic:

$$\chi^2 = \sum_{i=1}^{k} \frac{(y_i - e_i)^2}{e_i}$$

y_i = observed number of outcomes in ith category

$e_i = n\pi_{i_0}$ with $n = \sum_{i=1}^{k} y_i$

Region of rejection: $\chi^2 \geq \chi^2_{\alpha,k-1}$

EXERCISES

5.1.1. Use Table A.8 in the Appendix of Useful Tables to find the following:

a. $\chi^2_{0.01,7}$.

b. $\chi^2_{0.995,10}$.

c. $\chi^2_{0.025,70}$.

d. $P(\chi^2 > 31.410)$ if χ^2 is a chi-square random variable with 20 degrees of freedom.

e. $P(\chi^2 < 27.488)$ if χ^2 is a chi-square random variable with 15 degrees of freedom.

f. If $P(\chi^2 > b) = 0.05$ and χ^2 is a chi-square random variable with 10 degrees of freedom, find b.

g. If $P(\chi^2 \leq b) = 0.995$ and χ^2 is a chi-square random variable with 22 degrees of freedom, find b.

h. If $P(\chi^2 < 0.831) = 0.025$ and χ^2 is a chi-square random variable, find the degrees of freedom.

5.1.2. Computer programs for producing tables of random digits are often called "pseudo-random-number generators" because there is no way to prove that the digits are in random order. However, some properties of randomness can be tested. As an exercise, suppose that the 50 digits in row 1 of Table A.1 in the Appendix are a random sample.

a. State a null hypothesis about the proportion of even digits if the table is random.

b. State an alternative hypothesis that would indicate a lack of randomness.

c. Use a multinomial chi-square test with $\alpha = 0.05$ to test the above null hypothesis.

5.1.3. Assume the first three rows of Table A.1 are a random sample of size 150 and test that each of the digits 0, 1, ..., 9 is equally frequent in the whole table by means of a multinomial chi-square test ($\alpha = 0.05$).

5.1.4. The proportion of some Black populations who are carriers of the sickle-cell trait is estimated to be 30%. A public health officer on a Caribbean island wonders whether this estimate is correct for Blacks living on that island. Assuming that it will be a random sample, he requests that the next 150 blood tests performed for Blacks in a certain clinic also include a microscopic examination for this form of anemia. Given that there are 57 cases of sickle-cell anemia in the sample, perform a multinomial chi-square test to determine whether this proportion is correct. Use $\alpha = 0.05$.

5.1.5. When a certain red-flowering plant is self-fertilized, genetic theory indicates that the plants developed from the resulting seed should be in the ratio of three red-flowering plants to one white-flowering plant. If a random sample of 100 such seeds is collected and 68 produce red-flowering plants, 29 produce white-flowering plants, and 3 do not germinate, do these results agree with the theory? Use a multinomial chi-square test with $\alpha = 0.01$. What assumption must be made about the nongerminating seeds for this to be a valid test?

5.1.6. An experiment is conducted to study the effect of colors on the sale of certain products. Hamburger meat is packaged in equal quantities and displayed at the same price under red light, white light, and blue light. The display cases are kept full and the three colored areas are equally accessible and prominent. Records are kept of the number chosen from each area. If the experimental results are as follows,

Color	Number of Sales
Red	140
White	110
Blue	50

perform a test for a significance difference at the 5% level. What conclusion should be drawn about the best color of light to use for hamburger meat?

5.1.7. A manufacturer of automobile tires claims that 50% of his tires are driven 30,000 miles without a flat and only 20% of the rest have more than one flat during that mileage.

a. How many degrees of freedom would be used in a multinomial chi-square test to verify this claim?

b. If a random sample of 180 of these tires are driven 30,000 miles, what is the expected number having no flats? Only one flat? More than one flat?

c. If 72 tires have no flats, 60 have one flat, 31 have two flats, and 17 have three flats, test the claim at $\alpha = 0.05$.

5.1.8. A congressional representative circulates a questionnaire to all constituents to determine which national issue should be given the highest priority. A random sample of 500 gives the following:

Issue	Number Who Felt This Issue Deserves Highest Priority
Pollution	40
Energy	97
Poverty	31
Medical care	85
Foreign policy	53
Defense	71
Questionnaire not returned	123

The representative wants to know if there is a preference for one of the issues. Test the hypothesis that all of the issues are equally preferred against the hypothesis that some preference exists. What conclusion should the representative draw from this study? What assumption must be made about those who did not return the questionnaire in order for this analysis to be valid?

5.1.9. On the basis of size, blue crabs are categorized by marine biologists as "young," "juvenile," or "mature." In a healthy crab population that is being acceptably harvested by commercial fishermen, the percentage of each type is:

50% young   30% juvenile   20% mature

Deviations from these percentages usually indicate an unhealthy or "overfished" population. Fish and game biologists can dredge the bottom of a bay or estuary with nets to obtain a sample of crabs in an area close to commercial crabbing to determine if there is an unacceptable distribution of ages. Suppose that a small bay is dredged and the following categories of crab are netted:

58 young 33 juvenile 39 mature

a. Give the most logical null and alternative hypotheses for this study.
b. For this study, which is more serious, a Type I or Type II error? Why?
c. Perform a test of significance at $\alpha = 0.05$.
d. What is the experimental conclusion?
e. Suppose it is known that fishermen keep all mature and some juvenile crabs they net; all others are released unharmed. It is also known that young crabs are most susceptible to pollution, with juveniles the second most susceptible. Based on this information and the test of significance, which of the following is the appropriate action?
   i. Allow continued harvesting of crabs in the bay.
   ii. Close the bay to commercial crabbing because of overfishing.
   iii. Close the bay due to possible pollution.
   iv. Close the bay because of both overfishing and possible pollution.

5.1.10. In studying the genetic association between hair and eye color in human beings, a geneticist might hypothesize that the genes for hair color and eye color are located on the *same* chromosome. If a large group of dark-haired and brown-eyed people were to intermarry with another large group of light-haired and blue-eyed people, Mendel's law could be used to predict the characteristics of the second generation if the genes for hair color and eye color were on *different* chromosomes. The ratio of dark-haired and brown-eyed people to dark-haired and blue-eyed people to light-haired and brown-eyed people to light-haired and blue-eyed people would be 9:3:3:1. If the genes are on the same chromosome, this ratio does not appear.

a. What are the null and alternative hypotheses that should be used for this experiment?
b. If 1317 offspring of this type are located and classified with the following results:

Dark hair, brown eyes	782
Dark hair, blue eyes	234
Light hair, brown eyes	241
Light hair, blue eyes	60

what should the geneticist conclude?

5.1.11. At a certain state university, registration figures reveal that 55% of the American-born students are male and 20% are Black. The graduate dean wishes to determine whether there is any discrimination against female students or Black students in the distribution of the school's 100 graduate fellowships for U.S. citizens.

a. If there is no discrimination in awarding fellowships, what proportion of the fellowships will be awarded to non-Black males? Black males? Non-Black females? Black females?

b. If a multinomial chi-square procedure is used to test for discrimination, how many degrees of freedom would be involved?

c. Find the expected values for the chi-square procedure.

d. How large would the computed chi-square statistic have to be before the dean would conclude that there is discrimination if he wanted to have only a 1 in 100 chance of being misled by an unusual random sample?

e. The dean could also use his random sample to test for discrimination against Blacks. What would be the null hypothesis?

f. The dean could also test for discrimination against women. What would be the null hypothesis?

g. Give the critical value of chi-square if the tests in Parts e and f were done with $\alpha = 0.01$.

5.1.12. In a certain state the distribution of the population by age is as follows:

Age	Population (thousands)
Under 15	475
15–24	304
25–34	182
35–44	190
45–54	208
55–64	170
65–74	111
Over 74	72

a. Find the proportion of the population in each age group.

b. A certain planned city in this state claims that its inhabitants have the same proportion of people in each age group as the state as a whole. What null and alternative hypotheses should be used to test its claim?

c. If the city has a population of 12,500, compute the expected values for each age category if the null hypothesis is true.

d. If the city has the following distribution of ages, complete the test at the 5% significance level and state the conclusion.

Age	Population
Under 15	3016
15–24	2438
25–34	2037
35–44	2031
45–54	1253
55–64	977
65–74	585
Over 74	163

5.2. GOODNESS-OF-FIT TESTS

The multinomial chi-square test discussed in Section 5.1 is one type of *goodness-of-fit test*. It can be used to determine if the outcomes from a multinomial experiment fit a distribution with specified proportions of responses in certain categories.

A similar procedure can be used to determine whether a response variable for some population can be modeled by a certain probability distribution. For the case in which the parameters of the probability distribution are known, the test is very similar to the multinomial chi-square test. If the parameters are unknown and must be estimated, an adjustment in the degrees of freedom is necessary.

Example 5.3. Goodness-of-Fit Test with a Specified Parameter

Each day a salesman calls on five prospective customers and he records whether or not the visit results in a sale. For a period of 100 days his record is as follows:

Number of Sales:	0	1	2	3	4	5
Frequency:	15	21	40	14	6	4

A marketing researcher feels that a call results in a sale about 35% of the time, so he wants to see if this sampling of the salesman's efforts fits a theoretical binomial distribution for five trials with 0.35 probability of success, $b(y;5,0.35)$. This binomial distribution has the following probabilities and leads to the following expected values for 100 days of records.

y	$p(y)$	$e = 100p(y)$
0	0.1160	11.60
1	0.3124	31.24
2	0.3364	33.64
3	0.1812	18.12
4	0.0487	4.87
5	0.0053	0.53

Since the last category has an expected value of less than one, he combines the last two categories to perform the goodness-of-fit test.

Category A_i	Observed Frequency y_i	$P(A_i)$	Expected Frequency e_i	$y_i - e_i$	$(y_i - e_i)^2$	$(y_i - e_i)^2/e_i$
0	15	0.1160	11.60	3.40	11.5600	0.9966
1	21	0.3124	31.24	−10.24	104.8576	3.3565
2	40	0.3364	33.64	6.36	40.4496	1.2024
3	14	0.1812	18.12	−4.12	16.9744	0.9368
4 or 5	10	0.0540	5.40	4.60	21.1600	3.9185
						$\chi^2 = 10.4108$

In this goodness-of-fit test the hypotheses are:

H_0: *This sample is from* $b(y;5,0.35)$
H_a: *This sample is not from* $b(y;5,0.35)$

The degrees of freedom are $\nu = k - 1 = 5 - 1 = 4$. The critical value is $\chi^2_{0.05,4} = 9.488$. The null hypothesis is rejected if this value is exceeded. Thus the marketing researcher rejects the null hypothesis. The sales do not follow the pattern of this binomial distribution.

If the salesman has no idea of the proportion of the times he is successful, he could estimate π by dividing the total number of sales by the total number of visits, $187/500 = 0.374$. He could then test to see if his sales fit $b(y;5,0.374)$. The procedure is similar to the above, except now the degrees of freedom are $k - 2 = 5 - 2 = 3$. One additional degree of freedom is lost because of the estimated parameter. In general, $\nu = k - 1 - r$ where r is the number of parameters that are estimated.

A goodness-of-fit test for a Poisson distribution can be done in a similar manner.

Example 5.4. Goodness-of-Fit Test with an Unspecified Parameter
If the same typesetter sets all the copy for a book, the error rate should be approximately the same throughout the book. With this assumption, the

number of misprints per page may be a Poisson random variable. To check whether the Poisson model is correct, an efficiency expert collects the following data from a random sample of 100 pages.

Number of Mistakes per Page:	0	1	2	3	4	5	6
Observed Frequency y_i:	13	24	31	18	11	2	1

He wants to test

H_0: *This sample is from a Poisson distribution*

against

H_a: *This sample is not from a Poisson distribution*

To estimate λ, the average number of errors per page, he computes the total number of errors and divides by the number of pages, $200/100 = 2.00$. Thus 2.00 is an estimate of λ in the Poisson distribution. Looking at the Poisson distribution with $\lambda = 2.00$, he finds:

y	Probability
0	0.1353
1	0.2707
2	0.2707
3	0.1804
4	0.0902
5	0.0361
6	0.0120
Over 6	0.0045

If these eight categories are used for a goodness-of-fit test, the expected values for the last three categories will all be less than five. Since $3/8 = 0.375$, too many expected values are under five. To take care of this, he can combine the last three categories and compute the chi-square statistic as follows:

Category A_i	Observed y_i	$P(A_i)$	Expected e_i
0	13	0.1353	13.53
1	24	0.2707	27.07
2	31	0.2707	27.07
3	18	0.1804	18.04
4	11	0.0902	9.02
Over 4	3	0.0526	5.26
	100		

and

$$\chi^2 = \sum_{i=1}^{k} \frac{(y_i - e_i)^2}{e_i} = 2.345$$

The null hypothesis will be rejected if this computed chi-square value is greater than or equal to $\chi^2_{0.05,4} = 9.488$. There are four degrees of freedom because $\nu = k - 1 - 1 = 6 - 2 = 4$; the additional degree of freedom is lost because of the estimation of λ. The efficiency expert does not reject the null hypothesis in this study, and he concludes that the errors per page may be modeled by a Poisson distribution.

Both of the examples used in this section concern discrete probability distributions. It is also possible to do a chi-square goodness-of-fit test for continuous probability distributions. An example is given in Exercise 7.1.8.

■ ■ ■

Procedure. Chi-Square Goodness-of-Fit Test

H_0: *This sample is from distribution A*
H_a: *This sample is not from distribution A*

Significance level: α
Test statistic:

$$\chi^2 = \sum_{i=1}^{k} \frac{(y_i - e_i)^2}{e_i}$$

y_i = observed number of outcomes in category A_i

$e_i = nP(A_i) \qquad n = \sum_{i=1}^{k} y_i$

Region of rejection:

$\chi^2 \geq \chi^2_{\alpha,\nu} \qquad \nu = k - 1 - r$

r = number of parameters in distribution A estimated from the sample

■ ■ ■

EXERCISES

5.2.1. Sixty sample groups of four persons in each group contain the following distribution for the number of persons with type-O blood.

Number with Type-O:	0	1	2	3	4
Frequency:	8	18	21	8	5

Are these sample groups of four from the binomial distribution $b(y;4,0.40)$?

5.2.2. If the number of defects in a hundred 20-ft sections of wire are

Number of Defects:	0	1	2	3	4
Frequency:	88	10	1	0	1

does this fit a Poisson distribution with $\lambda = 0.10$?

5.2.3. A campground has five rustic campsites not accessible to campers on wheels. Some nights, some of these campsites are unoccupied because of the small number of campers with equipment for such campsites. The ranger keeps track of the number of unoccupied sites for 50 nights.

Number Unoccupied:	0	1	2	3	4	5
Frequency:	22	20	7	1	0	0

Do these data fit a binomial distribution?

5.2.4. If the following numbers of parasites are found on 80 hosts,

Number of Parasites:	0	1	2	3	4	5
Number of Hosts:	20	28	19	9	3	1

does this fit a Poisson distribution?

5.2.5. It seems that the history of the Supreme Court with respect to the occurrence of appointments within a year might be an example of a Poisson distribution (Kinney 1973, Wallis 1936). Test the following data for Poissonness, using a chi-square goodness-of-fit test at the 0.05 significance level.

Number of Appointments per Year	Number of Years (1790–1972)
0	108
1	55
2	19
3	1
4 or more	0

5.3. CONTINGENCY TABLE ANALYSIS

With goodness-of-fit tests, we can determine whether a single sample comes from a population that has a certain probability model. Sometimes we want to know whether or not *several samples* all come from the same population and perhaps we do not even know the appropriate model for the population. A *chi-square test of homogeneity* can often be used in this case.

For example, a speech pathologist might want to know whether the proportion of males among stammerers and the proportion of males among lispers are the same. Her null and alternative hypotheses are

$$H_0:\ \pi_S = \pi_L$$
$$H_a:\ \pi_S \neq \pi_L$$

in which π_S is the proportion of stammerers who are male and π_L is the proportion of lispers who are male. Note that the values of π_S and π_L are not specified in the null hypothesis. (The proportions for females could also be included in the null hypothesis, but this is unnecessary since there are only two classes, male and female, and the proportions must sum to one.)

The speech pathologist collects information from two random samples, one of stammerers and the other of lispers, and arranges the data in the form of a two-way table called a *contingency table*. (The following data are simplified in order to keep the arithmetic simple in this first example.)

	SAMPLES	
	STAMMER	LISP
MALE	32	28
FEMALE	18	22
TOTAL	50	50

The proportion of males in the sample of stammerers is 32/50 and the proportion of males in the sample of lispers is 28/50. Are these sample proportions so different that they indicate that the population proportions are not equal, $\pi_S \neq \pi_L$? To answer this, the speech pathologist computes the total number of males and females in the sample and uses these totals to find the expected value for each of the cells in the two-way layout if the null hypothesis is true.

	OBSERVED		TOTAL
	STAMMER	LISP	
MALE	32	28	60
FEMALE	18	22	40
TOTAL	50	50	100

	EXPECTED		
	STAMMER	LISP	TOTAL
MALE	30	30	60
FEMALE	20	20	40
TOTAL	50	50	100

The expected number of male stammerers is 30 because if the two populations are the same, 60/100 = 0.60 of the people with speech problems are males and 0.60(50) = 30, that is, there are 50 stammerers and 30 of them on the average should be males. There are two ways that the rest of the cells can be filled with expected values. Each expected value can be computed similarly to the one for the male stammerers; however, since the totals are known, the remaining cells can be filled by subtraction. For example, the expected number of male lispers is 60 − 30 = 30.

To find the expected value for a cell directly from the totals, we divide the product of the two corresponding marginal totals by the grand total. For the male stammerers this is (50)(60)/100 = 30. We can summarize this procedure by using the following symbols in which i identifies the row and j the column.

	OBSERVED		TOTAL
	y_{11}	y_{12}	$y_{1.}$
	y_{21}	y_{22}	$y_{2.}$
TOTAL	$y_{.1}$	$y_{.2}$	$y_{..}$

EXPECTED	
e_{11}	e_{12}
e_{21}	e_{22}

$$e_{ij} = \frac{(y_{i.})(y_{.j})}{y_{..}}$$

Once we have found the expected value, the χ^2 statistic is computed in the usual way.

Class	y_{ij}	e_{ij}	$y_{ij} - e_{ij}$	$(y_{ij} - e_{ij})^2$	$(y_{ij} - e_{ij})^2/e_{ij}$
Male, stammer	32	30	+2	4	0.133
Female, stammer	18	20	−2	4	0.200
Male, lisp	28	30	−2	4	0.133
Female, lisp	22	20	+2	4	0.200
					$\chi^2 = 0.666$

In a chi-square test of homogeneity, the degrees of freedom are $\nu = (r - 1)(c - 1)$ in which r is the number of rows and c is the number of columns. In this illustration $\nu = 1$. This corresponds to the fact that once we have computed one expected value from the totals in the two-by-two layout, all of the other values are determined.

The critical chi-square value for one degree of freedom is $\chi^2_{0.05,1} = 3.841$, and the null hypothesis is rejected if the chi-square statistic is greater than or equal to this value. The speech pathologist notes that the computed chi-square value is less than the critical value, and she decides that the proportion of males among stammerers may be the same as the proportion of males among lispers. She concludes that when males are tested for speech problems, they should not be tested for a specific problem such as stammering but should be given a general test that would identify both stammerers and lispers.

A chi-square test of homogeneity is used to determine whether two or more multinomial populations are the same. In the example just completed, the decision concerned two binomial populations. In the next example three multinomial population will be examined.

Example 5.5 Chi-square Test of Homogeneity

A political scientist is interested in how important defense spending is to the members of the different political parties. The three populations of interest are Democrats, Republicans, and Independents. He chooses a random sample of 100 from each of the parties, using party voter registration lists, and he asks the subjects to rate the importance of defense spending on a scale from 1 to 4. The results are as follows.

	VERY IMPORTANT 1	2	3	NOT IMPORTANT 4	TOTAL
DEMOCRATS	42	26	19	13	100
REPUBLICANS	55	21	14	10	100
INDEPENDENTS	38	30	22	10	100
TOTAL	135	77	55	33	300

In words, the hypotheses are

H_0: *Members of the three parties agree on the importance of defense spending* (homogeneity)

H_a: *Members of the three parties do not agree on the importance of defense spending* (lack of homogeneity)

Using the totals and the formula

$$e_{ij} = \frac{(y_{i.})(y_{.j})}{y_{..}}$$

the expected values are

	1	2	3	4	TOTAL
DEMOCRATS	45.0	25.7	18.3	11.0	$100 = y_{1.}$
REPUBLICANS	45.0	25.7	18.3	11.0	$100 = y_{2.}$
INDEPENDENTS	45.0	25.7	18.3	11.0	$100 = y_{3.}$
TOTAL	$135 = y_{.1}$	$77 = y_{.2}$	$55 = y_{.3}$	$33 = y_{.4}$	$300 = y_{..}$

The χ^2 statistic is computed.

Class	y_{ij}	e_{ij}	$(y_{ij} - e_{ij})^2/e_{ij}$
Democrats			
1	42	45.0	0.200
2	26	25.7	0.004
3	19	18.3	0.027
4	13	11.0	0.364
Republicans			
1	55	45.0	2.222
2	21	25.7	0.860
3	14	18.3	1.010
4	10	11.0	0.091
Independents			
1	38	45.0	1.089
2	30	25.7	0.719
3	22	18.3	0.748
4	10	11.0	0.091
			$\chi^2 = 7.425$

Since there are three rows and four columns in the contingency table

$$\nu = (r - 1)(c - 1) = (3 - 1)(4 - 1) = 6$$

At the 0.05 level of rejection, the null hypothesis is rejected if the computed chi-square value is greater than or equal to

$$\chi^2_{0.05,6} = 12.592$$

Since this is not the case in this study, the null hypothesis is accepted and the political scientist concludes that there is no evidence to indicate that the three populations are different with respect to their opinions on the importance of defense spending.

The chi-square test of homogeneity is applied to samples from two or more populations when the samples have been classified by one characteristic. There is a similar chi-square test that can be used to analyze data from a *single population* when the data have been classified by *two characteristics*. For example, in a state in which party affiliation is not declared at voter registration, a single sample of 300 registered voters could be selected at random and asked for their opinion on the importance of defense spending and also for their party preference. The contingency table would look similar to the table in Example 5.5 except that it is not likely that there would be exactly 100 from each party. The political scientist would be trying to determine whether party affiliation is related to opinion about defense spending, and the test procedure is called a *chi-square test of independence*.

H_0: *Party preference is independent of opinion about the importance of defense spending*
H_a: *Party preference is related to opinion about the importance of defense spending*

The test statistic and region of rejection is determined as in a test for homogeneity. A worked-out example follows.

Example 5.6. A Chi-Square Test of Independence
Football coaches feel that a football team has an advantage when it is playing a home game in its own stadium. The enthusiasm of the crowd, familiarity with the field, the lack of fatigue from travel all seem to contribute to this assumed advantage. A coach wants to test this theory at his school. If the theory is wrong, whether a game is won or lost is independent of whether the game is played at home or away. The hypotheses are

H_0: *The probability of winning is independent of where the game is played*
H_a: *The probability of winning depends on where the game is played*

The coach examines the records at his school over the past 31 years and classifies the results as follows (ties and bowl games are omitted).

OBSERVED

	HOME	AWAY	TOTAL
WON	97	69	166
LOST	42	83	125
TOTAL	139	152	291

Intuitively the data seem to confirm the coach's theory. Using the marginal totals, he computes the following expected values:

EXPECTED

	HOME	AWAY
WON	79.3	86.7
LOST	59.7	65.3

He then computes the chi-square statistic:

Class	y_{ij}	e_{ij}	$y_{ij} - e_{ij}$	$(y_{ij} - e_{ij})^2$	$(y_{ij} - e_{ij})^2/e_{ij}$
Won/home	97	79.3	17.7	316.3	3.99
Lost/home	42	59.7	−17.7	316.3	5.30
Won/away	69	86.7	−17.7	316.3	3.65
Lost/away	83	65.3	17.7	316.3	4.84
					$\chi^2 = 17.78$

Since $\chi^2_{0.05,1} = 3.841$, the null hypothesis is rejected and the coach concludes that if these 31 years are a random sample of this school's games, there is evidence of a home team advantage.

Since 2×2 contingency tables have one degree of freedom, the continuity correction can be used to improve the approximation of the discrete sampling distribution by the continuous theoretical chi-square distribution.

As in goodness-of-fit tests, contingency table tests do not work well for small expected values (below 5). In the 2×2 case, another test can be used when the expected values are small, *Fisher's exact test*. References to this test are given at the end of this chapter (Fisher 1973, Finney 1948, Latscha 1955).

■ ■ ■

Procedure. Contingency Table Analysis

Chi-square Test of Homogeneity

H_0: *The populations sampled are the same with respect to the categorization*

H_a: *The populations sampled are different with respect to the categorization*

and

Chi-square Test of Independence

H_0: *The row categories are independent of the column categories*

H_a: *The row categories and the column categories are dependent*

Significance level: α

Test statistic:

$$\chi^2 = \sum_i \sum_j \frac{(y_{ij} - e_{ij})^2}{e_{ij}}$$

$$y_{ij} = \text{number of occurrences in the } ij\text{th cell}$$

$$e_{ij} = \frac{(y_{i.})(y_{.j})}{y_{..}}$$

$$y_{i.} = \sum_j y_{ij}$$

$$y_{.j} = \sum_i y_{ij}$$

$$y_{..} = \sum_i \sum_j y_{ij}$$

Region of rejection:

$$\chi^2 \geq \chi^2_{\alpha,\nu} \qquad \nu = (r - 1)(c - 1)$$

r = number of rows

c = number of columns

■ ■ ■

EXERCISES

5.3.1. An English physician claims that people with last names beginning with the initials S through Z are more subject to stomach ulcers than those whose last names begin with initials earlier in the alphabet. Suppose a random sample of business executives is taken to test his claim. After selection, the executives are classified by their last names and by whether or not they have ulcers.

First Initial of Last Name	Evidence of Ulcers	
	Yes	No
A to *R*	18	42
S to *Z*	22	18

a. State the appropriate null and alternative hypotheses.

b. Perform the test at $\alpha = 0.05$.

c. What is the conclusion?

5.3.2. A serum thought to be effective in preventing colds is given to 300 persons. Their records for one year are compared with those of 200 untreated persons with the following results:

	No Colds	One Cold	More than One Cold
Treated	145	80	75
Untreated	80	70	50

Use a chi-square test of homogeneity to analyze these data.

5.3.3. A social scientist wants to determine if the feelings that parents have toward young people "living together" are affected by the age of their youngest child.

Age of Youngest Child	Parents' Feelings	
	Approve	Disapprove
Over 26	10	40
18 to 26	50	10
Under 18	60	30

a. State the null hypothesis verbally in terms of independence.
b. Perform a chi-square test of independence at the 0.05 level of significance.

5.3.4. It is reported that offspring produced by users of a certain drug may have a higher incidence of birth defects than the general population. To obtain information about a possible relationship between this drug and birth defects, 100 offspring of female rats fed the drug and 100 offspring from untreated female rats are examined. The results are given below:

	Progeny	
Females	Birth Defects	Normal
Treated	30	70
Untreated	20	80

Analyze these data. What do you conclude from the study? Is this a test of homogeneity or independence?

5.3.5. In an effort to reduce cost due to failure on a job training course, a brief screening examination is developed. It is given to the 300 candidates for the training program. The results are:

	Training Program	
Screening Exam	Passed	Failed
Passed	196	24
Failed	34	46

a. State the null and alternative hypotheses in words.
b. Give the critical value when $\alpha = 0.01$.
c. Perform the test and draw conclusions.
d. What assumption must be made about the sample for this test to be valid?

5.3.6. A consumer's union would like to compare three brands of flashlight batteries. Its testers randomly select 100 batteries of each brand and classify them into three groups depending on lifetimes.

Brand	Less than 5 Hours	5 to 10 Hours	Over 10 Hours	Total
X	30	60	10	100
Y	15	60	25	100
Z	30	30	40	100

a. State the null and alternative hypotheses to be tested.
b. Compute the chi-square statistic.
c. What are the statistical decision and the experimental conclusion?

5.3.7. An entomologist is interested in determining whether certain insecticides have a differential effect on black flies. The results of his experiment are:

Insecticide	Dead	Alive
A	165	35
B	172	28
C	173	27

a. What null hypothesis can be tested with these data?
b. If the entomologist sets the rejection level at 1%, how large must the chi-square statistic be in order for him to reject the null hypothesis?
c. Compute the statistic.
d. How likely is it that a sample as unusual as this will be obtained when the null hypothesis is true?
e. What decision should the entomologist make about the null hypothesis? What conclusion should be drawn?

5.3.8. A study was conducted on adult male cancer patients to determine whether there was any association between the kinds of work they performed and the kinds of cancer they had. The data are classified by the two categories as below:

	Site of Malignancy		
Occupation	Skin	Stomach	Prostate
Professional	25	58	37
Managerial	34	90	36
Laborer	41	52	27

a. State the null hypothesis verbally and in symbols.
b. Give the critical value of the test statistic for $\alpha = 0.05$.
c. Compute the expected value for the category laborer and stomach.
d. The computed value of chi-square is 10.49. Which of the following statements are appropriate to this survey?
   i. The type of work one does causes certain kinds of cancer.
   ii. The location of a cancer is independent of occupation.
   iii. There is a significant association between occupation and kind of cancer.

REVIEW EXERCISES

Decide whether each of the following statements is true or false. If a statement is false, explain why.

5.1. There is only one chi-square distribution.

5.2. The chi-square statistic does not have a continuous distribution, but the continuous distribution attributed to Helmert provides reliable probability statements.

5.3. If the computed value of chi-square is greater than the critical value, the null hypothesis is false.

5.4. H_0: $\pi = 0.7$ with H_a: $\pi \neq 0.7$ can be tested with either the binomial distribution or the chi-square distribution; if the sample size is large, the conclusion should be the same for the two tests.

5.5. If there are 75 observations from a binomial esperiment and in the multinomial chi-square test $(y_1 - e_1)^2 = 25$, then $(y_2 - e_2)^2 = 25$ also.

5.6. The greater the degrees of freedom, the less likely the sample values of chi-square will be in the region of rejection.

5.7. To say that a computed chi-square value is "significant" indicates that it is numerically smaller than the critical value against which it is compared.

5.8. In a multinomial experiment to test H_0: $\pi_1 = 0.25$, $\pi_2 = 0.50$, $\pi_3 = 0.25$, three degrees of freedom should be used.

5.9. If the sample size is less than 25, a correction for continuity should be made when testing a 1:2:1 ratio.

5.10. With random sampling, significant values of the chi-square statistic will be obtained only when the null hypothesis is false.

5.11. As the degrees of freedom for the chi-square distribution increase, the probability of rejecting a true null hypothesis decreases.

5.12. With random sampling, a computed chi-square value greater than the critical value can be obtained, even when the null hypothesis is true.

5.13. If there is close agreement between the observed and expected frequencies, the chi-square statistic should be relatively large.

5.14. The critical value at $\alpha = 0.05$ for a multinomial chi-square test about a 27:9:9:3:3:3:1 genetic ratio is 14.067.

5.15. The multinomial chi-square test is a goodness-of-fit test.

5.16. In order to test whether a set of samples can be modeled by a Poisson distribution, the experimenter must specify the Poisson parameter before sampling.

5.17. If the null hypothesis for a goodness-of-fit test is not rejected, we can conclude that the data are from a population with the specified probability distribution.

5.18. A chi-square contingency table analysis is not appropriate if it is suspected that the row and column categories are not independent.

5.19. To reject the null hypothesis in a chi-square test of independence is to decide that the categories in the rows are independent of those in the columns.

5.20. The chi-square test of homogeneity can be used if hypothetical ratios are unknown but may be equal for all population sampled.

5.21. A chi-square test of independence for a $k \times 2$ table has $k - 1$ degrees of freedom associated with it.

5.22. A chi-square test of homogeneity can be used to test the equality of the parameters in two binomial distributions.

5.23. A chi-square test of independence tests the null hypothesis that there is an association between row and column categories against the alternative that they are unrelated.

5.24. Many naturally occurring populations have the chi-square distribution as their model.

5.25. The expected value and the variance of a given chi-square distribution are equal.

SELECTED READINGS

Chapman, D. G., and R. C. Meng (1966). The power of chi-square tests for contingency tables. *Journal of the American Statistical Association,* **61,** 965–975.

Chase, G. R. (1972). On the chi-square test when the parameters are estimated independently of the sample, *Journal of the American Statistical Association,* **67,** 609–611.

Cochran, W. G. (1952). The chi-square test of goodness of fit, *The Annals of Mathematical Statistics,* **23,** 315–345.

——— (1954). Some methods for strengthening the common chi-square tests, *Biometrics,* **10,** 417–451.

Conover, W. J. (1974). Some reasons for not using the Yates correction on 2 × 2 contingency tables, *Journal of the American Statistical Association,* **69,** 374–382.

Finney, D. J. (1948). The Fisher–Yates test of significance in 2 × 2 contingency tables, *Biometrika,* **35,** 145–156.

Fisher, R. A. (1935). The logic of inductive inference, *Journal of the Royal Statistical Society,* Section A, **98,** 39–54.

——— (1973). *Statistical Methods for Research Workers,* 14th ed., Hafner, New York.

Good, I. J. (1973). What are degrees of freedom? *The American Statistician,* **27,** 227–228.

Grizzle, J. E. (1967). Continuity correction in the chi-square test for 2 × 2 tables, *The American Statistician,* **21** (Oct.), 28–32.

Guenther, W. C. (1977). Power and sample size for approximate chi-square tests, *The American Statistician,* **31,** 83–85.

Hoel, P. G. (1938). On the chi-square distribution for small samples, *The Annals of Mathematical Statistics,* **9,** 158–165.

Kinney, J. (1973). Poisson updated (letter to the editor), *The American Statistician,* **27,** 195.

Latscha, R. (1955). Tests of significance in a 2 × 2 contingency table: Extension of Finney's table, *Biometrika,* **40,** 74–86.

Liddell, F. D. K. (1972). Correcting the correction in the chi-square test in 2 × 2 tables, *Biometrics,* **28,** 268–269.

Plackett, R. L. (1964). The continuity correction in 2 × 2 tables, *Biometrika,* **51,** 327–337.

Roscoe, J. T., and J. A. Byars (1971). An investigation of the restraints with respect to sample size commonly imposed on the use of the chi-square statistic, *Journal of the American Statistical Association,* **66,** 755–759.

Wallis, W. A. (1936). The Poisson distribution and the Supreme Court, *Journal of the American Statistical Association,* **31,** 376–380.

Williams, C. A., Jr. (1950). On the choice of the number and width of classes for the chi-square test of goodness of fit, *Journal of the American Statistical Association,* **45,** 77–86.

Yarnold, J. K. (1970). The minimum expectation in chi-square goodness of fit tests and the accuracy of approximations for the null distribution, *Journal of the American Statistical Association,* **65,** 864–886.

Yates, F. (1934). Contingency tables involving small numbers and chi-square tests, *Journal of the Royal Statistical Society Supplement,* **1,** Series B, 217–235.

6

Sampling Distribution of Averages

In Chapters 3 through 5 we discussed techniques for analyzing certain types of data that are collected on the nominal scale. All of the procedures in those chapters dealt with data that are in the form of counts. This chapter is a transition to data that are collected on a numerical scale. The remainder of this book will deal with data that arise from measurements rather than counts.

6.1. POPULATION MEAN AND SAMPLE AVERAGE

As in the case of count data, researchers use statistical analysis of measurement data to make statements about populations that are not totally accessible from information obtained from properly chosen samples.

One of the parameters of a population that is often of interest is the *population mean*, because it is one way to describe the population's center or location. If the population were totally accessible, its mean would be computed by the formula

$$\mu = \frac{\sum y}{N}$$

in which μ (the lower-case Greek letter "mu") is the symbol for the population mean, $\sum y$ is the sum of all of the values of the variable of interest for the whole population, and N is the number of elements in the population.

We rarely have an opportunity to use this formula since most of the populations we study are not totally accessible; they are either too large, perhaps even infinite, or would be destroyed in the process of measurement.

Example 6.1. Computing a Population Mean

Historians often use the frequency of certain grammatical constructions to help identify the writings of a historical person. For example, a historian might determine the number of occurrences of a parallel series of adjectives such as

the worker was *tired* and *weary*

in 3000-word sections of a person's known writings. Imagine that the population of all of the known writings of the person can be arranged into 10 sections of 3000 words each, and the number of occurrences are

19 21 18 24 19 21 22 19 22 22

To find the population mean, the historian finds the sum of these data and divides by the number of observations:

$$\mu = \frac{\sum y}{N}$$

$$= \frac{19 + 21 + 18 + 24 + 19 + 21 + 22 + 19 + 22 + 22}{10}$$

$$= 20.7$$

That is, the mean number of parallel adjectives per 3000 words used by this author is 20.7.

If the population data are arranged in the form of a frequency distribution in which y is the value of the variable of interest and f is the number of occurrences, then the population mean can be computed by the formula

$$\mu = \frac{\sum yf}{N}$$

in which the summation is over the different values of y. To use this formula, a third column is added to the frequency table:

y	f	yf
18	1	18
19	3	57
21	2	42
22	3	66
24	1	24
	$N =$ 10	207 $= \sum yf$

and then

$$\mu = \frac{\sum yf}{N} = \frac{207}{10} = 20.7$$

If relative frequencies are given in the population table where

$$\text{relative frequency} = \mathbf{f} = f/N$$

then the computation of the population mean is simplified to

$$\mu = \sum y\mathbf{f}$$

Thus

y	**f**	y**f**
18	0.1	1.8
19	0.3	5.7
21	0.2	4.2
22	0.3	6.6
24	0.1	2.4
	$\mu = \sum y\mathbf{f} =$	20.7

We could represent the population by a graph (Figure 6.1), and then the mean μ can be interpreted as the balancing point of the distribution (Figure 6.2).

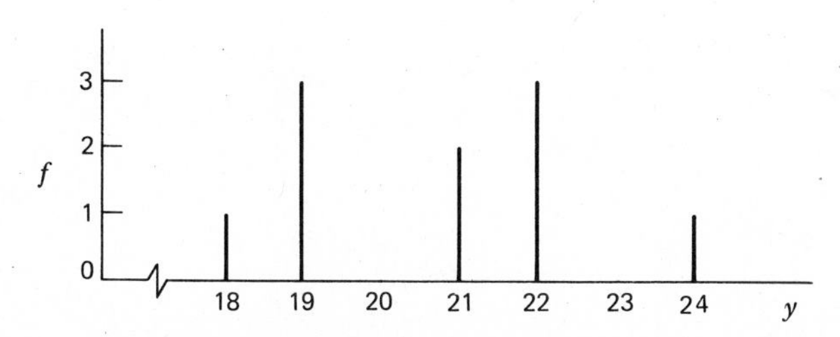

FIGURE 6.1. A population distribution.

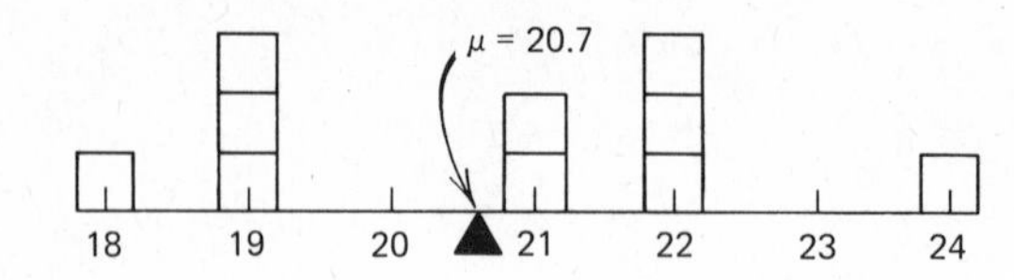

FIGURE 6.2. The population mean as the balancing point.

Since it is often impossible to obtain the population mean, statistical inference is used to estimate μ or to test a hypothesis concerning μ. The basic tool for these inferences (as in the case of count data) is a probability distribution that is a model of the population. We are already familiar with the concept of the expected value $E(y)$ of a probability distribution (see Section 2.4). If a certain probability distribution is the appropriate model for a population, then $E(y)$ will coincide with the population mean μ. Because of this, the expected value of a probability distribution is often called its mean, and we write $\mu = E(y)$. We should recall at this point that the expected value of a discrete probability distribution can be computed by the formula

$$E(y) = \sum yp(y)$$

This is analogous to the formula for a population mean if the values are arranged in a relative frequency distribution:

$$\mu = \sum y\mathbf{f}$$

Statistical inference about a population mean requires, in addition to a probability distribution to model the population, some information obtained from a sample of the population. A reasonable statistic to use is the *sample average*. The sample average is analogous to a population mean. If $\bar{y}$ is used as the symbol for a sample average, then

$$\bar{y} = \frac{\sum y}{n}$$

in which y is the value of the variable of interest for each of the members in the sample, the sum is over those values, and n is the number of observations in the sample. (The symbol $\bar{y}$ is read "y bar.") As in the case of population means, this formula can be modified for data arranged in a frequency table, then

$$\bar{y} = \frac{\sum yf}{n}$$

If the data are in a relative frequency table, then

$$\bar{y} = \sum y\mathbf{f}$$

Example 6.2. Computing a Sample Average
A random sample of 100 high-school students is taken prior to their senior year and the number of books they read that summer are recorded.

y	f
0	0.15
1	0.20
2	0.30
3	0.15
4	0.10
5	0.05
6	0.02
7	0.02
8	0.00
9	0.00
10	0.01

The sample average is computed by adding a third column to the relative frequency table.

y	f	yf
0	0.15	0.00
1	0.20	0.20
2	0.30	0.60
3	0.15	0.45
4	0.10	0.40
5	0.05	0.25
6	0.02	0.12
7	0.02	0.14
8	0.00	0.00
9	0.00	0.00
10	0.01	0.10

$$\bar{y} = \sum y\mathbf{f} = 2.26 \text{ books}$$

A sample average $\bar{y}$ is used as an estimator of the population mean μ. We write $\bar{y} = \hat{\mu}$ (which is read "mu hat") when we want to indicate that the sample average is an estimator of the population mean. The sample average is a maximum likelihood estimator. It is also unbiased and has a minimum variance among unbiased estimators (see Section 3.3).

■ ■ ■

Procedure. Measures of Location

Name	Formula: Ungrouped Data	Formula: Grouped Data — Frequency Distribution	Formula: Grouped Data — Relative Frequency Distribution
Population Mean	$\mu = \frac{\sum y}{N}$ N = population size	$\mu = \frac{\sum yf}{N}$ N = population size f = frequency	$\mu = \sum y\mathbf{f}$ $\mathbf{f}$ = relative frequency
Sample Average	$\bar{y} = \frac{\sum y}{n}$	$\bar{y} = \frac{\sum yf}{n}$ n = sample size f = frequency	$\bar{y} = \sum y\mathbf{f}$ $\mathbf{f}$ = relative frequency
Expected Value of a Discrete Probability Distribution		$E(y) = \sum yp(y)$	

■ ■ ■

EXERCISES

6.1.1. Find the population mean for the heights of the 50 male students given in Exercise 2.1.8.

6.1.2. Use the data in Exercise 2.1.8 for the following:

a. Arrange the heights into a population frequency distribution.

b. Compute the population mean from the population frequency distribution.

c. Find the population relative frequency distribution.

d. Compute the population mean from the relative frequency distribution.

6.1.3. The following data from a random sample of five-year-old children in the United States represent the number of cavities in their teeth.
4 0 1 0 3 2 1 0 4 3 2 3 4 2 2 3 2 1 1 2
a. Find the sample average from this ungrouped data.
b. Arrange the data into a frequency table.
c. Find the sample average from the frequency table.
d. Estimate the mean number of cavities for the population of all five-year-old children in the United States.

6.1.4. At a certain university a total census is made of all graduating seniors to determine how many courses they have failed during their undergraduate education. The population is as follows:

y:	0	1	2	3	4	5
f:	0.870	0.071	0.031	0.012	0.011	0.005

Find the population mean.

6.2. POPULATION VARIANCE AND SAMPLE VARIANCE

A second population parameter that is often of interest is σ^2, the *population variance*. Variance is a measure of the spread of the population. Suppose we want to choose between two investment plans and are told that both have mean earnings of 10% per annum; we might conclude that they were equally good. However, suppose we learn that Plan A has a variance twice as large as Plan B. This gives us additional information on which to base a choice. If we want to be relatively certain that our earnings are close to 10%, we would select Plan B. If we are willing to gamble that our earnings might be considerably in excess of 10% (or possibly considerably below 10%), we would choose Plan A.

A population variance can be computed from ungrouped data or from data that are grouped into a frequency or relative frequency distribution if the population is of the accessible variety.

For ungrouped data, a population variance is defined to be

$$\sigma^2 = \frac{\sum(y - \mu)^2}{N}$$

in which σ^2 is read "sigma squared" and represents the population variance. In practice, it is more convenient to use an equivalent computational form of this formula, especially when using a hand-held calculator:

$$\sigma^2 = \frac{\sum y^2 - (\sum y)^2/N}{N}$$

Example 6.3. Computing a Population Variance from Ungrouped Data

Consider again the small population of sections of all known writings of a historical person. The number of usages of parallel adjectives per 3000-word sections are

19 21 18 24 19 21 22 19 22 22

and the mean usage is $\mu = 20.7$. The population variance is the average squared deviation from the mean. In tabular form, the computations are as follows:

y	$y - \mu$	$(y - \mu)^2$
19	$19 - 20.7 = -1.7$	2.89
21	$21 - 20.7 = 0.3$	0.09
18	$18 - 20.7 = -2.7$	7.29
24	$24 - 20.7 = 3.3$	10.89
19	$19 - 20.7 = -1.7$	2.89
21	$21 - 20.7 = 0.3$	0.09
22	$22 - 20.7 = 1.3$	1.69
19	$19 - 20.7 = -1.7$	2.89
22	$22 - 20.7 = 1.3$	1.69
22	$22 - 20.7 = 1.3$	1.69
	$\sum(y - \mu)^2 =$	32.10

and

$$\sigma^2 = \frac{\sum(y - \mu)^2}{N} = \frac{32.10}{10} = 3.210$$

This process can be shortened by using the equivalent computational formula that is more adaptable to a calculator:

$$\sigma^2 = \frac{\sum y^2 - (\sum y)^2/N}{N}$$

$$\sum y = 207 \qquad \sum y^2 = 4317 \qquad N = 10$$

so

$$\sigma^2 = \frac{4317 - (207)^2/10}{10}$$

$$= 3.210$$

Sometimes population data are grouped into frequency or relative frequency tables. In these cases the formulas can be adapted. For a frequency table:

$$\sigma^2 = \frac{\sum (y - \mu)^2 f}{N} = \frac{\sum y^2 f - (\sum yf)^2/N}{N}$$

and for relative frequency tables:

$$\sigma^2 = \sum (y - \mu)^2 \mathbf{f} = \sum y^2 \mathbf{f} - (\sum y\mathbf{f})^2$$

This last formula is analogous to the computation of the variance of a discrete probability distribution:

$$\begin{aligned} V(y) &= \sum [y - E(y)]^2 p(y) \\ &= \sum y^2 p(y) - [\sum yp(y)]^2 \end{aligned}$$

If a probability distribution is used to represent a population and a certain probability distribution is an appropriate model, then σ^2, the variance of the population, will be the same as $V(y)$, the variance of the probability distribution. Because of this, σ^2 is often used when speaking of the variance of a probability distribution.

Usually we will be estimating the population variance by using a statistic from a random sample of the population. The statistic that is an estimator of the population variance is the *sample variance*, or s^2:

$$s^2 = \frac{\sum (y - \bar{y})^2}{n - 1} = \frac{\sum y^2 - (\sum y)^2/n}{n - 1}$$

Note that the denominator of s^2 is $n - 1$, an unusual way to "average" the squared deviations from the sample average. This modification is necessary so that the sample variance will be an unbiased estimator of the population variance. We write

$$s^2 = \hat{\sigma}^2$$

to indicate that the sample variance is an estimator of the population variance.

The formula for sample variance can be modified for data that are grouped into a frequency table:

$$s^2 = \frac{\sum (y - \bar{y})^2 f}{n - 1} = \frac{\sum y^2 f - (\sum yf)^2}{n - 1}$$

Example 6.4. Computing a Sample Variance from Grouped Data
In the high-school reading study (Example 6.2) of Section 6.1, the frequency table can be expanded to find $\sum yf$ in the third column and $\sum y^2 f$ in the fourth column:

y	f	yf	y^2f
0	15	0	0
1	20	20	20
2	30	60	120
3	15	45	135
4	10	40	160
5	5	25	125
6	2	12	72
7	2	14	98
8	0	0	0
9	0	0	0
10	1	10	100
	$n = 100$	$\sum yf = 226$	$\sum y^2f = 830$

Thus

$$s^2 = \frac{\sum y^2f - (\sum yf)^2/n}{n - 1}$$

$$= \frac{830 - (226)^2/100}{99}$$

$$= 3.22$$

A summary of the computational procedures for variances follows.

■ ■ ■

Procedure. Measures of Spread

Formula

Ungrouped Data	Grouped Data: Frequency Distribution	Grouped Data: Relative Frequency Distribution
Population Variance		
$\sigma^2 = \frac{\sum(y - \mu)^2}{N}$	$\sigma^2 = \frac{\sum(y - \mu)^2 f}{N}$	$\sigma^2 = \sum(y - \mu)^2\mathbf{f}$
$= \frac{\sum y^2 - (\sum y)^2/N}{N}$	$= \frac{\sum y^2 f - (\sum yf)^2/N}{N}$	$= \sum y^2\mathbf{f} - (\sum y\mathbf{f})^2$
N = population size	$N = \sum f$ f = frequency	$\mathbf{f}$ = relative frequency

Sample Variance

$s^2 = \dfrac{\sum(y - \bar{y})^2}{n - 1}$	$s^2 = \dfrac{\sum(y - \bar{y})^2 f}{n - 1}$	Convert relative frequencies to frequencies and use method to the left
$= \dfrac{\sum y^2 - (\sum y)^2/n}{n - 1}$	$= \dfrac{\sum y^2 f - (\sum yf)^2/n}{n - 1}$	
n = sample size	$n = \sum f$ f = frequency	

Variance of a Discrete Probability Distribution

$$V(y) = \sum[y - E(y)]^2 p(y)$$
$$= E(y^2) - [E(y)]^2$$
$$= \sum y^2 p(y) - [\sum y p(y)]^2$$

■ ■ ■

We might wonder at this point about the meaning of the numerical value of population and sample variances. Larger variances indicate a larger spread for the distribution, but can more than this be said? One approach is to use the result worked out by the Russian mathematician P. L. Chebyshev (1821–1894).

Chebyshev used the standard deviation, a measure related to the variance. A *population standard deviation* is the positive square root of the population variance

$$\sigma = \sqrt{\sigma^2}$$

and a *sample standard deviation* is the positive square root of the sample variance

$$s = \sqrt{s^2}$$

The standard deviation has the advantage of being in the same unit of measurement as the data, whereas the variance is in squared units that often have no intuitive meaning (as ''squared books'' in Example 6.4).

Chebyshev proved that in any collection of data, *at least* 3/4 of the values lie within two standard deviations of the mean (or average) and that at least 8/9 of the values lie within three standard deviations of the mean (or average).

TABLE 6.1. Chebyshev's Theorem for Some Values of $k > 1$

At least this proportion of the data:	Lies within this interval: Population	Sample
$1 - 1/2^2 = 3/4$	$\mu \pm 2\sigma$	$\bar{y} \pm 2s$
$1 - 1/3^2 = 8/9$	$\mu \pm 3\sigma$	$\bar{y} \pm 3s$
$1 - 1/4^2 = 15/16$	$\mu \pm 4\sigma$	$\bar{y} \pm 4s$
$1 - 1/k^2$	$\mu \pm k\sigma$	$\bar{y} \pm ks$

TABLE 6.2. The Empirical Rule

Approximately this proportion of the data:	Lies within this interval: Population	Large Sample
0.682	$\mu \pm 1\sigma$	$\bar{y} \pm 1s$
0.954	$\mu \pm 2\sigma$	$\bar{y} \pm 2s$
0.997	$\mu \pm 3\sigma$	$\bar{y} \pm 3s$

In general, the theorem states that for real numbers k, $k > 1$, at least $1 - 1/k^2$ of the values lie within k standard deviations of the mean (or average). Table 6.1 summarizes this result. Note that the theorem is true for *any* population or sample. Although this theorem gives only a lower bound for the proportion of the data within certain intervals, it is applicable to all data sets regardless of the shape of their distribution and regardless of their size.

If a population or a large sample is symmetrical and mound-shaped, an estimate is possible for the proportion of the data within certain intervals. The estimates in Table 6.2 are often called the *Empirical Rule*. (These proportions are determined from the standard normal distribution, see Section 7.1.)

EXERCISES

6.2.1. Find the population variance for the heights of the 50 males given in Exercise 2.1.8.

6.2.2. Use the height data and the tables found in Exercise 6.1.2. for the following.

a. Compute the population variance from the population frequency distribution.

b. Compute the population variance from the relative frequency distribution.

6.2.3. Use the sample data from Exercise 6.1.3. for the following.

a. Find the sample variance from the ungrouped data.

b. Find the sample variance from the frequency table.

6.2.4. Use the data from the population in Exercise 6.1.4 and find the population variance.

6.2.5. Consider the following three samples:

I. 1 2 2 3 3 3 4 4 5

II. 7 8 8 9 9 9 10 10 11

III. 1 1 1 2 2 3 4 4 5 5 5

a. Graph the frequency distribution for each of the three samples.
b. Compute the average of each sample.
c. Compute the variance of each sample.
d. Compare the average of samples I and II. What characteristic of the two data sets explains the difference in the averages?
e. Notice that the variances of set I and set II are equal. What geometric property of these two distributions accounts for this equality?
f. Note that set I and set III have the same average. Why is this possible for two data sets that seem so different?
g. Compare the shape of distributions I and III. Why would you expect the variance of I to be smaller than the variance of III?

6.2.6. Each mating season, birds of a certain species usually lay a "clutch" of six eggs in their nests. A biologist notices, however, that clutch number deviates from the usual when the birds feed on a certain kind of berry containing a narcotic alkaloid. He examines the nests of seven such birds and finds the following numbers of eggs:

8 2 5 7 4 10 6

a. Is there evidence that the alkaloid causes the birds to lay fewer eggs than usual?
b. Compute the variance of the sample.

6.2.7. Show that Chebyshev's theorem is true for the population in Exercise 6.1.4 for $k = 2$ and $k = 3$.

6.3. THE MEAN AND VARIANCE OF THE SAMPLING DISTRIBUTION OF AVERAGES

When dealing with binominal data, the useful statistic for inference is the number of occurrences in a certain category. This count summarizes the entire sample. Similarly, when dealing with numerical data there is a useful statistic which summarizes all of the measurements from the sample; this statistic is $\bar{y}$, the sample average. In many types of inference, we use the summary statistic $\bar{y}$ rather than the actual values obtained from the individuals in the sample. Since we use the sample average, it is necessary to further develop the properties of this statistic.

The first thing we should note is that $\bar{y}$ is a random variable, that is, it has a numerical value that is associated with the outcome of an experiment or survey. The sample average $\bar{y}$ depends upon the particular random sample chosen and varies for different samples.

Because $\bar{y}$ is a random variable, it has a probability distribution. The probability distribution associated with $\bar{y}$ is called the *sampling distribution of sample averages*. This sampling distribution consists of all possible values of $\bar{y}$ for a fixed sample size and the probabilities associated with these values of the random variable.

If the random variable is discrete and has a finite number of values, we can actually display the sampling distribution of averages. For example, if the population consists of ratings 1, 2, 3, 4, and all of these values are equally likely, then the population can be represented by the following probability distribution:

y:	1	2	3	4
$p(y)$:	1/4	1/4	1/4	1/4

The averages of all samples of size two are given in the body of the following table.

		OBSERVATION 2			
	$\bar{y}$	1	2	3	4
OBSERVATION 1	1	1	3/2	2	5/2
	2	3/2	2	5/2	3
	3	2	5/2	3	7/2
	4	5/2	3	7/2	4

If the random variable is continuous or has an infinite number of values, we cannot enumerate all of the averages but we can still think about them. In order to illustrate the properties of sampling distributions of averages, we will use the above small discrete example; however, the same properties are true for all sampling distributions of averages.

Since the sampling distribution of averages of all samples of a fixed size is a probability distribution, it has an expected value (mean) and a variance, and these parameters are related to the mean and variance of the underlying population.

In the discrete example concerning equally likely rankings, the mean of the population is

$$\begin{aligned}\mu_y = E(y) &= \sum yp(y)\\ &= (1 + 2 + 3 + 4)(1/4)\\ &= 5/2\end{aligned}$$

and the variance of the population is

$$\begin{aligned}\sigma_y{}^2 = V(y) &= \sum(y - 5/2)^2p(y)\\ &= 5/4\end{aligned}$$

To find the mean and the variance of the sampling distribution of averages of all samples of size $n = 2$, we first give the probability distribution in tabular form.

$\bar{y}$:	1	3/2	2	5/2	3	7/2	4
$p(\bar{y})$:	1/16	2/16	3/16	4/16	3/16	2/16	1/16

The graph of the sampling distribution of averages appears in Figure 6.3. The mean is

$$\begin{aligned}\mu_{\bar{y}} &= E(\bar{y}) = \sum \bar{y}p(\bar{y}) \\ &= 1(1/16) + (3/2)\,(2/16) + \cdots + 4(1/16) \\ &= 5/2\end{aligned}$$

and the variance is

$$\begin{aligned}\sigma_{\bar{y}}^2 &= V(\bar{y}) = \sum(\bar{y} - 5/2)^2 \\ &= 5/8\end{aligned}$$

We should note the following about this example of a sampling distribution of averages:

1. The sampling distribution of averages has the same mean as the underlying population.
2. The sampling distribution of averages has a smaller variance than the underlying population.
3. The sampling distribution of averages is symmetric and unimodal.

One particular illustration, of course, does not prove that these properties always hold. However, it can be proved mathematically that for all sampling

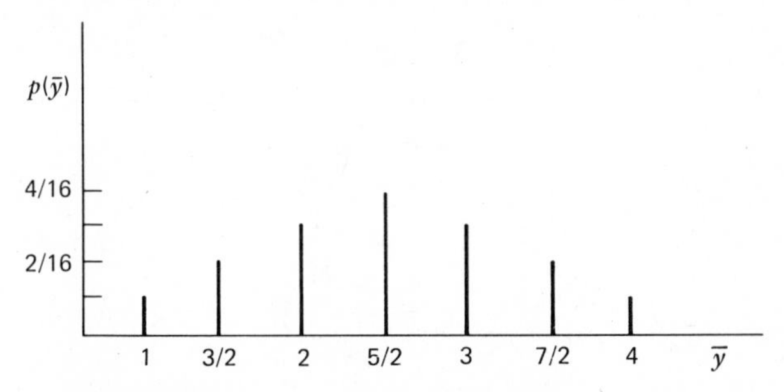

FIGURE 6.3. A sampling distribution of averages.

distributions of averages:

1. $\mu_{\bar{y}} = \mu_y$
2. $\sigma_{\bar{y}}^2 = \sigma_y^2/n$
3. If the sample size n is sufficiently large, then the distribution of $\bar{y}$ is almost symmetric and unimodal.

Another property of sampling distributions of averages is taken up in Chapter 7 after the discussion of normal distributions. In Chapters 7 and 8, the sampling distribution of averages is used for making an inference about the population mean.

In this section, as well as in the rest of this book, we assume that sampling is from an infinite population, or from a finite population and the sampling is with replacement. If the sampling is without replacement and from a finite population, we assume that the sample size is 5% or less of the population size. Many of the properties discussed in this text do not hold if sampling is without replacement from a finite population and the sample size is more than 5% of the population size.

EXERCISES

6.3.1. Let y be a discrete random variable with the following distribution:

$$p(y) = 1/3 \quad \text{for } y = 5, 7, 10$$
$$p(y) = 0 \quad \text{elsewhere}$$

a. Draw the graph of this probability distribution.
b. Find $E(y)$ and $V(y)$.
c. Find the sampling distribution of averages of all samples of size $n = 2$ from a population that is modeled by this distribution. Graph the sampling distribution of averages.
d. Compute $E(\bar{y})$ to show that it is equal to $E(y)$.
e. Compute $V(\bar{y})$ to show that it is equal to $V(y)/n$.

6.3.2. Let x and y be two independent random variables, each with the distribution described in Exercise 6.3.1. Show that:

a. $E(x + y) = E(x) + E(y)$.
b. $E(x - y) = E(x) - E(y)$.
c. $E(3y) = 3E(y)$.
d. $V(x + y) = V(x) + V(y)$.
e. $V(x - y) = V(x) + V(y)$.
f. $V(3y) = 9V(y)$.

6.3.3. The properties of expected value and variance illustrated in Exercise 6.3.2 are true in general:

$$E(x + y) = E(x) + E(y)$$
$$E(x - y) = E(x) - E(y)$$
$$E(ay) = aE(y), \text{ for a constant } a$$
$$V(x + y) = V(x) + V(y), \text{ if } x \text{ and } y \text{ are independent}$$
$$V(x - y) = V(x) + V(y), \text{ if } x \text{ and } y \text{ are independent}$$
$$V(ay) = a^2V(y), \text{ for a constant } a$$

Use these properties to show that in general, if $\bar{y} = \sum y/n$ in which the y's are independent, then:

a. $E(\bar{y}) = E(y)$.

b. $V(\bar{y}) = V(y)/n$.

6.3.4. For the population of heights given in Exercise 2.1.8:

a. What is $E(\bar{y})$ for all random samples of size 10?

b. What is $V(\bar{y})$ for all random samples of size 10?

6.3.5. Six female college students have heights (in inches) as follows: 62, 64, 65, 66, 65, 68. If these six students are considered to be a population from which sampling is done with replacement:

a. Draw the frequency distribution of the population.

b. Find the sampling distribution of averages for all samples of size two (with replacement) taken from this population. Draw its graph.

c. Find the population mean.

d. Find the mean of the sampling distribution of averages and confirm that it is the same as the population mean.

e. Find the variance of the population.

f. Find the variance of the sampling distribution of averages for samples of size $n = 2$ from the population variance.

REVIEW EXERCISES

Decide whether each of the following statements is true or false. If a statement is false, explain why.

6.1. It is appropriate to compute the average of a set of data collected on a nominal scale.

6.2. The sample average is always one of the values in the sample.

6.3. For any sample, $\sum(y - \bar{y}) = 0$.

6.4. If y is measured in feet, then $\bar{y}$ is also in feet.

6.5. If y is measured in inches, the unit of measurement for the standard deviation is squared inches.

6.6. If $\sum(y - c) = 0$ for a sample, then c is the sample average.

6.7. If for each value y in a sample $x = y + 10$, then $\bar{x} + 10 = \bar{y}$.

6.8. If for each value y in a sample $x = y + 10$, then the variance of y is equal to the variance of x.

6.9. If for each value y in a sample $x = ay$, then $\bar{x} = a\bar{y}$ and the variance of x is a^2 times the variance of y.

6.10. If y_1 and y_2 are random variables with the same probability distribution, then $E(y_1 - y_2) = 0$ and $V(y_1 - y_2) = 0$.

6.11. If two populations have the same mean, then they also have the same variance.

6.12. For many random samples the sample average $\bar{y}$ is not equal to the mean μ of the population from which the sample was chosen.

6.13. Because $\bar{y}$ is an unbiased estimator of μ, $\bar{y} = \mu$.

6.14. A sample average is computed in the same manner as a population mean.

6.15. A sample variance is computed in the same manner as a population variance.

6.16. If a population has a mean of 10 and a standard deviation of 2, then the sampling distribution of averages of samples of size $n = 2$ has a mean of 10 and a standard deviation of 1.

6.17. The variance of a sampling distribution of averages is larger than the variance of the underlying population because $\bar{y}$ has more distinct values than y.

6.18. Chebyshev's theorem shows that in all samples most of the data lie within three standard deviations of the average.

6.19. One of the advantages of using a sample average instead of a single observation to estimate the population mean is that the sample average is more likely to be close to the population mean.

6.20. The Empirical Rule cannot be applied to skewed distributions.

7

Normal Distributions

In Chapters 3 and 4 we discussed two types of discrete distributions, binomial and Poisson, that may be appropriate models for some discrete variables encountered in research. In Chapter 5 we discussed a continuous probability distribution, the chi-square distribution, which is not usually a direct model for a population but which can be used in an indirect way to answer questions about populations. In this chapter we discuss a second type of continuous probability distribution, the family of normal distributions. A normal distribution is sometimes the appropriate model for a population with a variable of interest that is continuous.

7.1 THE STANDARD NORMAL DISTRIBUTION

Some continuous variables can be modeled by a bell-shaped theoretical probability distribution called a *normal distribution*, also called a *Gaussian distribution* after Carl Friedrich Gauss (1777–1855), who investigated its mathematical properties.

For example, the sample of heights of 100 women measured to the nearest inch, as given in Table 7.1, can be grouped into a relative frequency distribution

y	f	y	f
60	0.01	67	0.14
61	0.04	68	0.08
62	0.03	69	0.01
63	0.07	70	0.01
64	0.26	71	0.01
65	0.19	72	0.01
66	0.14		

TABLE 7.1. Heights in a Sample of 100 Women

66	65	68	67	68	67	67	64	64	68
65	60	64	64	64	64	63	67	64	65
70	64	64	68	65	64	65	62	65	66
64	65	66	72	66	66	67	64	65	67
65	66	67	66	71	67	67	64	63	65
66	62	68	61	69	63	66	61	65	64
64	65	67	65	64	68	67	64	66	67
68	63	63	67	68	65	64	65	66	62
65	65	63	64	66	61	64	67	64	64
63	66	61	64	65	66	64	64	64	65

We should like to find a continuous probability distribution that can be used to model the population from which this sample was taken. Looking at the graph of the sample (Figure 7.1), we see that it is not perfectly bell-shaped, but the departures are not extreme. A sample of size 100 will resemble the population from which it was taken, but it will not be exactly like the population. It seems possible that the population of heights could be modeled by a theoretical normal distribution (Figure 7.2), with the following density function.

$$f(y) = \frac{1}{\sigma\sqrt{2\pi}} e^{-(y-\mu)^2/2\sigma^2}$$

The density function $f(y)$ gives the height of the curve above the y axis. In this density function, y is the random variable; y has all real numbers for its values. There are three constants in the density function: 2, π, and e. The constant π is the irrational number equal to approximately 3.14 (this use of π is not related to the binomial parameter), and the irrational e, approximately equal to 2.72, is the base of the natural logarithms. There are two independent parameters in the density, μ and σ^2; μ can be any real number and σ^2 can be any non-negative real number. In any particular

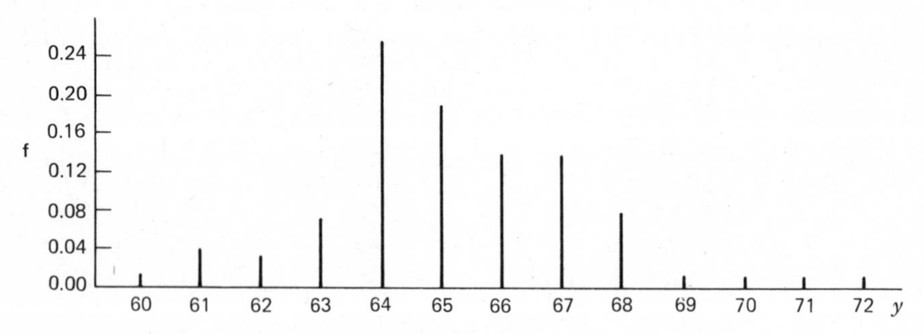

FIGURE 7.1. Heights in a sample of 100 women.

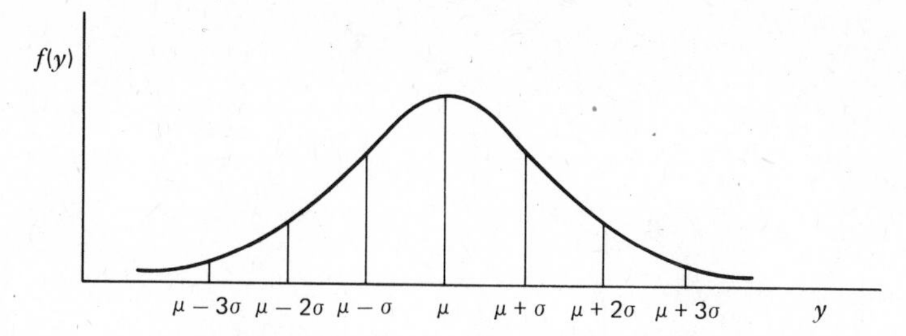

FIGURE 7.2. The normal distribution $N(\mu, \sigma^2)$.

normal density function, μ and σ^2 are fixed, thus there is a different normal distribution for each pair μ, σ^2.

The normal density function describes a curve that is:

1. Unimodal.
2. Symmetrical.
3. Asymptotic to the y axis.
4. Bell-shaped.

The normal distribution has:

1. $E(y) = \mu$.
2. $V(y) = \sigma^2$.
3. Inflection points at $\mu - \sigma$ and $\mu + \sigma$.
4. Total area between the curve and the y axis equal to one.
5. More than 99% of the area between $\mu - 3\sigma$ and $\mu + 3\sigma$.

In the sample of women's heights given above, the sample average $\bar{y}$ is 65.2 and the sample variance s^2 is 4.392. Thus, this sample might be from a population that can be modeled by a normal distribution with $E(y) = \mu = 65.2$ and $V(y) = \sigma^2 = 4.392$. We write $N(65.2, 4.392)$ to represent this theoretical distribution. (In Exercise 7.1.8 a goodness-of-fit test is described which can be used to check whether or not this is a good model; it is.)

Probabilities related to continuous random variables are represented by areas. Calculus (in particular, numerical integration) is necessary in order to find the areas of various sections under the normal curve. Tables, however, have been derived for the normal distribution $N(0,1)$, called the *standard normal distribution*. These tables can also be used to find the areas of sections under any normal curve by means of a standardization process.

The standard normal random variable is usually represented by z to distinguish it from other random variables. Table A.9 in the Appendix of Useful Tables gives the probabilities that the random variable z is greater than a

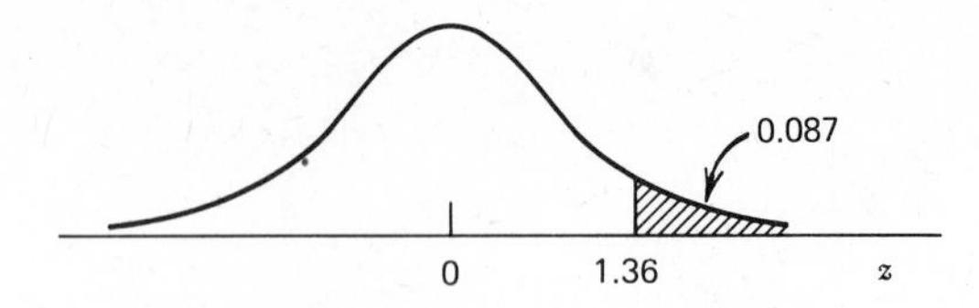

FIGURE 7.3. $P(z > 1.36) = 0.087$.

designated value between 0 and 3.09. For example, if $P(z > 1.36)$ is desired, the table is entered at row 1.30 and column 0.06, and the entry in the body of the table indicates that 0.087 of the area under the curve is to the right of $z = 1.36$ (Figure 7.3). To make this more practical, imagine that we have a freezer with temperatures that follow a standard normal distribution when measured on the Fahrenheit scale (the mean temperature is 0°F and the standard deviation is 1°F), then 8.7% of the time the temperature is above 1.36°F. Or we could say that the probability is 0.087 that the temperature is above 1.36°F. Areas relative to negative z values can be found by using the symmetry of the normal distribution. For example, $P(z < -1.36) = P(z > 1.36) = 0.087$.

If y is normally distributed with a mean of μ and variance σ^2, then y can be *standardized* by the formula

$$z = \frac{y - \mu}{\sigma}$$

Since z is the number of standard deviations y is from μ, z is sometimes called the *standard normal deviate*. If we want to find the probability that y is between 3 and 6 in $N(2,4)$, we compute:

$$z = \frac{3 - 2}{2} = 0.5 \quad \text{and} \quad z = \frac{6 - 2}{2} = 2$$

Then

$$\begin{aligned} P(3 \leq y \leq 6) &= P(0.5 \leq z \leq 2) \\ &= 0.309 - 0.023 \\ &= 0.286 \text{ (Figure 7.4)} \end{aligned}$$

Another example follows.

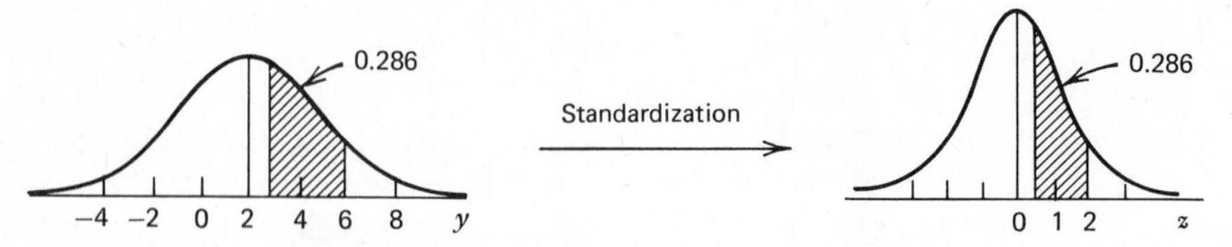

FIGURE 7.4. Standardization preserves area.

Example 7.1. Using the Standard Normal Distribution to Find Probabilities

Assume that an ecologist is studying the lungs of wild rabbits for possible contamination from a local power station. He has to build a trap to catch the rabbits, and he wants to make the door wide enough to catch a good percentage of them. Assume he knows that the mean width of rabbits' shoulders is $\mu = 3.80$ in. with a variance of $\sigma^2 = 0.36$ in.^{2} If he makes the door 5 in. wide, what percentage of rabbits will be able to go through the door? That is, what is $P(y < 5)$?

He finds that the standard normal deviate is

$$z = \frac{y - \mu}{\sigma} = \frac{5.0 - 3.8}{0.6} = 2.00$$

so the door is 2.00 standard deviations wider than the mean width of rabbits' shoulders. Using Table A.9, he finds that $P(z < 2.00) = 1 - 0.023 = 0.977$. This means that the area under the standard normal curve to the left of 2.00 is 0.977. It also means that in the normal distribution $N(3.80,\ 0.36)$, 0.977 of the area under the curve is to the left of 5, so 97.7% of the wild rabbits will fit through the door.

EXERCISES

7.1.1. Use Table A.9 to find:

a. $P(-1 \leq z \leq 2)$.
b. $P(-3.02 < z < 0)$.
c. $P(-0.5 < z < 0.5)$.
d. $P(z > 2.34)$.
e. $P(z > 0)$.
f. $P(z \geq -1.58)$.
g. $P(0.56 < z \leq 0.98)$.
h. $P(-2.44 < z < -0.12)$.
i. $P(|z| > 1)$.
j. $P(|z| > 2)$.
k. $P(|z| > 3)$.

7.1.2. Use Table A.9 to find:

a. $P(y < 4)$ if y is distributed as $N(5,0.64)$.
b. $P(10 < y < 13)$ if y is distributed as $N(12,4)$.

c. $P(y > 13)$ if y is distributed as $N(15,9)$.

d. $P(y < 0$ or $y > 3)$ if y is distributed as $N(1,9)$.

7.1.3. In $N(100,400)$, find:

a. The proportion of the values greater than 70.

b. The values of y within the central 90% of the distribution.

c. The smallest value of y that exceeds 85% of the distribution.

d. The largest value of y that is below 60% of the distribution.

7.1.4. Assume that Graduate Record Examination (GRE) scores follow a normal distribution with a mean of 1000 and a standard deviation of 200.

a. What percentage of graduates who take this exam have GRE scores greater than 750?

b. What GRE score separates the upper 30% of graduates from the other 70%?

c. Between what values are the scores of the central 90% of the graduates?

d. How likely is it that a randomly selected graduate will be one who has a GRE score greater than 1000?

e. How likely is it that a random sample of 10 graduates will contain more than seven who have GRE scores greater than 1000?

f. Suppose that a group of 10 graduates contains eight who have GRE scores greater than 1000.

i. Does this appear to be a random sample?

ii. Why?

7.1.5. The greater the sulfur content of coal, the less desirable it is as a heating fuel. Given that the variability among assays for sulfur in coal from a certain mine is $\sigma = 6$ lb per ton, and that they follow a normal distribution, answer the following:

a. Mines that assay 80 lb of sulfur per ton are considered worthless for heating fuel. How likely is it that a mine with mean sulfur content of $\mu = 62$ lb per ton will be placed in the worthless category on the basis of one random one-ton sample?

b. Some cities will not permit the sale of coal within the city limits if its assay for sulfur is as great as 34 lb per ton. How likely is it that coal with $\mu = 40$ lb per ton will be allowed to be sold within the city limits on the basis of one random one-ton sample?

7.1.6. The Food and Drug Administration has regulations governing the labeling of orange juice containers. The labeling depends on the amount of real orange juice per quart of drink. Assume the amounts of real orange juice per quart and the official designations of the drinks are:

Amount of Real Orange Juice Per Quart	Labeling To Be Used
0.0 to 3.2 oz	Orange-flavored drink
3.2 to 11.2 oz	Orange drink or orangeade
11.2 to 22.4 oz	Orange-juice drink

The juice canners claim that there is a standard deviation of 1.2 ounces of orange juice in their mixing, processing, and canning operation. Assume the distribution of real orange juice in quart containers follows a normal distribution.

a. If a canner plans to produce "orangeade" with a mean real orange juice content of $\mu = 4.0$ oz, how likely is it that a randomly chosen quart container will fail to meet the minimum standard for that drink?

b. If the "orangeade" producer wants at least 95% of his product to meet minimum standards, at which value should he set μ?

c. If a canner set the real orange-juice content at $\mu = 10.0$ oz and cheated on the labeling by calling it an "orange-juice drink," how likely is it that:

   i. A randomly chosen container will meet the specifications on the label?

   ii. Two randomly chosen containers will both meet the specifications on the label?

d. A company sets the amount of real orange juice at $\mu = 15$ ounces:

   i. What percentage of its product will contain between 14 and 16 ounces?

   ii. Between 13 and 17 ounces?

   iii. Within what range will be the central 95% of its product?

e. Suppose the company mentioned in Part d could reduce the variability of its product so that $\sigma = 0.6$, how would that change the answers to the questions in Part d about the distribution of orange-juice content in the containers?

7.1.7. A researcher in industrial relations notices that many men who receive high salaries are tall of stature. He decides to investigate the question whether height is related to salary. He wants to classify a man as "tall" if he is in the upper 10% of the heights of adult males. If adult male heights are normally distributed with a mean of 68 in. and a variance of 1.44 in.^{2}, what is the shortest height (to the nearest inch) that this researcher will classify as "tall"?

7.1.8. In the sample of women's heights given in this section, the sample average is $\bar{y} = 65.2$ in. and the sample variance is $s^2 = 4.392$, or $s = 2.1$ in. Use these sample values as estimates of μ and σ^2 in the normal distribution, and perform a chi-square goodness-of-fit test. Since two parameters are estimated, the degrees of freedom will be $k - 1 - 2$. Use the categories 59.5 to 60.5, 60.5 to 61.5, and so on. Expected values can be computed by finding the probability that a height is in such a section and multiplying by the sample size. If necessary, combine categories to prevent the expected values from becoming too small.

7.2. INFERENCE ABOUT A SINGLE OBSERVATION

Whenever possible we use samples consisting of several observations in order to make inference about a population; however, there are times when it is necessary to make a judgment about an unknown parameter from a single observation. Analogous to this is the reverse problem of placing limits on the value we expect to observe if we know the population parameters.

As an example of this reverse problem, suppose we want to know whether a particular flower is uncommonly large for its species. Assume the species has diameters that follow a normal distribution with a mean of 3 in. and a variance of 0.25 in.^{2} We can use the probabilities associated with the normal distribution to find the commonly occurring diameters of these flowers. In the standard normal distribution, 95% of the area is between -1.96 and 1.96. We write $z_{0.025} = 1.96$ to indicate that 2.5% of the area is to the right of 1.96. Thus $-1.96 = z_{0.975} = -z_{0.025}$ (Figure 7.5). Using the standardization procedure discussed in Section 7.1,

$$z = \frac{y - \mu}{\sigma}$$

we find that the lower limit y_L for the diameter of the flowers is

$$-1.96 = \frac{y_L - 3}{0.5}$$

$$y_L = 3 - 1.96(0.5) = 2.02$$

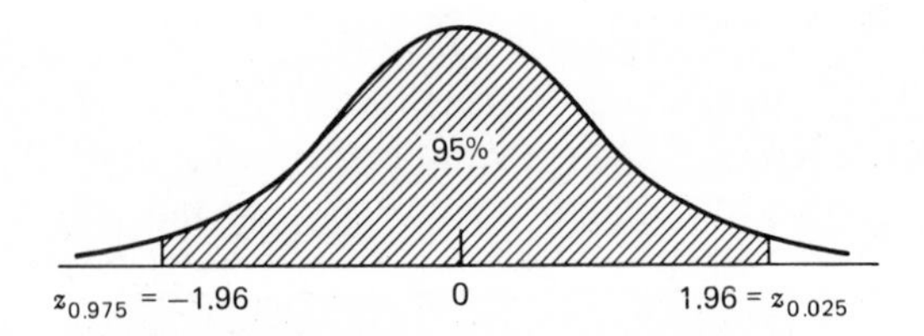

FIGURE 7.5. The standard normal distribution.

and the upper limit y_U is

$$1.96 = \frac{y_U - 3}{0.5}$$

$$y_U = 3 + 1.96(0.5) = 3.98$$

Thus the flowers of this species are commonly between 2.02 in. and 3.98 in. in diameter, $2.02 \le y \le 3.98$. These values are sometimes called *tolerance limits*, especially in engineering. A value within these limits is not a significant deviation from the mean. A value outside these limits is a significant deviation.

This procedure for finding tolerance limits is similar to the setting of confidence intervals. The interval is called a *prediction interval* (PI) for an individual observation when the population mean is known.

Example 7.2. A Prediction Interval for a Single Observation

An expectant mother is usually eager to predict the birthdate of the child she is carrying. The mean length of time from the beginning of the last menstrual period to the birth of the child is 280 days. Thus the expected birthdate of her child is $E(y) = \mu = 280$ days after onset of the last menses.

Anyone who has become a parent, however, knows that there is variability from the expected date of birth. In fact, the standard deviation is $\sigma = 10$ days. Assuming for this example that $N(280,100)$ is a good approximation of the distribution, one can determine the 95% most probable lengths of time. When standardized, only 0.025 of the lengths of full-term pregnancies have a z value less than $z_{0.975} = -1.96$, and only 0.025 have a z value greater than $z_{0.025} = +1.96$. Solving for the y value corresponding to these z values:

$$\begin{aligned} y_L &= \mu - 1.96\sigma \\ &= 280 - 1.96(10) \\ &= 260.4 \end{aligned}$$

and

$$\begin{aligned} y_U &= \mu + 1.96\sigma \\ &= 280 + 1.96(10) \\ &= 299.6 \end{aligned}$$

There is a 0.95 chance that any full-term pregnancy will result in birth $\mu \pm 1.96\sigma$ days after the start of the last menstrual period, or

$$\text{PI}_{0.95}\text{:} \quad 260 \le y \le 300$$

If the population mean is unknown, it is possible to carry out a test of hypothesis from a single observation (we stress, however, that whenever possible a larger number of observations should be used).

Example 7.3. Testing a Hypothesis About a Mean with a Sample of One Observation

Suppose a person showed many of the symptoms of hypothyroidism (an underactive thyroid gland). At one time his physician would have sent him to the hospital for a basal metabolism test. The test was fairly involved, somewhat lengthy, and required that the patient be in a fasting condition. Thus the decision whether or not to administer thyroid extract depended on a single observation of the patient's basal metabolism rate.

The mean basal metabolism rate for people with properly functioning glands is 40 calories per square meter per hour; a person suffering from hypothyroidism will have a reduced basal metabolism rate. Thus the null and alternative hypotheses are:

$$H_o\colon\ \mu = 40 \qquad \text{and} \qquad \mathrm{H}_a\colon\ \mu < 40$$

The variability in basal metabolism rate among people with properly functioning thyroids is also known, and for this example it is assumed that the population of such rates is distributed as $N(40,16)$. If the physician did not want more than a 0.05 probability of a misdiagnosis of a person with a properly functioning thyroid ($\alpha = 0.05$), he would compute the test statistic

$$z = \frac{y - \mu_0}{\sigma} = \frac{y - 40}{4}$$

in which μ_0 is the value of μ in the null hypothesis and σ is the known standard deviation. Evidence that the null hypothesis is false would be a large negative value of z since low basal metabolism rates are transformed to the left tail of the standard normal distribution (Figure 7.6). This z statistic is compared with the critical value of $z_{0.95} = -1.64$; if $z \leq -1.64$, H_0 is rejected.

If the physician did not understand how to carry out this test of hypothesis, he might ask a biostatistician to find the basal metabolism rate y that divides the area under the $N(40,16)$ curve into the lower 5% of the area and the

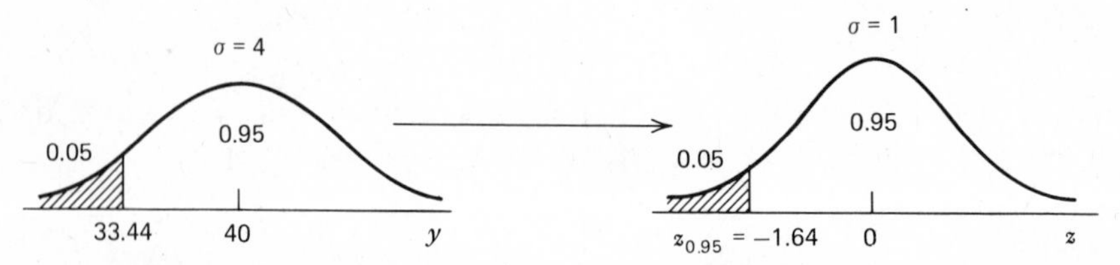

FIGURE 7.6. Low values in $N(40,16)$ which occur only 5% of the time.

upper 95% of the area. This is done by placing the critical value of z in the equation and solving for y. Thus

$$-1.64 = \frac{y - 40}{4}$$

$$y = 40 - 1.64(4) = 33.44$$

The physician would then make $y = 33.44$ his decision point. If the patient's basal metabolism was less than or equal to 33.44 calories, the diagnosis would be hypothyroidism and thyroid extract would be prescribed. In statistical terms, the null hypothesis of normal thyroid function would be rejected. If the patient's basal metabolism was greater than this value, the hypothesis would not be rejected, and the physician would investigate something other than the thyroid as the cause of the symptoms.

■ ■ ■

Procedure. Inference About a Single Observation from a Normal Distribution

Prediction Intervals

If μ and σ are known, then a single observation has a probability of $1 - \alpha$ of falling within the following tolerance limits.

$$\text{PI}_{1-\alpha}\text{:} \quad \mu - z_{\alpha/2} \leq y \leq \mu + z_{\alpha/2}$$

Test of Hypothesis

H_0: $\mu = \mu_0$
H_a: $\mu \neq \mu_0$ or $\mu > \mu_0$ or $\mu < \mu_0$

Significance level: α

Test statistic:

$$z = \frac{y - \mu_0}{\sigma}$$

Region of rejection: $|z| \geq z_{\alpha/2}$ or $z \geq z_\alpha$ or $z \leq -z_\alpha$, respectively

■ ■ ■

EXERCISES

7.2.1. Use Table A.9 in the Appendix to find:

a. $z_{0.05}$

b. $z_{0.95}$

c. $z_{0.01}$
d. $z_{0.99}$
e. $z_{0.005}$
f. $z_{0.995}$

7.2.2. Assume that the temperatures of healthy infants follow an $N(99,1)$ distribution when measured on a Fahrenheit scale.

a. If a particular infant has a temperature of 100.5°F, should his temperature be considered "normal"? That is, test the hypothesis H_0: $\mu = 99$ against H_a: $\mu \neq 99$ at $\alpha = 0.05$.
b. Find the 95% tolerance limits for a single observation.

7.2.3. Legend has it that Archimedes made his discovery concerning specific gravity (Archimedes' principle) while trying to determine whether the king's crown was made of pure gold or an alloy. Working with metal samples which he knew to be pure gold or alloys, he found that his device for measuring specific gravity produced a mean determination of $\mu = 19.3$ for pure gold, whereas all alloys tested yielded lower mean specific gravities. For the sake of this problem, suppose Archimedes' measuring device followed an $N(\mu,0.09)$ distribution.

a. What would be a suitable null hypothesis for such an experiment?
b. What would be the most logical alternative hypothesis?
c. If $\alpha = 0.05$, what should be the region of rejection for this experiment?
d. How likely is it that a random sample of an alloy with a specific gravity determination of 18.7 would be mistakenly called pure gold in this experiment?

7.2.4. A dairy farmer buys a heifer (female calf) from a Holstein–Friesian herd that is thought to be genetically superior to others in the region. The quantity of milk production among mature cows in the herd is normally distributed with $\mu = 18{,}000$ lb per year and $\sigma = 2500$ lb per year. Assuming the new owner can provide feed, shelter, and other environmental factors equivalent to those for the herd from which the calf was bought:

a. Give the numerical value of $E(x)$, the expected milk production of the calf when it reaches maturity.
b. Place a 95% prediction interval of the calf's milk production at maturity.
c. What is the probability that the calf will produce at a greater rate than the mean of the herd from which it was bought?
d. What is the probability that it will produce at a rate greater than the breed mean of $\mu = 14{,}000$?

7.3. THE CENTRAL LIMIT THEOREM

Normal distributions occur frequently in experiments; however, many random variables are not normally distributed, and it would be inappropriate to use a normal distribution as the model. In spite of this, if the samples are large enough, a normal distribution can often still be used to find certain probabilities associated with the experiment because of some results that are known from the mathematical theory of statistics. The theory relevant to this use concerns the properties of sampling distribution of averages.

In Section 6.3 we noted that the sampling distribution of averages has the following properties:

1. $\mu_{\bar{y}} = \mu_y$, that is, the mean of the sampling distribution of averages is the same as the mean of the underlying population.
2. $\sigma_{\bar{y}}^2 = \sigma_y^2/n$, that is, the variance of the sampling distribution of averages is equal to the variance of the underlying population divided by the sample size.
3. If n is sufficiently large, then the sampling distribution of averages is almost symmetrical and unimodal.

The third property can now be made more explicit. If a population is normal, then the sampling distribution of averages is normal. If a population is not normal, then the sampling distribution of averages is approximately normal for large n.

This last property is known as the *central limit theorem*. It is because of this property that normal distributions come into play in many statistical analyses. With very few exceptions,† no matter what form the underlying population distribution takes, as n increases the sampling distribution of averages approaches a normal distribution, thus the normal distribution can be used to approximate probabilities in cases of reasonably large samples ($n \geq 30$) from non-normal distributions.

Usually in statistics we observe a sample and use the data collected to make decisions about the population. If we compute the sample average we have one value from the sampling distribution of averages. Using the three properties just discussed, we can answer probability questions about sample averages. If the underlying population is normally distributed, then the sampling distribution of averages is also normally distributed and has the same expected value as the population distribution and a variance that is $1/n$ of the population variance. If the underlying distribution is not normal, then the sampling distribution of averages for large n is approximately normal and has the same expected value as the population distribution and a variance of $1/n$ times the population variance.

† It is sufficient that the distribution have a finite variance.

Example 7.4. Probabilities Associated with a Sample Average
An educational psychologist is working with a random sample of five adults. They are going to take a standardized IQ test which has scores that are normally distributed with a mean of 105 and a standard deviation of 15. The psychologist wants to know how likely it is that the average score of the five subjects will be greater than 108, that is, $P(\bar{y} > 108)$.

Since he is working with a sample average, he has a single value from the sampling distribution of averages that is normally distributed with a mean of 105 and a variance of $\sigma_{\bar{y}}^2 = \sigma_y^2/n = 15^2/5 = 45$. Thus

$$P(\bar{y} > 108) = P(z > 0.45) = 0.326$$

because

$$z = \frac{\bar{y} - \mu_{\bar{y}}}{\sigma_{\bar{y}}} = \frac{\bar{y} - \mu_y}{\sigma_y/n} = \frac{108 - 105}{\sqrt{45}} = 0.45$$

The psychologist concludes that the probability is 0.326 that the average scores of his five subjects will be above 108.

EXERCISES

7.3.1. If the basal metabolism rate for people with properly functioning thyroid glands can be modeled by a normal distribution with mean 40 calories per square meter per hour and a standard deviation of 4, find:

a. The probability that a healthy person chosen at random will have a rate less than 35.

b. The probability that five healthy persons chosen at random will all have a rate less than 35.

c. The probability that the average rate of five healthy persons chosen at random is less than 35.

7.3.2. A certain aptitude test for job trainees follows a normal distribution with a mean of 80 and a standard deviation of 16.

a. What is the probability that a random sample of four trainees will all have scores above 88?

b. What is the probability that the average score for a random sample of four trainees will be above 88?

7.4. INFERENCES ABOUT A POPULATION MEAN AND VARIANCE

Although it is sometimes necessary to make decisions on the basis of a single observation (as in Section 7.2), in general this is not the preferred procedure. Larger samples yield more information on which to base decisions. If we are interested in making a decision about μ or an estimate of μ, using $\bar{y}$ with $n > 1$ instead of a single observation has the advantage that $\bar{y}$ is less variable than y. A smaller variance increases the probability of obtaining a sample value close to the true population mean. Another advantage of using averages of samples is that even if the original population does not have a normal distribution, the sampling distribution of averages for large n is approximately normal (central limit theorem).

Tests of hypotheses based on averages are analogous to the procedure for an individual observation. For a single observation, the standardization procedure is

$$z = \frac{y - \mu}{\sigma}$$

For averages of samples of size n, the standardization procedure is

$$z = \frac{\bar{y} - \mu}{\sigma/\sqrt{n}}$$

because the mean of the sampling distribution of averages is the same as the original mean and the standard deviation of the sampling distribution is $\sigma/\sqrt{n}$. (This denominator is sometimes called the *standard error*. "Error" in this context does not imply a mistake, but variability due to sampling.)

Example 7.5. Using the Standard Normal Distribution to Test a Hypothesis About μ

Assume that an ecologist is studying the effects of DDT on wild birds. DDT sprayed on insects and vegetation subsequently eaten by female birds seems to affect the thickness of their eggshells and thus affects the protection of the embryonic birds. Assume the ecologist knows that the mean eggshell thickness in untreated birds (birds that have not eaten matter sprayed with DDT) is 330 microns and the variance is 2500 microns^2. Assume also that the thicknesses follow a normal distribution. He would like to test H_0: $\mu = 330$ against the alternative H_a: $\mu \neq 330$. If the sample size is $n = 16$ and $\bar{y} = 290$, then for a test at the 5% level of significance, he will reject H_0 if $|z| \geq z_{0.025} = 1.96$;

$$z = \frac{\bar{y} - \mu_0}{\sigma/\sqrt{n}} = \frac{290 - 330}{50/4} = -3.2$$

Since $|-3.2| > 1.96$, this is a significant deviation from the hypothesized mean. The ecologist rejects the null hypothesis and concludes that DDT decreases eggshell thickness.

Confidence intervals on μ can also be determined from samples with $n > 1$.

Example 7.5. Using the Standard Normal Distribution to Find a Confidence Interval on μ

Assume that a researcher at an agricultural experiment station knows that the variance in butterfat production for Holstein–Friesian dairy cattle is $\sigma^2 = 6400$ (lb per year)^2. He treats a group of dairy cattle by adding inorganic nitrate to their diet, and he wants to know the mean butterfat production for this treatment group, that is, the value of μ. He would perform a test of hypothesis to get some information about μ, the mean for the treatment group. If the null and alternative hypotheses are

$$H_0: \quad \mu = \mu_0$$

$$H_a: \quad \mu \neq \mu_0$$

and $\alpha = 0.05$, he would use the formula

$$z = \frac{\bar{y} - \mu_0}{\sigma/\sqrt{n}}$$

He would not reject the null hypothesis if

$$-1.96 \le \frac{\bar{y} - \mu_0}{\sigma/\sqrt{n}} \le 1.96\dagger$$

or, the equivalent, if

$$\bar{y} - 1.96\frac{\sigma}{\sqrt{n}} \le \mu_0 \le \bar{y} + 1.96\frac{\sigma}{\sqrt{n}}$$

Thus the 95% confidence interval on μ is:

$$\text{CI}_{0.95}: \quad \bar{y} \pm 1.96\frac{\sigma}{\sqrt{n}}$$

and if $\bar{y} = 465$ and $n = 25$, then

$$\text{CI}_{0.95}: \quad 465 - 1.96(80/5) \le \mu \le 465 + 1.96(80/5)$$

$$433.64 \le \mu \le 496.36$$

for the treatment group.

If the population variance σ^2 is unknown (as is commonly the case) it can be estimated by the sample variance

$$s^2 = \frac{\sum(y - \bar{y})^2}{n - 1} = \frac{\sum y^2 - (\sum y)^2/n}{n - 1}$$

† Strictly speaking, we do not reject the null hypothesis if $-1.96 < z < 1.96$. Since this is a continuous distribution, however, $P(z = 1.96) = 0$ and the two types of inequalities are equivalent.

If the sample size is large ($n \geq 30$), s^2 can be used in place of σ^2 in inferences concerning μ.

■ ■ ■

Procedure. Inferences About a Population Mean

Assumptions: 1. $n \leq 30$, population normal, and σ known, or
2. $n \geq 30$

Confidence Intervals

$$\text{CI}_{1-\alpha}: \quad \bar{y} - z_{\alpha/2}\frac{\sigma}{n} \leq \mu \leq \bar{y} + z_{\alpha/2}\frac{\sigma}{n}$$

if σ is known. If σ is unknown and $n \geq 30$ estimate σ by s.

Test of Hypothesis

H_0: $\mu = \mu_0$
H_a: $\mu \neq \mu_0$ or $\mu > \mu_0$ or $\mu < \mu_0$
Significance level: α
Test statistic:

$$z = \frac{\bar{y} - \mu_0}{\sigma/\sqrt{n}}$$

if σ is known. If σ is unknown and $n \geq 30$, estimate σ by s.
Region of rejection: $|z| \geq z_{\alpha/2}$ or $z \geq z_\alpha$ or $z \leq -z_\alpha$, respectively.

■ ■ ■

Sometimes the parameter of interest is not only the population mean, but also the population variance. Several examples follow. A teacher is interested in both the mean performance of his class and the variability of the grades; a large variance may indicate that although the class as a whole is performing well, some individuals may not be performing at an acceptable level. During the manufacturing of drugs, the variance of the potency is of concern and also the variance of the purity level. During the machine filling of boxes or bottles with a product, the variance of the quantity put into the container is of concern. Variability of sentence length has been used to establish authorship. These are only some of the areas in which the investigator needs information about the variance.

It is possible to test hypotheses and determine confidence intervals for a population variance if the population is normal. These procedures make use of the fact that

$$\frac{\sum(y - \bar{y})^2}{\sigma^2} = \frac{(n - 1)s^2}{\sigma^2}$$

is distributed as a chi-square distribution with $n - 1$ degrees of freedom if y is normally distributed.

Example 7.6. Inference About the Variance of a Normal Population
In a certain city, the mean electric consumption for residences is 7.2 thousand kilowatt hours with a variance of 2.25 thousand (kilowatt hours)^2. Differences in home consumption are due to the energy efficiency of the house and the life-style of the occupants.

In a sample of 101 homes from an area in which all of the residences are of equal size and equal energy efficiency, the sample variance is 1.21 thousand (kilowatt hours)^2. Does this indicate that uniform energy-efficient homes significantly lower the variance of electric consumption?

The null and alternative hypotheses are

$$H_0\colon\ \sigma^2 = 2.25 \qquad H_a\colon\ \sigma^2 < 2.25$$

The test statistic is

$$\chi^2 = \frac{(n - 1)s^2}{\sigma_0^{\,2}}$$

with $n - 1 = 100$ degrees of freedom. At $\alpha = 0.05$ the region of rejection is

$$\chi^2 \leq \chi^2_{0.95,100} = 77.929$$

The value of the test statistic is

$$\chi^2 = \frac{100(1.21)}{2.25} = 53.778$$

Thus the null hypothesis is rejected and there is evidence that uniform housing significantly reduces the variability of electric consumption. This result suggests that a program to encourage persons to make their homes more energy efficient might be worthwhile.

If desired, a central confidence interval can be determined for σ^2 for the population of uniform residences of the type sampled.

$$\mathrm{CI}_{0.95}\colon\ \frac{(n - 1)s^2}{\chi^2_{0.025,n-1}} \leq \sigma^2 \leq \frac{(n - 1)s^2}{\chi^2_{0.975,n-1}}$$

$$\frac{100(1.21)}{129.561} \leq \sigma^2 \leq \frac{100(1.21)}{74.222}$$

$$0.93 \leq \sigma^2 \leq 1.63$$

The inferences relative to the variance of a normal population can be summarized as follows.

■ ■ ■

Procedure. Inferences About a Population Variance

Assumption: Normality

Confidence Intervals

$$\mathrm{CI}_{1-\alpha}: \quad \frac{(n-1)s^2}{\chi^2_{\alpha/2,n-1}} \leq \sigma^2 \leq \frac{(n-1)s^2}{\chi^2_{1-\alpha/2,n-1}}$$

Test of Hypothesis

H_0: $\sigma^2 = \sigma_0{}^2$
H_a: $\sigma^2 \neq \sigma_0{}^2$ or $\sigma^2 > \sigma_0{}^2$ or $\sigma^2 < \sigma_0{}^2$
Significance level: α
Test statistic:

$$\chi^2 = \frac{(n-1)s^2}{\sigma_0{}^2}$$

Region of rejection: $\chi^2 \leq \chi^2_{1-\alpha/2,n-1}$ and $\chi^2 \geq \chi^2_{\alpha/2,n-1}$, or $\chi^2 \geq \chi^2_{\alpha,n-1}$, or $\chi^2 \leq \chi^2_{1-\alpha,n-1}$, respectively

■ ■ ■

EXERCISES

7.4.1. On an IQ test which is distributed as $N(100,225)$, the average IQ score for a certain second grade in a private school in Victoria, Texas, is $\bar{y} = 106$. If $\alpha = 0.05$, how often might a deviation this large or larger occur by chance in a random sample of 25?

7.4.2. A certain intelligence test has an $N(100,100)$ distribution. To see whether intelligence is inherited, tests are given to the eldest child of each of a random sample of 16 acclaimed scholars. The average score of the children is 10[illegible]

a. Give the null hypothesis to be tested.
b. Give the alternative hypothesis.
c. Perform the test.
d. How likely is it that data like these represent a sample from a population in which the null hypothesis is true?

7.4.3. A synthetic female hormone (DES) has been used to fatten livestock. If this substance appears in the meat, it affects the sexual maturity of young animals eating the meat. Biological assays can be used to test for the presence of DES in meat. Young female rats are fed the suspected meat, and if they mature earlier than expected, it is probably because of DES in the meat. Suppose for a given strain of rat, time until sexual maturity in the females follows an essentially normal distribution with a mean of 90 days and a variance of 144.

a. What is the probability that a randomly selected female rat will reach sexual maturity before 90 days? Before 86 days?

b. What is the probability that the average time until sexual maturity for a random sample of nine female rats will be less than 90 days? Less than 86 days?

c. A random sample of nine female rats is fed a diet including meat suspected of containing DES.

   i. What are the most logical null and alternative hypotheses?

   ii. If $\alpha = 0.05$, which values of the sample average will lead to the rejection of the null hypothesis?

   iii. Suppose for female rats on a diet containing DES, sexual maturity follows an $N(86,144)$ distribution; what is the probability of making a Type II error?

7.4.4. It is believed that concentration can improve one's performance on a standardized examination. To evaluate this belief, a random sample of 10 students are hypnotized and told that their minds are clear of all distractions just before taking a standard $N(50,160)$ statistics examination. If $\bar{y} = 57$ for the 10 students, what conclusion would you draw from this experiment?

7.4.5. A coal research scientist has discovered that West Virginia coal contains an ore rich in aluminum. Although it is present in coal only as a trace mineral, it may be economically practical to recover the ore from the ash left when coal is burned in large boilers of power plants. In order to estimate the quantity of the ore in coal, the scientist takes a random sample consisting of 100 observations and computes the following:

$$\sum y = 8{,}400 \text{ ppm}$$

$$\left(\sum y\right)^2 = 70{,}560{,}000 \text{ ppm}^2$$

$$\sum y^2 = 715{,}500 \text{ ppm}^2$$

a. What is the best estimate of the mean content of aluminum ore in West Virginia coal?

b. Show that the sample standard deviation is 10 ppm.

c. A coal economist calculates that the recovery of the ore will be profitable if it is present to an extent greater than 82.3 ppm in the coal burned in the boilers. On the basis of these data, would you recommend attempting to recover the ore?

7.4.6. A random sample of size 16 is drawn from a normal distribution with an unknown mean and $\sigma^2 = 6400$. If $\bar{y} = 200$, give the 95% most probable $N(\mu,6400)$ distributions from which these results could have come.

7.4.7. The following are weights in ounces of 30 apples of a particular variety:

5	7	9	6	4	5	3	4	4	7
5	6	8	5	4	6	7	9	5	7
3	6	5	9	4	7	8	6	5	8

a. Compute s^2.

b. Find a 95% confidence interval for σ^2.

c. Perform a test of hypothesis at the 5% level of significance to determine whether or not this sample came from a population that has a variance of 3.2.

d. Find a 95% confidence interval for μ using s^2 to approximate σ^2.

7.4.8. Many organic phosphorus compounds are effective insecticides, but they are also chemically stable and likely to get into the human food chain. They have even been detected in the digestive tracts of recently born infants, but it is not known to what extent this is via mother's milk and to what extent these compounds pass through the placental membrane prior to birth. To get answers to these questions, a medical research team draws samples of amniotic fluid from the wombs of 64 pregnant women and performs chemical analysis for a certain organic phosphorus insecticide. The following data are obtained:

$$\sum y = 320.00 \text{ ppm}$$

$$\sum y^2 = 1{,}761.28 \text{ ppm}^2$$

a. Estimate the mean ppm of the compound found in amniotic fluid.

b. Show that the sample variance is 2.56 ppm^2.

c. Place a 95% confidence interval on the mean.

d. Place a 95% confidence interval on the variance.

7.4.9. It can be illustrated that $s^2 = \sum(y - \bar{y})^2/(n - 1)$ is an unbiased estimator of σ^2 by the following special case. Let the population be an equally likely distribution of 1, 2, 3, 4. This population was discussed in Section 6.3.

a. List all possible samples (with replacement) of size two.
b. Compute the sample variance of each sample.
c. Find the relative frequency of each different sample variance found in Part b.
d. Find $E(s^2)$ and show that $E(s^2) = \sigma^2$.

7.5. USING A NORMAL DISTRIBUTION TO APPROXIMATE OTHER DISTRIBUTIONS

A normal distribution can sometimes be used to approximate the probabilities associated with response variables that follow a binomial or a Poisson distribution.

In the case of a binomial distribution, the central limit theorem implies that if n is fairly large ($n \geq 25$) and π is fairly close to 0.5 ($0.2 \leq \pi \leq 0.8$), then the binomial random variable y can be transformed into a random variable that is distributed approximately as the standard normal random variable

$$z \cong \frac{y - n\pi}{\sqrt{n\pi(1 - \pi)}}$$

Note that $n\pi = \mu$, the mean of the binomial distribution, and $\sqrt{n\pi(1 - \pi)}$ is the standard deviation.

Example 7.7. Using a Normal Distribution to Approximate Probabilities for a Binomial Random Variable

A sociologist studying families headed by a single parent would like to know the probability of finding 40 or more such families in a random sample of 100 families if 30% of families are of this type.

Since $E(y) = n\pi = 100(0.30) = 30$, and $V(y) = n\pi(1 - \pi) = 100(0.30)(0.70) = 21$, then

$$\begin{aligned} P(y \geq 40) &\cong P\left(z \geq \frac{40 - 30}{\sqrt{21}}\right) \\ &= P(z \geq 2.18) \\ &= 0.015 \end{aligned}$$

Thus, if the sociologist needs at least 40 cases for a study, a sample of 100 families will probably not be sufficient.

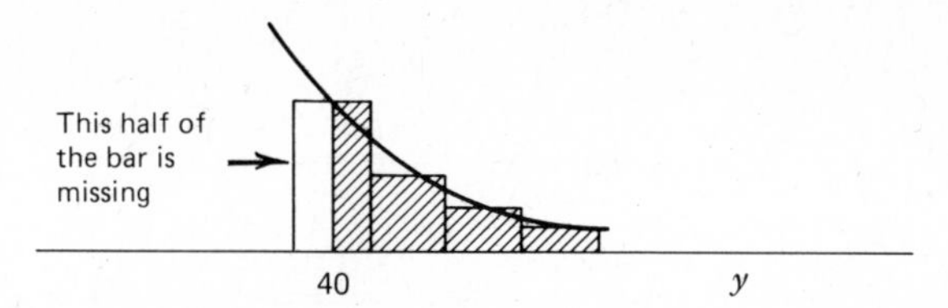
This half of
the bar is
missing
40
y

FIGURE 7.7. Approximating a binomial distribution by a normal distribution.

Since the binomial distribution is discrete and the normal distribution is continuous, the approximation will be poor in the case of small sample sizes. To compensate for this, a continuity correction of 0.5 is often made. If we represent the binomial probabilities by bars of unit width so that the area of the bar over y is the probability of y, and we represent the normal distribution by a smooth curve, we can see (Figure 7.7) that using 40 as the cutoff point in the above example does not take into consideration half of the bar above 40. Thus, instead of finding $P(y \geq 40)$, we should find $P(y \geq 39.5)$. The sociologist above would then find

$$P(y \geq 39.5) \cong P\left(z \geq \frac{39.5 - 30}{\sqrt{21}}\right)$$

$$= P(z \geq 2.07)$$

$$= 0.019$$

The additional accuracy may be important in some experiments.

A test of hypothesis can also be done about the binomial parameter, making use of the fact that $(y - n\pi)/\sqrt{n\pi(1 - \pi)}$ is approximately standard normal. This procedure is especially helpful for large sample sizes, since exact binomial tables may not be available.

Example 7.8. Using a Normal Distribution to Test a Hypothesis About π

Most people have a dominant eye which looks directly ahead while the other eye adjusts to it in order to bring a viewed object into focus. A reading specialist wants to determine whether there is any tendency for one eye to be dominant in children with a certain reading problem. She takes a random sample of 225 children with the reading problem and determines the dominant eye for each of them. Suppose she finds that for 144 of the children the right eye is dominant. The null and alternative hypotheses are:

$$H_0: \ \pi = 0.5 \qquad \text{and} \qquad H_a: \ \pi \neq 0.5$$

The test statistic is:

$$z \cong \frac{y - n\pi_0}{\sqrt{n\pi_0(1 - \pi_0)}} = \frac{144 - 225(0.5)}{\sqrt{225(0.5)(0.5)}} = 4.2$$

At $\alpha = 0.05$, she will reject the null hypothesis if $|z| \geq 1.96$. Since $|4.2| > 1.96$, she rejects the null hypothesis and concludes that more than half the children with this reading problem have a dominant right eye.

If the specialist in the above example would like to find a confidence interval for π, she could make use of the fact that

$$z \cong \frac{y - n\pi}{\sqrt{n\pi(1 - \pi)}} = \frac{y/n - \pi}{\sqrt{\dfrac{\pi(1 - \pi)}{n}}}$$

and that y/n is the best point estimate of π. Analogous to confidence intervals on μ, the confidence interval on π would be:

$$\text{CI}_{1-\alpha}: \quad y/n \pm z_{\alpha/2}\sqrt{\frac{\pi(1 - \pi)}{n}}.$$

However, since π is unknown, it must be estimated in the standard error by y/n, giving

$$\text{CI}_{1-\alpha}: \quad y/n \pm z_{\alpha/2}\sqrt{\frac{(y/n)(1 - y/n)}{n}}$$

In the sample, since $y = 144$, she would find

$$\text{CI}_{0.95}: \quad 144/225 \pm 1.96\sqrt{\frac{(144/225)(1 - 144/225)}{225}}$$

$$0.640 \pm 1.96(0.0320)$$

$$0.640 \pm 0.0637$$

$$0.576 \leq \pi \leq 0.704$$

If desired, the statistic

$$z \cong \frac{y/n - \pi_0}{\sqrt{\dfrac{\pi_0(1 - \pi_0)}{n}}}$$

can be used for tests of hypothesis. This is equivalent to the method illustrated in the example.

■ ■ ■

Procedure. Normal Approximation of a Binomial Distribution

Assumption: $n \geq 25$ and $0.2 \leq \pi \leq 0.8$

Confidence Intervals

$$\text{CI}_{1-\alpha}:\ y/n - z_{\alpha/2}\sqrt{\frac{(y/n)(1 - y/n)}{n}} \leq \pi \leq y/n + z_{\alpha/2}\sqrt{\frac{(y/n)(1 - y/n)}{n}}$$

Tests of Hypotheses

H_0: $\pi = \pi_0$
H_a: $\pi \neq \pi_0$ or $\pi > \pi_0$ or $\pi < \pi_0$
Significance level: α
Test statistic:

$$z \cong \frac{y - n\pi_0}{\sqrt{n\pi_0(1 - \pi_0)}} = \frac{y/n - \pi_0}{\sqrt{\dfrac{\pi_0(1 - \pi_0)}{n}}}$$

Region of rejection: $|z| \geq z_{\alpha/2}$ or $z \geq z_\alpha$ or $z \leq -z_\alpha$, respectively

■ ■ ■

The normal distribution can also be used to approximate probabilities related to variables that follow a Poisson distribution. This approximation arises from the central limit theorem. If y is a Poisson random variable and λ is large, y can be transformed into a random variable that is distributed approximately as the standard normal random variable

$$z \cong \frac{y - \lambda}{\sqrt{\lambda}}$$

Note that λ is the mean and $\sqrt{\lambda}$ the standard deviation of the Poisson distribution.

Example 7.9. Using a Normal Distribution to Approximate Probabilities for a Poisson Random Variable

A traffic-control specialist wants to know the probability that more than 30 vehicles will pass a given intersection in a three-minute period at 3:00 PM if the expected number of vehicles to pass that intersection in three minutes at that time is 25.

$$P(y > 30) \cong P\left(z > \frac{30.5 - 25}{\sqrt{25}}\right)$$
$$= P(z > 1.1)$$
$$= 0.136$$

This computation is much simpler than working with the exact Poisson distribution. Note that a continuity correction is used because the discrete Poisson distribution is being approximated by the continuous normal distribution.

Tests of hypotheses about λ can also be done with a z statistic using the fact that $(y - \lambda)/\sqrt{\lambda}$ is approximately standard normal for large λ.

■ ■ ■

Procedure. Normal Approximation of a Poisson Distribution

Test of Hypothesis

H_0: $\lambda = \lambda_0$

H_a: $\lambda \neq \lambda_0$ or $\lambda > \lambda_0$ or $\lambda < \lambda_0$

Significance level: α

Test statistic:

$$z = \frac{y - \lambda_0}{\sqrt{\lambda_0}}$$

Region of rejection: $|z| \geq z_{\alpha/2}$ or $z \geq z_\alpha$ or $z \leq -z_\alpha$

■ ■ ■

EXERCISES

7.5.1. A physical education professor claims that 35% of third-grade children can do a handstand. If this claim is true:

a. Find the probability that 10 or more third-grade children out of a random sample of 25 can do a handstand.

   i. Use the exact binomial distribution.
   ii. Use the normal distribution without a continuity correction.
   iii. Use the normal distribution with a continuity correction.

b. Find the probability that 40 or more third-grade children out of a random sample of 100 can do a handstand.
   i. Use the normal distribution without a continuity correction.
   ii. Use the normal distribution with a continuity correction.

c. Based on the results of Parts a and b, is the correction for continuity more important in large or in small samples?

7.5.2. A customer relations bureau located in a large eastern city claimed that 80% of the complaints registered with it were settled to the satisfaction of the customers. The local newspaper, doubting whether the percentage was really that large, takes a random sample of 40 complainants and asks them whether they had received satisfaction. Only 12 indicate that they had. Use the normal approximation to make a test of significance at $\alpha = 0.01$.

7.5.3. In a certain Midwestern community, 25% of the population consists of third-generation descendants of one Finnish immigrant family. Within the community there is a remittant nervous disorder which may be transmitted genetically. There are 75 cases of the disorder on which to base studies.

a. If the disorder is *not* genetic nor in any way associated with racial origin, what percentage of those with the disorder are likely to be third-generation descendants of that family?

b. What are the most logical null and alternative hypotheses to test whether the disorder is genetically controlled?

c. If 28 of the 75 cases are third-generation descendants of the Finnish family, carry out the test at the 0.05 level of significance.

7.5.4. A random sample of 100 high-school dropouts in Pittsburgh aged 17 to 19 revealed that 20% of them were unemployed.

a. Place a 95% confidence interval on the percentage of all similar people in that area who are unemployed.

b. The average unemployment rate for the entire work force in Pittsburgh is 7.0%. Is the unemployment rate among high-school dropouts significantly higher than for the entire work force? Justify your answer.

7.5.5. Many people claim they can distinguish the difference in taste between fish that has been frozen and fish that is prepared fresh. In an experiment, a random sample of 100 consumers is presented with two portions of cooked fish, one of each kind. Of these consumers, 64 can correctly distinguish between the fresh and the frozen fish.

a. Estimate by means of a point estimate the proportion of people in the population who can make this distinction.

b. The answer to Part a is an estimate and thus subject to variability. What is the estimated variance of this estimate?

c. Use the normal approximation to the binomial distribution in order to place a 95% confidence interval on the proportion.

d. Is there evidence that some people can distinguish fresh fish and are not just guessing? Explain.

7.5.6. The President has difficulty obtaining energy legislation because many people do not believe that there is a national natural gas shortage. To determine the thinking of his own political party on this, assume a random sample is taken of 225 county chairmen, and 90 of them indicate that they do not believe there is a shortage.

a. Make a point estimate of the proportion of county chairmen who do not believe the report of a shortage.

b. Place a 90% confidence interval on that proportion.

c. Which of the following statements are proper statistical inferences from this study?

i. There is no evidence that a majority of the county chairmen doubt the existence of a shortage.

ii. The margin of error is so wide that it is impossible to conclude whether or not a majority of chairmen doubt the shortage.

iii. If another random sample had been taken, there is a 0.90 probability that the sample proportion would have fallen within the confidence interval computed above.

7.5.7. The theory of radioactive decay predicts that a certain material is expected to emit 40 radioactive particles in 10 milliseconds.

a. What is the probability that at least 35 particles will be emitted in 10 milliseconds?

b. What is the probability that between 30 and 35 particles (inclusive) will be emitted?

7.5.8. A nuclear physicist suspects that a counter is missing some radioactive particles because it has a certain "dead" period as it counts; that is, if two particles are emitted very close together the counter misses the second one. Assume that the theory correctly states that the expected number of radioactive particles emitted in 10 milliseconds from a certain material is 40. If a counter counts 26 particles in 10 milliseconds, does the physicist have evidence that the counter is giving undercounts?

REVIEW EXERCISES

Decide whether each of the following statements is true or false. If a statement is false, explain why.

7.1. Neither of the parameters of a normal distribution can be negative.

7.2. For random sampling, $P(y > \mu) = 0.5$ only if y is from a normal distribution.

7.3. If grades in a statistics class follow the normal distribution $N(70,81)$, about 95% of the class will have grades within the interval 61 to 79.

7.4. Extensive investigation has shown that most data are distributed according to a normal distribution irrespective of the research field from which they come.

7.5. All bell-shaped distributions are normal distributions.

7.6. In a normal distribution, if μ has a large numerical value, then σ^2 will also tend to be large.

7.7. In a normal distribution, about 95% of the values lie within -2 to $+2$.

7.8. If the variance of a population that follows a normal distribution is known, then if necessary a test of hypothesis concerning the mean can be performed from a sample of size $n = 1$.

7.9. Confidence intervals concern parameters; prediction intervals concern observations.

7.10. If possible, samples of size larger than one should be used for purposes of inference.

7.11. According to the central limit theorem, if n is large, the sampling distribution of averages is closely approximated by a normal distribution.

7.12. The central limit theorem can only be applied to symmetrical distributions.

7.13. A test of hypothesis involving the z statistic is frequently used because most experimental populations follow normal distributions with known variances.

7.14. The normal distribution is a valuable statistical tool because of the central limit theorem.

7.15. If a population has variance $\sigma^2 = 12$, then the variance among the averages of all samples of size three drawn at random from the population will be $\sigma_{\bar{y}}^2 = 4$.

7.16. For a test of hypothesis using a z statistic, the region of rejection is uniquely determined by the alternative hypothesis and the sample size.

7.17. The region of rejection is based on the null hypothesis.

7.18. The danger in misusing a one-tailed test when a two-tailed test should be used is that it makes α larger than for the proper test.

7.19. The danger in misusing a two-tailed test when a one-tailed test should be used is that it takes β larger than for the proper test.

7.20. The power $1 - \beta$ of a test can never be less than α.

7.21. When the null hypothesis is suspected of being false, it would be wise to decrease α from 0.05 to 0.01 to be less likely to reject it by chance.

7.22. A random sample from any population is normally distributed with a mean of μ and variance σ^2.

7.23. Other things being equal, in a test of hypothesis the larger the sample size, the smaller the α level.

7.24. Other things being equal, in a confidence interval the larger the sample size, the narrower the interval.

7.25. If a population distributed as $N(\mu,\sigma^2)$ is randomly sampled and $(\bar{y} - \mu)/(s/\sqrt{n})$ is used to compute a z statistic, the probabilities will be reliable only if n is large.

7.26. If the $1 - \alpha$ central confidence interval on μ does not contain the value of μ in the null hypothesis, then a two-tailed test would lead to rejection of the null hypothesis at the α level of significance.

7.27. If the variance of a normal distribution is unknown and is estimated by s^2, then two separate random samples of the same size could produce two confidence intervals of different widths.

7.28. A hypothesis about the binomial parameter π tested by the exact binomial distribution and by the normal approximation give exactly the same probabilities.

7.29. When n is large and π is near 0.5, the binomial distribution is approximately a normal distribution.

7.30. In the standard normal distribution, $\mu = n\pi$ and $\sigma^2 = n\pi(1 - \pi)$.

SELECTED READINGS

Adams, W. J. (1974). *The Life and Times of the Central Limit Theorem*, Kaedman, New York.

Pearson, K. (1924). Historical note on the origin of the normal curve of errors, *Biometrika*, **16**, 402–404.

Tate, R. F., and G. W. Klett (1959). Optimal confidence intervals for the variance of a normal distribution, *Journal of the American Statistical Association*, **54**, 674–682.

8

Student's *t* Distribution

In most experimental situations, the population variance is unknown. In Chapter 7 we noted that if a population variance is unknown and the sample size is 30 or more, the population variance can be estimated by the sample variance and then the standard normal distribution can be used for inference. If the sample size is below 30, this procedure will not give reliable probabilities. We discuss the appropriate procedure for such situations in this chapter.

8.1. THE NATURE OF *t* DISTRIBUTIONS

At the beginning of this century, William Sealy Gosset was an employee of the Guinness brewery in Dublin where he interpreted data and planned barley experiments. In 1906 and 1907 he was sent to University College, London, to study statistics with Karl Pearson. In 1908 he published a paper in which he noted that if random samples of size less than 30 are taken from a normal distribution and the samples used to estimate the variance, then the statistic

$$\frac{\bar{y} - \mu}{s/\sqrt{n}}$$

is not normally distributed. The probabilities in the tails of this distribution are greater than for the standard normal distribution (Figure 8.1). This is reasonable since

$$z = \frac{\bar{y} - \mu}{\sigma/\sqrt{n}}$$

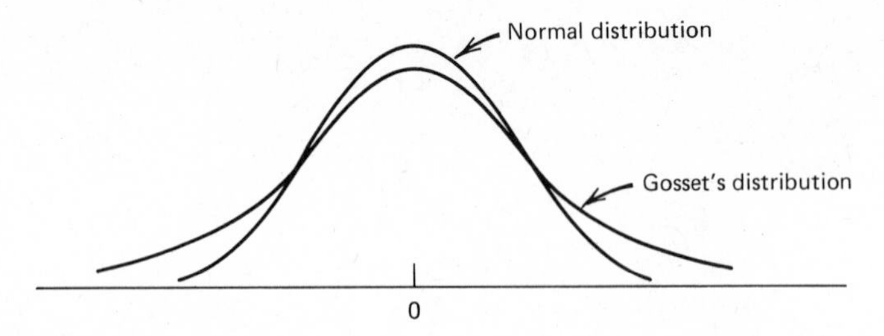

FIGURE 8.1. Comparison of the standard normal distribution and a t distribution.

contains only one random variable $\bar{y}$, while

$$\frac{\bar{y} - \mu}{s/\sqrt{n}}$$

contains two random variables $\bar{y}$ and s. Gosset also noticed that as n increases, this new distribution approaches the standard normal distribution.

Gosset published his findings under the pseudonym "Student" because of the Guinness company's restrictive policy on publication by its employees. The sampling distributions he studied are called *Student's t distributions*, and we write

$$t = \frac{\bar{y} - \mu}{s/\sqrt{n}}$$

The density functions for Student's t distributions are known, and a description of the curve may be helpful (see Figure 8.2).

Student's t distributions are:

1. Unimodal.
2. Asymptotic to the horizontal axis.
3. Symmetrical about zero; $E(t) = 0$.
4. Dependent on ν, the degrees of freedom. (For the statistic under discussion, $\nu = n - 1$.)

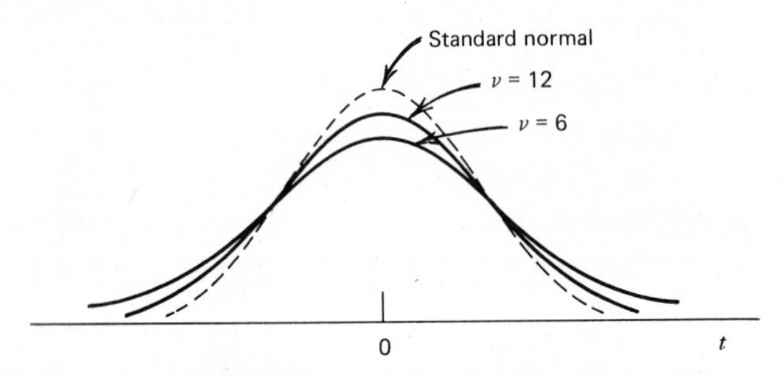

FIGURE 8.2. Student's t distributions.

5. More variable than the standard normal distribution; $V(t) = \nu/(\nu - 2)$ for $n > 2$.
6. Approximately standard normal if ν is large.

Table A.10 in the Appendix of Useful Tables gives many of the critical values of the t distributions needed for inference. The t distributions are listed by degrees of freedom. In the table, α corresponds to the probability that t exceeds the tabular value; thus $P(t > 1.721 \text{ if } \nu = 21) = 0.05$. We write $t_{0.05,21} = 1.721$.

Since the t distribution is symmetrical, critical values for the lower tail can be obtained from the upper tail,

$$t_{1-\alpha,\nu} = -t_{\alpha,\nu}$$

Thus

$$t_{0.95,16} = -t_{0.05,16} = -1.746$$

It should be emphasized that the t statistic arises only when we are sampling from a population with a normal distribution and when σ^2 is estimated by s^2. Whether the sample size is large or small,

$$\frac{\bar{y} - \mu}{s/\sqrt{n}}$$

has a t distribution. However, since the t distribution is quite close to the standard normal for $n \geq 30$, it is common to approximate the probabilities in the t distribution by the standard normal for large sample sizes. If more accuracy is desired and the appropriate table is available, the t distribution can be used.

It is permissible to use the t distribution to estimate probabilities when we are sampling from a distribution that is not normal if the distribution is at least symmetrical, unimodal, and with a variance that is not inordinately large. In this case, the t distribution is a good estimate of the actual sampling distribution.

EXERCISES

8.1.1. Use Table A.10 to find:

a. $t_{0.01,10}$.

b. $t_{0.99,10}$.

c. $t_{0.025,7}$.

d. $t_{0.975,7}$.

e. $t_{0.005,23}$.

f. $t_{0.995,23}$.

8.1.2. Use Table A.10 to find:

a. $P(t > 2.145 \text{ if } \nu = 14)$.
b. $P(t < 2.518 \text{ if } \nu = 21)$.
c. $P(t < -1.782 \text{ if } \nu = 12)$.
d. $P(t > -1.363 \text{ if } \nu = 11)$.
e. $P(-2.120 \leq t \leq 2.120 \text{ if } \nu = 16)$.
f. $P(|t| \geq 2.831 \text{ if } \nu = 21)$.

8.1.3. A random sample is taken of 16 women who are the sole support of their families, and information is obtained about their annual income (in dollars):

$$\sum y = 128{,}000$$

$$\sum y^2 = 1{,}177{,}600{,}000$$

Assume that the distribution of incomes is normal.

a. Find the best point estimate of the mean income of all women who are the sole support of their families.
b. Estimate the population variance.
c. If μ is actually \$6400, compute

$$t = \frac{\bar{y} - \mu}{s/\sqrt{n}}$$

d. How likely is it that a t statistic of this magnitude or larger will arise when choosing random samples of size 16 from this population?

8.2. INFERENCE ABOUT A SINGLE MEAN

Under the following conditions, t distributions may be used for inference about μ:

1. The population distribution is normal (or at least symmetrical and unimodal).
2. The population variance is unknown and estimated by the sample variance.
3. The sample is random.

Tests of hypothesis about a population mean μ and confidence intervals for μ using t distributions are analogous to using the standard normal distribution.

Example 8.1. Using a *t* Distribution to Find a Confidence Interval for μ

After running about 17 miles, marathon runners encounter a form of physiological stress which they call "hitting the wall." To better pinpoint where in a race to expect this phenomenon, a sports physiologist has 12 male marathon runners race until each feels this stress. The variable of interest is the number of miles run until the stress occurs. These are:

15.8	16.5	15.3	16.2	17.1	16.4
17.5	17.3	16.9	16.6	17.0	17.7

The physiologist would like to use a t distribution to find a 95% confidence interval on the mean distance a marathon runner covers before "hitting the wall." He finds that $\sum y = 200.4$ miles and $\sum y^2 = 3{,}352.08$. He computes a point estimate for the mean

$$\bar{y} = \frac{200.4}{12} = 16.70$$

and the sample variance is

$$s^2 = \frac{\sum y^2 - (\sum y)^2/n}{n-1} = \frac{3352.08 - (200.4)^2/12}{11} = 0.4909$$

The sample standard deviation is $s = 0.70$ and the standard error of the mean is $s/\sqrt{n} = 0.70/\sqrt{12} = 0.20$. Since there are 12 subjects, the degrees of freedom are $n - 1 = 12 - 1 = 11$.

$$\text{CI}_{0.95}: \quad \bar{y} \pm t_{0.025,11} \frac{s}{\sqrt{n}}$$

$$16.70 \pm 2.201(0.20)$$

$$16.70 \pm 0.44$$

$$16.26 \leq \mu \leq 17.14$$

In order for this to be valid, the physiologist must be able to assume that the variable of interest is normally distributed, or at least approximately so. If he has been observing this phenomenon for some time in the course of his other investigations of marathon runners, he may have accumulated enough rough measurements to draw a graph and check on the symmetry and unimodality; or it is possible that theoretical knowledge of the physiological changes that occur during running could justify this assumption. If he is unable to justify the assumption, he will have to be cautious about how much faith he has in the accuracy of the interval.

Another condition for the validity of this confidence interval (as well as for other inferences) is that the subjects are a random sample from the population of interest. To literally obtain a completely random sample of 12 runners from the population of all male marathon runners in this country is not feasible. Often the investigator must rely on local volunteers. It would be better if he could find a list of runners from across the country and try to obtain a sample of distance runners from this group. If only local runners are feasible, the generalization to all runners is not as credible. There could be some local condition that affect the variable of interest, for example, altitude.

At a later stage in the experimentation, the physiologist may want to test a hypothesis about the distance until stress occurs. For example, he might decide to extend his investigation to female runners. An immediate question would be whether the distance until stress for women is also 17 miles.

Example 8.2. Using a *t* Distribution to Test a Hypothesis About μ
The sports physiologist would like to test H_0: $\mu = 17$ against H_a: $\mu \neq 17$ for female marathon runners. In a random sample of 8 female runners, he finds

$$\bar{y} = 15.8 \qquad \text{and} \qquad s^2 = 0.65$$

Since $n = 8$, the degrees of freedom are $\nu = 7$ and at $\alpha = 0.05$, the null hypothesis will be rejected if $|t| \geq t_{0.025,7} = 2.365$. The test statistic is

$$t = \frac{\bar{y} - \mu_0}{s/\sqrt{n}} = \frac{15.8 - 17}{\sqrt{0.65/8}} = -4.21$$

Thus he rejects the null hypothesis and concludes that for women the distance until stress is less than 17 miles.

A two-tailed test was used in the above example. If the physiologist had some previous information that stress occurs earlier in women, then a one-tailed test in the lower tail would have been appropriate. Using H_a: $\mu < 17$, at $\alpha = 0.05$ the region of rejection is $t \leq -t_{0.05,7} = -1.895$.

It is possible to make inference about another type of mean, the mean of the *difference between two matched groups*. For example, the mean difference between pretest scores and post-test scores for a certain course might be desired, or the mean difference in reaction time when the same subjects have received a certain drug or have not received the drug. In situations of this type, the experimenter will have two sets of sample data (in the examples just given, pretest/post-test or received/did not receive); however, both sets are obtained from the same subjects. Sometimes the matching is done in other ways, but the object is always to remove extraneous variability from the experiment. For example, identical twins might be used to control for genetically caused variability, or two types of seeds are planted in identical plots of soil under identical conditions to control for the effect of environment on plant growth.

If the experimenter is dealing with two matched groups, the two sets of sample data contain corresponding members—thus he has, essentially, one set consisting of pairs of data. Inference about the mean difference between these two dependent groups can be made by working with the differences within the pairs and using a t distribution with $n - 1$ degrees of freedom in which n is the number of pairs.

Example 8.3. Matched-Pair *t* Test

Two types of calculators are compared to determine if there is a difference in the time required to perform a certain common statistical calculation. Twelve students chosen at random are given drills with both calculators so that they are familiar with the operation of each type. Then the time they take to complete the calculation on each device is measured in seconds (which calculator they are to use first is determined by some random procedure to control for any additional learning during the first calculation). The data are as follows:

Student	Calculator A	Calculator B	Difference y_d	(Difference)^2 y_d^2
1	23	19	4	16
2	18	18	0	0
3	29	24	5	25
4	22	23	−1	1
5	33	31	2	4
6	20	22	−2	4
7	17	16	1	1
8	25	23	2	4
9	27	24	3	9
10	30	26	4	16
11	25	24	1	1
12	27	28	−1	1
			$\sum y_d = 18$	$\sum y_d^2 = 18$

The null hypothesis is H_0: $\mu_d = 0$ and H_a: $\mu_d \neq 0$ in which μ_d is the population mean for the difference in time on the two devices.

$$\bar{y}_d = \frac{\sum y_d}{n} = \frac{18}{12} = 1.5$$

$$s_d^2 = \frac{\sum y_d^2 - (\sum y_d)^2/n}{n - 1} = \frac{82 - (18)^2/12}{11} = 5$$

The test statistic is

$$t = \frac{\bar{y}_d - \mu_{d_0}}{s_d/\sqrt{n}} = \frac{1.5 - 0}{\sqrt{5/12}} = 2.325$$

Using $\alpha = 0.05$ and $\nu = 12 - 1 = 11$, $t_{0.025,11} = 2.201$, and since $t > 2.201$, the test is significant and the two calculators differ in the time necessary to perform the calculation. Looking at the data, since $\bar{y}_d$ is positive, the experimenter concludes that the calculation is faster on machine *B*.

In the above example, the experimenter was interested in whether or not there is a difference in time required on the two calculators, thus $\mu_d = 0$ was tested. The population mean specified in the null hypothesis need not be zero; it could be some other specified amount. For example, in an experiment about reaction time the experimenter might hypothesize that after taking a certain drug, reaction times are slower by two seconds; then H_0: $\mu_d = 2$ would be tested, with $y_d = y_{\text{after}} - y_{\text{before}}$. The alternative hypothesis may be one-tailed or two-tailed, as appropriate for the experimental question.

Using a matched-pair design is a way to control extraneous variability. If the study of the two calculators involved a random sample of 12 students who used calculator *A* and another random sample of 12 students who used calculator *B*, additional variability would be introduced because the two groups are made up of different people. Even if they were to use the same calculator, the means of the two groups would probably be different. If the differences among people are large, they interfere with our ability to detect any difference due to the calculators. If possible, a design involving two dependent samples that can be analyzed by a matched-pair t test is preferable to two independent samples. The analysis proper for two independent samples is discussed in Section 8.3.

If confidence intervals are desired for the mean of the difference between two dependent samples, they can also be computed:

$$\text{CI}_{1-\alpha}: \quad \bar{y}_d \pm t_{\alpha/2,n-1} \frac{s_d}{\sqrt{n}}$$

■ ■ ■

Procedure. Inference About a Mean Using a *t* Distribution

Assumptions: normality, or at least symmetry and unimodality; unknown population variance

Confidence Intervals

$$\text{CI}_{1-\alpha}: \quad \bar{y} - t_{\alpha/2,n-1} \frac{s}{\sqrt{n}} \le \mu \le \bar{y} + t_{\alpha/2,n-1} \frac{s}{\sqrt{n}}$$

Test of Hypothesis

H_0: $\mu = \mu_0$
H_a: $\mu \neq \mu_0$ or $\mu > \mu_0$ or $\mu < \mu_0$
Significance level: α
Test statistic:

$$t = \frac{\bar{y} - \mu_0}{s/\sqrt{n}}$$

Region of rejection: $|t| \geq t_{\alpha/2,n-1}$ or $t \geq t_{\alpha,n-1}$ or $t \leq -t_{\alpha,n-1}$, respectively

■ ■ ■

EXERCISES

8.2.1. From a random sample of 16 applicants for certain graduate fellowships, the following statistics are obtained about their GRE scores:

$$\sum y = 16{,}000$$

$$(\sum y)^2 = 256{,}000{,}000$$

$$\sum y^2 = 18{,}400{,}000$$

a. Give the best point estimate of the population mean.
b. Find the standard error of this estimate.
c. Place a 95% confidence interval on this population mean.

8.2.2. The mean pulse rate for active males of college age is 72 beats per minute, but it is thought to be greater for less active men of the same age. A physician at a student health center questions his male patients on whether they participate in leisure-time sports and measures the pulse rates of a random sample of 10 who do not. The following pulse rates are obtained: 65, 77, 66, 83, 73, 75, 68, 73, 91, 77.

a. Criticize the sample on the basis of the population it may represent.
b. Assuming some valid inference can be made, prepare for a test of hypothesis by giving:
   i. The most logical null and alternative hypotheses.
   ii. The critical region of the test statistic for $\alpha = 0.05$.
c. Conduct the test of significance by computing:
   i. The sample average and variance.
   ii. The value of the test statistic.

d. Assume the inference is valid; what would you conclude from this study?

8.2.3. Distance runners are known to have lower pulse rates than their contemporaries. Suppose pulse rates are measured on a random sample of 25 runners five minutes after they have completed a 10-kilometer run. The data yield $\bar{y} = 38.2$ beats per minute and $s^2 = 72.25$.

a. Compute the standard error of the average.

b. Use the standard error to set a 95% confidence interval for the mean pulse rate of distance runners.

8.2.4. Fruit flies (*Drosophila melanogaster*) are attracted to light. This phenomenon is called positive phototaxis, and it may be an inherited behavior. Suppose a geneticist measures the phototactic response of all flies for one generation and finds a mean response time of 80 seconds. He then mates the male and female that showed the fastest response times. The following data are obtained on the phototactic response times of their offspring:

$$n = 30$$

$$\sum y = 2136 \text{ seconds}$$

$$\sum y^2 = 155{,}225.2$$

a. If phototactic behavior is inherited, then should the offspring of the male and female that showed the most rapid response have an average response time greater or less than that of the previous generation?

b. Use the answer to Part a to set up the most logical null and alternative hypotheses.

c. Perform the test of significance and state the conclusion.

8.2.5. Organic phosphorus insecticides are very stable chemically and are known to collect in the soil and water and eventually to enter the food chain of human beings. In a study made in an agricultural region in the Orient, the milk of 40 nursing mothers was examined and found to have an average of 4.2 ppm of organic phosphorus insecticides. The sample standard deviation was 1.2 ppm.

a. Place a two-sided 99% confidence interval on the mean level of these compounds in mothers' milk in the region.

b. Place a one-sided 99% confidence limit on the worst the mean contamination might be.

8.2.6. The mean score on the Graduate Record Exam is 1000 for all students who take the exam. No extensive study has been made to determine whether higher or lower mean scores are attained by students 30 years of age or older. A pilot study is done, and the following data are obtained:

$$n = 18$$

$$\sum y = 18{,}972$$

$$\sum(y - \bar{y})^2 = 435{,}200$$

a. Prepare for a test of significance by giving:
   i. The most logical null and alternative hypotheses.
   ii. The critical value for the test statistic for $\alpha = 0.05$.
b. Compute the average and variance.
c. Conduct the test of significance and state the conclusion.

8.2.7. At a certain university, an English proficiency test must be passed before undergraduates can receive their degrees. Some students have been known to take the test twice before passing it. A random sample of 25 such students was taken, and the number of "comma errors" was counted on the first and second tests. The average difference on the two tests was a decrease of 2.4 errors. The standard deviation was 6.0.

a. If a college administrator wants to test to show that there was no improvement, what are the null and alternative hypotheses?
b. Perform the test.

8.2.8. One side of the brain is dominant over the other. A psychologist wishes to determine whether the reaction time for voluntary movement is more rapid for the hand controlled by the dominant side of the brain. Fifteen random subjects are given five instructions for each hand in random order and the difference in total reaction time for each hand is recorded for each subject.

a. Give the most logical null and alternative hypotheses.
b. What is the test statistic?
c. Give the degrees of freedom and the critical value at $\alpha = 0.05$.

8.2.9. Agronomists have identified seven different geographical areas with respect to raising corn in West Virginia and have managed to obtain an experimental farm in each area. To see if a single variety of corn can be recommended for the entire state, the two leading varieties are compared for yield at all seven localities. The following yields in bushels per acre are obtained:

	Geographical Area						
Variety	1	2	3	4	5	6	7
A	45	41	58	60	42	32	57
B	47	44	62	63	46	35	59
$(B - A)$	2	3	4	3	4	3	2
$(B - A)^2$	4	9	16	9	16	9	4

a. Why is it a good design to compare the two varieties at each location?
b. What is the average difference in the yields?
c. Show that the standard error of this difference is 0.309.
d. The seed company that sells variety B claims it will exceed variety A in yield by more than two bushels per acre. Test this claim at $\alpha = 0.05$.
e. What is your conclusion about the seed company's claim?
f. Find a 95% central confidence interval on the mean difference in yield of the two types of seed. How is this confidence interval related to the test in Part d?

8.2.10. An industrial psychologist devises a 50-point questionnaire to measure a worker's attitude toward his job; the higher the score, the more favorably the worker views it. The industrial psychologist is concerned that attitude may be affected by the relationship of the day questioned to payday, with a worker responding more favorably if he has been recently paid. To evaluate the effect of payday, he draws a random sample of 16 workers and gives them all the same questionnaire the day before (with score y_1) and the day after (with score y_2) they are paid. The difference in each worker's two scores ($y_d = y_1 - y_2$) is the variable analyzed.

a. Give the most logical null and alternative hypotheses.
b. Use the following sample data,

$$\sum y_1 = 512$$

$$\sum y_2 = 608$$

$$\sum(y_d - \bar{y}_d)^2 = 1500$$

and $\alpha = 0.05$ to give the critical value of the test statistic. Make the test of significance.
c. Is there a payday effect?

8.2.11. Listed below are the gains in pounds of a random sample of pairs of twin lambs in which one member of each pair is treated with an antibiotic and the other remains untreated (control).

Pair	1	2	3	4	5	6	7	8	9	Total
Treated	33	29	29	20	30	33	15	15	21	225
Control	30	34	18	16	25	19	15	18	23	198
y_d	3	−5	11	4	5	14	0	−3	−2	27

a. If $\sum y_d^2 = 405$, compute s_d^2.
b. If you had no knowledge, before this experiment of the effect of

antibiotics on weight gain, give the most logical null and alternative hypotheses.

c. Conduct the test at $\alpha = 0.05$, stating your decision about the null hypothesis and your experimental conclusion.

d. Place a 95% confidence interval on the mean difference in weight gain and explain how this confidence interval could be used to test the null hypothesis.

8.3. INFERENCE ABOUT TWO MEANS

At the end of Section 8.2 we discussed a matched-pair t procedure for two dependent samples. In this section we discuss the appropriate procedure for two *independent* random samples that meet the following conditions:

1. The experimenter is interested in the difference of two population means, $\mu_1 - \mu_2$.
2. The two samples, one from each population, are independent.
3. Both populations are normal, or at least approximately so.
4. The population variances are unknown, but are the same for both populations, $\sigma_1^2 = \sigma_2^2 = \sigma^2$.

Example 8.4. Group Comparison *t* Test

Chemical compounds that are carcinogenic to mammals also commonly cause genetic mutations in lower organisms. Preliminary screening of possible cancer-producing compounds can be performed by testing whether these compounds increase the mutation rate of microorganisms.

Suppose an experimenter uses this procedure as the first safety screening of an aromatic hydrocarbon that could be used as an industrial solvent. He adds the compound to a medium of an Ascomycetes fungus in several petri dishes and compares the mutation rate of this group (the treatment group) with the control group (untreated group).

The variable measured is the number of mutant colonies per petri dish. The experimenter realizes that this discrete random variable probably is not normally distributed but rather has a Poisson distribution. Since he would like to use a t test to make the comparison, he first transforms his counts, x, by letting $y = \log_{10}x$. (If there are any zero counts, he would use $y = \log_{10}(x + 1)$.) Experience has shown him that in this situation this transformation will yield distributions that, although discrete, are approximately normal. After the transformation, his data are summarized as follows:

	Control Group	Treatment Group
Sample sizes	$n_1 = 7$	$n_2 = 8$
Sample means	$\bar{y}_1 = 1.52$	$\bar{y}_2 = 1.68$
Sample variances	$s_1^2 = 0.12$	$s_2^2 = 0.14$

From his previous work he believes that the variances of the two populations, although unknown, are in fact equal. The closeness of the sample variances seems to confirm this. (If he were in doubt, he could apply the test to be described in Section 8.4 to the sample variances in order to test the hypothesis $\sigma_1^2 = \sigma_2^2$.) Since he believes the two variances are equal, the best point estimate of this common variance will be an average of the two sample variances weighted by the degrees of freedom. This weighted average is called the *pooled sample variance* and is computed as follows:

$$s_p^2 = \frac{\sum(y_1 - \bar{y}_1)^2 + \sum(y_2 - \bar{y}_2)^2}{(n_1 - 1) + (n_2 - 1)}$$

$$= \frac{(n_1 - 1)s_1^2 + (n_2 - 1)s_2^2}{n_1 + n_2 - 2}$$

In this experiment,

$$s_p^2 = \frac{6(0.12) + 7(0.14)}{7 + 8 - 2} = 0.131$$

He would like to test

$$H_0\colon \ \mu_1 - \mu_2 = 0 \qquad \text{against} \qquad H_a\colon \ \mu_1 - \mu_2 < 0$$

In other words,

$$H_0\colon \ \mu_1 = \mu_2 \qquad \text{against} \qquad H_a\colon \ \mu_1 < \mu_2$$

The test statistic has $\nu = n_1 + n_2 - 2 = 13$ degrees of freedom, corresponding to the denominator of the pooled sample variance, and

$$t = \frac{(\bar{y}_1 - \bar{y}_2) - (\mu_1 - \mu_2)_0}{\sqrt{\dfrac{s_p^2}{n_1} + \dfrac{s_p^2}{n_2}}} = \frac{(1.52 - 1.68) - 0}{\sqrt{\dfrac{0.131}{7} + \dfrac{0.131}{8}}} = -0.85$$

The critical value at $\alpha = 0.05$ is $t_{0.95,13} = -1.771$. Thus the null hypothesis is not rejected, and the experimenter concludes that there is no evidence that this aromatic hydrocarbon increases the mutation rate of the fungus.

Note that the t statistic, although different from the statistic used for one sample or matched-pair tests, is still of the same form:

$$t = \frac{\begin{pmatrix}\text{estimate}\\ \text{of the parameter}\end{pmatrix} - \begin{pmatrix}\text{hypothesized value}\\ \text{of the parameter}\end{pmatrix}}{\begin{pmatrix}\text{standard error}\\ \text{of the estimator}\end{pmatrix}}$$

The estimator of $\mu_1 - \mu_2$ is $\bar{y}_1 - \bar{y}_2$. Since the variances of the two groups are equal ($\sigma_1^2 = \sigma_2^2 = \sigma^2$) and the samples are independent,

$$V(\bar{y}_2 - \bar{y}_2) = V(\bar{y}_1) + V(\bar{y}_2)$$
$$= \frac{\sigma^2}{n_1} + \frac{\sigma^2}{n_2}$$

This is estimated by

$$\frac{s_p^2}{n_1} + \frac{s_p^2}{n_2}$$

and the standard error of the estimator is estimated by

$$\sqrt{\frac{s_p^2}{n_1} + \frac{s_p^2}{n_2}}$$

A caution about this procedure: The test is not reliable if the variances of the two groups are unequal. If there is doubt, this should be checked by the method to be described in the next section. If the variances prove to be unequal and the sample sizes are small ($n_1 < 30$ or $n_2 < 30$), then there is no exact test available and an approximation procedure such as the one in the next section should be used.

The test in this section is the appropriate one for two *independent* samples. Two independent samples should *not* be analyzed by means of a matched-pair procedure, for the degrees of freedom will be lower, increasing the magnitude of the critical value and reducing the power of the test.

If the combined sample size is large ($n_1 + n_2 \geq 30$), the critical value may be estimated by a z-value for convenience. If both samples are large ($n_1 \geq 30$ and $n_2 \geq 30$), the test statistic may be replaced by

$$z = \frac{(\bar{y}_1 - \bar{y}) - (\mu_1 - \mu_2)_0}{\sqrt{\frac{s_1^2}{n_1} + \frac{s_2^2}{n_2}}}$$

eliminating the need to pool the sample variances. Whether or not the population variances are equal, this z statistic is valid for two large samples. If the actual population variances are known, then

$$z = \frac{(\bar{y}_1 - \bar{y}_2) - (\mu_1 - \mu_2)_0}{\sqrt{\frac{\sigma_1^2}{n_1} + \frac{\sigma_2^2}{n_2}}}$$

is the appropriate statistic for all sample sizes.

Confidence intervals for $\mu_1 - \mu_2$ may also be computed. For $n_1 < 30$ or $n_2 < 30$ with $\sigma_1^2 = \sigma_2^2$ and σ_1^2, σ_2^2 unknown, use

$$\text{CI}_{1-\alpha}: \quad \bar{y}_1 - \bar{y}_2 \pm t_{\alpha/2, n_1+n_2-2} \sqrt{\frac{s_p^2}{n_1} + \frac{s_p^2}{n_2}}$$

For $n_1 \geq 30$ and $n_2 \geq 30$ with σ_1^2, σ_2^2 unknown, use

$$\text{CI}_{1-\alpha}: \quad \bar{y}_1 - \bar{y}_2 \pm z_{\alpha/2} \sqrt{\frac{s_1^2}{n_1} = \frac{s_2^2}{n_2}}$$

If σ_1^2 and σ_2^2 are known, use

$$\text{CI}_{1-\alpha}: \quad \bar{y}_1 - \bar{y}_2 \pm z_{\alpha/2} \sqrt{\frac{\sigma_1^2}{n_1} + \frac{\sigma_2^2}{n_2}}$$

regardless of sample size.

■ ■ ■

Procedure. Inference About Two Independent Means

Assumptions: normality or at least symmetry and unimodality

σ_1^2, σ_2^2 unknown, $\quad \sigma_1^2 = \sigma_2^2$, $\quad$ and $\quad n_1$ or $n_2 < 30$

Confidence Interval on $\mu_1 - \mu_2$

$$\text{CI}_{1-\alpha}: \quad \bar{y}_1 - \bar{y}_2 \pm t_{\alpha/2, n_1+n_2-2} \sqrt{\frac{s_p^2}{n_1} + \frac{s_p^2}{n_2}}$$

with

$$s_p^2 = \frac{(n_1 - 1)s_1^2 + (n_2 - 2)s_2^2}{n_1 + n_2 - 2}$$

Test of Hypothesis

H_0: $\mu_1 - \mu_2 = (\mu_1 - \mu_2)_0$
H_a: $\mu_1 - \mu_2 \neq (\mu_1 - \mu_2)_0$ or $\mu_1 - \mu_2 > (\mu_1 - \mu_2)_0$
or $\mu_1 - \mu_2 < (\mu_1 - \mu_2)_0$

Significance level: α
Test statistic:

$$t = \frac{\bar{y}_1 - \bar{y}_2 - (\mu_1 - \mu_2)_0}{\sqrt{\dfrac{s_p^2}{n_1} + \dfrac{s_p^2}{n_2}}} \qquad \text{with } s_p^2 \text{ as above}$$

Region of rejection: $|t| \geq t_{\alpha/2, n_1+n_2-2}$ or $t \geq t_{\alpha, n_1+n_2-2}$ or $t \leq -t_{\alpha, n_1+n_2-2}$, respectively

---

Assumptions: n_1 and $n_2 \geq 30$

Confidence Interval on $\mu_1 - \mu_2$

$$\text{CI}_{1-\alpha}: \quad \bar{y}_1 - \bar{y}_2 \pm z_{\alpha/2} \sqrt{\frac{\sigma_1^2}{n_1} + \frac{\sigma_2^2}{n_2}}$$

Use s_1^2 and s_2^2 to estimate σ_1^2 and σ_2^2 if the population values are unknown

Test of Hypothesis

H_0: $\mu_1 - \mu_2 = (\mu_1 - \mu_2)_0$
H_a: $\mu_1 - \mu_2 \neq (\mu_1 - \mu_2)_0$ or $\mu_1 - \mu_2 > (\mu_1 - \mu_2)_0$ or $\mu_1 - \mu_2 < (\mu_1 - \mu_2)_0$

Significance level: α
Test statistic:

$$z = \frac{\bar{y}_1 - \bar{y}_2 - (\mu_1 - \mu_2)_0}{\sqrt{\frac{\sigma_1^2}{n_1} + \frac{\sigma_2^2}{n_2}}}$$

Use s_1^2 and s_2^2 to estimate σ_1^2 and σ_2^2 if the population values are unknown
Region of rejection: $|z| \geq z_{\alpha/2}$ or $z \geq z_\alpha$ or $z \leq -z_\alpha$, respectively

EXERCISES

8.3.1. After an extended dry period, measurements are taken on atmospheric pollution in urban and rural locations. The data are summarized as follows:

	Urban	Rural
n	7	5
$\bar{y}$	26.0 ppm	12.2 ppm
s^2	91	126

a. Compute the pooled variance.
b. What are the null and alternative hypotheses if the experimenter is looking for evidence of higher pollution in the urban locations?
c. Perform the test of significance at $\alpha = 0.05$ assuming that the variables meet the assumptions for a group comparison t test.
d. Place a 95% confidence interval on the maximum difference between the two means.

8.3.2. A study is done on insecticide residues on fruit. Normal spraying practices are followed in an apple orchard. After the fruit is picked, a random sample of 16 apples is washed individually by hand. A second sample of 16 is washed mechanically. The experimenter is unsure which method would be more effective in removing insecticide residues. The level of insecticide present on each fruit is determined chemically, yielding the following data:

By Hand	Mechanically
$\bar{y} = 3.5$ ppm	$\sum y = 48$ ppm
$\sum y^2 = 200.5$	$\sum(y - \bar{y})^2 = 5.1$

Test for a significant difference of insecticide residue at the 0.01 level of significance.

8.3.3. A certain industrial solvent absorbs atmospheric moisture very rapidly. The absorbed moisture dilutes the solvent and lessens its usefulness. Two types of containers are used in an effort to find a method of storage that will retard moisture absorption. After two months of storage, 10 containers are chosen at random from each kind and are examined for moisture content.

	Container *A*	Container *B*
$\sum y$	100	120
$\sum y^2$	1012	1450.5

Place a 99% central confidence interval on the difference in the moisture content of the two types of containers.

8.3.4. In a study of the effect of protein quality in the diet, two groups of juvenile female rats are fed diets of the same caloric content, but they differ in the quality of the protein. The experimenter believes that by the end of the experiment the rats on a high-quality protein diet will gain on the average more than five grams more than those on a low-quality protein diet. The experiment begins with equal numbers of rats on each diet, but some are mistakenly assigned to another experiment and have to be eliminated from the protein experiment. Data on the weight gain (in grams) of the remaining rats are collected and summarized:

	High Quality	Low Quality
Sample size	12	7
Sample average	119.7	101.2
Sample standard deviation	21.4	20.6

a. Give the most appropriate null and alternative hypotheses for this experiment.

b. What assumptions are necessary in order to apply a *t* test for two independent groups?

c. Assuming the two populations have the same variance, test the null hypothesis.

d. What do you conclude about the diets?

8.3.5. At a certain university, Graduate Record Exam scores are compared for doctoral students who completed their Ph.D. work within seven years of their bachelor's degree and those who did not complete their work within that time. Random sampling provides the following results:

	Completed Work	Did Not Complete Work
Sample size	25	25
Average score	1056	912
Standard deviation	295	270

Is there any evidence that those who finish their Ph.D. work within seven years score higher on the GRE than those who do not finish within that time? Do you believe that lower GRE scores can be used to predict those who will have difficulty completing their doctoral work on time? Why or why not?

8.3.6. In order to study the effect on reading caused by a family move between first and second grades, second-grade children are classified as having completed first grade in the city of the study or as having transferred to the city after completing the first grade elsewhere. A separate random sample is taken from each group, and all children are given the same standard reading test. The results are:

	First Grade Completed	
Sample Value	In City of Study	Elsewhere
n	14	10
$\sum y$	840	650
$\sum(y - \bar{y})^2$	770	385

a. What hypotheses can be tested about the effect on reading due to changing schools between the first and second grade?

b. What assumptions must be made in order to perform a t test on these data?

c. Find the pooled sample variance.

d. Perform the t test and draw a conclusion.

8.3.7. Two experimental methods of controlling acid drainage from coal mines are compared. The data are as follows, with greater numerical values indicating the more effective method.

	Method A	Method B
Average	5.60	6.70
Variance	0.98	0.85
Sample size	6	9

a. Place a 95% confidence interval on the difference between the means for the two methods.
b. Using the confidence interval, what decision would you make about the equality of the means for the two methods?

8.3.8. An educator thinks that engineers, although known to be equal to physical scientists in quantitative skills, have less verbal ability. To test this, GRE verbal scores are compared for large random samples of engineering and physical-science seniors.

	Engineering	Physical Science
Average	414	422
Standard deviation	30	40
Sample size	100	100

a. State the most logical null and alternative hypotheses.
b. Take advantage of the large sample sizes and perform the appropriate z test.
c. What conclusion should be drawn from this study?

8.4. INFERENCE ABOUT TWO VARIANCES

In Section 8.3 we described procedures for analyzing data from two populations having equal variances. There are situations, of course, in which the variances of the two populations under consideration are different. The variability in the weights of elephants is certainly different from the variability in the weights of mice, and in many experiments even though we do not have these extremes the treatments may affect the variances as well as the means.

The null hypothesis H_0: $\sigma_1^2 = \sigma_2^2$ is tested by using a statistic that is in the form of a ratio rather than a difference; the statistic is s_1^2/s_2^2. Intuitively, if the variances are equal this ratio should be approximately equal to one, so values that differ greatly from one indicate inequality.

It has been found that the statistic s_1^2/s_2^2 from two normal populations with equal variances follows a theoretical distribution known as an F distribution. The density functions for F distributions are known, and we can get some understanding of their nature by listing some of their properties. Let us call a random variable that follows an F distribution F, then

1. $F > 0$.
2. The density function of F is not symmetrical.
3. F depends on an ordered pair of degrees of freedom ν_1 and ν_2; that is, there is a different F distribution for each ordered pair ν_1, ν_2. (ν_1 corresponds to the degrees of freedom of the numerator of s_1^2/s_2^2 and ν_2 corresponds to the denominator.)
4. If α is the area under the density curve to the right of the value F_{α,ν_1,ν_2}, then

$$F_{\alpha,\nu_1,\nu_2} = 1/F_{1-\alpha,\nu_2,\nu_1}$$

5. The F distribution is related to the t distribution:

$$F_{\alpha,1,\nu_2} = (t_{\alpha/2,\nu_2})^2$$

Table A.11 in the Appendix gives upper critical values for F if $\alpha = 0.050$, 0.025, 0.010, 0.005, and 0.001. Lower values can be found using Property 4 above.

Example 8.5. Testing for the Equality of Two Variances

Both rats and mice carry ectoparasites that can transmit disease organisms to man. To determine which of the two rodents presents the greater health hazard in a certain area, a public health officer traps (presumably at random) both and counts the number of ectoparasites each carries. He finds:

	n	s^2	$\bar{y}$
Rats	31	62	16.3
Mice	9	18	11.4

He wants to test for the equality of means with a group comparison t test. He assumes that these discrete counts are approximately normally distributed, but he has some doubts about the equality of the variances in the two populations. Thus he first must test

$$H_0\colon\ \sigma_1^2 = \sigma_2^2 \qquad \text{against} \qquad H_a\colon\ \sigma_1^2 \neq \sigma_2^2$$

with the test statistic $F = s_1^2/s_2^2 = 62/18 = 3.44$. Since $n_1 = 31$ and $n_2 = 9$, the degrees of freedom for the numerator are $\nu_1 = n_1 - 1 = 30$ and for the denominator $\nu_2 = n_2 - 1 = 8$. In Table A.11 he finds

$$F_{0.05,30,8} = 3.079 \qquad \text{and} \qquad F_{0.05,8,30} = 2.266$$

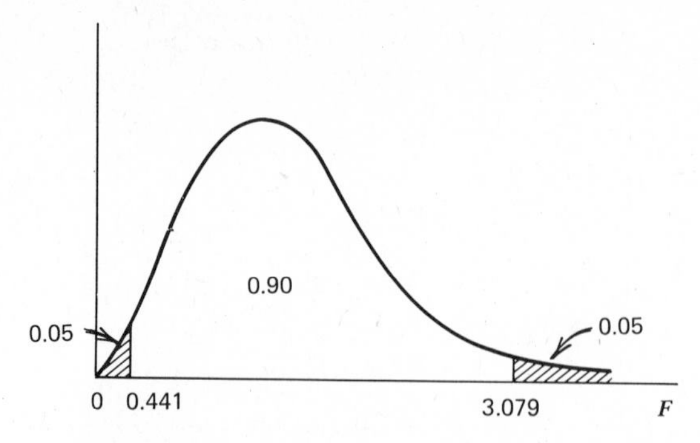

FIGURE 8.3. Regions of rejection in an F distribution.

thus the region of rejection (Figure 8.3) at $\alpha = 0.10$ is

$$F \geq F_{0.05,30,8} = 3.079 \qquad \text{and} \qquad F \leq F_{0.95,30,8} = 1/F_{0.05,8,30}$$

$$= 1/2.266 = 0.441$$

Since the computed F equals 3.44, the null hypothesis is rejected, and the public health officer concludes that the variances are unequal. Since one of the sample sizes is small, he may not perform the usual t test for two independent samples.

One-tailed tests of hypotheses involving the F distribution can also be performed, if desired, by putting the entire probability of a Type I error in the appropriate tail.

Central confidence intervals on σ_1^2/σ_2^2 are found as follows:

$$\text{CI}_{1-\alpha}: \quad \frac{s_1^2}{s_2^2} \frac{1}{F_{\alpha/2,\nu_1,\nu_2}} \leq \sigma_1^2/\sigma_2^2 \leq \frac{s_1^2}{s_2^2} F_{\alpha/2,\nu_2,\nu_1}$$

Although the public health officer cannot perform the usual t test for two independent samples because of the unequal variances and the small sample size, there are approximation methods available. One such test is called the Behrens–Fisher, or the t' test for two independent samples and uses adjusted degrees of freedom.

Example 8.6. Testing $\mu_1 - \mu_2$ if $\sigma_1^2 \neq \sigma_2^2$

To test H_0: $\mu_1 = \mu_2$ against H_a: $\mu_1 \neq \mu_2$ at $\alpha = 0.05$, the health officer uses the test statistic

$$t' = \frac{(\bar{y}_1 - \bar{y}_2) - (\mu_1 - \mu_2)_0}{\sqrt{\dfrac{s_1^2}{n_1} + \dfrac{s_2^2}{n_2}}} = \frac{(16.3 - 11.4) - 0}{\sqrt{\dfrac{62}{31} + \dfrac{18}{9}}} = 2.45$$

with adjusted degrees of freedom

$$\nu \cong \frac{\left(\frac{s_1^2}{n_1} + \frac{s_2^2}{n_2}\right)^2}{\frac{\left(\frac{s_1^2}{n_1}\right)^2}{n_1 - 1} + \frac{\left(\frac{s_2^2}{n_2}\right)^2}{n_2 - 1}} = \frac{\left(\frac{62}{31} + \frac{18}{9}\right)^2}{\frac{\left(\frac{62}{31}\right)^2}{30} + \frac{\left(\frac{18}{9}\right)^2}{8}} = 25.26$$

or $\nu = 25$. H_0 will be rejected if $|t'| \geq t_{0.025,25} = 2.060$. Since $|t'| = 2.45 > 2.060$, the null hypothesis is rejected, and the public health officer concludes that on the average there are more ectoparasites on rats than on mice.

The adjusted degrees of freedom may be rounded to the closest integer, as in the example, or interpolation may be used in the t table for a more accurate critical value. Since this t' test is only an approximate procedure and is usually very conservative (rejection is difficult), it should be avoided if possible. Instead, larger samples sizes should be obtained when feasible.

A summary of several test statistics in the form of a flow-chart for making a decision about the appropriate procedure is given in Figure 8.4. Degrees of freedom involved in the t, F, and χ^2 procedures are indicated by subscripts; for example, t_{n-1} means that the test has $n - 1$ degrees of freedom. Since a matched-pair t test is essentially a one-sample procedure (the set of differences is a single sample), this test does not appear explicitly in the flowchart.

EXERCISES

8.4.1. Use Table A.11 to find:

a. $F_{0.01,11,7}$.

b. $F_{0.01,7,11}$.

c. $F_{0.05,20,15}$.

d. $F_{0.95,15,20}$.

e. $F_{0.99,8,3}$.

8.4.2. The writings of different authors can be partially characterized by the variability in the lengths of their sentences. Two manuscripts, A and B, are found by a historian and he wants to know whether they have the same author. Several sentences from each are chosen at random, and word counts are taken; the variable of interest y is the number of words per sentence.

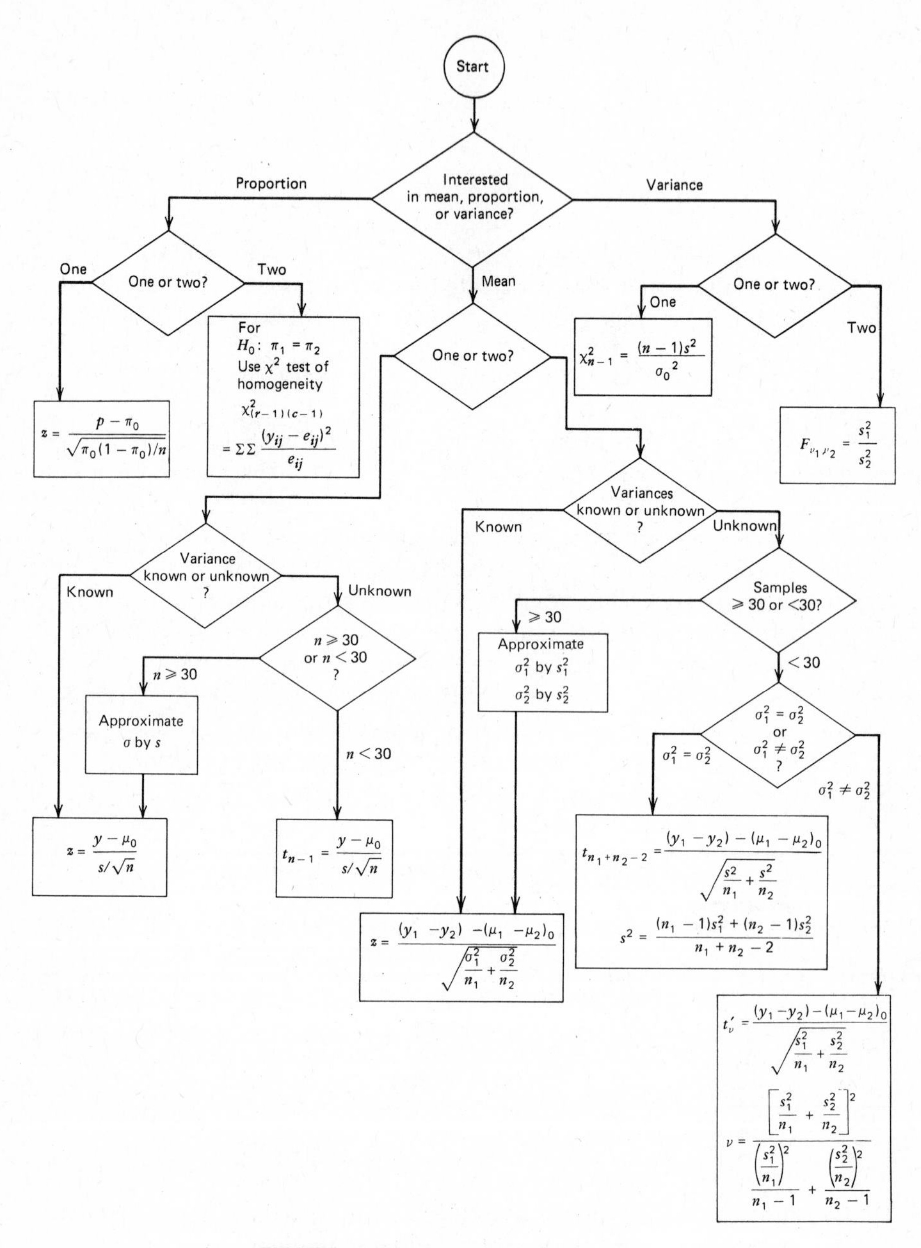

FIGURE 8.4. Flowchart of test statistics.

	Manuscript A	Manuscript B
n	15	15
$\sum y$	141	210
$\sum y^2$	1327	2942

Is there evidence of different authorship at the 0.02 level of significance?

8.4.3. A highway engineer wishes to compare the resin content of asphalt from a Caribbean source with those from a North American source. The following statistics are obtained:

	Average	Sample Value Variance	Size
Caribbean	21.4	0.44	10
North American	22.0	0.11	8

Given only this information, perform the appropriate test of hypothesis to determine if there is a difference in the mean resin content from the two sources (use $\alpha = 0.10$).

8.4.4. A nutritionist wishes to study vitamin B production by bacteria in the caecum (a portion of the digestive tract) and wishes to use either mice or meadow voles, whichever have the larger mean caecum volume. The sample data on which he must make his decision are:

	Mice	Voles
Number of observations	16	11
Average caecum volume	6.5	8.9
Variance	4.6	13.1

a. Should he use a t test or a t' test? (Use $\alpha = 0.10$.)
b. Test to see if there is a significant difference in the average caecum volumes. (Use $\alpha = 0.10$.)
c. What would you suggest to the nutritionist?

8.4.5. The following values were computed from the length of life of two brands of light bulbs (in hours):

	Brand A	Brand B
n	9	16
$\bar{y}$	1560	1573
$\sum(y - \bar{y})^2$	440	1860

a. Is there a difference in the variability of lifetimes for the two brands of bulbs? ($\alpha = 0.02$.)

b. Find a 98% confidence interval on the ratio of the two variabilities.

REVIEW EXERCISES

Decide whether each of the following statements is true or false. If a statement is false, explain why.

8.1. The t distribution is appropriate for small sample sizes irrespective of whether or not the variance is known.

8.2. For each positive-integer degree of freedom, there is a different t distribution.

8.3. The "standard error of averages" is the standard deviation of the averages of all samples of a fixed size from some population.

8.4. The denominator of a t test is sometimes referred to as the "standard error."

8.5. Other things being equal, in order to reduce the standard error of the averages by one-half, it is necessary to double the sample size.

8.6. When the variance of a normal distribution is known, a t distribution should not be used for purposes of inference.

8.7. The proportion of area under the t distribution to the right of a specified value of t in the right tail is greater than the proportion beyond the same value in the standard normal distribution.

8.8. Gosset discovered that when n is small, s^2 tends to overestimate σ^2.

8.9. For a one-sample t test, the region of rejection is uniquely determined by the alternative hypothesis and the sample size.

8.10. For a test of hypothesis about μ, the t distribution is used in place of the standard normal distribution when σ^2 is unknown and n is small.

8.11. Whether a one-tailed or a two-tailed test is used depends on the question being asked by the experimenter.

8.12. One-tailed tests are usually chosen to reduce the probability of a Type I error.

8.13. For a fixed α level, as the degrees of freedom increase in a t test, the absolute value of the critical value increases.

8.14. $CI_{0.95}$: $\bar{y} \pm t_{0.025}\, s/\sqrt{n}$ contains 95% of all population means.

8.15. $\bar{y} \pm t_{\alpha/2,\nu} s/\sqrt{n}$ is narrower than the corresponding interval based on the standard normal distribution $\bar{y} \pm z_{\alpha/2}\, s/\sqrt{n}$.

8.16. If two samples consist of pairs of data, the experimenter may choose between the matched-pair t test or the t test for two independent samples.

8.17. Since paired t tests are more powerful than two-sample t tests, the experimenter should try to plan a matched-pair design if at all possible.

8.18. In the matched-pair t test, the parameter in the null hypothesis must equal zero.

8.19. In a paired comparison t test involving 20 pairs of twins, there are 38 degrees of freedom.

8.20. Even when $\mu_1 = \mu_2$, due to random sampling the calculated value of t may be large enough to cause a Type I error.

8.21. If a t test fails to indicate significance for the difference between two sample means, a Type II error results.

8.22. Other things being equal, it is easier to detect that $\mu_1 = \mu_2$ is false when $\mu_1 - \mu_2 = 1.5$ than when $\mu_1 - \mu_2 = 4.5$.

8.23. The pooled variance s_p^2 is a weighted average of the two sample standard deviations.

8.24. A paired comparison t test should always be used when $\sigma_1^2 = \sigma_2^2$.

8.25. If a t test determines that the difference between two sample averages is significant, then the experimenter should conclude that two different populations were sampled.

8.26. If in a two-sample t test $\mu_1 = \mu_2$, then the computed value of t will be exactly zero.

8.27. If for two populations $\sigma_1^2 = \sigma_2^2$, the best estimate of the common variance is $(s_1^2 + s_2^2)/2$ irrespective of other considerations.

8.28. If the experimenter is unsure that $\sigma_1^2 = \sigma_2^2$, this hypothesis can be tested prior to performing the two-sample t test.

8.29. If H_0: $\mu_1 = \mu_2$ is true, then for the group comparison t test the t statistic should be close to 0.

8.30. If $\sigma_1^2 = \sigma_2^2$ is true, then the F statistic should be close to 0.

8.31. The F test is a procedure for testing hypotheses about variances only when means are unequal.

8.32. The t' test is a test for equality of variances rather than equality of means.

8.33. When σ_1^2 and σ_2^2 are unequal and unknown, and the samples are small, there is no exact test for a hypothesis of equality of means from the two populations.

8.34. The t' test is an approximate procedure for testing hypotheses about means only when variances are equal.

8.35. There are many F distributions, one for each ordered pair of degrees of freedom.

8.36. The F distributions are symmetrical.

8.37. $F_{\alpha,\nu_1,\nu_2} = -F_{1-\alpha,\nu_1,\nu_2}$.

8.38. All values of F are non-negative.

8.39. $1/F_{0.005,6,8} = F_{0.995,8,6}$.

8.40. The only alternative hypothesis possible for an F test for equality of variance using this book is H_a: $\sigma_1^2 > \sigma_2^2$ because only the upper critical values of F are given in Table A.11 in the Appendix.

SELECTED READINGS

Boneau, C. A. (1960). The effects of violations of assumptions underlying the *t* test, *The Psychological Bulletin,* **57,** 49–64.

Box, J. F. (1981). Gosset, Fisher, and the *t* distribution, *The American Statistician,* **35,** 61–66.

Eisenhart, C. (1979). On the transition from "Student's" *z* to "Student's" *t, The American Statistician,* **33,** 6–10.

Gayen, A. K. (1949). The distribution of "Student's" *t* in random samples of any size drawn from non-normal universes, *Biometrika,* **36,** 353–369.

Geary, R. C. (1947). Testing for normality, *Biometrika,* **34,** 209–242.

Grunow, D. G. C. (1951). Test for the significance of the difference between means in two normal populations having unequal variance, *Biometrika,* **38,** 252–256.

Guenter, W. C. (1981). Sample size formulas for normal theory *t* tests, *The American Statistician,* **35,** 243–244.

Neyman, J. (1938). Mr. W. S. Gosset, *Journal of the American Statistical Association,* **33,** 226–228.

Owen, D. B. (1965). The power of Student's *t* test, *Journal of the American Statistical Association,* **60,** 320–333.

Scheffé, H. (1943). On solutions of the Behrens–Fisher problem based on the *t*-distribution, *The Annals of Mathematical Statistics,* **14,** 35–44.

——— (1944). A note on the Behrens–Fisher problem, *The Annals of Mathematical Statistics,* **15,** 430–432.

"Student" [William Sealy Gosset] (1908). The probable error of a mean, *Biometrika,* **6,** 1–25.

Walsh, J. E. (1947). On the power efficiency of a *t*-test formed by pairing sample values, *The Annals of Mathematical Statistics,* **18,** 601–604.

Welch, B. L. (1937). The significance of the difference between two means when the population variances are unequal, *Biometrika,* **29,** 350–362.

9

Distributions of Paired Variables

Thus far our discussion of inference has focused on the values of a *single* variable of interest obtained from a random sample. We saw in Chapter 2, however, that it is possible to consider more than one variable associated with a given population. For example, *two* variables from the same population that might be considered are age and blood pressure. Other examples are height and weight, and calories intake and weight loss. In this chapter we consider pairs of variables and possible relationships between these variables. It is also possible to study the relationship among several variables—for example, blood pressure is related to age, weight, and exercise. Relationships among more than two variables are discussed in Chapter 14.

9.1. SIMPLE LINEAR REGRESSION

A question often asked about a pair of variables x and y is, ''How do changes in x affect the value of y?'' For example, as a man ages five years, how will this affect his blood pressure? Or we might ask a related question, ''What is the expected value of y for a certain value of x?'' For example, if a man is 30 years old, what is his expected blood pressure?

The x variable age is called the *independent variable* or the *predictor variable*, and the y variable blood pressure is called the *dependent variable* or the *response variable*. If x and y have a relationship with each other, in order to predict y from x we have to be able to find a model for the relationship. The simplest model of a relationship is a straight line. If a straight-line model is appropriate, the line is called the *regression line* and we say that we are regressing y on x. This type of regression is called *simple linear regression*; ''simple'' indicates that there is only one independent variable, and ''linear'' indicates that the model is a straight line.

When dealing with pairs of variables we have the same difficulty as with a single variable, namely, we usually are unable to measure all possible members of the population. In the single variable case, we solved this difficulty by using a random sample to make inference about the population. We do the same for pairs of variables. For example, if we are interested in studying a possible linear relationship between age and blood pressure in adult males, we use a random sample of men, obtain sample data about age and blood pressure, and then see if a straight line fits the data.

Say a random sample of 10 adult males yields the following data:

Age x:	28	23	52	42	27	29	43	34	40	28
Systolic Blood Pressure (mm) y:	70	68	90	75	68	80	78	70	80	72

We begin our analysis by plotting the pairs x,y as points (Figure 9.1). This graph is called a *scatter plot*. The points certainly do not fall exactly on a straight line, but there does appear to be a general linear upward trend such that higher ages are associated with higher systolic blood pressure. Regression is used to fit a straight line to this data in a unique way so that the line can be used to predict systolic blood pressure from age.

It is possible, of course, that two variables are related in some other manner than by a straight-line relationship, or perhaps they are not related to each other at all. Thus our discussion of simple linear regression must include a method for determining whether or not a straight line is the appropriate model for a given set of data (Section 9.2).

Since the simplest possible relationship between two variables is a straight line, it is natural to try to use this model before considering more complex models. Sometimes even if the true relationship is something other than a straight line (as in Figure 9.2), a straight line may be close enough to the true relationship for a preliminary analysis. A straight line is convenient to use because the mathematics involved is relatively simple.

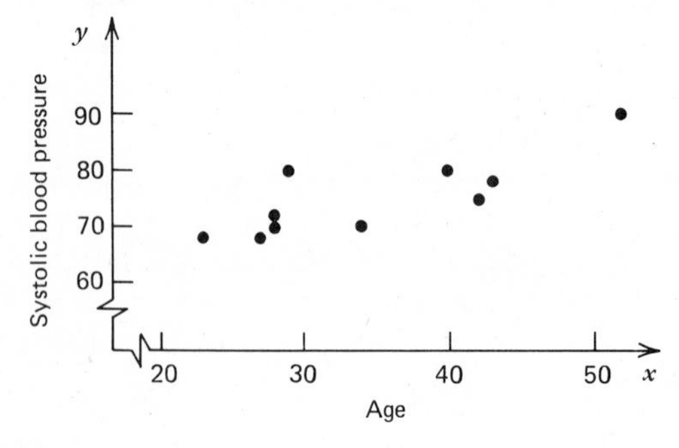

FIGURE 9.1. A scatter plot of age and systolic blood pressure.

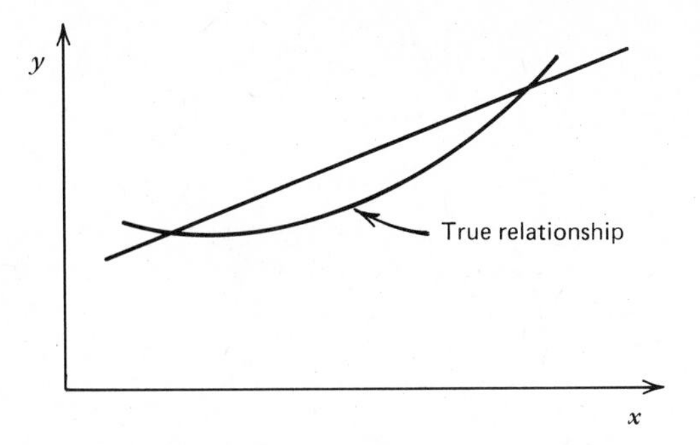

FIGURE 9.2. A relationship that is approximated by a straight line.

Sometimes the true relationship is definitely not linear and a straight line is a very poor model of the relationship. One example is the relationship between the amount of nitrogen fertilizer used on a field and the yield of the crop. The true relationship is quadratic and would be represented by a parabola. In this example, however, economy limits the amount of fertilizer that the farmer would consider using, and in the economical range the relationship might be approximated by a straight line (Figure 9.3). Unfortunately, not every curvilinear relationship will have such a subset of x values that are the main interest of the investigator. Curvilinear relationships are discussed in Sections 14.5 and 14.6.

In order to understand how a straight line is fitted to a set of data that consists of pairs of values obtained for two variables, we consider an overly simplified example. Imagine that an efficiency expert is investigating a possible linear relationship between the number of hours of instruction employees receive about a certain assembly procedure in a factory and the number of units they are able to produce per hour. The following data are

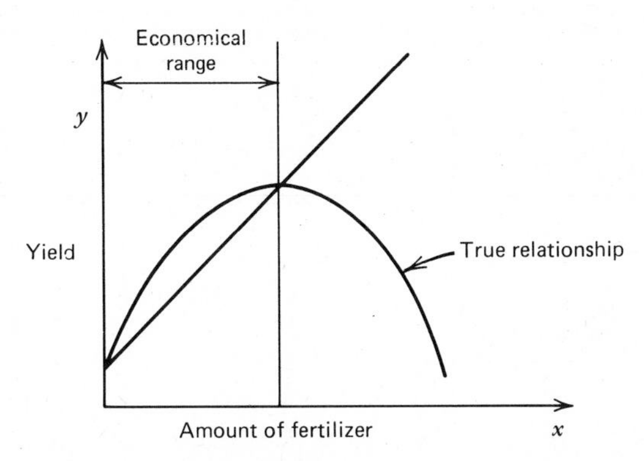

FIGURE 9.3. A relationship that is approximated by a straight line in a certain region of the independent variable.

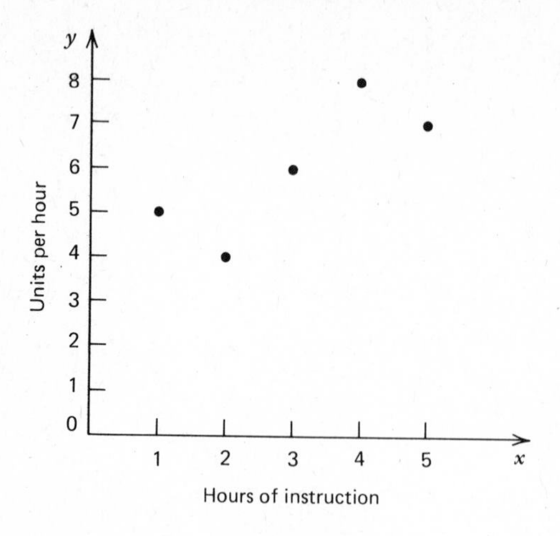

FIGURE 9.4. Scatter diagram for the production study.

collected from five employees:

Hours of Instruction x	Units per Hour y
1	5
2	4
3	6
4	8
5	7

In a real study the investigator would take a random sample of several employees from the groups of employees with the different levels of instruction. However, to keep this illustration simple, we imagine a random sample of just one employee at each level. The approach is the same for several employees at each level.

The first thing the investigator does is graph the scatter diagram (Figure 9.4). If there are enough points in the scatter diagram, it may indicate the general shape of the curve or line that can possibly be used as a model for the variables. A generalized random scatter may indicate that there is no relationship between the variables.

Even if the relationship is linear, not all of the points will lie exactly on the line. The model (Figure 9.5) is of the form

$$y = \alpha + \beta x + \epsilon$$

The regression line is given by the function

$$f(x) = \alpha + \beta x$$

in which α is the y intercept and β is the slope† (the change in y per unit

† Note that this use of α and β is entirely different from the use of these symbols in connection with Type I and Type II error.

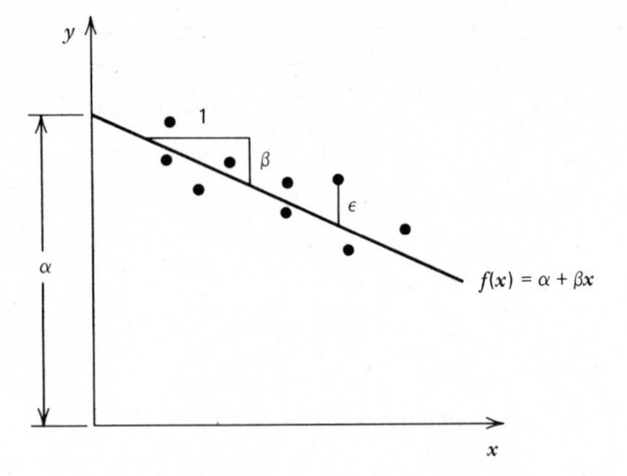

FIGURE 9.5. A regression line.

increase in x). The term ϵ indicates the vertical deviation of a particular point from the line, that is, the line represents the mean y response at a given x value, but individuals will deviate from the mean response due to random variability.

Returning now to the factory example, if the investigator thinks the relationship is linear, the problem is to specify the line that characterizes the relationship by finding the equation of the line. Since only a sample is available, the parameters α and β must be approximated. One approach is simply to draw a line that seems to fit the data; however, this would not be a unique solution. Another approach is to draw a line that has an equal number of points above and below; this is not unique either. Or the line might be drawn such that the vertical deviations would sum to zero; but again, this is not unique.

The problem of approximating the true regression line is solved by using the *least-squares trend line*, also called the *sample regression line*. The least-squares trend line is that unique line for which the sum of the squares of the vertical distances of the sample points from the line is as small as possible (Figure 9.6). Assume that the least-squares line is of the form

$$\hat{y} = a + bx$$

in which a is the y intercept and b is the slope. We minimize the function

$$f(a,b) = \sum (y - \hat{y})^2$$

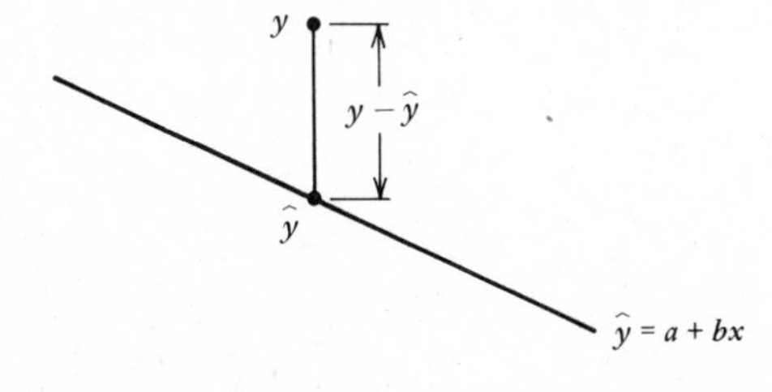

FIGURE 9.6. A vertical deviation from a least-squares line.

in which y is an observed value and $\hat{y}$ is the value predicted by the line for the corresponding x. That is, we find a and b such that this sum is as small as possible. This is done using calculus and leads to two simultaneous equations called the *normal equations*

$$an + b\sum x = \sum y$$

$$a\sum x + b\sum x^2 = \sum xy$$

Solving these two equations simultaneously, the slope is

$$b = \frac{\sum xy - (\sum x)(\sum y)/n}{\sum x^2 - (\sum x)^2/n}$$

and

$$a = \bar{y} - b\bar{x}$$

The denominator of the slope should be familiar; it is similar to the computational form for the sum of squared deviations that appears in a sample variance,

$$\sum (x - \bar{x})^2 = \sum x^2 - (\sum x)^2/n$$

The numerator of the slope can be shown to be a sum of products

$$\sum (x - \bar{x})(y - \bar{y}) = \sum xy - (\sum x)(\sum y)/n$$

Because expressions of this type are used so frequently in regression, it is convenient to use some brief symbols to represent them. We use

$$S_{xx} = \sum (x - \bar{x})^2 = \sum x^2 - (\sum x)^2/n$$

and

$$S_{xy} = \sum (x - \bar{x})(y - \bar{y}) = \sum xy - (\sum x)(\sum y)/n$$

for the sum of the squared x deviations and for the sum of the products of deviations. Then the estimated slope is

$$b = \frac{S_{xy}}{S_{xx}}$$

The least-squares line has the property of containing the point $(\bar{x},\bar{y})$, in which $\bar{x}$ is the sample average of the x values and $\bar{y}$ is the sample average of the y values. This point may or may not be one of the sample points; in this example it happens to be a data point (Figure 9.7). Since one of the points on the line is known, $(\bar{x},\bar{y})$, the line can be determined once we know its slope. The slope is given by the formula

$$b = \frac{S_{xy}}{S_{xx}} = \frac{\sum xy - (\sum x)(\sum y)/n}{\sum x^2 - (\sum x)^2/n}$$

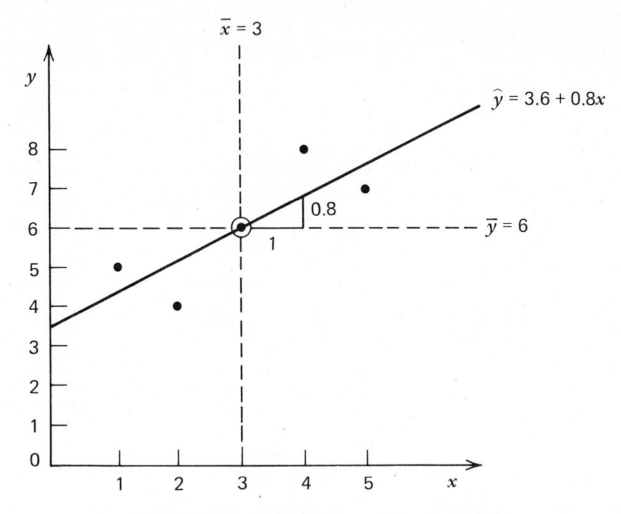

FIGURE 9.7. The least-squares trend line.

so it can be computed as follows:

x	y	x^2	xy
1	5	1	5
2	4	4	8
3	6	9	18
4	8	16	32
5	7	25	35
15	30	55	98

$$b = \frac{98 - (15)(30)/5}{55 - (15)^2/5} = \frac{8}{10} = 0.8$$

The slope indicates that as x increases one unit, y increases 0.8 units. An additional hour of instruction increases mean productivity by 0.8 units per hour. Using the slope and starting at $(\bar{x},\bar{y}) = (3,6)$, we move one unit to the right and 0.8 units up to locate a second point on the line (if the slope had been negative we would move down). Since two points determine a unique straight line, the least-squares trend line can now be drawn.

The y intercept can be found from the formula

$$\begin{aligned} a &= \bar{y} - b\bar{x} \\ &= 6 - 0.8(3) \\ &= 3.6 \end{aligned}$$

Thus the equation of the line is

$$\hat{y} = 3.6 + 0.8x$$

This is the sample regression line, and assuming that it is the proper model for the investigation, it is used to predict y for a given x, that is, it can predict the number of units per hour that would be produced if an employee had a certain number of hours of training. Only values between 1 and 5 may be specified for the independent variable x, since data were collected only for that range. Extrapolation outside the range of the x variable is not reliable since the relationship may not be linear in other regions.

Remember that a sample regression line may be used for prediction only if the model is appropriate. It is always possible to compute the least-squares line; its usefulness for prediction is a different question, which will be dealt with in the next section.

The slope of the least-squares line gives us some information about the nature of the relationship. If b is close to zero, it may be approximating a true slope of $\beta = 0$. A slope of $\beta = 0$ indicates that there is no relationship between x and y, or that y has a constant value, or it could indicate a nonlinear relationship (however, not all nonlinear relationships have $\beta = 0$). If x and y are linearly related and increase together, then b approximates $\beta > 0$. If y decreases as x increases, then b approximates $\beta < 0$ (Figure 9.8).

Note that the slope of the least-squares line is not a pure number, but it is expressed in certain units of measurement. For example, if the variables are x, height in inches, and y, weight in pounds, then b is expressed in

$$\frac{\text{(inches)(pounds)}}{\text{(inches)}^2} = \frac{\text{pounds}}{\text{inch}}$$

that is, in pounds per inch. If the same subjects were measured in centimeters and kilograms, b would have a different value because it would be in different units of measurement. Because of this, the magnitude of the slope cannot be used as a measure of the strength of the linear relationship. A measurement used to express the degree of association between x and y is the *correlation coefficient*. This is discussed in Section 9.4.

Further, we should note that the equation

$$\hat{y} = a + bx$$

is the sample regression line for the *regression of y on x*. The *regression of x on y* is usually a different line. Thus if x is hours of sleep per night and y is pounds overweight, we might regress pounds overweight on hours of sleep, that is, we would want to predict pounds overweight from hours of sleep (if in fact there was a linear relationship). On the other hand, we might be interested in the regression of hours of sleep on pounds overweight, that is, we would want to predict hours of sleep from pounds overweight. In most studies, the two lines would be different.

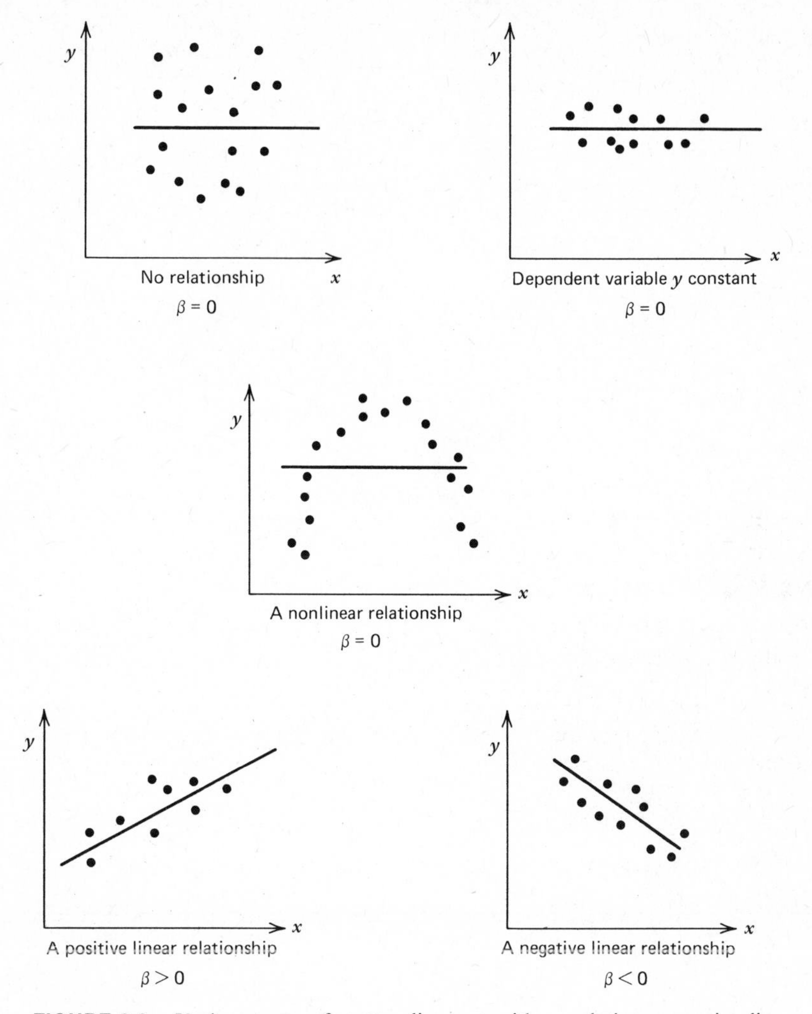

FIGURE 9.8. Various types of scatter diagrams with population regression lines.

■ ■ ■

Procedure. The Least-Squares Trend Line

Given n pairs of observations x,y, the least-squares trend line or sample regression line for the regression of y on x is

$$\hat{y} = a + bx$$

To find this line, compute

$$\sum x, \quad \sum x^2, \quad \sum y, \quad \text{and} \quad \sum xy$$

and then compute

$$\bar{x} = \sum x/n$$

$$\bar{y} = \sum y/n$$

$$S_{xx} = \sum x^2 - (\sum x)^2/n$$

$$S_{xy} = \sum xy - (\sum x)(\sum y)/n$$

The slope is

$$b = S_{xy}/S_{xx}$$

and the y intercept is

$$a = \bar{y} - b\bar{x}$$

EXERCISES

9.1.1. Which of the following completes the statement correctly? In the equation $\hat{y} = a + bx$, the value of a:

a. Can never be negative.

b. Determines the slope of the trend line.

c. Determines the point at which the trend line intersects the y axis.

d. Determines the point at which the trend line intersects the x axis.

9.1.2. Draw a scatter diagram and find the least-squares trend line for the following sample data.

Number of Hours of Study x:	4	5	6	7	8	9	10	11	12
Grade on Exam y:	55	60	50	70	70	70	80	90	85

9.1.3. If x is measured in pounds and y is measured in days, what are the units of measurement for the slope of the least-squares trend line?

9.1.4. In each case below, use the information given to obtain the numerical value of the slope of the least-squares trend line.

a. $\hat{y} = 5$ if $x = 10$, and $\hat{y} = 10$ if $x = 20$.

b. $\sum(x - \bar{x})(y - \bar{y}) = 30$, $\sum(y - \bar{y})^2 = 10$, and $\sum(x - \bar{x})^2 = 5$.

c. $\hat{y} = -3 + 15x$.

d. $\bar{y} = 10$, $\bar{x} = 13$, and $\hat{y} = 15$ if $x = 15$.

9.1.5. A botanist studying *Arabadopsis thaliana* notes a relationship between the number of branches on the plant and the number of seed pods it produces. A preliminary analysis yields the following data:

Branches x:	14	15	16	17	18
Seed Pods y:	50	60	70	100	120

a. Find $\sum(x - \bar{x})(y - \bar{y})$.
b. Compute the slope of the trend line.
c. Give the equation of the trend line.
d. What is the predicted number of seed pods on a plant with 16 branches?

9.1.6. Obesity in mice is inherited. For every gram above mean mature weight that a female mouse is in her generation, the mean of her daughters' mature weights is 2/5 gram above the mean weight in their generation.

a. What is the slope of the regression line?
b. Predict the mature weight of a daughter if her mother's weight is 28 grams, the mean for the mother's generation is 23 grams, and the mean for the daughter's generation is 20 grams.
c. Predict the mature weight of a daughter if her mother's weight is 23 grams, the mean for the mother's generation is 20 grams, and the mean for the daughter's generation is 22 grams.

9.1.7. A study of nursing activities is conducted in a 100-bed hospital in Kansas. The nursing staff remains constant throughout the study, but the patient load varies, so it is possible to observe how nurses allocate their duty time with different patient loads. One of the nursing activities observed and measured is patient care, and another is the time spent on records and reports. A separate study is made for each hospital ward, and the data below represent the minutes per staff duty hour spent on these activities by the nurses in the surgery ward under varying patient loads.

Patient Load:	2	3	4	6	7	8
Patient care	44.7	53.0	71.7	111.3	129.4	159.9
Records and reports:	15.8	16.0	13.3	10.4	7.2	9.3

a. Examine the relationship between patient load and time spent in patient care.
   i. What sort of linear relationship seems logical, positive or negative?
   ii. Do the data tend to support the experimental hypothesis?
   iii. Compute the slope of the least-squares trend line that shows

how an increase in patient load affects staff time allocated to patient care.

iv. What are the units of measurement for the slope of the trend line?

v. Find the equation that would allow surgery-ward nurses to predict the amount of time they have to allocate per staff duty hour for a given number of patients in their ward.

vi. Use the equation to estimate the amount of time required for patient care if there were only one patient in the ward. (Since one patient is outside the range of the data collected, this may be a poor estimate.) Use it to estimate the time required for five patients.

b. Examine the relationship between patient load and time spent on records and reports.

i. Does the linear relationship appear to be positive or negative?

ii. Does such a relationship seem intuitively logical prior to the survey, or is the relationship one that can be rationalized after the data are collected?

iii. Compute the least-squares trend line that shows how an increase in patient load affects the staff time allocated to records and reports.

iv. Suppose that a minimum of five minutes per staff duty-hour is required for necessary records and reports. Assume that the trend can be extrapolated, and estimate the point at which patient load becomes so heavy that the surgical nursing staff no longer has adequate time for record keeping.

9.1.8. When a straight line is fitted to data that follow a binomial distribution, a special procedure known as *probit analysis* is employed. This procedure takes into account such conditions as the relationship between the mean and the variance of the binomial distribution and the fact that the trend is rarely linear over the full range of π. However, the first step in probit analysis is to fit a "provisional" line to the data, and this can be done by employing the least-squares procedure developed in this section. Suppose an advertising firm wants to determine the relationship between the number of times a commercial is shown on national television and the percentage of viewers who have seen the commercial.

Times Commercial Shown x:	10	15	20	25	30
Percentage of Viewers y:	13	32	35	53	67

a. Use least-squares procedures to find the slope of the trend line.

b. Give the equation of the "provisional" line.

c. Use the equation to estimate how many times a television commercial must be shown before 50% of the viewers have seen it. (This is called the "50% effective dose" or "ED_{50}" in probit analysis.)

9.1.9. Francis Galton extended least squares techniques by employing them in a study of the relationship between mature heights of fathers and their sons. He collected hundreds of observations, plotted them on graph paper, and noted a straight-line relationship among average heights. Some of his data might be as follows:

Fathers' Height:	65	66	67	68	69	70	71
Average Height of Sons:	66.9	67.8	68.0	67.9	69.6	69.2	70.1

a. What is the average height of the fathers' generation?
b. What is the average height of the sons' generation?
c. If a group of fathers are each 1 in. above average height for their generation, what is the expected average deviation of their sons from the average height of their respective generation?

9.1.10. A study is made to determine the rate of disappearance from the environment of radioactive chemicals after a nuclear accident. Strontium 85 is released in an alfalfa field in a simulated accident. Twenty goats are allowed to graze the field, and at 30-day intervals the level of strontium 85 is measured in dried samples of alfalfa as well as in the goats' milk. The alfalfa data are given below:

Days After Release x:	30	60	90	120	150
ppm in Dried Alfalfa y:	1.85	1.43	1.21	1.19	1.37

a. Compute the least-squares trend line.
b. What are the units of measure for the slope? For the y intercept?
c. The measured level of strontium 85 in alfalfa on day 150 seems somewhat contrary to the trend shown in the other data. Compute the predicted level for $x = 150$. Compute the deviation of the observed value from this point on the trend line.

9.1.11. Fit a straight line to the age and blood pressure data given in this section.

9.2. MODEL TESTING

The least-squares line can always be computed for any set of two or more points with different x values. It may not be appropriate however, to predict from this line. For prediction, two conditions are necessary:

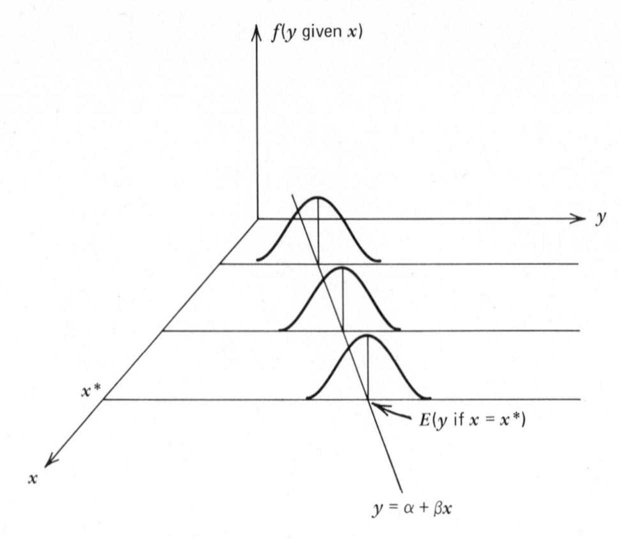

FIGURE 9.9. The regression model.

1. The straight line model fits the data.
2. The straight line being estimated is not horizontal ($\beta \neq 0$), that is, the regression line is a better predictor of y than $\bar{y}$.

In this section we discuss each of these conditions in turn.

First we need to be more precise as we speak of a regression line being a model for a certain research situation. Two variables x,y (Figure 9.9) meet the conditions for the regression of y on x if:

1. The x values are fixed by the experimenter and are measured with negligible error.†
2. For each x value there is a normal distribution of y values. (This assumption is necessary for inference.)
3. The distribution of y for each x has the same variance σ^2.
4. The expected values of y for each x lie on a straight line.

Another way to express these conditions is to say that the variables satisfy the model

$$y = \alpha + \beta x + \epsilon$$

in which the ϵ's are normally distributed with a mean of 0 and a variance of σ^2, and the ϵ's are independent of the x's and independent of each other.

One way to test for violations of these assumptions is by an examination

† Regression analysis is also possible in cases where x is a random variable (see Section 9.4).

of the residuals $y - \hat{y} = e$ that result from fitting the least-squares line to the sample data.

In the small example about employee training used for illustration purposes in Section 9.1, the residuals could be computed as follows.

x	y	$\hat{y}$	$y - \hat{y} = e$
1	5	$3.6 + 0.8(1) = 4.4$	0.6
2	4	$3.6 + 0.8(2) = 5.2$	−1.2
3	6	$3.6 + 0.8(3) = 6.0$	0.0
4	8	$3.6 + 0.8(4) = 6.8$	1.2
5	7	$3.6 + 0.8(5) = 7.6$	−0.6

Since the e's estimate the ϵ's in the model, to check for normality an overall plot of the residuals can be drawn as a dot diagram (Figure 9.10). In this unrealistically small example it is difficult to check for departures from normality because of the small number of points. Some patterns that appear with larger samples are illustrated in Figure 9.11.

Linearity can be checked by plotting the residuals e against the predicted values $\hat{y}$ (Figure 9.12). A linear relationship is reflected in a random scatter about a horizontal line at $e = 0$. If the relationship is nonlinear, it usually results in a systematic plot that has some pattern. A systematic pattern could also indicate that another independent variable is affecting y.

Equality of variances can be checked by plotting the residuals e against the predicted values $\hat{y}$ or the independent variable x (Figure 9.13). Equal variances result in a horizontal band of points, whereas variances that depend on the magnitude of x will result in a fan-shaped distribution.

The regression model assumes independence of the ϵ's. This means that the random error in one observation does not affect the random error in another observation. This assumption is sometimes violated. If the observations have a natural sequence in time or space, the lack of independence is called autocorrelation.

Autocorrelation may occur for several reasons: The dependent variable may follow economic trends; an instrument may be drifting out of calibration; batch processes in a reactor system may leave some of the product to be carried over to the next batch; observations may be from adjacent experimental plots that have similar conditions. These are only some examples. Diagnosis is difficult, but this type of dependence can sometimes be detected by plotting the residuals against the time order or the spatial order of the observations (Figure 9.14).

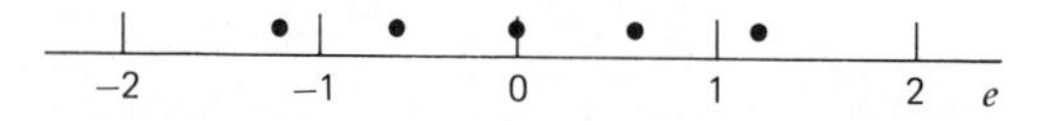

FIGURE 9.10. An overall plot of residuals.

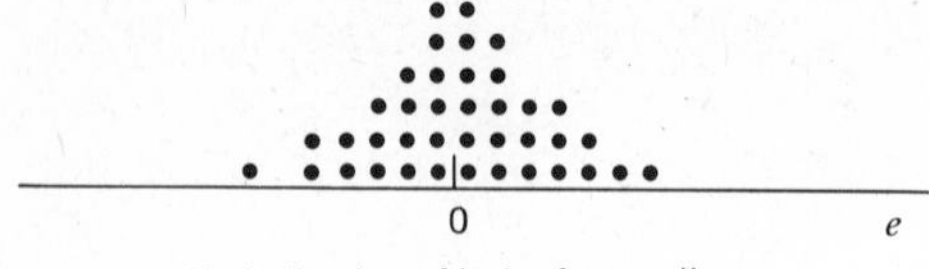

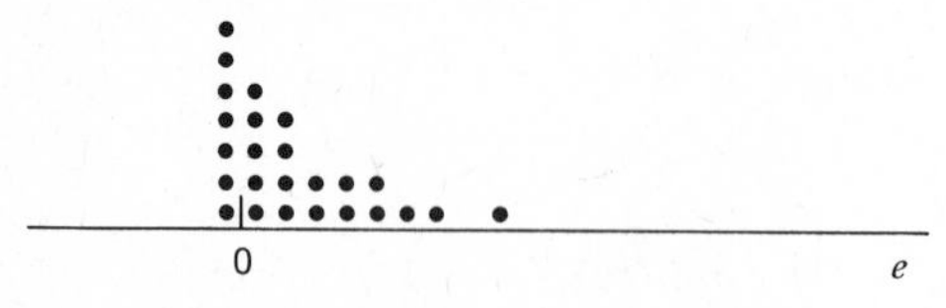

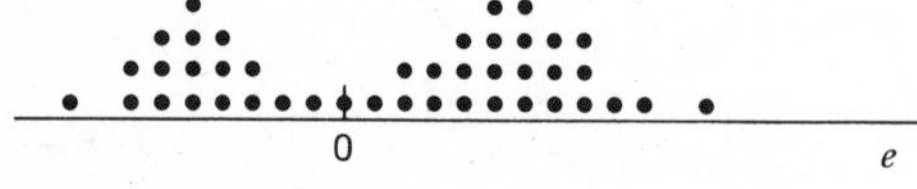

FIGURE 9.11. Checking overall plots of residuals for violations of normality.

The visual inspection of the original scatter diagram of the data and the various types of residual plots is an important first step in any regression analysis and should not be omitted. Statistical programs on computers make it possible to inspect these diagrams with little labor. If the diagrams reveal any departures from the assumptions required for regression, a different model may be necessary, or perhaps a transformation can be used on the data before the regression analysis (Sections 14.5 and 14.6). If the visual inspection does not turn up any departures from assumptions, we have not proved that the model is correct but at least there is no overwhelming evidence that it is wrong.

Besides these visual checks of the assumptions, there is a statistical test that can be performed to see if there is a significant lack of fit with a straight line. Repeated observations are necessary at each x value to carry out such a test (see Draper and Smith 1981, pp. 33–40).

If we decide that a straight line seems to be a reasonable model, then we need to determine that the line is not horizontal. A horizontal line indicates that x does not make a significant contribution to the prediction of y, that is, there is no linear relationship. To test whether the line is horizontal, we test

$$H_0: \ \beta = 0$$

in which β is the slope of the population regression line. Rejection of this hypothesis is evidence that the line explains a significant portion of the variability in y. Acceptance of this hypothesis means that there is no advantage to considering the values of x as we attempt to predict y. We could do just as well by using the model $\hat{y} = \bar{y}$.

The test statistic is a t statistic in which b is the estimator of the parameter β. To estimate the standard error of the estimator b for the denominator of the t test, we first must consider the variability of the y values about the sample regression line. We use the residuals and compute the sum of the squared residuals, and then we divide this sum by the degrees of freedom that are $n - 2$ for simple linear regression (thus a minimum of three points is required for this test).

For example, in the employee training example, the variability of the

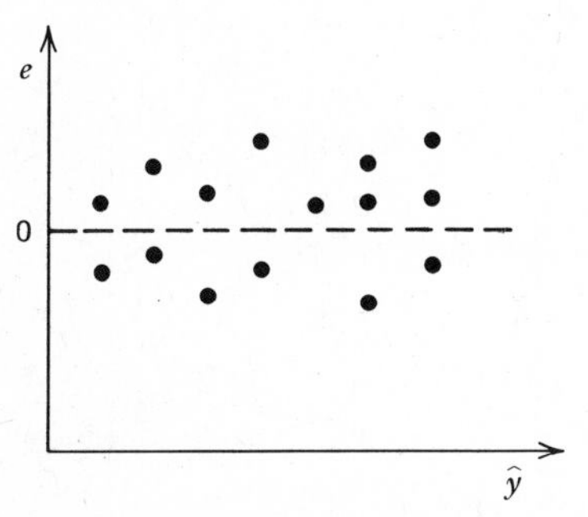

A linear model is appropriate

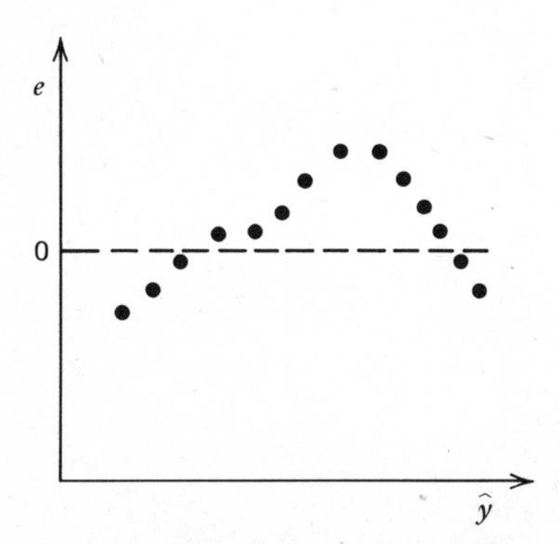

A nonlinear relationship (or a second independent variable is involved)

FIGURE 9.12. Residuals plotted against predicted values to check for a linear relationship.

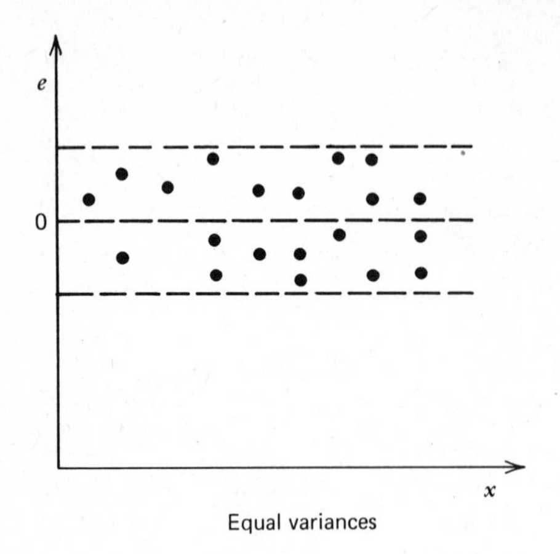

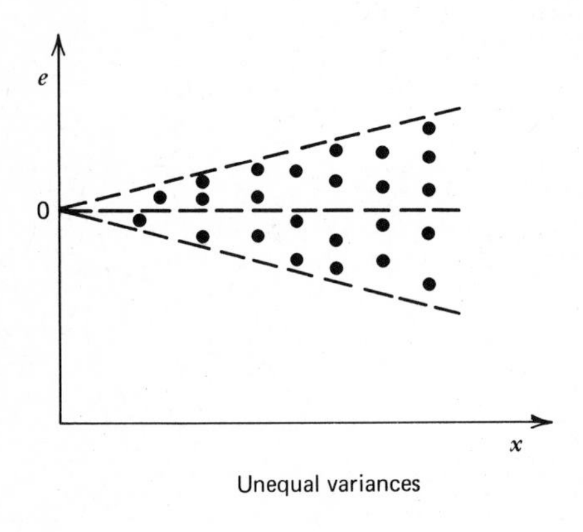

FIGURE 9.13. Residuals plotted against the independent variable to check for equality of variances.

observations about the least-squares line is computed as follows:

$y - \hat{y}$	$(y - \hat{y})^2$
0.6	0.36
−1.2	1.44
0.0	0.00
1.2	1.44
−0.6	0.36
	3.60

and

$$s_{y \cdot x}^2 = \frac{\sum(y - \hat{y})^2}{n - 2} = \frac{3.60}{5 - 2} = 1.2$$

in which n is the number of pairs of data. Variability about the trend line is the variability in y when we have removed the effect of the x variable. In the employee training example, *before* we have removed the effect of the x variable the variability in y is

$$s_y^2 = \frac{\sum(y - \bar{y})^2}{n - 1} = \frac{10}{4} = 2.5$$

This represents the variability of the data points about $\bar{y}$. In contrast, $s_{y \cdot x}^2$ is the variability about the trend line and is the variability in y independent of x. Note that 2.5 is reduced to 1.2 when the effect of x is removed (Figure 9.15).

In practice it is usually easier to use the short computational formula

$$\sum(y - \hat{y})^2 = \sum(y - \bar{y})^2 - [\sum(x - \bar{x})(y - \bar{y})]^2/\sum(x - \bar{x})^2$$
$$= S_{yy} - S_{xy}^2/S_{xx}$$
$$= S_{yy} - bS_{xy}$$

in which

$$S_{yy} = \sum(y - \bar{y})^2 = \sum y^2 - (\sum y)^2/n$$

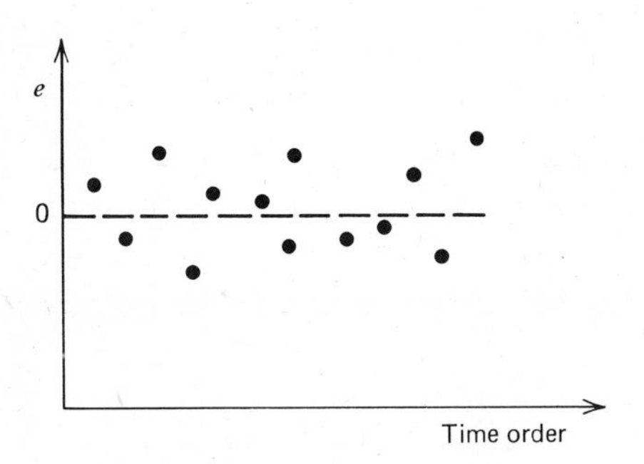

Independence of ϵ's with each other

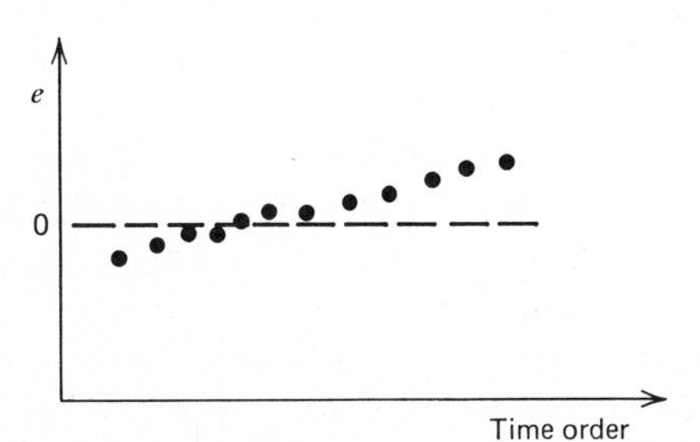

Autocorrelation

FIGURE 9.14. Residuals plotted against the order in which they were observed.

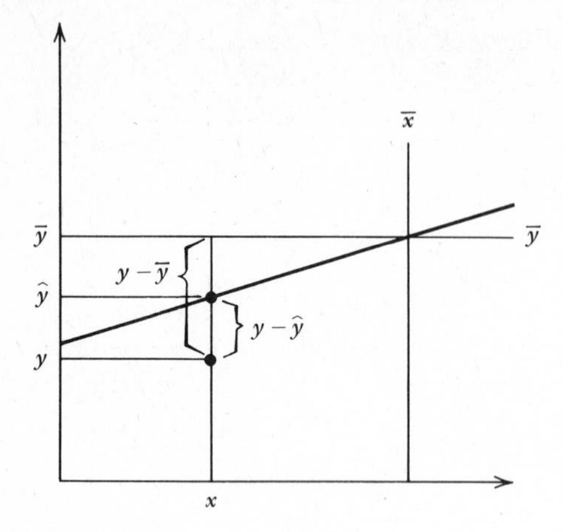

FIGURE 9.15. Deviation of an observed y value from the average y value and from a predicted y value.

Using $s_{y \cdot x}{}^2$ the standard error of b can be shown to be

$$s_{y \cdot x}/\sqrt{S_{xx}}$$

and the t statistic for a test of H_0: $\beta = 0$ is

$$t = \frac{b - \beta_0}{s_{y \cdot x}/\sqrt{S_{xx}}}$$

with $n - 2$ degrees of freedom.

In the training example, to test H_0: $\beta = 0$ against H_a: $\beta > 0$ at $\alpha = 0.05$, we would reject the null hypothesis if $t \geq t_{0.05,3} = 2.353$. A one-tailed test is used because additional training is expected to increase productivity if it is of any effect at all. Then

$$t = \frac{0.8 - 0}{\sqrt{1.2}/\sqrt{10}} = 2.31$$

and the null hypothesis is not rejected. Thus the line seems to be horizontal and the equation of the trend line should not be used for prediction. Note that the t statistic of 2.31 is very close to the critical value, so it is possible that a larger sample size might provide evidence that the line does contribute significant information about y. We repeat again that the small sample size here is unrealistic and is used only to keep the computations to a minimum.

If it is possible to reject $\beta = 0$, then prediction from the least-squares line is appropriate. Prediction may be done only for values of x within the range of the collected data. Extrapolation outside of that range is not reliable.

Values other than zero may be used in the null hypothesis when testing the slope parameter if this is reasonable for the experiment. The test procedure is analogous.

■ ■ ■

Procedure. Testing the Slope Parameter

Assumption: $y = \alpha + \beta x + \epsilon$ with the ϵ's independently normally distributed with a mean of 0 and a variance σ^2

Test of Hypothesis

H_0: $\beta = \beta_0$
H_a: $\beta \neq \beta_0$ or $\beta > \beta_0$ or $\beta < \beta_0$

Significance level: α
Test statistic:

$$t = \frac{b - \beta_0}{s_{y \cdot x}/\sqrt{S_{xx}}}$$

with

$$b = S_{xy}/S_{xx} \qquad \text{and} \qquad s_{y \cdot x}{}^2 = (S_{yy} - bS_{xy})/(n - 2)$$

Region of rejection: $|t| \geq t_{\alpha/2,n-2}$ or $t > t_{\alpha,n-2}$ or $t < -t_{\alpha,n-2}$, respectively

EXERCISES

9.2.1. For the data in Exercise 9.1.2:
   a. Carry out a residual analysis.
   b. Show that $s_{y \cdot x}{}^2 = 33.57$.
   c. To test the significance of the least-squares line:
      i. Give the most logical null and alternative hypotheses.
      ii. Give the critical value.
      iii. Compute the test statistic and state the conclusion.

9.2.2. Explain the difference between y and $\hat{y}$.

9.2.3. If y is the number of fish caught in x hours of fishing, give the units of measurement for:
   a. The slope of the trend line.
   b. A predicted y value.
   c. The point in which the trend line meets the y axis.

9.2.4. Some species of tropical fish bear their young alive rather than lay eggs. An aquarium keeper wants to determine whether the number of young increases with each parity (time when young are produced).

The following data are available for study:

Order of Parity:	1	2	3	4	5
Number of Young:	7	11	9	13	15

a. Find the slope of the sample regression line.
b. Compute the sample variance about the trend line.
c. What are the most logical null and alternative hypotheses about the slope of the regression line?
d. Why is a two-sided alternative inappropriate?
e. Complete the test. What conclusion should be drawn?

9.2.5. Review Exercise 9.1.7 of this chapter, in which there is a discussion of the effect of patient load on nursing activities in a hospital.

a. Conduct a test of hypothesis to see if patient load can be used to predict the time spent on patient care.
   i. Give the null hypothesis in symbols and in a complete sentence.
   ii. Why should the alternative hypothesis be one-sided?
   iii. Give the critical value of the test statistic for $\alpha = 0.05$.
   iv. Perform the test of significance.
b. Conduct a test of hypothesis about patient load as a predictor of the time available for records and reports.
   i. Give the null hypothesis.
   ii. Why should the alternative hypothesis be two-sided?
   iii. Perform the test of significance at $\alpha = 0.01$.

9.2.6. When experimentation with lysergic acid diethylamide (LSD) first began, the hallucinogenic effect was noted as so similar to the symptoms of schizophrenia that medical scientists thought they had discovered a chemical cause of the mental disorder. Because an increase in the level of copper in the blood is frequently (but not always) associated with schizophrenia, a study was made to see whether the level of blood copper increased with the administration of increasing dosages of LSD.

a. What null hypothesis would be used in an analysis of this experiment?
b. What would be the alternative hypothesis?
c. Dosages were calibrated according to the percentage of those receiving the dosage who hallucinate. The level of blood copper measured at each dosage. The data obtained were as follows:

Effective Dosage (%):	0	25	50	75	100
Copper (mg/liter):	0.87	0.98	0.70	0.90	1.05

i. Compute the slope of the least-squares trend line.
ii. Test the $\beta = 0$ at $\alpha = 0.05$.
iii. Draw conclusions, answering the following questions: Do increasing dosages of LSD cause significant increases in blood copper level? Because increased blood copper is a common condition in schizophrenia, is there significant evidence that LSD may be a chemical cause of schizophrenia?

9.2.7. Review Exercise 9.1.10 in which a nuclear accident is simulated by releasing strontium 85 in an alfalfa field.

a. Compute $\sum(y - \hat{y})^2$ by using the short computational formula.
b. Compute $\sum(y - \hat{y})^2$ by finding the expected value on the trend line for each value of x and subtracting it from the observed value.
c. In performing a test of significance of the least-squares trend line:
   i. What is the null hypothesis?
   ii. Why is the alternative H_a: $\beta < 0$?
   iii. What is the critical value of the test statistic for $\alpha = 0.05$?
   iv. What is the decision about the null hypothesis? What should be concluded?

9.2.8. In Exercise 9.1.9 involving the relationship between fathers' and sons' heights:

a. Compute the expected height of sons y of fathers of each height x given in the experiment.
b. Compare observed height y with expected height $\hat{y}$ and compute:
   i. The sum of the deviations from the trend line, $\sum(y - \hat{y})$.
   ii. The sum of the squared deviations from the trend line, $\sum(y - \hat{y})^2$.
c. Compare observed height y and expected height $\hat{y}$ in terms of how they deviate from the average; compute:
   i. The sums of the deviations from the average, $\sum(y - \bar{y})$ and $\sum(\hat{y} - \bar{y})$.
   ii. The sums of the squared deviations from the average, $\sum(y - \bar{y})^2$ and $\sum(\hat{y} - \bar{y})^2$.
d. Use the above computations to empirically verify the following mathematical identities:
   i. The sum of squares from the average equals the sum of squares due to the linear trend plus the sum of squares from the trend line: $\sum(y - \bar{y})^2 = \sum(\hat{y} - \bar{y}) + \sum(y - \hat{y})^2$.
   ii. The sum of squares due to the linear trend is $\sum(\hat{y} - \bar{y})^2 = [\sum(y - \bar{y})(x - \bar{x})]^2/\sum(x - \bar{x})^2 = S_{xy}{}^2/S_{xx} = bS_{xy}$.
   iii. The sum of squares from the trend line is $\sum(y - \hat{y})^2 = \sum(y - \bar{y})^2 - [\sum(y - \bar{y})(x - \bar{x})]^2/\sum(x - \bar{x})^2 = S_{yy} - bS_{xy}$.

9.3. INFERENCES RELATED TO REGRESSION

The term "regression" originated with the work of Francis Galton. The studies of inheritance inspired by Darwin's work led Galton to believe that everything could be studied quantitatively. One of his studies involved the linear trend between the heights of fathers and their sons. The slope of the trend line in this particular study was positive but less than one, so Galton called the relationship a "regression toward the mean." The term "regression" was then applied to any linear trend. It is unfortunate, however, since the slope of a least-squares trend line need not be less than one.

Several types of inference are possible in relation to the regression line. Confidence intervals and tests of hypotheses are possible for parameters α and β and for $\mu_y^* = E(y \text{ if } x = x^*)$, the expected value of y for a specific value x^* of x. These procedures are summarized in Table 9.1.

The following example will illustrate the use of some of these procedures.

Example 9.1. Inferences Related to Regression

If the efficiency expert in Section 9.1 had obtained the following data instead of that previously given,

x:	1	1	2	4	4	5	6	6	7
y:	3	6	4	3	6	5	9	10	8

he could organize the regression analysis as follows:

$$n = 9 \qquad \sum x = 36 \qquad \sum x^2 = 184$$
$$\sum xy = 248 \qquad \sum y = 54 \qquad \sum y^2 = 376$$

$$\bar{x} = \sum x/n = 36/9 = 4.0$$
$$\bar{y} = \sum y/n = 54/9 = 6.0$$

$$S_{xx} = \sum(x - \bar{x})^2 = \sum x^2 - \left(\sum x\right)^2/n = 184 - (36)^2/9 = 40$$
$$S_{yy} = \sum(y - \bar{y})^2 = \sum y^2 - \left(\sum y\right)^2/n = 376 - (54)^2/9 = 52$$

$$S_{xy} = \sum(x - \bar{x})(y - \bar{y}) = \sum xy - \left(\sum x\right)\left(\sum y\right)/n = 248 - (36)(54)/9 = 32$$

The estimated slope is

$$b = \frac{\sum(x - \bar{x})(y - \bar{y})}{\sum(x - \bar{x})^2} = \frac{32}{40} = 0.80$$

The y intercept is

$$a = \bar{y} - b\bar{x} = 6 - 0.8(4.0) = 2.8$$

The least-squares trend line is

$$\hat{y} = 2.8 + 0.8x$$

TABLE 9.1 Inferences Related to Regression

Parameter	Test Statistic $\nu = n - 2$	$1 - \alpha$ Central Confidence Interval
α	$t = \dfrac{a - \alpha_0}{s_{y \cdot x}\sqrt{\dfrac{1}{n} + \dfrac{\bar{x}^2}{S_{xx}}}}$	$a \pm t_{\alpha/2, n-2}\, s_{y \cdot x}\sqrt{\dfrac{1}{n} + \dfrac{\bar{x}^2}{S_{xx}}}$
β	$t = \dfrac{b - \beta_0}{\dfrac{s_{y \cdot x}}{\sqrt{S_{xx}}}}$	$b \pm \dfrac{t_{\alpha/2, n-2}\, s_{y \cdot x}}{\sqrt{S_{xx}}}$
$\mu_y^* =$ $E(y \text{ if } x = x^*)$	$t = \dfrac{\hat{y} - (\mu_y^*)_0}{s_{y \cdot x}\sqrt{\dfrac{1}{n} + \dfrac{(x^* - \bar{x})^2}{S_{xx}}}}$	$\hat{y} \pm t_{\alpha/2, n-2} s_{y \cdot x}\sqrt{\dfrac{1}{n} + \dfrac{(x^* - \bar{x})^2}{S_{xx}}}$

Assuming that a residual analysis uncovers no deviations from the assumptions, it is valid to predict from this line because testing

$$H_0: \quad \beta = 0 \qquad \text{against} \qquad H_a: \quad \beta > 0$$

at $\alpha = 0.05$, we find

$$s_{y \cdot x}^2 = \frac{\sum(y - \hat{y})^2}{n - 2} = \frac{S_{yy} - S_{xy}^2/S_{xx}}{n - 2}$$

$$= \frac{52 - (32)^2/40}{7} = 3.78$$

$$s_{y \cdot x} = \sqrt{3.78} = 1.95$$

and

$$t = \frac{b - 0}{s_{y \cdot x}/\sqrt{S_{xx}}} = \frac{0.8}{1.95/\sqrt{40}} = 2.595$$

with

$$t_{0.05,7} = 1.895.$$

The 95% central confidence interval on β is

$$\text{CI}_{0.95}: \quad b + t_{0.025,7} s_{y \cdot x}/\sqrt{S_{xx}}$$

$$0.8 \pm 2.365(1.95)/\sqrt{40}$$

$$0.8 \pm 0.73$$

If the researcher wants to find the average productivity under 3.5 hours of instruction, he finds

$$\hat{y} = 2.8 + 0.8x = 2.8 + 0.8(3.5) = 5.6$$

This is the estimate of the average productivity for 3.5 hours of instruction, $E(y \text{ if } x = 3.5)$. The 95% central confidence interval on this parameter is

$$\text{CI}_{0.95}: \quad \hat{y} \pm t_{0.025,7} s_{y \cdot x} \sqrt{\frac{1}{n} + \frac{(x^* - \bar{x})^2}{S_{xx}}}$$

$$5.6 \pm 2.365(1.95)\sqrt{\frac{1}{9} + \frac{(3.5 - 4)^2}{40}}$$

$$5.6 \pm 1.58$$

If an experimenter is interested in predicting the next y observation at a given level x^* of x, the point estimate is the same as for the expected y value at that level:

$$\hat{y} = a + bx^*$$

However, the formula for the prediction interval on the next observation is slightly different than the formula for the confidence interval on the expected value:

$$\text{PI}_{1-\alpha}: \quad \hat{y} \pm t_{\alpha/2,n-2} s_{y \cdot x} \sqrt{1 + \frac{1}{n} + \frac{(x^* - \bar{x})^2}{S_{xx}}}$$

These prediction intervals are wider than the corresponding confidence intervals, and this seems logical because we are trying to predict a single value rather than the population mean for all values of y with a common x^*. Both types of intervals are narrowest at $x^* = \bar{x}$ (Figure 9.16).

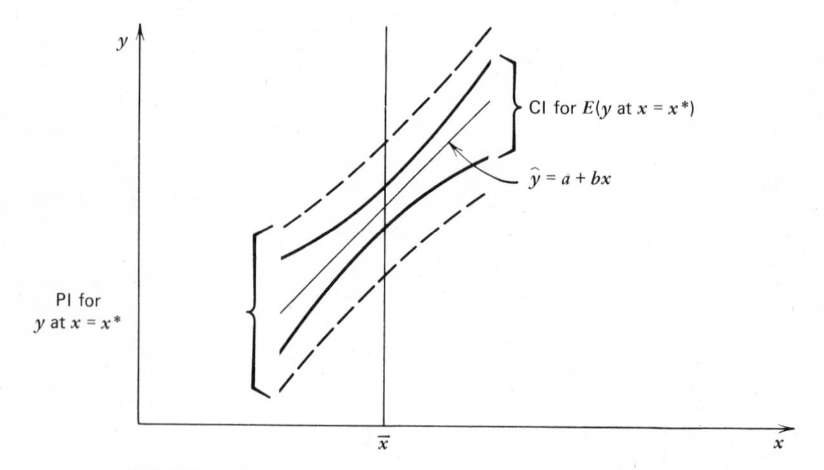

FIGURE 9.16. Prediction intervals and confidence intervals.

EXERCISES

9.3.1. The linear relationship between weight y (in grams) and age x (in days) has been studied in a strain of inbred guinea pigs. The following values have been computed. The guinea pigs ranged from 8 to 14 days of age.

$$n = 16, \quad b = 5.0, \quad \bar{x} = 11, \quad \bar{y} = 87$$
$$\sum(x - \bar{x})(y - \bar{y}) = 200, \quad \sum(y - \bar{y})^2 = 1{,}126$$

a. Find $\sum(x - \bar{x})^2$.

b. Compute the variability about the least-squares trend line.

c. Place a 95% confidence interval on the mean weight of eight-day-old guinea pigs.

9.3.2. A random sample of 27 college men yields the following data in a study of the relationship between arm length x (in inches) and leg length y (in inches).

$$\sum x = 675 \quad \sum y = 810 \quad b = 1.2$$
$$\sum(x - \bar{x})^2 = 25 \quad \sum(y - \bar{y})^2 = 136$$

a. Compute the variance around the sample regression line.

b. Make a test of significance of this line against the most logical alternative.

c. Find a 95% confidence interval for β.

d. Predict the leg length of a man with arms 25 in. long.

e. Find the 95% prediction interval for this length.

9.3.3. In an effort to find a method of predicting the dental work required by army recruits, an army dentist studies the dental records of a random sample of 10 men completing their service. He computes the relationship between the number of cavities filled during the first two years of service y with the number of cavities filled in the two years before service x.

a. State the null hypothesis that should be used to test for the usefulness of the regression line.

b. Give the alternative hypothesis you would suggest to the dentist and the reason for that alternative.

c. Give the critical value.

9.3.4. Suppose the following statistics are computed for the dental study in Exercise 9.3.3:

$$\sum x = 50, \quad \sum y = 52, \quad \sum xy = 321$$
$$\sum(x - \bar{x})^2 = 68, \quad \sum(y - \bar{y})^2 = 75.6$$

a. Find the estimate of the slope of the trend line.

b. Find the standard error of the estimate of the slope.

c. Find 95% central confidence intervals for:

  i. The slope of the trend line.

  ii. The average number of cavities a draftee will have filled during his first two years of service.

d. Find the 95% prediction interval for the number of cavities to be filled in the teeth of a new draftee who in the previous two years had three new fillings.

9.3.5. In an experiment involving 12 female mice and their first litters, a study is made of the relationship between the rate of weight gain (gain divided by original weight) of the female during pregnancy x and the birth weight y of her litter. The following statistics are computed:

$$\bar{x} = 0.10, \quad \bar{y} = 20.00, \quad \sum xy = 24.48$$
$$\sum(x - \bar{x})^2 = 0.16, \quad \sum(y - \bar{y})^2 = 15.84$$

a. Find b.

b. Find the sample variance about the trend line.

c. Test the significance of the trend line against the most logical one-sided alternative hypothesis.

d. Estimate the average birth weight of a litter for a mouse that gained 0.12 during pregnancy.

e. Place a 95% confidence interval on this estimate.

f. Find the intersection of the trend line with the y axis.

g. Place a 90% confidence interval on α.

h. Comment on the validity of Parts d through g.

9.3.6. Data from 23 boys between the ages of 14 and 16 are used to estimate the regression of weight on height. The following results are reported:

average weight = 122 lb
average height = 68 in.
95% confidence interval on β: 6.60 ± 1.56

a. What are the units of measurement for the slope of the regression line?

b. What is the expected weight of boys 2 in. taller than the average height?

c. At which point does the regression line intersect the y axis?

d. Is there a significant linear relationship between the weights and heights of boys in this age range?

9.3.7. Refer again to Exercise 9.1.7, which discusses the effect of patient load on nursing activities.

a. Place a one-sided 95% confidence interval on lowest value of the slope of the trend line that relates time spent on patient care with patient load.

b. Place a two-sided 95% confidence interval on the slope of the trend line relating time spent on records and reports with patient load.

9.3.8. For Exercise 9.2.6, which examines the relationship between LSD dosage and blood copper level:

a. Compute a 90% two-sided confidence interval on the slope.

b. Compute a 90% central confidence interval on the y-intercept.

c. Compute a 90% confidence interval for the lowest mean copper level of those receiving a 50% dosage.

d. Find the 90% prediction interval for the lowest copper level of an individual who would receive a 70% dosage.

e. Is it valid to use these intervals?

9.3.9. For Exercise 9.1.10, which involves a simulated nuclear accident:

a. Place a 95% central confidence interval on the mean ppm of all alfalfa samples that could be taken on the 150th day.

b. Place a 95% central prediction interval on the ppm of a single sample that could be taken that day.

c. How does the observed sample correspond to these intervals?

d. The data do not record the amount of strontium 85 released and immediately available to the alfalfa at the start of the experiment.

i. Estimate this from the data available.

ii. Place a 99% confidence interval on this estimate.

iii. Would you have any hesitation about using these estimates?

9.4. CORRELATION

The main use of regression is for prediction. If the slope of the regression line is zero, there is no significant linear relationship and prediction is not valid. It is difficult to judge without a test of hypothesis whether b, the estimator of β, is significantly different from zero because b depends on the units of measurement employed in the experiment. Thus $b = 0.01$ may be significantly different from zero, while $b = 100$ may not be significantly different from zero.

If we could remove the units of measurement from b, it would be easier to get a feeling for the strength of the linear relationship. We compute

$$r = b\frac{\sqrt{S_{xx}}}{\sqrt{S_{yy}}} = \frac{S_{xy}}{\sqrt{S_{xx}S_{yy}}}$$

The statistic r is called the *sample correlation coefficient.* In all cases $-1 \leq r \leq 1$. If $r = -1$, there is a perfect negative relationship and all of the data points are on the sample regression line. If $r = +1$, the relationship is a perfect positive one with all of the points on the sample regression line. As r gets closer to zero, there is less association between the variables. The amount of association can be judged to some degree by simply looking at the magnitude of r.

Another useful statistic related to the sample correlation coefficient is the *coefficient of determination,* r^2. The coefficient of determination has the following interpretation.

$$\left.\begin{array}{l}\text{The proportion of variability}\\ \text{in } y \text{ unexplained by the}\\ \text{linear relationship}\end{array}\right\} = \frac{\sum(y - \hat{y})^2}{\sum(y - \bar{y})^2}$$

$$= \frac{\sum(y - \bar{y})^2 - [\sum(x - \bar{x})(y - \bar{y})]^2/\sum(x - \bar{x})^2}{\sum(y - \bar{y})^2}$$

$$= 1 - \frac{S_{xy}{}^2}{S_{xx}S_{xy}}$$

$$= 1 - r^2$$

and so

$$r^2 = 1 - \text{the proportion of unexplained variability in } y$$

$$= \text{the proportion of variability in } y \text{ explained by the linear relationship}$$

Thus r^2 indicates the proportion of the variability in y explained by the linear relationship to x. If r^2 is large (close to one), most of the variability is accounted for by the relationship and y is a useful predictor. If r^2 is close to zero, the regression equation is not very useful. The r^2 statistic is often the most meaningful statistic that can be computed, since it gives us a measure of the usefulness of the prediction.

In the revised training example of Section 9.3, we find that

$$r = \frac{\sum(x - \bar{x})(y - \bar{y})}{\sqrt{\sum(x - \bar{x})^2 \sum(y - \bar{y})^2}} = \frac{32}{\sqrt{(40)(52)}} = 0.70$$

and $r^2 = 0.49$. Thus, 49% of the variability in the number of units produced is accounted for by the relationship between productivity and training. Management could then decide whether the cost of additional training would be offset by the increased production.

Care should be taken in the interpretation of regression and correlation. If a significant linear relationship exists, this in itself does not indicate that changes in the x variable *cause* changes in the y variable. There may or may not be causality involved. In this efficiency example, it is possible that increased instruction causes increased productivity; however, the significance of the regression line alone does not prove this. Causality must be demonstrated by an argument outside the statistical analysis. In many cases there may be no causality involved. For example, if it is found that in the group of people who sleep less than four hours a night there is a negative linear relationship between the number of hours of sleep per night and the rate of heart attacks, this does not in itself imply that a decrease in sleep causes an increase in the heart attack rate. It could be that a third factor causes both the increase in the heart attack rate and the decrease in sleep. For example, an increase in stress may decrease sleep and increase the heart attack rate.

The foregoing discussion of correlation and regression indicates that they are different, but not mutually exclusive, techniques. Roughly, regression is used for prediction, whereas correlation is used to determine the degree of association.

Besides the different functions served by regression and correlation, different assumptions are used to develop the theory behind these procedures (see Table 9.2 and Figure 9.17).

As a result of these models, the following guidelines should be used. All regression procedures (Sections 9.1, 9.2, 9.3) may be applied to both models. Also, the computation of the sample correlation coefficient and the coefficient of determination may be applied to both models. However, inference about the population correlation coefficient should only be made if the experimenter believes the variables are bivariate normal (fit the correlation

TABLE 9.2 Difference Between Regression and Correlation

Regression Model	Correlation Model
1. x is fixed at levels chosen by the experimenter. (Scientists call this an "independent variable.") At each fixed x level, subjects are chosen at random and y is measured. (Scientists call y the "dependent variable.")	1. Subjects are sampled at random and the x, y measurements are recorded.
2. x is measured without error, that is, there is no sampling variability in x. Only y contains sampling variability.	2. Both x and y contain sampling variability.
3. For each value of x there is a normal distribution of y.	3. For each value of x there is a normal distribution of y, and for each value of y there is a normal distribution of x.
4. Each distribution of y has the same variance.	4. The x distributions have the same variance. The y distributions have the same variance.
5. The expected value of the normal y distribution lie on a straight line.	5. The joint distribution of x and y is the bivariate normal distribution.

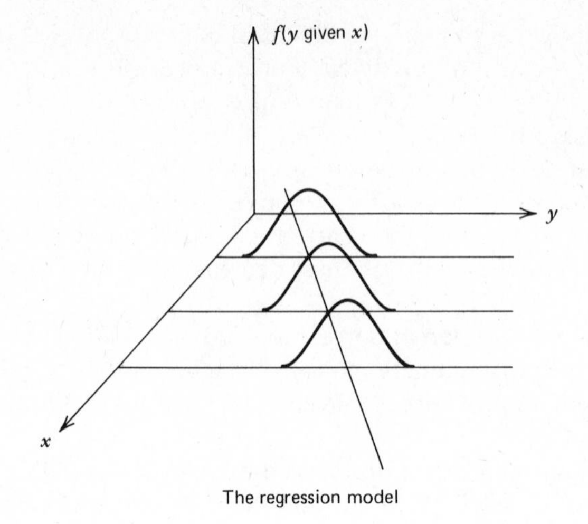

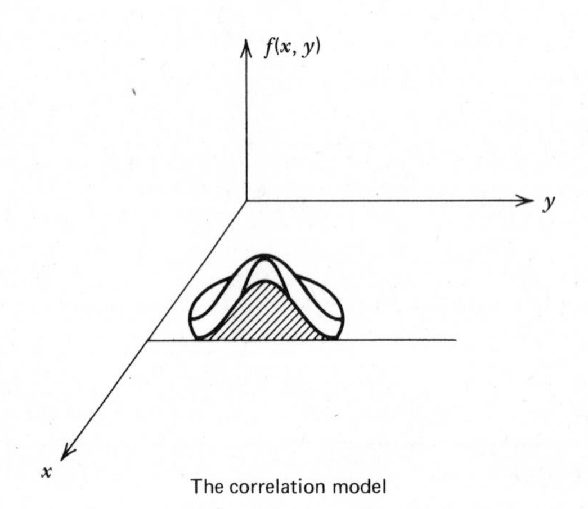

FIGURE 9.17. The different assumptions for regression and correlation.

model); for example, the statistic r may be used as an estimate of the population correlation coefficient ρ. If $\rho = 0$ for a bivariate normal distribution, then there is no useful linear relationship and we can also conclude that x and y are *independent* in the statistical sense. (Recall that in regression analysis, if $\beta = 0$ it is still possible that x and y are related by some type of relationship other than a linear one.)

The hypothesis H_0: $\rho = 0$ is tested with a t statistic having $n - 2$ degrees of freedom:

$$t = \frac{r}{\sqrt{\dfrac{1 - r^2}{n - 2}}}$$

Sometimes in research papers we find that the correlation coefficient and its test of significance are used in instances when the x and y variables do not have a bivariate normal distribution. The t test is valid with the usual assumptions (random sampling, normality, and independence) made only about the y variable at each level of x, but the interpretation is different. If there is a bivariate normal population and an investigator wants to learn more about the relationship between the two variables (perhaps height and weight) in that population, he draws a random sample of members of the population and computes r as an estimate of ρ. In contrast to this, an agronomist may select six increasing levels of fertilizer x and then compute the correlation with yield of corn y. He is using the correlation coefficient as the square root of the coefficient of determination, or as an index of how well a linear relationship fits the experimental data. He can use the t test to determine whether the levels of fertilizer explain a significant portion of the variability in corn yield, but the value of r is not an estimate of correlation between yield and levels of fertilizer.

The experimenter who wishes to use correlation procedures needs to be aware of an unusual feature about ρ. This t test is valid only to decide whether x and y are independent or whether there is a useful linear relationship between x and y, that is, the specific null hypothesis $\rho = 0$. It cannot be used to test a hypothesis such as $\rho = 0.5$. Furthermore, the analogy between the t test and confidence interval, which we have observed in other situations, does not hold true with regard to the correlation coefficient.

This situation arises because the correlation coefficient is bounded between -1 and $+1$, and therefore the distribution of the sample estimates, the r's, is symmetrical only when $\rho = 0$. If the value of ρ is very close to $+1$, then the range of overestimates is small but the range of underestimates is relatively large. The opposite is true if ρ is closer to -1. Thus when ρ is not zero, the sampling distribution will be skewed to the right or left depending upon whether ρ is negative or positive, respectively. Furthermore, the sample correlation coefficient r is a biased estimate of the parameter ρ when the latter is nonzero. Thus it is obvious that the sampling distribution of r is not a normal distribution when $\rho \neq 0$, and therefore a t test cannot be used because, as we have seen, such a test requires that the sampling distribution be normal.

A solution to the difficulty was first presented by R. A. Fisher, whose early theoretical research in statistics involved the sampling distribution of the correlation coefficient. Three of Fisher's findings are of particular use to us:

1. Although we assume a bivariate normal distribution of the x,y data points when we estimate the population correlation parameter ρ, when this parameter has a value of zero the distribution of r does not depend on the distribution of x but only on that of y. This is important here,

because it means that since y has a normal distribution, the two tests for a useful linear relationship are equivalent:

$$t = \frac{r}{\sqrt{\dfrac{1 - r^2}{n - 2}}} \quad \text{and} \quad t = \frac{b}{\sqrt{\dfrac{s_{y \cdot x}^2}{S_{xx}}}}$$

Thus we may use whichever is more convenient when testing $\rho = 0$.

2. No matter what the value of ρ, there is a transformation

$$\mathbf{z}_r = \log_e \sqrt{(1 + r)/(1 - r)}$$

that provides a near-normal sampling distribution and permits the use of procedures involving the normal distribution.

3. The variance of the transformed value $\mathbf{z}_r$ is practically independent of ρ and r and can be considered a known parameter $\sigma^2 = 1/(n - 3)$. Because the variance is known, we use the normal distribution rather than the t distribution when dealing with the $\mathbf{z}_r$ transformation.

As a consequence of points 2 and 3, we can make the following kinds of statistical inference about the correlation coefficient.

Example 9.2. Confidence Interval for ρ

In a study of obesity, the sample correlation coefficient for weights of 28 mature obese brother–sister pairs is computed to be $r = 0.64$. A nutritionist wishes to place a 95% confidence interval on the population correlation coefficient ρ.

A confidence interval is first found on the transformed parameter $\mathbf{z}_\rho$ using $\mathbf{z}_r$, and then the confidence limits are transformed back to r values.

$$\text{CI}_{1-\alpha}: \quad \mathbf{z}_r - z_{\alpha/2}(1/\sqrt{n-3}) \leq \mathbf{z}_\rho \leq \mathbf{z}_r + z_{\alpha/2}(1/\sqrt{n-3})$$

Since $r = 0.64$ is transformed to $\mathbf{z}_r = \log_e \sqrt{(1 + 0.64)/(1 - 0.64)} = 0.758$ (see Table A.12a in the Appendix),

$$\text{CI}_{0.95}: \quad 0.758 - 1.96(1/5) \leq \mathbf{z}_\rho \leq 0.758 + 1.96(1/5)$$

$$0.366 \leq \mathbf{z}_\rho \leq 1.150$$

Using Table A.12b, the corresponding r values are:

$$\mathbf{z}_r = 0.366 \rightarrow r = 0.350$$

$$\mathbf{z}_r = 1.150 \rightarrow r = 0.818$$

Thus

$$\text{CI}_{0.95}: \quad 0.350 \leq \rho \leq 0.818$$

A similar approach is used to test whether the population correlation coefficient is some nonzero value.

Example 9.3. Test of H_0: $\rho = \rho_0$ with $\rho_0 \neq 0$
The nutritionist in the previous example wants to test H_0: $\rho = 0.5$ against H_a: $\rho \neq 0.5$ because of some prior theory or available evidence. The test is a z test with statistic

$$z = \frac{\mathbf{z}_r - \mathbf{z}_{\rho_0}}{1/\sqrt{n-3}}$$

Since $r = 0.64$, it follows that $\mathbf{z}_r = 0.758$, and $\rho_0 = 0.5$ is transformed to $\mathbf{z}_{\rho_0} = 0.549$ (Table A.12a). Thus

$$z = \frac{0.758 - 0.549}{1/5} = 1.048$$

The null hypothesis is rejected at $\alpha = 0.05$ if $|z| > 1.96$, so the nutritionist concludes that ρ may be 0.5.

Fisher's transformation can also be used to compare two correlation coefficients.

Example 9.4. Testing $\rho_1 = \rho_2$
Suppose that the nutritionist has data on 23 brother–sister pairs of conventional mature weight in addition to the data above for obese pairs where $r_1 = 0.64$. For the conventional sample, $r_2 = 0.38$. To test whether the correlation is the same for both populations at $\alpha = 0.05$, the following test is used.

$$H_0: \rho_1 = \rho_2 \quad \text{against} \quad H_a: \rho_1 \neq \rho_2$$

is tested with

$$z = \frac{\mathbf{z}_{r_1} - \mathbf{z}_{r_2}}{\sqrt{\dfrac{1}{n_1 - 3} + \dfrac{1}{n_2 - 3}}}$$

Thus

$$z = \frac{0.758 - 0.400}{\sqrt{\dfrac{1}{25} + \dfrac{1}{20}}} = 1.193$$

Since $z_{\alpha/2} = 1.96$, there is no significant difference between the two correlation coefficients. The correlation between weights of brother–sister pairs may be the same for obese siblings as for those of conventional weight.

The various types of inference about correlation coefficients are summarized below.

■ ■ ■

Procedure. Inferences About Correlation Coefficients

Assumption: bivariate normal distribution

Tests of Hypotheses (Significance level: α)

1. H_0: $\rho = 0$
   H_a: $\rho \neq 0$ or $\rho > 0$ or $\rho < 0$
   Test Statistic:

$$t = \frac{r}{\sqrt{\dfrac{1 - r^2}{n - 2}}}$$

Reject H_0 if $|t| \geq t_{\alpha/2,n-2}$ or $t \geq t_{\alpha,n-2}$ or $t \leq -t_{\alpha,n-2}$, respectively

2. H_0: $\rho = \rho_0$ with $\rho_0 \neq 0$
   H_a: $\rho \neq \rho_0$ or $\rho > \rho_0$ or $\rho < \rho_0$
   Test statistic:

$$z = \frac{\mathbf{z}_r - \mathbf{z}_{\rho_0}}{1/\sqrt{n - 3}}$$ using Table A.12a for $\mathbf{z}_r$ and $\mathbf{z}_{\rho_0}$

Reject H_0 if $|z| \geq z_{\alpha/2}$ or $z \geq z_\alpha$ or $z \leq -z_\alpha$, respectively

3. H_0: $\rho_1 = \rho_2$
   H_a: $\rho_1 \neq \rho_2$ or $\rho_1 > \rho_2$ or $\rho_1 < \rho_2$
   Test statistic:

$$z = \frac{\mathbf{z}_{r_1} - \mathbf{z}_{r_2}}{\sqrt{\dfrac{1}{n_1 - 3} + \dfrac{1}{n_2 - 3}}}$$

Reject H_0 if $|z| \geq z_{\alpha/2}$ or $z \geq z_\alpha$ or $z \leq -z_\alpha$, respectively

Confidence Interval on $\boldsymbol{\rho}$

Compute $\text{CI}_{1-\alpha}$: $\mathbf{z}_r \pm z_{\alpha/2}(1/\sqrt{n-3})$ then use Table A.12b to transform the lower and upper limits back to r values.

■ ■ ■

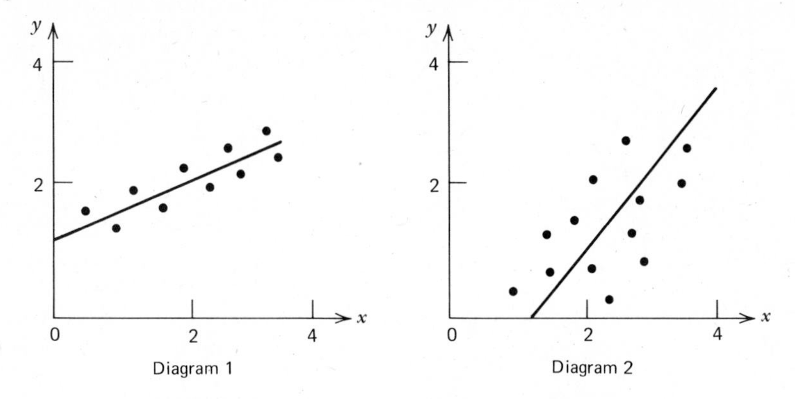

FIGURE 9.19. Scatter diagrams for Exercise 9.4.5.

9.4.8. If the sample correlation coefficient for the relationship between the number of cavities of 19 pairs of fathers and daughters is 0.57, test to see whether there is a significant difference between this correlation and the sample correlation coefficient for the relationship between the number of cavities of fathers and sons as computed in Exercise 9.4.6.

REVIEW EXERCISES

Decide whether each of the following statements is true or false. If a statement is false, explain why.

9.1. The sample regression line is called the least-squares trend line because for it $\sum(y - \hat{y})^2$ is smaller than for any other straight line fitted to the sample points.

9.2. The trend line always passes through the origin (0, 0).

9.3. If the independent variable is measured in radians and the dependent variable is measured in millimeters, then the slope of the least-squares trend line is in radians per millimeter.

9.4. If $b = 1$, all of the sample points fall on the least-squares trend line.

9.5. If $b = 0$, none of the sample points fall on the least-squares trend line.

9.6. If the slope of the regression line relating cake volume to amount of baking powder is 3.22 cc/gm, this means that for each additional gram of baking powder, the mean increase in the volume of the cake will be 3.22 cc.

9.7. $\sum(x - \bar{x})(y - \bar{y})$ is always positive.

9.8. If the trend line passes through the points (0, 0) and $(\bar{x}, \bar{y})$, then the slope is $b = 1$.

9.9. For a given sample value x, the corresponding sample value of y may or may not fall on the sample regression line.

9.10. It is possible to fit a line other than the least-squares trend line so that $\sum(y - \hat{y}) = 0$.

9.11. In a sample for which all the x values are equal, $b = 0$.

9.12. If the sign of the slope of the trend line is negative, the slope is probably not significantly different from zero.

9.13. The experimenter would test H_0: $\beta > 0$ if he thought that the slope of the trend line was positive.

9.14. Since $\sum(y - \hat{y})^2 \leq \sum(y - \bar{y})^2$, it follows that $s_{y \cdot x}^2 \leq s_y^2$.

9.15. If there are 30 sample points, then there will be 28 degrees of freedom associated with a t test for the validity of predicting from the least-squares trend line.

9.16. The better the line fits the sample points, the smaller $\sum(x - \bar{x})^2$ will be.

9.17. As the scatter of the points becomes larger, the standard error of the slope of the trend line becomes smaller.

9.18. If the numerical value of b is greater than 10.0, then b is probably significantly different from zero.

9.19. Units of measurement can affect both the magnitude of the slope and the significance of the slope of the least-squares trend line.

9.20. There can be a strong dependent relationship between y and x that will not be detected by linear regression analysis.

9.21. Along a significant least-squares line, $\hat{y}_1 \neq \hat{y}_2$ if $x_1 \neq x_2$.

9.22. The vertical deviations of the sample y values from the least-squares trend line will sum to zero.

9.23. The variance around the trend line is estimated with $n - 1$ degrees of freedom.

9.24. $s_{y \cdot x}^2/\sum(x - \bar{x})^2$ is to b as s^2/n is to $\bar{x}$.

9.25. H_0: $\beta = 0$ may be accepted and $\hat{y} = a + bx$ may still be a valid prediction equation.

9.26. The phrase "regression of y on x" indicates a negative relationship between the y and x variables.

9.27. For a given x, the prediction interval for the next y will be wider than the confidence interval for the average y value at that x.

9.28. The confidence interval for $E(y$ if $x = x^*)$ will be greater at $x^* = \bar{x}$ than for any $x^* \neq \bar{x}$.

9.29. Confidence intervals can be set for the true slope of the regression line, the true intercept on the y axis, and the true mean of y for any given value of x.

9.30. When computing a correlation coefficient, the experimenter assumes that there is a cause-and-effect relationship between x and y.

9.31. If $\sum(y - \hat{y})^2$ is large relative to $\sum(y - \bar{y})^2$, this indicates that a large portion of the variability in y is attributed to the linear relationship between y and x.

9.32. The greater the magnitude of r, the stronger the relationship between x and y.

9.33. $\sum(y - \hat{y})^2$ measures the variability that is not explained by the trend line.

9.34. $\sum(y - \hat{y})^2$ can never be greater than $\sum(y - \bar{y})^2$.

9.35. If the slope of the sample regression line is a large numerical value, the relationship between x and y is strong.

9.36. If all of the sample points fall on the least-squares line, $\sum(y - \bar{y})^2 = 0$.

9.37. One of the assumptions made in regression analysis is that the dependent variable follows a normal distribution.

9.38. In testing b for significance, it is assumed that y has the same variance for each fixed value of x.

9.39. For regression analysis, the experimenter needs to assume a cause-and-effect relationship between the two variables.

9.40. As the strength of the relationship between two variables increases, the regression line becomes a better fit for the points.

9.41. To accept H_0: $\beta = 0$ is to decide there is no relationship between x and y.

9.42. To accept H_0: $\rho = 0$ is to decide there is no relationship between x and y.

9.43. Other things being equal, the greater $\sum(y - \hat{y})^2$ the more likely that the slope of the trend line is significantly different from zero.

9.44. In making statistical inference about the correlation coefficient, x values may be fixed or may arise from random sampling, but the y values must arise from random sampling.

9.45. If the roles of x and y are interchanged, the regression equation will be the same.

9.46. If the roles of x and y are interchanged, the correlation coefficient will be the same.

SELECTED READINGS

Anscombe, F. J., and J. W. Tukey (1963). The examination of residuals, *Technometrics,* **5,** 141–160.

Bartlett, M. S. (1949). Fitting a straight line when both variables are subject to error, *Biometrics,* **5,** 207–212.

Behnken, D. W., and N. R. Draper (1972). Residuals and their variance patterns, *Technometrics,* **14,** 101–111.

Box, G. E. P. (1966). Use and abuse of regression, *Technometrics,* **8,** 625–629.

Daniel, C., and F. S. Wood (1971). *Fitting Equations to Data*, Wiley, New York.

Draper, N., and H. Smith (1981). *Applied Regression Analysis*, 2nd ed., Wiley, New York.

Gillingham, R., and D. Heien (1971). Regression through the origin, *The American Statistician*, **25,** 54–55.

Joiner, B. L. (1981). Lurking variables: Some examples, *The American Statistician,* **35,** 227–233.

Jurečková, J. (1971). Nonparametric estimate of regression coefficients, *The Annals of Mathematical Statistics,* **42,** 1328–1338.

Kendall, M. G. (1970). *Rank Correlation Methods*, 4th ed., Griffin, London.

Kruskal, W. H. (1958). Ordinal measures of association, *Journal of the American Statistical Association,* **53,** 814–861.

Madansky, A. (1959). The fitting of straight lines when both variables are subject to error, *Journal of the American Statistical Association,* **54,** 173–205.

Olkin, I., and J. W. Pratt (1958). Unbiased estimation of certain correlation coefficients. *The Annals of Mathematical Statistics,* **29,** 201–211.

Prescott, P. (1975). An approximate test for outliers in linear regression, *Technometrics,* **17,** 129–132.

Sampson, A. R. (1974). A tale of two regressions, *Journal of the American Statistical Association,* **69,** 682–689.

10

Techniques For One-Way Analysis of Variance

In Chapter 8 we discussed a group comparison test for *two* independent samples that came from normal populations with possibly different means but with the same variance. The hypothesis H_0: $\mu_1 = \mu_2$ was tested. In this chapter we test similar hypotheses for *three, four,* or *more* independent samples taken from normal populations with possibly different means but a common variance.

10.1. THE ADDITIVE MODEL

A psychologist studying factors that influence the amount of time mice require to solve a new maze might be observing four groups of three mice each. Each group has had a different amount of previous experience at maze solving, and the psychologist is looking for evidence of learning. The mice in the first group have had one previous experience in maze solving; those in the second group have solved two mazes; the third group has solved three; and the fourth group has solved four. Each mouse is now placed in a new maze, and the amount of time (in minutes) required to solve the maze is recorded.

The data (simplified for this example) might be as follows:

Group			
1	2	3	4
11	7	6	5
9	9	5	3
10	8	7	4

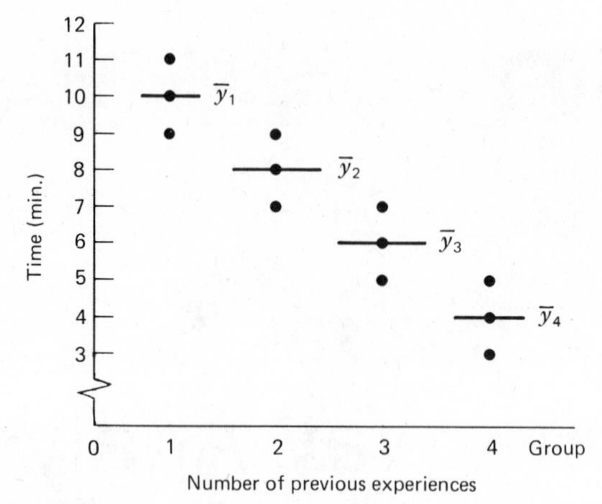

FIGURE 10.1. Data on time required to solve the maze.

Before a formal analysis of these data, we plot the values as in Figure 10.1, and add the sample averages ($\bar{y}_1$, $\bar{y}_2$, $\bar{y}_3$, $\bar{y}_4$,) to the graph.

Learning would be indicated by a decrease in the time required to solve the maze. The graph does seem to indicate a decrease in time for increased experience. However, the apparent differences in the graph could be due to sampling variability rather than learning. We need a method for deciding whether the differences in the sample averages are significant. If there is no learning, the four populations from which the samples were taken will all have the same means, $\mu_1 = \mu_2 = \mu_3 = \mu_4$. The analysis of variance is a formal method for testing this hypothesis.

In order to be able to speak more precisely about these data, in this text the symbol y_{ij} is used for the jth observation from the ith group. The first subscript i is reserved for the treatment group number irrespective of whether the groups are displayed in columns or in rows. Experimenters differ in how they display and label their data, so four groups of three observations each may be displayed as in Table 10.1 or as in Table 10.2. When reading books and articles, be careful to check how the subscripts are being used since the notation is not consistent.

TABLE 10.1 Treatment Groups Displayed in Columns

	Group				
	y_{1j}	y_{2j}	y_{3j}	y_{4j}	
	$y_{11} = 11$	$y_{21} = 7$	$y_{31} = 6$	$y_{41} = 5$	
	$y_{12} = 9$	$y_{22} = 9$	$y_{32} = 5$	$y_{42} = 3$	
	$y_{13} = 10$	$y_{23} = 8$	$y_{33} = 7$	$y_{43} = 4$	
Total:	30	24	18	12	$\sum_i \sum_j y_{ij} = 84$

TABLE 10.2 Treatment Groups Displayed in Rows

Group				Total
y_{1j}	$y_{11} = 11$	$y_{12} = 9$	$y_{13} = 10$	30
y_{2j}	$y_{21} = 7$	$y_{22} = 9$	$y_{23} = 8$	24
y_{3j}	$y_{31} = 6$	$y_{32} = 5$	$y_{33} = 7$	18
y_{4j}	$y_{41} = 5$	$y_{42} = 3$	$y_{43} = 4$	12
				$\sum_i \sum_j y_{ij} = 84$

In the example under consideration, the number of groups is $a = 4$ and the number of observations within each group is $n = 3$. We assume in all of the examples (until stated otherwise) that each group contains the same number of observations, n observations.

The psychologist in the present example wants to know if the amount of previous experience changes the time required to solve a maze. He wants to test H_0: $\mu_1 = \mu_2 = \mu_3 = \mu_4$ (that is, each of the samples comes from a population with the same mean) against H_a: *At least one inequality* (that is, $\mu_1 \neq \mu_2$ or $\mu_1 \neq \mu_3$ or $\mu_1 \neq \mu_4$ or $\mu_2 \neq \mu_3$ or $\mu_2 \neq \mu_4$ or $\mu_3 \neq \mu_4$). He is assuming that the four populations have a common variance σ^2.

It would be possible to test the equality of each pair of means by a t test; however, $\binom{4}{2} = 6$ separate t tests would be required for the null hypothesis under consideration. Besides being tedious, six separate t tests on the same data would have an α level much higher than the α used in each t test. A possible alternative procedure involves comparing the sample variability among groups with the sample variability within groups. This test is possible because if the null hypothesis is true, both of these statistics are estimates of σ^2.

In order to understand why the test is based on variability, it will be helpful if we consider the different types of averages associated with these data.

The grand average:	$\bar{y} = \sum_i \sum_j y_{ij}/an = 84/12 = 7$
The group averages:	$\bar{y}_1 = \sum_j y_{1j}/n = 30/3 = 10$
	$\bar{y}_2 = \sum_j y_{2j}/n = 24/3 = 8$
	$\bar{y}_3 = \sum_j y_{3j}/n = 18/3 = 6$
	$\bar{y}_4 = \sum_j y_{4j}/n = 12/3 = 4$

The average of the group averages $=$ The grand average $= \bar{y} = 7$

If we consider the population parameters related to these sample averages, each observation can be thought of in terms of an additive model consisting of three terms,

$$y_{ij} = \mu + \alpha_i + \epsilon_{ij}$$

in which μ (estimated by $\bar{y}$) is the mean time for all mice, α_i (estimated by $\bar{y}_i - \bar{y}$) is the mean treatment effect, or adjustment, for all mice in the ith group, and ϵ_{ij} is a random effect due to sampling. The data could then be written:

Group 1	Group 2
$11 = 7 + (10 - 7) + 1$	$7 = 7 + (8 - 7) + (-1)$
$9 = 7 + (10 - 7) + (-1)$	$9 = 7 + (8 - 7) + 1$
$10 = 7 + (10 - 7) + 0$	$8 = 7 + (8 - 7) + 0$

Group 3	Group 4
$6 = 7 + (6 - 7) + 0$	$5 = 7 + (4 - 7) + 1$
$5 = 7 + (6 - 7) + (-1)$	$3 = 7 + (4 - 7) + (-1)$
$7 = 7 + (6 - 7) + 1$	$4 = 7 + (4 - 7) + 0$

In terms of the additive model, the null hypothesis can be written in a different manner now:

$$H_0: \quad \alpha_1 = \alpha_2 = \alpha_3 = \alpha_4 = 0$$

with

$$H_a: \quad \textit{At least one inequality}$$

The development of the F test that follows, comparing the variability among groups with the variability within groups to test the above hypothesis, assumes this additive model. It also assumes that all treatments of interest to the experimenter are being used, that each treatment group is normally distributed, that all groups have the same variance, and that the experimental units are randomly assigned to the treatment group. For example, in this experiment the 12 mice should be chosen at random from those available and randomly assigned to groups 1, 2, 3, and 4. This type of analysis of variance is called a *one-way completely randomized ANOVA* (analysis of variance). In symbols, the assumptions are written:

$$y_{ij} = \mu + \alpha_i + \epsilon_{ij}$$

with

$$\sum_i \alpha_i = 0$$

and

$$\epsilon_{ij} \text{ IND}(0, \sigma^2),$$

that is, the ϵ_{ij} are *independently normally distributed* with a mean of zero and a variance of σ^2.

Returning now to the three types of sample averages, there are three types of sample variability that can be obtained by considering deviations from these sample averages. A sample variance is an average squared deviation from a sample average in which the averaging is achieved by dividing by the corresponding degrees of freedom. Thus the three types of sample variances are as given in Table 10.3.

The within-group variability is a pooled variance as in Chapter 8. The multiplication by n in the among-group variability is necessary if this variability is to be compared with the within-group variability. The among-group variability estimates the dispersion in the sampling distribution of averages of all samples of size n (that is, σ^2/n), so the among-group variability must be multiplied by n to estimate the dispersion of the original distribution.

The three types of deviations considered above are illustrated in Figure 10.2. The straight lines at right angles indicate the deviations of the observations from the grand average; these will be used for the total variability. The braces indicate the deviations of the observations from their respective group average; these will be used for the within-group variability. The dotted lines indicate the deviations of the group averages from the grand average and these will be used for the among-group variability. If the null hypothesis is true, $\bar{y}_1$, $\bar{y}_2$, $\bar{y}_3$, and $\bar{y}_4$ are not significantly different from $\bar{y}$, and the within-group variability will be approximately the same as the among-group variability. However, if the null hypothesis is false, then the among-group variability will be larger because of the significant deviations of the group averages from the grand average.

In the maze example, the sum of squares (SS) or numerator of the var-

TABLE 10.3 Three Types of Variability

Type of Variance	Formula	Meaning
Total variability	$\dfrac{\sum_i \sum_j (y_{ij} - \bar{y})^2}{na - 1}$	The average squared deviation of the observations from the grand average
Within-group variability	$\dfrac{\sum_i \sum_j (y_{ij} - \bar{y}_i)^2}{a(n - 1)}$	The average squared deviation of the observations from their respective group average (the pooled variance)
Among-group variability	$n\left[\dfrac{\sum_i (\bar{y}_i - \bar{y})^2}{a - 1}\right]$	The average squared deviation of the group averages from the grand average multiplied by the number of observations in each group

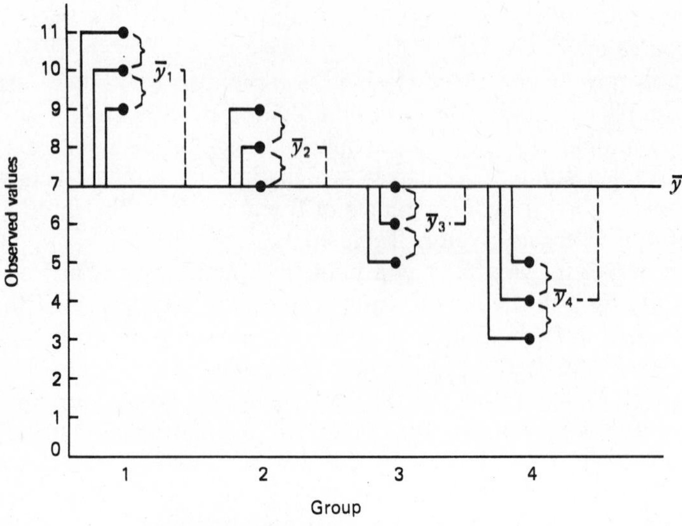

FIGURE 10.2. Three types of deviation.

iability in each case is as follows:

Total SS $\sum_i \sum_j (y_{ij} - \bar{y})^2 = 4^2 + 2^2 + 3^2 + \cdots + (-4)^2 + (-3)^2$

$= 68$

Within SS $\sum_i \sum_j (y_{ij} - \bar{y}_i)^2 = [1^2 + (-1)^2 + 0^2] + \cdots + [1^2 + (-1)^2 + 0^2]$

$= 8$

Among SS $n \sum_i (\bar{y}_i - \bar{y})^2 = 3[3^2 + 1^2 + (-1)^2 + (-3)^2]$

$= 60$

This example illustrates that the total sum of squares can be partitioned into two parts, the among-group sum of squares and the within-group sum of squares.

Total-SS	=	Among-SS	+	Within-SS
68	=	60	+	8

This relationship among the total, among-group, and within-group sum of squares leads to a shorter computational method, to be developed later. For now, the computation of the sum of squares just given will be used for the test. In order to change the sums of squares into variability (mean squares, or MS) they must be divided by their degrees of freedom.

The degrees of freedom are also partitioned as the sums of squares.

Total-df	=	Among-df	+	Within-df
$na - 1$	=	$a - 1$	+	$a(n - 1)$
11	=	3	+	8

A conventional form used is a work table, as follows:

Source	df	SS	MS
Among groups	$a - 1 = 3$	60	$60/3 = 20$
Within groups	$a(n - 1) = 8$	8	$8/8 = 1$
Total	$an - 1 = 11$	68	

If the null hypothesis H_0: $\mu_1 = \mu_2 = \mu_3 = \mu_4$ is true, the among-MS and the within-MS are both estimates of σ^2. This is because we are sampling from the same population (Figure 10.3). The variance among the averages estimates σ^2/n so n times the variance among the averages, or the among-group variance, estimates σ^2.

The test of the hypothesis about the equality of means is therefore an F test for the equality of two variances.

$$F = \frac{\text{among-MS}}{\text{within-MS}} = \frac{20}{1} = 20$$

is computed. This F statistic is compared with the critical value $F_{0.05,3,8}$ and leads to rejection if $F \geq 4.066$. This is a one-sided F test since if the null hypothesis is false, the among-MS is greater than the within-MS. In this example, $F \geq 4.066$, so the null hypothesis is rejected and it is concluded that the sample came from four populations among which there is at least

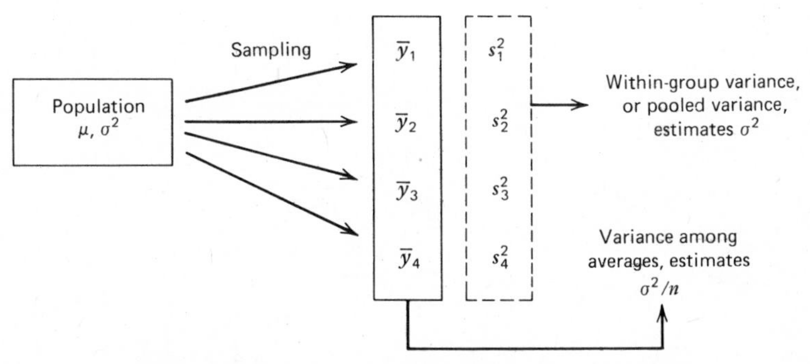

FIGURE 10.3. Within-group and among-group variances.

one inequality; that is, prior experience does affect the time required for the mice to solve a new maze.

EXERCISES

10.1.1. Compute the total-sum of squares, among-sum of squares, and within-sum of squares for the following data:

Group		
1	2	3
1	2	3
1	1	2
0	1	2
0	0	3
0	1	1

Show that the total-SS = among-SS + within-SS.

10.1.2. Four groups, each comprising four randomly selected persons, are asked to perform a simple mechanical task. Prior to the task, Group A is given a strong depressant, Group B a mild depressant, Group C a mild stimulant, and Group D a strong stimulant. The times (in seconds) required to complete the task are:

Group				
A	4	2	3	2
B	2	3	3	2
C	2	2	3	1
D	1	2	1	1

a. Graph these data and add the group averages to the graph.

b. Do the drugs seem to affect the time required to complete the task?

c. Test the hypothesis H_0: $\mu_A = \mu_B = \mu_C = \mu_D$ using an F test.

10.1.3. Four pea plants of a certain variety are grown without fertilizer, and four plants of the same variety are grown with fertilizer. The mature heights (in feet) are recorded below:

Without:	0.9	1.0	0.8	1.2
With:	1.5	1.2	1.6	1.3

a. Test H_0: $\mu_1 = \mu_2$ by the analysis of variance technique described in this section.
b. Test H_0: $\mu_1 = \mu_2$ by a two-sample t test.
c. What is the relationship between the F statistic and the t statistic?

10.1.4. In the maze example developed in this section, show that the average of the group averages is equal to the grand average. Why is this always true?

10.2. ONE-WAY ANALYSIS OF VARIANCE PROCEDURE

The procedure explained in Section 10.1 is a one-way analysis of variance. In this section, we develop a shorter computational method for this procedure.

This short method depends on the fact already noted:

$$\text{Total-SS} = \text{Within-SS} + \text{Among-SS}$$

This fact is used with an approach similar to the computational formula for the sample variance (Section 6.2)

$$s^2 = \frac{\sum y^2 - (\sum y)^2/n}{n - 1}$$

In the computational formula, the sum of squares (the numerator) is found by considering the sum of the squared deviations from the origin, $\sum y^2$, and subtracting the correction factor, $(\sum y)^2/n$, in order to get the sum of the squared deviations from the sample average. This method is used because it is simpler to compute with the deviations from the origin (the actual values) than with deviations from the average.

In the analysis of variance, a similar computational approach is used. We illustrate this using the mouse study of Section 10.1.

	y_{1j}	y_{2j}	y_{3j}	y_{4j}	
	11	7	6	5	
	9	9	5	3	
	10	8	7	4	
Totals	30	24	18	12	Grand total 84

When analyzing these data, we can consider three types of totals:

1 total of 12 observations	$\sum_i \sum_j y_{ij}$: 84
4 totals of 3 observations	$\sum_j y_{ij}$: 30, 24, 18, 12
12 totals of 1 observation	y_{ij}: 11, 9, ..., 3, 4

For the short computational method, these totals will be squared, divided by the number of observations per total, and summed. Table 10.4 summarizes this procedure.

The analysis of variance can then be computed from these uncorrected sums of squares as follows:

Source	df	SS	MS
Among groups	$a - 1 = 3$	$A - CF = 60$	$60/3 = 20$
Within groups	$a(n - 1) = 8$	$T - A = 8$	$8/8 = 1$
Total	$an - 1 = 11$	$T - CF = 68$	

In articles in professional journals, the sums of squares column is not usually given, nor is the row for the total.

To aid memory, it should be noted that the degrees of freedom and the number of squared values (totals) can be used to determine the sum of squares in the analysis of variance table. For example, the among-SS has $a - 1$ degrees of freedom, and among-SS $= A - CF$ in which A contains a squared values and CF contains 1 squared value. The within-SS has $a(n - 1) = an - a$ degrees of freedom, and within-SS $= T - A$, with T containing an squared values and A containing a squared values. Similarly for the total-SS.

Another, more realistic, example of a one-way analysis of variance follows.

TABLE 10.4 Uncorrected Sums of Squares for Equal-sized groups

Name	Symbol	Number of Totals	Observations/ Total	Formula	Numerical Value
Uncorrected total SS	T	$na = 12$	1	$\sum_i \sum_j y_{ij}^2$	$11^2 + 9^2 + \cdots + 4^2 = 656$
Uncorrected group SS	A	$a = 4$	$n = 3$	$\sum_i (\sum_j y_{ij})^2/n$	$30^2/3 + 24^2/3 + 18^2/3 + 12^2/3 = 648$
Correction factor	CF	1	$na = 12$	$(\sum_i \sum_j y_{ij})^2/na$	$84^2/12 = 588$

Example 10.1. One-Way Completely Randomized Analysis of Variance with Equal Sample Sizes

In a study of the physiological stress resulting from operating hand-held chain saws, experimenters measured the kickback that occurs when a saw is used to cut a 3-in. thick synthetic fiber board. The variable of interest was the angle (in degrees) to which the saw is deflected when it begins to cut the board. Below are the angles of deflection recorded for five random saws from each of four different manufacturers' models. A graph of the data and group averages appears in Figure 10.4.

	Chain Saw Model				
	A	*B*	*C*	*D*	*Totals*
	42	28	57	29	
	17	50	45	40	
	24	44	48	22	
	39	32	41	34	
	43	61	54	30	
$\sum_j y_{ij}$	165	215	245	155	780
$\sum_j y_{ij}^2$	5,999	9,965	12,175	4,981	33,120
$(\sum_j y_{ij})^2$	27,225	46,225	60,025	24,025	157,500

The hypothesis to be tested is:

$$H_0: \quad \alpha_A = \alpha_B = \alpha_C = \alpha_D$$

against

$$H_a: \quad \textit{At least one inequality}$$

$$T = 33{,}120$$

$$A = 157{,}500/5 = 31{,}500$$

$$CF = 780^2/20 = 30{,}420$$

Source	df	SS	MS
Among groups	$a - 1 = 3$	$SS_a = A - CF$ $= 1080$	$MS_a = SS_a/(a - 1)$ $= 360$
Within groups (error)	$a(n - 1) = 16$	$SS_e = T - A$ $= 1620$	$MS_e = SS_e/a(n - 1)$ $= 101.25$

The test statistic is $F = 360/101.25 = 3.56$ and $F_{0.05,3,16} = 3.239$. The null

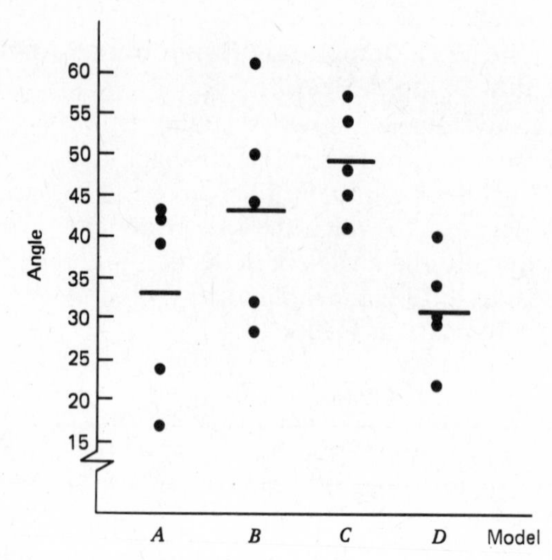

FIGURE 10.4. Angles of deflection for four types of chain saw.

hypothesis is rejected. There is a significant difference among the average kickbacks of the four types of saws.

In order to determine which of the models are different with respect to kickback, a follow-up procedure will be needed. This procedure is developed in the next section.

We can summarize the one-way analysis of variance procedure for equal group sizes as follows. The symbol SS_e is used for the within-group sum of squares because this quantity represents the variability due to random sampling, that is the sampling *error*.

■ ■ ■

Procedure. One-Way Completely Randomized Analysis of Variance with Equal Sample Sizes

H_0: $\alpha_1 = \alpha_2 = \cdots = \alpha_a$ (or $\mu_1 = \mu_2 = \cdots = \mu_a$)
H_a: *At least one inequality*

y_{ij} = jth observation in the ith treatment group
$i = 1, \ldots, a \quad j = 1, \ldots, n$

Compute:

$$T = \sum_i \sum_j y_{ij}^2$$

$$A = \sum_i (\sum_j y_{ij})^2/n$$

$$CF = (\sum_i \sum_j y_{ij})^2/an$$

Source	df	SS	MS	F
Among groups	$a - 1$	$SS_a = A - CF$	$MS_a = SS_a/(a - 1)$	MS_a/MS_e
Within groups (error)	$a(n - 1)$	$SS_e = T - A$	$MS_e = SS_e/a(n - 1)$	
Total	$an - 1$	$SS_t = T - CF$		

Reject H_0: if $F \geq F_{\alpha, a-1, a(n-1)}$

■ ■ ■

Many times the experimenter has no control over sample size, and an unbalanced design is necessary. This can happen in a genetics experiment in which the experimenter has no control over the number of offspring, in wildlife experiments that depend on the number of animals trapped, in a botany experiment in which some plants die (due to causes extraneous to the experiment), or in situations where cost restricts equalizing the sample sizes. The one-way analysis of variance can also be used if the sample sizes are unequal, although there may be some loss of power. The sums of squares needed for the computations are as in Table 10.5.

Example 10.2. One-Way Completely Randomized Analysis of Variance with Unequal Groups

A psychologist is studying several types of behavioral disorders in children and has reached a stage where he can classify children as belonging to one of seven types, depending on certain behavioral characteristics. He has a feeling that the mean level of intelligence may differ in some of these groups, so he begins to examine the IQ scores of children in these seven categories. In his files he finds cases of all seven types. There is some question in his

TABLE 10.5 Uncorrected Sums of Squares for Unequal-sized Groups

Source	Symbol	Number of Squared Values	Observations/ Square	Formula
Uncorrected total SS	T	N	1	$\sum_i \sum_j y_{ij}^2$
Uncorrected group SS	A	a	n_i	$\sum_i (\sum_j y_{ij})^2/n_i$
Correction factor	CF	1	N	$(\sum_i \sum_j y_{ij})^2/N$

NOTE: n_i is the number of observations in the ith group, and $N = \sum_i n_i$.

mind about the randomness of these data and also whether they meet the other assumptions of an analysis of variance. However, as a preliminary investigation he would like to test

$$H_0: \mu_1 = \mu_2 = \cdots = \mu_7$$

that is, that there is no difference among the mean IQ of children in the different categories. Since the psychologist has no control over the number of cases in his file, the groups have unequal sizes.

	Disorder						
	1	2	3	4	5	6	7
	105	115	103	124	115	85	79
	98	109	96	127	112	106	87
	110	121	105	118		98	
		130	107			111	
			112				
$\sum_j y_{ij}$	313	475	523	369	227	400	166
n_i	3	4	5	3	2	4	2
$\frac{(\sum_j y_{ij})^2}{n_i}$	32,656.3	56,406.2	54,705.8	45,387.0	25,764.5	40,000.0	13,778.0
$\sum_j y_{ij}^2$	32,729	56,647	54,843	45,429	25,769	40,386	13,810

$\sum_i n_i = 23 \qquad \sum_i \sum_j y_{ij} = 2473 \qquad \sum_i \sum_j y_{ij}^2 = 269{,}613$

$$T = \sum_i \sum_j y_{ij}^2 = 269{,}613.00$$

$$A = \sum_i (\sum_j y_{ij})^2/n_i = 268{,}697.80$$

$$CF = (\sum_i \sum_j y_{ij})^2/N = 265{,}901.26$$

Source	df	SS	MS	F	Critical Value $\alpha = 0.05$
Among groups	$a - 1 = 6$	2,796.54	466.09	8.14	2.741
Within groups	$N - a = 16$	915.20	57.20		

The null hypothesis is rejected, and the psychologist concludes that there seems to be a difference among the mean IQ of the children in the different categories.

The procedure for unequal groups can be summarized as follows.

■ ■ ■

Procedure. One-Way Completely Randomized Analysis of Variance with Unequal Sample Sizes

H_0: $\alpha_1 = \alpha_2 = \cdots = \alpha_a = 0$
H_a: *At least one inequality*

y_{ij} = jth observation in the ith treatment group
$i = 1, \ldots, a \quad j = 1, \ldots, n_i \quad \sum_i n_i = N$

Compute:

$$T = \sum_i \sum_j y_{ij}^2$$

$$A = \sum_i (\sum_j y_{ij})^2 / n_i$$

$$CF = (\sum_i \sum_j y_{ij})^2 / N$$

Source	df	SS	MS	F
Among groups	$a - 1$	$SS_a = A - CF$	$MS_a = SS_a/(a - 1)$	MS_a/MS_e
Within groups (error)	$N - a$	$SS_e = T - A$	$MS_e = SS_e/(N - a)$	
Total	$N - 1$	$SS_t = T - CF$		

Reject H_0 if $F \geq F_{\alpha, a-1, N-a}$

■ ■ ■

EXERCISES

10.2.1. Five groups of four men each are randomly assigned diets. At the end of a week, the following changes in weight (in pounds) are observed.

Diet 1	2	3	4	5
+3	+2	+4	+3	+1
−2	0	0	0	−1
0	+2	+1	−1	−2
−2	+1	+2	+1	−1

Perform an analysis of variance to see if there is any difference among the effects of these diets.

10.2.2. Five brands of lawnmowers are compared on the basis of hours of trouble-free operation. Eight randomly chosen mowers of each type are used in the study. Complete the following analysis of variance table.

Source	df	SS	MS
Among brands	—	140	—
Within brands	—	—	11

Give the null and alternative hypotheses to be tested by these data. Draw conclusions concerning the hypotheses.

10.2.3. Given the information below about the life (in months) of three types of light bulbs, graph the data, and complete the analysis of variance table.

	Brand		
	A	B	C
	7.0	13.4	9.5
	11.8	15.0	13.6
	10.5	14.6	10.6
	12.6	17.3	13.5
$\sum_j y_{ij}$	41.9	60.3	47.2
$\bar{y}_i$	10.5	15.1	11.8
$\sum_j (y_{ij} - \bar{y}_i)^2$	18.35	7.99	12.86

$41.9 + 60.3 + 47.2 = 149.4$

$(41.9)^2 + (60.3)^2 + (47.2)^2 = 7{,}619.54$

$(149.4)^2 = 22{,}320.36$

What is the hypothesis about the means of the brands? Would the hypothesis be accepted? What conclusion do you draw about the light bulbs?

10.2.4. Tomato plants are treated with five different fertilizers, and the sum of the weight (in pounds) of the ripe fruit is recorded for each plant that matures.

Fertilizer	*A*	*B*	*C*	*D*	*E*
Number of mature plants	4	7	6	5	6
$\sum_j y_{ij}$	81	111	138	96	101
$\sum_j y_{ij}^2$	1649	1775	3184	1850	1715

Perform an analysis of variance to test for equality of means. What assumptions are necessary for this analysis to be valid?

10.2.5. Three different methods of processing orange juice are compared. The amount of vitamin C per 8-oz serving is the variable of interest (in milligrams). Six servings are chosen at random from each process.

	Processing Method			
	A	*B*	*C*	Totals
	96	123	76	
	87	115	78	
	85	122	79	
	92	118	77	
	90	122	80	
$\sum_j y_{ij}$	450	600	390	1,440
$(\sum_j y_{ij})^2$	202,500	360,000	152,100	714,600
$\sum_j y_{ij}^2$	40,574	72,046	30,430	143,050

What null hypothesis can be tested? Graph the data. Does the null hypothesis appear to be true? If $\alpha = 0.05$, what is the critical value of the test statistic? Show that the correction factor for these data is 138,240. Complete the analysis of variance table. Should the null hypothesis be rejected? What conclusion do you draw?

10.2.6. Given the following information, complete the analysis of variance to test for equality of group means.

Source	Number of Squared Values	Observations per Squared Value	Numerical Value
$\sum_i \sum_j y_{ij}^2$	30	1	1565
$\sum_i (\sum_j y_{ij})^2/n$	6	5	1325
$(\sum_i \sum_j y_{ij})^2/an$	1	30	1200

10.2.7. Live traps are set to capture samples of rabbits at five different locations in a large wooded area. The weights (in ounces) are as follows:

		Area		
1	2	3	4	5
37	29	49	40	50
40	33	47	38	46
46	34		42	49
	31		39	
			41	

Graph the data and the group averages. Is there evidence that the mean weights of the rabbits differ at some of the locations? Test at the 1% level of significance. What assumptions are necessary about the rabbits?

10.2.8. A dean at a small college believes there may be a difference in the mean age of his faculty in different departments. He obtains the following information about faculty ages.

Mathematics	28	35	31	32	
English	45	37	42	38	36
Foreign languages	27	32	29		
History	43	39			

a. Are there significant differences in the average ages for these four departments?

b. What assumptions must be made in order for ANOVA techniques to be valid for this study?

10.2.9. A forest entomologist has isolated seven insecticides that are reasonably safe to the rest of the environment when used to control gypsy moths. She wants to determine whether any one of them produces significantly greater mortality than the others when applied topically to adult gypsy moths. Using standard bioassay techniques, she applies a given insecticide to the abdomen of each of 100 moths. This procedure is repeated five times for each insecticide, with new solutions being prepared each time. Mortality is recorded after 24 hours for each insecticide trial. Assume that the data, although distributed in a binomial fashion, will approximate the normal distribution adequately for ANOVA procedures.

a. Although 3500 moths are used, why are there only 34 degrees of freedom associated with the experiment?

b. In using the y_{ij} notation, does the j subscript refer to the insecticide or the trial?
c. What are the assumptions for an analysis of variance?
d. Use this information to complete the accompanying ANOVA table:

$$\left(\sum_i \sum_j y_{ij}\right)^2 / an = 143{,}360 \qquad \sum_i \sum_j y_{ij}^2 = 144{,}334$$

Source	df	SS	MS
Insecticides	—	—	55
Trials within insecticides	—	—	—

e. Give the null and alternative hypotheses.
f. Give the critical value ($\alpha = 0.05$) of the above hypothesis and draw conclusions about the experiment.

10.2.10. The following linear model is used in a study involving five artists and four paintings per artists:

$$u_{ij} = \lambda + \nu_i + \delta_{ij}$$

in which $i = 1, \ldots, a = 5$ and $j = 1, \ldots, n = 4$. The data below give the number of smudges per picture:

	Artist				
	A	*B*	*C*	*D*	*E*
	7	2	4	11	2
	6	4	6	7	0
	8	4	6	8	3
	7	4	2	4	5
Total	28	14	18	30	10

a. In order to perform an analysis of variance on these data, what must be assumed about ν_i and δ_{ij}?
b. What is the numerical value of u_{23} and $\sum_j u_{3j}$?
c. Given that $7^2 + 6^2 + 8^2 + \cdots + 5^2 = 630$ and $28^2 + 14^2 + 18^2 + 30^2 + 10^2 = 2304$, complete a table for the uncorrected sums of squares giving the number of squared values, the number of observations per squared value, and the numerical value of the uncorrected sum of squares.

10.2.11. Suppose that a building contractor wants to test three types of wooden beams for weight-bearing capacity. Five beams of each type are broken by stacking lead weights on them, and the weight required to break each beam is recorded.

a. Given the mathematical model

$$z_{hi} = \psi + \delta_{\_} + \xi_{\_}$$

in which z_{hi} = the breaking strength of beam i within type h
$\delta_{\_}$ = the symbol of the type effect
$\xi_{\_}$ = the symbol of the beam within type effect,
fill in the blanks with the appropriate subscripts.

b. What assumptions must be made about $\delta_{\_}$ and $\xi_{\_}$?

c. What is the largest numerical value that can be taken by each subscript in the model?

d. If the following computations are made on the experimental data

$$\sum_h \sum_i z_{hi}^2 = 3{,}620{,}000$$

$$\sum_h \left(\sum_i z_{hi}\right)^2 = 18{,}040{,}000$$

$$\left(\sum_h \sum_i z_{hi}\right)^2 = 54{,}000{,}000$$

complete the analysis of variance.

10.3. MULTIPLE COMPARISON PROCEDURES

In Sections 10.1 and 10.2 of this chapter, the analysis of variance procedure is developed to test H_0: $\mu_1 = \mu_2 = \cdots = \mu_a$ (or H_0: $\alpha_1 = \alpha_2 = \cdots = \alpha_a$). If the null hypothesis is rejected, we conclude that there is at least one inequality among the means of the treatment groups (or among the treatment effects).

If the treatment groups under consideration exhaust the cases that are of interest to the experimenter (as we have been assuming in this chapter) and the F test is significant, the experimenter may want to draw some further conclusions. He may want to decide which pairs of treatments are different, or he may want to contrast one treatment effect with the average of some other treatment effects, or he may want to estimate some of the parameters in the experiment.

In this section we discuss several procedures for deciding which pairs of means are different. In general these techniques are called *multiple comparison procedures*. Contrasts and estimation are discussed in Sections 10.4 and 10.5.

Several multiple comparison procedures are available to researchers. We discuss five different approaches and their relative merits for various experimental situations. In all cases we assume equal sample sizes for the treatment groups.

Some Multiple Comparison Procedures

1. Fisher's Least Significant Difference
2. Duncan's New Multiple Range Test
3. The Student–Newman–Keuls' Procedure
4. Tukey's Honestly Significant Difference
5. Scheffé's Method

*1. **Fisher's Least Significant Difference.*** R. A. Fisher's multiple comparison procedure is known as the least significant difference. It is based on a t test. If the treatment groups are all of equal size n, then two sample averages, $\bar{y}_1$ and $\bar{y}_2$ for example, can be tested for a significant difference by the statistic

$$t = \frac{\bar{y}_1 - \bar{y}_2}{\sqrt{\frac{s_p^2}{n} + \frac{s_p^2}{n}}} = \frac{\bar{y}_1 - \bar{y}_2}{\sqrt{\frac{2s_p^2}{n}}}$$

in which s_p^2 is the pooled sample variance as in Chapter 8. Thus Fisher said the difference $\bar{y}_i - \bar{y}_j$ is significant if

$$|y_i - y_j| \geq t_{\alpha/2, a(n-1)} \sqrt{\frac{2\text{MS}_e}{n}}$$

since MS_e in the analysis of variance is a pooled estimate of the common variance of the treatment groups and MS_e has $a(n - 1)$ degrees of freedom.

In order to protect the overall Type I error rate for the experiment, Fisher's procedure requires a prior significant F test in the analysis of variance. With this condition, the overall error rate has been shown by simulation to be approximately the α level of the F test.

Example 10.3. Fisher's Least Significant Difference

In the chain saw study, Example 10.1 of Section 10.2, the sample averages are:

$$\bar{y}_A = 165/5 = 33 \qquad \bar{y}_C = 245/5 = 49$$
$$\bar{y}_B = 215/5 = 43 \qquad \bar{y}_D = 155/5 = 31$$

The experimenter wants to test $\binom{a}{2} = \binom{4}{2} = 6$ hypotheses

$$H_0\colon \ \mu_A = \mu_B \qquad H_0\colon \ \mu_B = \mu_C$$
$$H_0\colon \ \mu_A = \mu_C \qquad H_0\colon \ \mu_B = \mu_D$$
$$H_0\colon \ \mu_A = \mu_D \qquad H_0\colon \ \mu_C = \mu_D$$

to locate the specific difference or differences he believes exist because of the prior significant F test.

If Fisher's test is used, the differences between all pairs of sample averages must be compared with

$$t_{\alpha/2, a(n-1)} \sqrt{\frac{2\text{MS}_e}{n}} = t_{0.025,16} \sqrt{\frac{2(101.25)}{5}}$$
$$= 2.120(6.36)$$
$$= 13.5$$

at $\alpha = 0.05$.

To keep track of all possible differences between sample averages, he arranges them in order according to size, from the smallest to the largest

31 33 43 49

and forms a table listing the ordered averages on the left omitting the largest, and across the top omitting the smallest.

	33	43	49
31			
33			
43			

If the top average is larger than one on the left, he subtracts the average on the left from the average on the top and enters the difference in the table.

	33	43	49
31	2	12	18
33		10	16
43			6

These differences are then compared with the least significant difference 13.5, which was computed earlier. He begins at the right of the top row of differences. There he finds 18, which is greater than 13.5, so he marks 18 with an asterisk and concludes that $\mu_D \neq \mu_C$. The next entry in the top row is 12, which is less than 13.5, so he goes no further in that row. He then treats the second and third rows in the same manner. The final table has the following form:

		$\bar{y}_A$ 33	$\bar{y}_B$ 43	$\bar{y}_C$ 49
$\bar{y}_D$	31	2	12	18*
$\bar{y}_A$	33		10	16*
$\bar{y}_B$	43			6

The only pairs of means that are different are $\mu_D \neq \mu_C$ and $\mu_A \neq \mu_C$.

In a journal, in order to save space he would report that at the $\alpha = 0.05$ level by Fisher's least significant difference, any two averages not underlined by the same line-segment are significantly different.

31 33 43 49

Since the middle line is already indicated by the first line, it can be omitted:

31 33 43 49

Fisher's test has a drawback; it requires that the null hypothesis be rejected in the ANOVA procedure. It is possible that the F test will fail to detect a single significant difference among several treatment groups. In a case like this, Fisher's least significant difference cannot be used. The other four multiple comparison procedures do not require a significant F test; they protect the Type I error rate by different approaches.

2. ***Duncan's New Multiple Range Test.*** We will not go into the details of Duncan's method for protecting the error rate. Briefly, he considers the error rate for each pairwise comparison (rather than an overall rate) and allows a higher rate for pairs of sample averages that are further apart when ordered by size. Thus, if

$$\bar{y}_1 \qquad \bar{y}_2 \qquad \bar{y}_3$$

are three sample averages arranged from smallest to largest, a test of $\mu_1 = \mu_3$ would have a higher error rate than the test of $\mu_1 = \mu_2$. Because of this, Duncan's procedure will involve several different critical differences in contrast to Fisher's single least significant difference.

To reject H_0: $\mu_i = \mu_j$ when $\bar{y}_i$, $\bar{y}_j$ span r ranked sample averages, it is necessary that

$$|\bar{y}_i - \bar{y}_j| \geq d_{\alpha,r,a(n-1)}\sqrt{\frac{\text{MS}_e}{n}}$$

in which $d_{\alpha,r,a(n-1)}$ is found in Tables A.13a and A.13b in the Appendix. The α is the significance level set by the experimenter; Duncan makes the necessary adjustments in his table. Note also that the radical does not contain the factor of 2 found in the t test; it has been absorbed into the d value. If we are dealing with adjacent sample averages,

$$d_{\alpha,2,\nu} = t_{\alpha/2,\nu}\sqrt{2}$$

Example 10.4. Duncan's New Multiple Range Test
Using the table of differences of sample averages for the power saw data, we see that the lowest diagonal consists of differences of adjacent rank-

	33	43	49	
31	2	12	18	
				↖ spans 4 ranked averages
33		10	16	
				↖ span 3 ranked averages
43			6	
				↖ span 2 ranked averages

ed averages, that is, a span of two ranked averages. The second diagonal consists of differences of averages separated by one average, that is, the difference spans three ranked averages. The remaining difference spans four ranked averages. Using Table A.13a in the Appendix, the experimenter finds

$$d_{0.05,2,16}\sqrt{\frac{\text{MS}_e}{n}} = 2.998(4.50) = 13.5$$

$$d_{0.05,3,16}\sqrt{\frac{\text{MS}_e}{n}} = 3.144(4.50) = 14.1$$

$$d_{0.05,4,16}\sqrt{\frac{\text{MS}_e}{n}} = 3.235(4.50) = 14.6$$

Comparing the differences with these critical values, he finds two significant differences:

	33	43	49	
31	2	12	18*	Compare with
				↖ 14.6
33		10	16*	
				↖ 14.1
43			6	
				↖ 13.5

His conclusion would be identical with the one reached with Fisher's procedure.

Duncan's test is slightly more conservative than Fisher's, that is, it will sometimes find fewer significant differences. However, there is about 95% agreement between the two procedures.

*3. **The Student–Newman–Keuls' Procedure.*** The Student–Newman–Keuls' procedure is still more conservative than Duncan's. Like Duncan's test, different critical values are used depending on the span of the two ranked averages being compared. However, this test protects the error rate using a constant level for each diagonal.

Two sample averages which span r ranked averages are significantly different if

$$|\bar{y}_i - \bar{y}_j| \geq q_{\alpha,r,a(n-1)}\sqrt{\frac{\text{MS}_e}{n}}$$

in which the q values are found in Tables A.14a and A.14b in the Appendix, the Studentized Range.

Example 10.5. Student–Newman–Keuls' Procedure

Using the chain saw data of Example 10.3 and Table A.14a in the Appendix, the investigator finds:

$$q_{0.05,2,16}\sqrt{\frac{\text{MS}_e}{n}} = 2.998(3.50) = 13.5$$

$$q_{0.05,3,16}\sqrt{\frac{\text{MS}_e}{n}} = 3.649(3.50) = 16.4$$

$$q_{0.05,4,16}\sqrt{\frac{\text{MS}_e}{n}} = 4.046(3.50) = 18.2$$

The table of differences is

	33	43	49	
31	2	12	18	Compare with
				↖ 18.2
33		10	16	
				↖ 16.4
43			6	
				↖ 13.5

Thus, none of the differences are significant using this procedure.

This procedure is so conservative that it located no differences, whereas the F test in the ANOVA indicated that a difference exists.

4. Tukey's Honestly Significant Difference. Tukey's procedure is still more conservative. It uses a single critical difference:

$$q_{\alpha,a,a(n-1)}\sqrt{\frac{\text{MS}_e}{n}}$$

that is, the largest critical difference in the Student–Newman–Keuls' procedure. The error rate is for the entire experiment.

Example 10.6. Tukey's Honestly Significant Difference
For the chain saw data (see Example 10.3), two averages $\bar{y}_i$, $\bar{y}_j$ are significantly different if

$$|\bar{y}_i - \bar{y}_j| \geq q_{0.05,4,16}\sqrt{\frac{101.25}{5}}$$

$$= 4.046(4.50)$$

$$= 18.2$$

Thus, none of the pairs of averages are significantly different.

---

5. Scheffé's Method. Scheffé's method can be used to compare means and also to make other types of contrasts. For example, we might want to test

$$H_0: \quad \mu_1 = \frac{\mu_2 + \mu_3}{2}$$

that is, that Treatment 1 is the same as the average of Treatments 2 and 3. The error rate α in Scheffé's procedure applies to all possible contrasts.

To compare two means using this method, $\bar{y}_i$ and $\bar{y}_j$ are significantly different if

$$|\bar{y}_i - \bar{y}_j| \geq \sqrt{(a-1)F_{\alpha,a-1,a(n-1)}}\sqrt{\frac{2\text{MS}_e}{n}}$$

Example 10.7. Scheffé's Method for Comparing Means
In the chain saw study, the critical difference is

$$\sqrt{3F_{0.05,3,16}}\sqrt{\frac{2(101.25)}{5}} = \sqrt{3(3.239)}\sqrt{40.5} = 19.8$$

Again this yields no significant difference.

TABLE 10.6 Comparison of Multiple Comparison Procedures

Multiple Comparison Procedure	Power	Type I Error Rate
Fisher's	Highest	Highest
Duncan's	↓ More conservative, less likely to detect real differences	↑ More likely to indicate false differences
Student–Newman–Keuls'		
Tukey's		
Scheffé's	Lowest	Lowest

Scheffé's is the most conservative of the five methods we have discussed. It is very likely to miss detecting a real difference that exists. Scheffé's approach is used more often for the other contrasts; in these cases an adjustment is needed in the standard error. For example, to test

$$H_0\colon \ \mu_1 = \frac{\mu_2 + \mu_3}{2} \qquad \text{or the equivalent,} \qquad H_0\colon \ \mu_1 - \frac{\mu_2}{2} - \frac{\mu_3}{2} = 0$$

the standard error is $\sqrt{3\text{MS}_e/2n}$. The coefficient 3/2 is the sum of $1^2 + (-1/2)^2 + (-1/2)^2$, that is, the sum of the squares of the coefficients in the linear combination of the μ's in the null hypothesis.

The five procedures we have just outlined are only some of the multiple comparisons available to the researcher. Which procedure should be used depends upon which type of error is more serious. In the chain saw example, assume the prices are approximately the same. Then a Type I error is not serious; it would imply that we decide one model has less kickback than another when in fact the two models have the same amount of kickback. A Type II error would imply that a difference in kickback actually exists but we fail to detect it, a more serious error. Thus, in this experiment we want maximum power and we would probably use Fisher's least significant difference. The experimenter should decide before the experimentation which method will be used to compare the means.

Table 10.6 lists the five tests indicating decreasing power and increasing error rate. The five procedures can be summarized as follows.

■ ■ ■

Procedure. Multiple Comparison Procedures

$H_0\colon \ \mu_1 = \mu_2, \qquad H_0\colon \ \mu_1 = \mu_3,$ and so on, for all pairs of treatment means.

$H_a\colon \ \mu_1 \neq \mu_2, \qquad H_a\colon \ \mu_1 \neq \mu_3, \ldots .$

Compute $\bar{y}_1, \bar{y}_2, \ldots, \bar{y}_a$, the a sample averages, and arrange them in order from the smallest to the largest:

$$\bar{y}_{(1)}, \bar{y}_{(2)}, \ldots, \bar{y}_{(a)}$$

Form a table of differences:

	$\bar{y}_{(2)}$	$\bar{y}_{(3)}$	$\cdots$	$\bar{y}_{(a)}$
$\bar{y}_{(1)}$	$\bar{y}_{(2)} - \bar{y}_{(1)}$	$\bar{y}_{(3)} - \bar{y}_{(1)}$	$\cdots$	$\bar{y}_{(a)} - \bar{y}_{(1)}$
$\bar{y}_{(2)}$		$\bar{y}_{(3)} - \bar{y}_{(2)}$	$\cdots$	$\bar{y}_{(a)} - \bar{y}_{(2)}$
$\vdots$			$\ddots$	
$\bar{y}_{(a-1)}$				$\bar{y}_{(a)} - \bar{y}_{(a-1)}$

Determine the critical difference of differences:

Fisher's	$t_{\alpha/2,a(n-1)}\sqrt{\dfrac{2\mathrm{MS}_e}{n}}$	Apply to all differences
Duncan's	$d_{\alpha,2,a(n-1)}\sqrt{\dfrac{\mathrm{MS}_e}{n}}$	Apply to bottom diagonal
	$d_{\alpha,3,a(n-1)}\sqrt{\dfrac{\mathrm{MS}_e}{n}}$	Apply to second lowest diagonal
	$\vdots$	$\vdots$
	$d_{\alpha,a,a(n-1)}\sqrt{\dfrac{\mathrm{MS}_e}{n}}$	Apply to top diagonal
Student–Newman–Keuls'	$q_{\alpha,2,a(n-1)}\sqrt{\dfrac{\mathrm{MS}_e}{n}}$	Apply to bottom diagonal
	$q_{\alpha,3,a(n-1)}\sqrt{\dfrac{\mathrm{MS}_e}{n}}$	Apply to second lowest diagonal
	$\vdots$	$\vdots$
	$q_{\alpha,a,a(n-1)}\sqrt{\dfrac{\mathrm{MS}_e}{n}}$	Apply to top diagonal

Tukey's	$q_{\alpha,a,a(n-1)}\sqrt{\dfrac{\text{MS}_e}{n}}$	Apply to all differences
Scheffé's	$\sqrt{(a-1)F_{\alpha,a-1,a(n-1)}}\sqrt{\dfrac{2\text{MS}_e}{n}}$	Apply to all differences

Only Fisher's procedure requires a prior significant ANOVA.
In each procedure, reject H_0 if $|\bar{y}_i - \bar{y}_j| \geq$ critical difference.

It is possible to modify Fisher's and Scheffé's procedures for unequal sample sizes. The standard error becomes

$$\sqrt{\frac{\text{MS}_e}{n_i} + \frac{\text{MS}_e}{n_j}}$$

For Duncan's, Student–Newman–Keuls', and Tukey's procedures an approximation approach is possible by letting n be

$$\tilde{n} = \frac{a}{\dfrac{1}{n_1} + \dfrac{1}{n_2} + \cdots + \dfrac{1}{n_a}}$$

EXERCISES

10.3.1. An analysis of variance is conducted to compare the yields of several different varieties of blight-resistant corn.

Source	df	SS	MS
Among varieties	—	—	598
Within varieties	20	3600	—

Variety:	C	A	D	B	E
Average yield:	60	80	82	85	93

a. Complete the analysis of variance table.

b. Show that the standard error of a sample average is 6.0.

c. Would it be appropriate to use Fisher's least significant difference to compare variety means in this experiment?

d. Perform Fisher's test at $\alpha = 0.05$.

10.3.2. Five kinds of insecticide are used in effort to control insect damage to a certain crop. Damage is measured in terms of square centimeters of leaf area destroyed. The data are summarized as follows:

Insecticide	1	2	3	4	5	Totals
Plants examined	4	4	4	4	4	20
$\sum_j y_{ij}$	24	19	29	67	34	173
$\sum_j y_{ij}^2$	178	97	237	1313	342	2167
$(\sum_j y_{ij})^2$	576	361	841	4489	1156	7423
$(\sum_j y_{ij})^2/n$	144.00	90.25	210.25	1122.25	289.00	1855.75

a. Show that the correction factor is 1496.45.
b. Perform an analysis of variance at $\alpha = 0.05$.
c. Use Fisher's procedure to test for differences among the means

10.3.3. A behavioral biologist subjected spiders to different stressful conditions and then measured the number of gaps in their webs.

Condition 1	2	3	4
11	13	21	10
4	9	18	4
6	14	15	19
21	36	54	33

$$\sum_i \sum_j y_{ij}^2 = 2086$$

$$\sum_i (\sum_j y_{ij})^2 = 5742$$

$$(\sum_i \sum_j y_{ij})^2 = 20736$$

a. Complete the analysis of variance at $\alpha = 0.01$.
b. Would it be valid to use Fisher's procedure to test for a difference between group means? Why or why not?
c. Use Scheffé's procedure to test for a difference between means.

10.3.4. Five male students are selected at random from each of five colleges in a study to determine whether there is an association between sentimentality and the selected field of study. They are shown a movie about a little crippled orphan, his blind dog, and a senile grandfather who is trying to care for them in his cabin which is in the path of a strip-mine operation. Polygraph equipment is used to record emotional response to the picture. The F test for differences among colleges is

$$F = \frac{\text{among-college MS}}{\text{within-college MS}} = \frac{50.00}{11.25}$$

a. Show that the standard error of a college average is 1.5.
b. Use Duncan's procedure to test for differences in emotional response among the college means.

College:	Law	Business	Agriculture	Arts and Sciences	Engineering
Sample average:	3	7	14	15	21

10.3.5. In order to see whether three commonly used weed killers may have differential effects on the yield of rye, each is sprayed on six different plots of rye at the seedling stage. The within-spray MS is 96, and the average yields are:

Weed killers:	I	II	III
Averages:	10	20	30

a. Use the Student–Newman–Keuls' procedure to determine whether there are any differences in the mean yields.
b. If the agronomist conducting the experiment wants to use Fisher's least significant difference, how large would the F value have to be in order for him to be justified in using the procedure? Does the computed F value exceed this critical value?
c. How could the experimenter test whether the plot sprayed with weed killer III produces an average yield that is significantly different from the average of the other two?

10.3.6. Consider a significant analysis of variance in which $a = 6$, $n = 5$, $MS_e = 33.78$ and the treatment averages are:

Treatment Average:	39.3	45.2	48.4	50.4	55.5	58.2
Treatment:	A	B	C	D	E	F

Use all five multiple comparison procedures at $\alpha = 0.05$ on these data and form a table indicating the different conclusions reached by each test.

10.4. ONE DEGREE OF FREEDOM COMPARISONS

The multiple comparison procedures in Section 10.3 are known as *a posteriori* tests, that is, they are after the fact. After the experiment is completed, the investigator decides to look for possible pairwise differences.

There is also an *a priori* approach, that is, contrasts that are planned before the experiment. The experimenter believes prior to the investigation that certain factors may be related to differences in treatment groups. For example, in the chain saw experiment (Example 10.1 of Section 10.2), suppose that Models A and D are light-weight chain saws for home use and that B and C are heavy-duty industrial types. The investigator might want to know if the kickback from the home type is the same as the kickback from the industrial type. In addition he might also be interested in any differences in kickback within types.

Comparison	H_0 to Be Tested
1. Home vs. industrial	$\frac{\mu_A + \mu_D}{2} - \frac{\mu_B + \mu_C}{2} = 0$
2. Home model A vs. home model D	$\mu_A - \mu_D = 0$
3. Industrial model B vs. industrial model C	$\mu_B - \mu_C = 0$

Each of the null hypotheses is a linear combination of the treatment means:

Linear Combinations
1. $(1/2)\mu_A - (1/2)\mu_B - (1/2)\mu_C + (1/2)\mu_D$
2. $(1)\mu_A + (0)\mu_B + (0)\mu_C - (1)\mu_D$
3. $(0)\mu_A + (1)\mu_B - (1)\mu_C + (0)\mu_D$

A set of linear combinations of this type is called a set of *orthogonal contrasts* or orthogonal comparisons. A set of linear combinations must satisfy two mathematical properties in order to be orthogonal contrasts:

A. The sum of the coefficients in each linear combination must be zero; this makes the linear combination a contrast.

In 1: $1/2 - 1/2 - 1/2 + 1/2 = 0$

In 2: $1 + 0 + 0 - 1 = 0$

In 3: $0 + 1 - 1 + 0 = 0$

B. The sum of the products of the corresponding coefficients in any two contrasts must equal zero; this makes the contrasts orthogonal.

In Contrasts 1 and 2:

$$(1/2)(1) + (-1/2)(0) + (-1/2)(0) + (1/2)(-1) = 0$$

In Contrasts 1 and 3:

$$(1/2)(0) + (-1/2)(1) + (-1/2)(-1) + (1/2)(0) = 0$$

In Contrasts 2 and 3:

$$(1)(0) + (0)(1) + (0)(-1) + (-1)(0) = 0$$

In general, if

$$L = a_1\mu_1 + a_2\mu_2 + \cdots + a_a\mu_a$$

and

$$M = b_1\mu_1 + b_2\mu_2 + \cdots + b_a\mu_a$$

are two linear combinations, then L and M are orthogonal contrasts if

$$\sum_i a_i = 0, \quad \sum_i b_i = 0, \quad \text{and} \quad \sum_i a_i b_i = 0$$

A set of contrasts is mutually orthogonal if every pair of contrasts is orthogonal. An experiment involving a treatments can have several different sets of mutually orthogonal contrasts, but each set consists of at most $a - 1$ orthogonal contrasts.

If the experimenter is able to plan reasonable comparisons of this type prior to the experiment, then the tests can be done within the analysis of variance procedure. If contrasts are not incorporated into the design of the experiment but are suggested during the data gathering or analysis, Scheffé's procedure can be used instead of the procedure discussed here. Also, Scheffé's procedure can be used when the contrasts of interest are not orthogonal. Generally, however, such tests will not be as powerful as those for planned orthogonal contrasts, and it seems reasonable that experiments which are well designed and which test specific hypotheses will have the greatest statistical power.

Example 10.8. One Degree of Freedom Comparisons

Five toothpastes are being tested for their abrasiveness. The variable of interest is the time in minutes until mechanical brushing of a material similar to tooth enamel exhibits wear. The five toothpastes are all the same except for the absence or presence of certain additives. The material is assigned randomly to the treatments.

Toothpaste	Additive
I	Whitener
II	None
III	Fluoride
IV	Fluoride with freshener
V	Whitener with freshener

Group totals and the basic ANOVA table are as follows for four observations per treatment group.

Toothpaste:	I	II	III	IV	V
$T_i = \sum_j y_{ij}$:	197.4	199.0	211.3	215.8	186.5

Source	df	SS	MS	F
Among toothpastes	4	136.8	34.20	39.8
Within toothpastes	15	13.0	0.86	

The investigator deliberately chose these five toothpastes so that the following $a - 1$ orthogonal contrasts could be made.

Comparison	H_0 to Be Tested
Additive vs. no additive	$\frac{\mu_1 + \mu_3 + \mu_4 + \mu_5}{4} - \mu_2 = 0$
Whitener vs. fluoride	$\frac{\mu_1 + \mu_5}{2} - \frac{\mu_3 + \mu_4}{2} = 0$
Whitener vs. whitener with freshener	$\mu_1 - \mu_5 = 0$
Fluoride vs. fluoride with freshener	$\mu_3 - \mu_4 = 0$

To test these comparisons within the ANOVA procedure, the among-SS is partitioned into $a - 1$ components which are each sums of squares for a one degree of freedom F test. The sum of squares for *additive vs. no additive* is found as follows. The null hypothesis is rewritten as

$$H_0: \quad \mu_1 + \mu_3 + \mu_4 + \mu_5 - 4\mu_2 = 0$$

by multiplying by 4. The contrast is then in an equivalent form without fractions:

$$L_1 = \mu_1 + \mu_3 + \mu_4 + \mu_5 - 4\mu_2$$

The coefficients are:

$$a_1 = a_3 = a_4 = a_5 = 1 \quad \text{and} \quad a_2 = -4$$

The sum of squares is

$$SS_{L_1} = \frac{[\sum_i a_i T_i]^2}{n \sum_i a_i^2} = \frac{[197.4 + 211.3 + 215.8 + 186.5 - 4(199)]^2}{4[1^2 + 1^2 + 1^2 + 1^2 + (-4)^2]} = 2.8$$

Similarly, the sum of squares can be found for the other three contrasts:

Whitener vs. fluroide

$$H_0: \quad L_2 = \mu_1 + \mu_5 - \mu_3 - \mu_4 = 0$$

$$SS_{L_2} = \frac{(197.4 + 186.5 - 211.3 - 215.8)^2}{4[1^2 + 1^2 + (-1)^2 + (-1)^2]} = 116.6$$

Whitener vs. whitener with freshener

$$H_0: \quad L_3 = \mu_1 - \mu_5 = 0$$

$$SS_{L_3} = \frac{(197.4 - 186.5)^2}{4[1^2 + (-1)^2]} = 14.9$$

Fluoride vs. fluoride with freshener

$$H_0: \quad L_4 = \mu_3 - \mu_4 = 0$$

$$SS_{L_4} = \frac{(211.3 - 215.8)^2}{4[1^2 + (-1)^2]} = 2.5$$

The ANOVA table is then enlarged as follows:

Source	df	SS	MS	F	$F_{0.05}$
Among toothpastes	4	136.8	34.20	39.8*	3.056
Add. vs. No Add.	1	2.8	2.8	3.3	4.543
Wh. vs. Fl.	1	116.6	116.6	135.6*	4.543
Wh. vs. Wh. & Fr.	1	14.9	14.9	17.4*	4.543
Fl. vs. Fl. & Fr.	1	2.5	2.5	2.9	4.543
Within toothpastes	15	13.0	0.86		

These comparisons show a significant difference between the abrasiveness of the whitener and the fluoride; the whitener is more abrasive. There is also a significant difference between the whitener alone and the whitener with freshener, the latter being still more abrasive.

It should be noted in the above example that the among-SS has been partitioned, that is, divided into nonoverlapping parts, by the orthogonal contrasts. This has an advantage over the multiple comparison procedures of the previous section in that the partition can be used to determine the percentage of variability that is due to the different factors. In this example, the difference between the whitener and the fluoride is responsible for 116.6/136.8 = 85% of the variability among the contrasts.

A significant F test is not a prerequisite for these one degree of freedom tests. In fact, the ANOVA procedure need not be carried out. It is essential, however, that the contrasts be planned before examining the data, otherwise the investigator may be biased by what he sees.

A priori tests of this type are not always possible because there may be insufficient information to set up reasonable contrasts. The experimenter needs a great deal of information to be able to choose treatment groups in such a way that a set of orthogonal contrasts relevant to the experiment will exist. When possible, these contrasts usually answer more relevant questions than multiple comparisons.

The one degree of freedom comparisons can be summarized as follows.

■ ■ ■

Procedure. One Degree of Freedom Comparisons

To test a set of $a - 1$ mutually orthogonal comparisons, write each contrast in the form $L = a_1\mu_1 + a_2\mu_2 + \cdots + a_a\mu_a$ with integer coefficients. Then the sum of squares for each contrast is found by the formula

$$\mathrm{SS}_L = \frac{[\sum_i a_i T_i]^2}{n \sum_i a_i^2}$$

in which T_i is the ith treatment group total and n is the number of observations in each group. This sum of squares has one degree of freedom. The contrast is tested with the statistic

$$F = \frac{\mathrm{MS}_L}{\mathrm{MS}_e}$$

and the comparison is significant if $F \geq F_{\alpha,1,a(n-1)}$.

■ ■ ■

The procedure described in this section applies only to groups of equal sample sizes.

If desired, the sums of squares for the one degree of freedom tests can be computed from the group averages instead of the group totals. In that case the formula becomes

$$\mathrm{SS}_L = \frac{[\sum_i a_i \bar{y}_i]^2\, n}{\sum_i a_i^2}$$

EXERCISES

10.4.1. In the chain saw experiment, test the three comparisons proposed at the beginning of this section by means of one degree of freedom F tests.

10.4.2. Certain people convicted of crimes return to prison over and over again, while others seem to be rehabilitated. To determine whether this may be related to the nature of the first offense, a sociologist sampled prison records of former inmates of the same age. He recorded the nature of the first offense and the total number of times they were imprisoned.

Nature of Crime:	Assault	Rape	Fraud	Embezzlement
Average number of imprisonments:	7.5	5.5	4.5	2.5

a. Make the following orthogonal comparisons if $n = 10$ and $\text{MS}_e = 15$.

Assault vs. rape
Fraud vs. embezzlement
Violent vs. nonviolent

b. What conclusions can be drawn from this analysis?

10.4.3. A study is done on the effectiveness of various types of analgesics. There are six treatment groups, one of which is a control group and receives a placebo. Five persons who have pain are chosen at random for each treatment. All patients take the medication in capsule form and do not know which of the six groups they are in. The capsules that contain aspirin (with or without something else) all contain the same amount of aspirin. The variable of interest is the amount of time (in hours) until relief from pain is felt.

Group	Treatment	$\sum_j y_{ij}$	$[\sum_j y_{ij}]^2$	$\sum_j y_{ij}^2$	$\bar{y}_i$
1	Placebo	20	400	105	4.0
2	Aspirin, Brand 1	5	25	6	1.0
3	Aspirin with caffeine	10	100	19	2.0
4	Aspirin, Brand 2	6	36	7	1.2
5	Aspirin with buffer	8	64	10	1.6
6	Aspirin with buffer and caffeine	11	121	22	2.2
	Totals	60	746	169	

a. State the null and alternative hypotheses.
b. Perform the analysis of variance at $\alpha = 0.01$.
c. Make the following orthogonal comparisons:
   Placebo vs. analgesic
   Pure aspirin vs. aspirin with additives
   Aspirin 1 vs. aspirin 2
   Aspirin with caffeine (alone) vs. aspirin with buffer (with or without caffeine)
   Aspirin with buffer vs. aspirin with buffer and caffeine
d. Show that the set of comparisons in Part c are mutually orthogonal.
e. What part of the variability among groups is caused by the difference between pure aspirin and aspirin with additives?
f. What should the experimenter conclude from the above analyses?

10.5. ESTIMATION

Often an investigator wants to obtain one or more estimates of parameters after an analysis of variance. He may want to estimate μ (the overall mean), $\mu + \alpha_i$ (the ith treatment mean) or α_i (the ith treatment effect). He might also be interested in the difference of two parameters as $\alpha_1 - \alpha_2$ or some other linear combination of parameters as $\mu_1 - (\mu_2 + \mu_3)/2$. Usually he wants the estimate in the form of a confidence interval.

The following table summarizes the point estimators and standard errors needed to form these confidence intervals.

$$\text{CI}_{1-\alpha}\text{: Point Estimator} \pm t_{\alpha/2, N-a}(\text{Standard Error})$$

Parameter	Symbol	Point Estimator	Standard Error
Mean	μ	$\bar{y}$	$\sqrt{\text{MS}_e/N}$
Treatment mean	$\mu_i = \mu + \alpha_i$	$\bar{y}_i$	$\sqrt{\text{MS}_e/n_i}$
Treatment effect	α_i	$\bar{y}_i - \bar{y}$	$\sqrt{\text{MS}_e\left(\dfrac{N - n_i}{n_i N}\right)}$
Difference between treatment means	$\mu_i - \mu_{i'}$ or $\alpha_i - \alpha_{i'}$	$\bar{y}_i - \bar{y}_{i'}$	$\sqrt{\dfrac{\text{MS}_e}{n_i} + \dfrac{\text{MS}_e}{n_{i'}}}$

$\text{CI}_{1-\alpha}$: Point Estimator $\pm$ $t_{\alpha/2,N-a}$(Standard Error) (*Continued*)

Parameter	Symbol	Point Estimator	Standard Error
A linear combination of means	$\sum_i a_i \mu_i$ with $\sum_i a_i = 0$	$\sum_i a_i \bar{y}_i$	$\sqrt{\text{MS}_e \left[\sum_i \left(\frac{a_i^2}{n_i} \right) \right]}$

All of the standard errors except the one for the treatment effect can be seen to follow from the properties of the variance of a linear combination of random variables. The standard error for the treatment effect is different because $\bar{y}_i$ and $\bar{y}$ are dependent.

Example 10.9. Confidence Intervals Related to ANOVA
In the chain saw study, Example 10.1 of Section 10.2, the averages are:

$\bar{y}_A$	$\bar{y}_B$	$\bar{y}_C$	$\bar{y}_D$	$\bar{y}$
33	43	49	31	39

$n = 5$ and $\text{MS}_e = 101.25$. Some of the possible point estimates are given in Figure 10.5.

The experimenter wants to find 95% confidence intervals for the overall mean, for the mean of Model B, for the Model B effect, for the difference between Model A and Model D, and for the difference between the oldest model, Model A, and the average of the three newer models.

Overall mean, μ

$$\text{CI}_{0.95}: \quad \bar{y} \pm t_{0.025,a(n-1)} \sqrt{\frac{\text{MS}_e}{N}}$$

$$39 \pm 2.120 \sqrt{\frac{101.25}{20}}$$

$$39 \pm 4.77$$

Mean of Model B, μ_B

$$\text{CI}_{0.95}: \quad \bar{y}_B \pm t_{0.025,a(n-1)} \sqrt{\frac{\text{MS}_e}{n}}$$

$$43 \pm 2.120 \sqrt{\frac{101.25}{5}}$$

$$43 \pm 9.5$$

Model B effect, α_B

$$\text{CI}_{0.95}:\quad \bar{y}_B - \bar{y} \pm t_{0.025,a(n-1)}\sqrt{\text{MS}_e\frac{(N - n_i)}{n_iN}}$$

$$(43 - 39) \pm 2.120\sqrt{101.25\,\frac{(20 - 5)}{5(20)}}$$

$$4 \pm 8.27$$

Since this interval contains zero, Model B does not differ significantly from the overall mean of all four models.

The difference between Model A and Model D, $\mu_A - \mu_D$

$$\text{CI}_{0.95}:\quad \bar{y}_A - \bar{y}_D \pm t_{0.025,a(n-1)}\sqrt{\frac{\text{MS}_e}{n} + \frac{\text{MS}_e}{n}}$$

$$(33 - 31) \pm 2.120\sqrt{\frac{2(101.25)}{5}}$$

$$2 \pm 13.49$$

Since this interval contains zero, Models A and D do not differ significantly with respect to kickback.

The difference between Model A and the average of the other three models, $\mu_A - (\mu_B + \mu_C + \mu_D)/3$

$$a_A = 1, \qquad a_B = a_C = a_D = -1/3 \qquad \sum a_i = 0$$

$$\text{CI}_{0.95}:\quad \bar{y}_A - \frac{\bar{y}_B + \bar{y}_C + \bar{y}_D}{3} \pm t_{0.025,a(n-1)}\sqrt{\text{MS}_e \sum_i \frac{a_i^2}{n}}$$

$$33 - \frac{43 + 49 + 31}{3} \pm 2.120\sqrt{101.25\left[\frac{1^2 + 3(-1/3)^2}{5}\right]}$$

$$-8 \pm 11.0$$

Thus the older one does not seem to be significantly different from the average of the three newer ones.

The investigator should remember that repeated estimates within the same experiment will not preserve the original α level. By chance alone, one or more of the intervals may fail to cover the parameter. One way to avoid

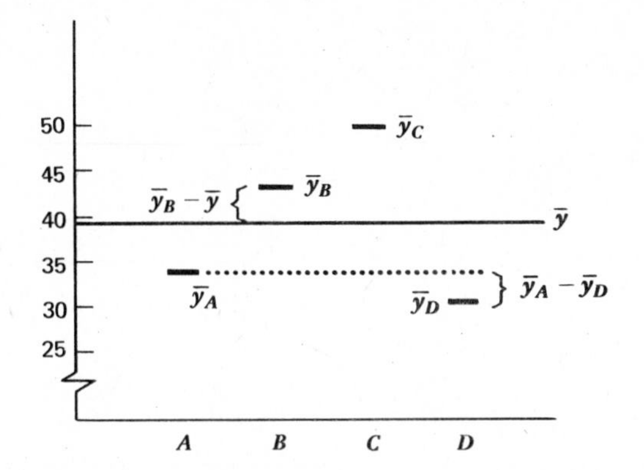

FIGURE 10.5. Point estimators of parameters in ANOVA.

this difficulty is to set the α level for the entire set of confidence intervals desired. If m confidence intervals are involved, then $t_{\alpha/2m,N-a}$ is used for each individual confidence interval. The set of intervals is then called *multiple-t confidence intervals*. A t table that lists very small values of α is necessary to find most multiple-t confidence intervals.

EXERCISES

10.5.1. In the insecticide study of Exercise 10.3.2,

a. Place a 95% confidence interval on the overall experimental mean.
b. Place a 99% confidence interval on the effect of the third insecticide.
c. Place a 90% confidence interval on the difference between the second and the fourth insecticide.
d. Place a 95% confidence interval on the fifth treatment mean.

10.5.2. In the spider study of Exercise 10.3.3,

a. Place a 95% confidence interval on the mean of the second treatment.
b. Place a 95% confidence interval on the difference between the mean of the first and the third treatments.
c. Place a 95% confidence interval on the difference between the first and second treatment effects.

10.5.3. Four normal populations with homogeneous variances give rise to the following data from random samples.

Group			
1	2	3	4
52	40	38	48
41	28	33	36
52		27	38
39		33	38
		39	48
			49
			36
			38
			47

a. Perform an analysis of variance.
b. Estimate $\mu_1 - \mu_3$ with a 90% confidence interval.
c. Estimate μ with a 90% confidence interval.
d. Estimate α_3 with a 90% confidence interval.
e. Estimate $\frac{\mu_1 + \mu_4}{2} - \frac{\mu_2 + \mu_3}{2}$ with a 90% confidence interval.

REVIEW EXERCISES

Decide whether each of the following statements is true or false. If a statement is false, explain why.

10.1. In an analysis of variance (ANOVA), there is a degree of freedom associated with each squared total in the uncorrected sums of squares.

10.2. The standard deviation among sample averages is called the standard error and is computed from an analysis of variance procedure by (within-MS)/n.

10.3. Either a t test or an analysis of variance may be used if only two treatment groups are being compared.

10.4. In ANOVA the uncorrected total sum of squares will be equal to or greater than any other corrected or uncorrected sum of squares.

10.5. An analysis of variance uses both sides of the F distribution for critical values because the alternative hypothesis contains $\neq$.

10.6. An analysis of variance cannot be done if the treatment groups are unequal in size.

10.7. An analysis of variance requires that all treatment groups have the same variance, and this variance is estimated by MS_e.

10.8. If the null hypothesis is rejected in an analysis of variance, we can conclude that the group with the smallest sample average has a mean that is different from all of the other group means.

10.9. In an analysis of variance, the data from a control group are handled in a manner different from the treatment groups.

10.10. Fisher's least significant difference requires equal treatment group sizes.

10.11. Fisher's procedure is the only multiple comparison procedure available to the researcher.

10.12. A confidence interval on the difference between two treatment means is the same as a confidence interval on the difference between two treatment effects.

10.13. The method of one degree of freedom comparisons is an example of a multiple comparison procedure.

10.14. The correction factor is the average variability from the overall average.

10.15. Multiple comparison procedures and orthogonal contrasts are both methods for drawing conclusions from experiments in which H_0 is not true.

10.16. It is common to imbed a set of multiple comparisons into the design of an experiment for which ANOVA will be used.

10.17. A set of mutually orthogonal contrasts can be used to make all pairwise contrasts among a set of group means.

10.18. Although the F test involves variances, when it is used in ANOVA it is to test hypotheses about means.

10.19. An F test is used to decide whether Duncan's test should be used to find significant differences among group means.

10.20. Orthogonal comparisons can be used to divide the treatment mean square into independent parts the sum of which equals the treatment mean square.

SELECTED READINGS

Anderson, R. L. (1965). Negative variance estimates, *Technometrics,* **7,** 75–76.

Andrews, H. P., and R. D. Snee (1980). Graphical display of means, *The American Statistician,* **34,** 195–199.

Bernhardson, C. S. (1975). Type I error rates when multiple comparison procedures follow a significant F test of ANOVA, *Biometrics,* **31,** 229–232.

Carmer, S. G., and M. R. Swanson (1971). Detection of differences between means: A Monte Carlo study of five pairwise multiple comparison procedures, *Agronomy Journal,* **63,** 940–945.

——— (1973). Evaluation of ten pairwise multiple comparison procedures by Monte Carlo methods, *Journal of the American Statistical Association,* **68,** 66–74.

Chew, V. (1977). *Comparisons Among Treatment Means in an Analysis of Variance*, Agricultural Research Service, U.S. Department of Agriculture, Washington, D.C.

Duncan, D. B. (1955). Multiple range and multiple F tests, *Biometrics*, **11**, 1–42.

Dunn, O. J. (1961). Multiple comparisons among means, *Journal of the American Statistical Association*, **56**, 52–64.

Dunnett, C. W. (1955). A multiple comparison procedure for comparing several treatments with a control, *Journal of the American Statistical Association*, **50**, 1096–1121.

Keuls, M. (1952). The use of the "Studentized range" in connection with an analysis of variance, *Euphytica*, **1**, 112–122.

Kramer, C. Y. (1956). Extension of multiple range tests to group means with unequal numbers of replications, *Biometrics*, **12**, 307–310.

Light, R. J., and B. H. Margolin (1971). An analysis of variance for categorical data, *Journal of the American Statistical Association*, **66**, 534–544.

Scheffé, H. (1953). A method for judging all contrasts in the analysis of variance, *Biometrika*, **40**, 87–104.

——— (1959). *The Analysis of Variance*, Wiley, New York.

Shaffer, J. P. (1977). Multiple comparisons emphasizing selected contrasts: An extension and generalization of Dunnett's procedure, *Biometrics*, **33**, 293–303.

Sirotnik, K. (1971). On the meaning of the mean in ANOVA (or, The case of the missing degree of freedom), *The American Statistician*, **25**(Oct.), 36–37.

Tukey, J. W. (1949). Comparing individual means in the analysis of variance, *Biometrics*, **5**, 99–114.

11

The Analysis of Variance Model

Now that we are familiar with the basic ANOVA procedure, we need to look more closely at the underlying model and its assumptions.

11.1. RANDOM EFFECTS AND FIXED EFFECTS

The one-way analysis of variance discussed in Chapter 10 can be applied to many different experiments. For example, it could be used to pick the least corrosive chemical from among six chemicals that are all effective for melting ice. Or it could be used to test whether there is significant variability among the achievements of introductory economics classes when they use the same method and materials but are taught by different teachers.

In Chapter 10 we assumed experimental situations similar to the ice-melting chemical example. That is, we assumed that all treatments of interest, the six chemicals, were included in the experiment. This type of ANOVA is based on a model called the *fixed effects model.* In this model the experimenter—usually in the latter stages of experimentation—narrows down the possible treatments to several in which he has a special interest. In the case of the chemicals, for example, tests would already have been completed to determine that these six were all available, suitable for melting ice, and economically feasible. Now a final choice is to be made on the basis of corrosiveness. In the fixed effects model we are usually trying to pick the best of several possibilities. The inference made is restricted to the treatments used in the experiment.

The fixed effects model is sometimes called Model I. It is referred to as "fixed" because if the investigator decided to repeat the experiment he would use the same treatments in the repetition.

The achievement of economics classes taught by different teachers is an example of Model II, or the *random effects model*; it is also called the components of variance model. The random effects model assumes that the treatments are a random sample of all of the treatments of interest. It does not look for differences among the group means of the treatments being tested, but rather asks whether there is significant variability among all possible treatment groups. For example, if five teachers were used in the study, these five teachers would be the treatments and the grades of their students on some standardized test might be the variable of interest. The investigator would be interested in the variability among *all* economics teachers using this method and these materials. The five teachers in the experiment are a random sample from all of the treatments of interest. If the experiment were to be repeated, five different teachers chosen at random would be used.

When the random effects model is used, the investigator is interested in σ_A^2, the variability among all possible treatment groups. The ANOVA procedure can be used to test H_0: $\sigma_A^2 = 0$. If this null hypothesis is rejected, there is evidence of variability among groups. In the teacher example, if the null hypothesis is rejected, teachers do have an effect on the achievements of introductory economics classes. The inference is to all economics teachers, not just the five involved in the study.

In Chapter 10 we did not consider examples that follow the random effects model. The assumptions for the underlying mathematical additive model $y_{ij} = \mu + \alpha_i + \epsilon_{ij}$ differ for fixed effects and random effects. However, the numerical procedure for the one-way ANOVA is identical for both models.

The following table summarizes the two models.

$$y_{ij} = \mu + \alpha_i + \epsilon_{ij}$$

$$i = 1, 2, \ldots, a$$
$$j = 1, 2, \ldots, n$$

	Fixed Effects Model (FEM)		Random Effects Model (REM)
H_0:	$\mu_1 = \mu_2 = \cdots = \mu_a$	H_0:	$\sigma_A^2 = 0$
H_a:	At least one inequality	H_a:	$\sigma_A^2 > 0$
μ:	A constant, the mean of all possible experiments using the *a* designated treatments	μ:	A constant, the population mean for all experiments involving all possible treatments of the type being considered
α_i:	A constant for the *i*th treatment group, the deviation from the mean due to the *i*th treatment: $\sum_i \alpha_i = 0$	α_i:	A constant for the *i*th treatment group, a random deviation from the population mean. The α_i's are normal, with $E(\alpha_i) = 0$ and $V(\alpha_i) = \sigma_A^2$

	Fixed Effects Model (FEM)		Random Effects Model (REM)
ϵ_{ij}:	A random effect containing all uncontrolled sources of variability. The ϵ_{ij}'s are IND$(0,\sigma^2)$, that is, they are normally distributed with a mean of zero and a variance σ^2 and they are independent of each other and of the α_i's	ϵ_{ij}:	Same as for FEM
MS_a:	Estimates $\sigma^2 + n \sum_i \alpha_i^2/(a-1)$	MS_a:	Estimates $\sigma^2 + n\sigma_A^2$
MS_e:	Estimates σ^2	MS_e:	Estimates σ^2

In both models we assume that the experimental units are chosen at random from the population and assigned at random to the treatments. Frequently these assumptions are not completely met. Sometimes it is almost impossible to obtain a random sample from the entire population of interest. For example, the investigator may want to make inference about all white mice but must use a random sample of the white mice received from distributors. Or a researcher may be studying the effect of exercise on blood pressure in human males and may want to make inference to all males but may have to use volunteers with no opportunity to choose subjects at random. In both of these examples, however, it is possible to assign the subjects at random to the treatments. In some other investigations, even this second stage of randomization is not possible. For example, in a study of the effect of different teaching methods on the learning of college students, the investigator may have to utilize for the treatment groups the classes in which the students have enrolled. In this example there is no opportunity for a random choice of students nor for a random assignment of the students to the treatments.

The ANOVA procedure is reliable if the assumptions are met. The more the experiment deviates from the assumptions, the less reliable are the conclusions. An investigator should mention any shortcomings of this type in the report of the study.

The follow-up procedures after ANOVA will differ depending upon whether the fixed effects or random effects model is being used. For the FEM we use multiple comparisons, orthogonal contrasts, or estimation of parameters (or linear combinations of parameters). For REM we are interested in the *intraclass correlation*, an estimate of the percentage of the total variability that is due to the differences among the treatments.

The intraclass correlation serves a function similar to that of the coefficient of determination which we examined in our study of linear trend. The intraclass correlation gives the percentage of the variability that is explained by the groups or treatments in the model. If the effects α_i are on the numerical scale, we could compute the coefficient of determination r^2 but it would never be greater than the intraclass correlation r_I. That is because r^2 gives the percentage of variability explained by a linear relationship, whereas r_I provides the percentage explained by any relationship. Another advantage of r_I is that the groupings or α_i effects can be on the nominal scale and we can still obtain a statement of relationship of the treatments to the y variable and have an estimate of the variability explained by the method of groupings employed in the experiment.

Example 11.1. One-way ANOVA for the Random Effects Model

An educational psychologist is looking for evidence of the heritability of intelligence. He selects 30 identical twins at random and administers an IQ test to each twin.

The 30 pairs of twins are his treatment groups. Each group has a sample size of two. These 30 pairs are a random sample of all possible pairs of twins, so this is the random effects model.

The ANOVA is carried out as in Chapter 10 except that the null hypothesis is H_0: $\sigma_A^2 = 0$.

Source	df	SS	MS	F
Among twin pairs	$a - 1 = 29$	25,921	894	4.43
Within pairs	$a(n - 1) = 30$	6,050	202	

Since $F_{0.05,29,30} = 1.847$, H_0 is rejected and there is significant variability among the IQ scores of the twin pairs, that is, there is some evidence of the heritability of intelligence.

Since MS_a estimates $\sigma^2 + n\sigma_A^2$ and MS_e estimates σ^2, the investigator computes the intraclass correlation r_I as follows:

$$\hat{\sigma}^2 = \text{MS}_e = 202$$

$$\hat{\sigma}_A^2 = \frac{\text{MS}_a - \text{MS}_e}{n} = \frac{894 - 202}{2} = 346$$

$$r_I = \frac{\hat{\sigma}_A^2}{\hat{\sigma}_A^2 + \hat{\sigma}^2} = \frac{346}{346 + 202} = 0.631$$

that is, 63.1% of the total variability in IQ's is due to the differences among the twin pairs.

There are many experimental situations in which the random effects model is used and the intraclass correlation is calculated. For example, in an environmental study on the amount of lung damage in wild animals in a heavily industrial region, the region is divided into sections, random sections chosen, and traps set to capture a sample of animals. The random sections are the treatments and the intraclass correlation indicates the amount of variability in lung damage due to the different sections.

For another example, the random effects model would be used in a preliminary study to see if bees are attracted to color. Alfalfa blossoms range in color from dark purple to yellow to white. A random sample of alfalfa plants with different colored blossoms is chosen. The number of visits of bees to the different plants is the variable of interest. If the null hypothesis is rejected, plans can be made to conduct experiments that would reveal the specific color or colors that attract bees.

When the intraclass correlation is computed, the investigator is interested in the percentage of the total variability due to the treatments. The specific percentage that is meaningful depends on the experiment. If the investigator is looking for evidence of repeatability, as in a lab test to measure blood sugar where the treatment groups are different samples of blood, he will want a high intraclass correlation, perhaps 95%. In many other situations a lower value is meaningful, for example, in a study to see if obesity runs in families in which the families are the treatment groups. It is possible that the ANOVA procedure leads to a significant F value, but at the same time r_I is so small for the purpose of experiment that the variability that has been uncovered is not useful.

The intraclass correlation procedure can be summarized as follows.

■ ■ ■

Procedure. Intraclass Correlation

Perform the ANOVA as in Chapter 10.
Estimate $\sigma_A{}^2$ and σ^2 as follows:

$$\hat{\sigma}^2 = \text{MS}_e$$

$$\hat{\sigma}_A{}^2 = \frac{\text{MS}_a - \text{MS}_e}{n}$$

Then r_I, the intraclass correlation, is:

$$r_I = \frac{\hat{\sigma}_A{}^2}{\hat{\sigma}_A{}^2 + \hat{\sigma}^2}$$

$$0 \leq r_I \leq 1$$

The intraclass correlation can be interpreted as the proportion of the total variability due to the differences in all possible treatments of this type.

■ ■ ■

EXERCISES

11.1.1. Decide whether each of the following is using the fixed effects model or the random effects model.

a. A professor is trying to select a textbook for a sociology course from four different ones which are available. He divides his students at random into four groups and assigns the textbooks to the groups at random. After using the different books for the course, all students still enrolled take the same examination. ANOVA is used to analyze the results.

b. A manufacturer builds a piece of equipment to turn out machined parts. In order to study the performance of his machines, he selects eight machines at random and then selects ten parts at random from the production of each of these machines. He measures the lengths of the 80 pieces and performs an analysis of variance.

c. An educator wishes to study the competence in algebra of all New York City students who have just completed the ninth grade. Five junior high schools are selected at random, and within each school a random sample of ninth-grade students are given examinations. Using these scores, the hypothesis that there is variability among the schools is tested.

d. Worms are classified into three groups by a structural characteristic: small, medium, or large ventral flap. Three random samples of 11 worms are taken from each group and the weight of each worm is recorded. The hypothesis is tested that the mean weight of each group is the same.

e. A psychologist devises an examination in such a way that the final score depends almost entirely upon the ability of the subject to follow instructions. The test is given to 40 students who have been divided into four equal groups at random. The instructions are given in the following four ways:

Group I	written and brief
Group II	oral and brief
Group III	written and detailed
Group IV	oral and detailed

An analysis of variance is performed.

11.1.2. An analysis of variance is used to study the effect of seam differences on variability in the sulfur content of coal. Seams and samples from seams are taken at random.

Source	df	SS	MS
Among coal seams	24	2400	100
Within coal seams	125	5000	40

a. Do differences among seams contribute significantly to the variability in the sulfur content of coal?
b. What percentage of the variability in the sulfur content of coal is attributable to seam differences?
c. Would you advise coal producers in search of low-sulfur coal to seek low-sulfur seams or to seek other factors that might affect variability? Justify your answer on the basis of the above analyses.

11.1.3. Given the following data from a (fictional) study of obesity:

	Brothers			
Family	Pounds Overweight			Total
A	59	66	83	208
B	70	87	90	247
C	67	83	92	242
D	83	78	77	238
E	82	95	90	267
F	96	75	78	249
G	101	78	66	245
H	79	79	84	242
I	85	72	89	246
J	84	79	83	246

$\sum_i \sum_j y_{ij}^2 = 199{,}508 \qquad \sum_i (\sum_j y_{ij})^2 = 592{,}392 \qquad \sum_i \sum_j y_{ij} = 2430$

a. Complete the analysis of variance.
b. Compute the intraclass correlation.
c. What conclusion do you draw?

11.1.4. Given the following analysis of variance, compute the intraclass correlation.

Source	df	SS	MS
Among treatments	10	4368	436.8
Within treatments	33	4320	130.9

11.1.5. Suppose a physiologist is working on a new method to measure blood sugar. Blood samples are taken from 10 people, and two assays are done on each sample.

Source	df	SS	MS
Among persons	9	1710	190
Within persons	10	100	10

a. Which model is being used?
b. What is the null hypothesis?
c. Should the null hypothesis be rejected?
d. Compute the intraclass correlation.
e. Does this new method seem to be reliable?

11.1.6. Fifteen varieties of corn are chosen at random from all available varieties, and plots are planted of each variety. At maturity, five random plants are chosen from each plot and the yield is measured, leading to the following analysis.

Source	df	SS	MS
Among corn varieties	14	4368	—
Within corn varieties	—	—	72

a. Complete the analysis of variance.
b. Compute the intraclass correlation.
c. Interpret the intraclass correlation.

11.2. TESTING THE ASSUMPTIONS FOR ANOVA

In both the fixed effects and random effects models we assume the observations fit the additive model

$$y_{ij} = \mu + \alpha_i + \epsilon_{ij}$$

in which the ϵ_{ij}'s are IND$(0,\sigma^2)$. In practice, this means:

1. The treatment groups are normally distributed (this is required so that the ϵ_{ij}'s will be normally distributed).
2. The treatment groups all have the same variance (this is required so that the ϵ_{ij}'s will have the same variance for each i).
3. The experimental units are picked at random and assigned at random to the treatment groups (this is required so that the ϵ_{ij}'s are independent of each other and the α_i's).

We discuss each of these conditions in turn.

Normality. The normality of the treatment groups can be roughly checked by constructing histograms of the sample from each treatment group. Histograms reveal skewness and bimodality. Another approach is to plot the cumulative frequencies on normal probability paper; a normal distribution leads to a straight line. Unfortunately, a large number of observations are needed for both of these procedures. The analysis of variance, however, leads to valid conclusions in some cases where there are departures from normality. For small sample sizes the treatment groups should be symmetric and unimodal. For large samples, more radical departures are acceptable since the central limit theorem comes into play. Thus if there is doubt about normality, one solution is to use a large number of observations.

Some traditionally small experiments lead to non-normal distributions:

1. Data composed of small counts, even into the hundreds, such as the number of parasites on wildlife.
2. Data composed of very large counts, such as bacterial counts.
3. Proportions, or percentage data.
4. Arbitrary scales, such as a 10-point taste test.
5. Weights of very small things.

In the first three cases, not only is the assumption of normality invalid but the variances of the treatment groups may be unequal and there may be a lack of independence between the random effect and the treatment effect. One approach in these cases is to transform the data and perform the ANOVA on the transformed values; this is discussed in Section 11.3.

In experiments involving arbitrary scales, as the taste test, normality can be approximately achieved by using several tasters (five or more) and recording their average ratings.

Weights of very small things are often not normally distributed because of the limits of the accuracy of the weighing process. Weighing objects in groups can sometimes overcome this difficulty.

Equality of Variances. An analysis of variance assumes homogeneity of variances (*homoscedasticity*), that is, all of the treatment groups have the same variance. F tests are robust with respect to departures from homogeneity, that is, moderate departures from equality of variances do not greatly effect the F statistic. If the experimenter fears a large departure from homogeneity, several procedures are available to test

$$H_0\colon \sigma_1^2 = \sigma_2^2 = \cdots = \sigma_a^2$$

Unfortunately, most of these tests rely on the assumption of normality.

We discuss here only one test for homogeneity of variances, the F-max test developed by Hartley (1950). Hartley's test is one of the simplest; it may be used when all treatment groups are the same size and involves comparing the largest sample variance with the smallest sample variance.

Example 11.2. *F*-max Test for Homogeneity of Variances
In the chain saw study (Example 10.1), the investigator wants to test

$$H_0\colon\ \sigma_A^2 = \sigma_B^2 = \sigma_C^2 = \sigma_D^2$$

He first computes the sample variance for each treatment group

	Group			
	D	A	B	C
$\sum_j y_{ij}$	155	165	215	245
$\sum_j y_{ij}^2$	4981	5999	9965	12,175
$(\sum_j y_{ij})^2/n$	4805	5445	9245	12,005
s_i^2	44.0	138.5	180.0	42.5

$$F_{\max} = \frac{\text{largest treatment variance}}{\text{smallest treatment variance}}$$

$$= \frac{180}{42.5} = 4.24$$

$F_{\max}$ is significant if it exceeds the value given in the table computed by Hartley, Table A.15 in the Appendix of Useful Tables. This table is entered by a, the number of treatment groups, and $\nu = n - 1$, in which n is the number of observations per treatment group. In this example

$$F_{\max_{0.05,a,\nu}} = F_{\max_{0.05,4,4}} = 20.6$$

Thus the null hypothesis of homogeneity of variances is accepted.

Hartley's procedure can be summarized as follows.

■ ■ ■

Procedure. Hartley's Test for Homogeneity of Variances

To test:

$H_0\colon\ \sigma_1^2 = \sigma_2^2 = \cdots = \sigma_a^2$ against H_a: *At least one inequality*

when each of the a populations is normal and there is a random sample of size n from each population, compute:

$$s_1^2, s_2^2, \ldots, s_a^2$$

and calculate:

$$F_{\max} = \frac{\text{largest } s_i^2}{\text{smallest } s_i^2}$$

$F_{\max}$ is significant if it equals or exceeds the value $F_{\max_{\alpha,a,\nu}}$ in Hartley's table, Table A.15 in the Appendix, with a the number of populations and $\nu = n - 1$.

■ ■ ■

Because of the sensitivity of this test to departures from normality, if $F_{\max}$ is significant it indicates either unequal variances or a lack of normality.

Two other commonly used tests of homogeneity of variances are those of Cochran and Bartlett. In most situations, Cochran's test is equivalent to Hartley's. Bartlett's test has a more complicated test statistic but has two advantages over the other two: It can be applied to groups of unequal sample sizes, and it is more powerful. Scheffé has a test that is less sensitive to departures from normality. For a discussion of these tests see Winer (1971), pages 205–220.

If the experimenter finds that only one or two of the treatment groups have a different variance, he might discard these samples and work only with the remaining ones. However, if discarding these treatment groups makes it impossible to answer the experimental questions, another approach may be needed. One possibility is to transform the data as described in Section 11.3; another would be a nonparametric technique in place of ANOVA.

Independence. The random effects (ϵ_{ij}'s) in the additive model must be:

1. Independent of each other
2. Independent of the treatment effects (α_i's)

If these conditions are missing, it will be difficult to detect real differences that may exist.

The first condition is usually satisfied if the experimental units are randomly chosen and randomly assigned to the treatments. If the treatment groups already exist, such as members of a certain profession, the experimenter does not have the opportunity to assign the subjects at random to the treatments. In such cases he uses random samples from each treatment group.

It is not usually acceptable to use ANOVA on repeated observations on the same subject unless precautions are taken to avoid a systematic effect caused by the repetition of the experiment, for example, learning by the subject who repeats the same task. Sometimes lack of independence occurs because of instrument wear or drift. This type of dependence within groups can be detected by plotting the data in the order in which they were collected.

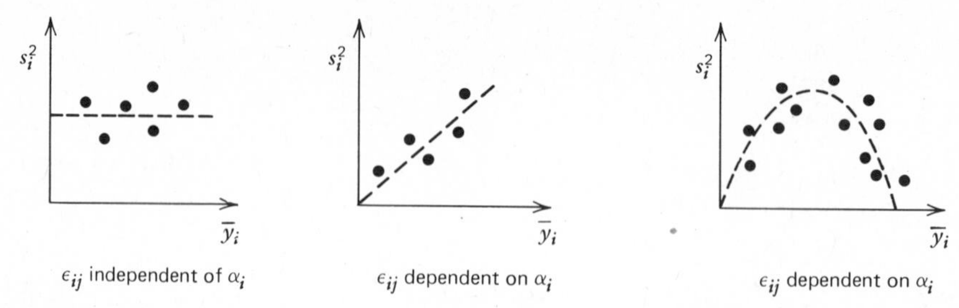

FIGURE 11.1. Visual test for the independence of the error term and the treatment effect.

The second condition, that the random effect is independent from the treatment effect, can be checked by plotting the sample means against the sample variances (Figure 11.1). Independence will lead to an unpatterned scatter around an horizontal line, while dependence usually takes the form of some curve. A transformation can sometimes be used to remove this type of dependence.

EXERCISES

11.2.1. Given below are the calculations from an experiment involving the breaking strengths of six different fabrics.

	Nylon	Rayon	Linen	Dacron	Cotton	Silk
$\sum_j y_{ij}$	144	96	119	168	98	140
n	10	10	10	10	10	10
$\sum_j y_{ij}^2$	2080.8	1063.8	1449.4	2904.4	1018.0	1979.8
$(\sum_j y_{ij})^2/n$	2073.6	921.6	1416.1	2822.4	960.4	1960.0

a. Test to decide whether the different fabrics have a common variance for breaking strength.

b. Which variances are significantly different from each other? (*Hint:* Test all pairs of variances by using a two-way table similar to the table for multiple comparisons; however, use the ratios of the variances and F-max tests along each diagonal.)

11.2.2. In the light bulb experiment, Exercise 10.2.3, show that the variances of the three brands are equal.

11.2.3. In the orange-juice experiment, Exercise 10.2.5, show that there is no evidence that the variances of vitamin C are different among the three methods of processing orange juice.

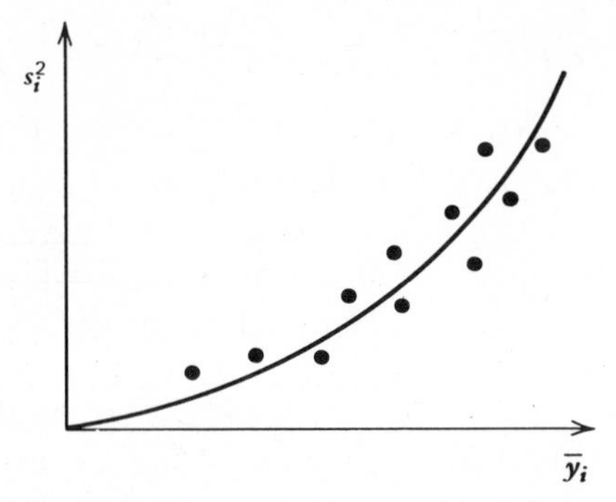

FIGURE 11.2. Data that may be improved by a log transformation.

11.3. TRANSFORMATIONS

If we find that the variances are not homogeneous, or if we find a lack of normality, or if there is a dependence between the treatment effects and the random effects, it is sometimes possible to use a transformation to get the data into a form for which the analysis of variance is valid. A transformation replaces each observed value y_{ij} by another value x_{ij} according to a certain rule, for example, $x_{ij} = \log y_{ij}$. It is essential that any transformation preserve the order of the data values; thus, if y_1 and y_2 are transformed to x_1 and x_2 respectively and $y_1 < y_2$, then $x_1 < x_2$. Since the order of the observations is not changed by the transformations we use, any conclusion about differences in the transformed data are true for the original data. This technique, however, has the disadvantage that we must report results in unusual units of measure, as the log of a length or the square root of the number of fish.

Various transformations are available, and there are sometimes clues to help the experimenter decide which one to use. One approach is to plot the sample averages against the sample variances. If the graph has a parabolic shape (Figure 11.2), the log transformation usually helps. This shape often

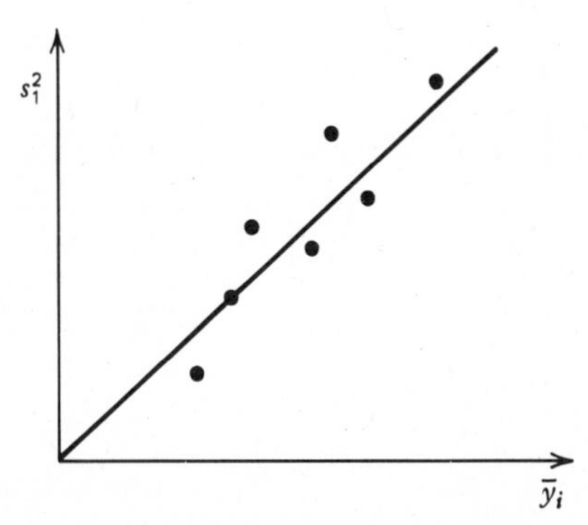

FIGURE 11.3. Data that may be improved by a square-root transformation.

occurs when the data arise from large counts (as blood counts or bacterial counts). Each observation y_{ij} is transformed to $x_{ij} = \log y_{ij}$ or to $x_{ij} = \log(y_{ij} + c)$ with $c > 0$ if zero or negative numbers are in the data. Either logs base 10 or logs base e may be used. Table A.16 in the Appendix is a table of logs base 10. These log transformations preserve the order of the data, the order of the means, and the order of the variances. They reduce the amount of difference among the variances. The analysis of variance is carried out as usual, except that the x_{ij} replaces each y_{ij}. Before beginning the ANOVA procedure, however, it is wise to check the transformed data to see that it has the properties of normality, homogeneity of variances, and independence.

A graph that frequently appears when sample averages are plotted against sample variances for small counts is a straight line at a 45° angle (Figure 11.3). This graph indicates a Poisson distribution in which $\mu = \sigma^2$. The transformation that often helps is to replace y_{ij} with $x_{ij} = \sqrt{y_{ij}}$ or $x_{ij} = \sqrt{y_{ij} + c}$.

If the data are from a population with a binomial distribution (percentage or proportion data), the mean and the variance are not independent,

$$\mu = n\pi \qquad \text{and} \qquad \sigma^2 = n\pi(1 - \pi)$$

The diagram in this case has the form found in Figure 11.4. A transformation often used in this case, especially if $\pi < 0.2$ or $\pi > 0.8$, is arc sin $\sqrt{y_{ij}}$ in which y_{ij} is expressed as a percentage. Tables are available for this transformation. Table A.17 in the Appendix is one such table.

Since the analysis of variance was designed for *continuous* variables and proportions arise from *discrete* variables, the investigator should remember that ANOVA may not be the best way to analyze data of this type. In fact, an F test with or without a transformation may be less powerful than the appropriate procedure. Sometimes the investigator may decide to use ANOVA because of its convenience or for reporting results in a uniform way when ANOVA is being used on other variables in the study. This approach, however, is at most second best.

Many transformations are available in addition to the ones discussed in

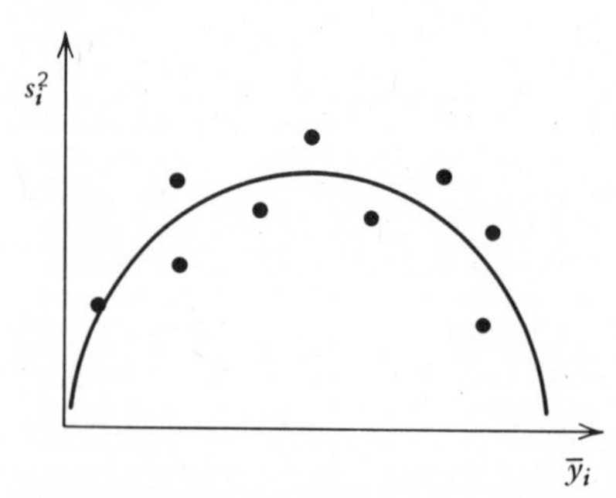

FIGURE 11.4. Data that may be improved by an angular transformation.

this section. Some computer packages offer several to the investigator. It is invalid to transform the data by each available transformation and perform ANOVA in order to pick out the transformation that leads to significant results. However, several transformations can be used on the data, and the one that equalizes the ranges of the samples the most could be used for ANOVA since the ranges are closely related to the variances. If the ranges are not very different, then the variances may be homogeneous.

EXERCISES

11.3.1. In a certain experiment in graph reading, subjects take the following amounts of time (in seconds divided by 10) to answer a set of questions.

	Group		
	A	*B*	*C*
	28	16	31
	17	13	22
	18	16	16
	21	12	21
	13	13	13
	29	12	16
$\sum_j y_{ij}$	126	82	119
$\sum_j y_{ij}^2$	2848	1138	2567

a. Show that the variances of the groups are unequal using Hartley's F-max test.

b. Use a square root transformation on the times.

c. Does the transformation correct the lack of homogeneity of variances?

d. Perform ANOVA on the transformed data.

e. How would the results of the ANOVA be reported?

11.3.2. Four groups of subjects were given a certain task to perform. The number of mistakes out of 18 trials is recorded.

Group	Errors out of 18 Trials							
1	0	0	0	0	1	3	0	0
2	5	3	2	11	3	0	0	0
3	1	1	0	0	2	3	0	0
4	3	0	1	0	4	1	1	2

a. Convert the number of errors to percentage of errors.
b. Show that the groups have unequal variances when the variable is percentage of errors.
c. Use arc sin $\sqrt{\%}$ to transform the data.
d. Check the transformed data for homogeneity of variance.

REVIEW EXERCISES

Decide whether each of the following statements is true or false. If a statement is false, explain why.

11.1. The random effects model could be called the component of variance model because the experimenter is more interested in causes of variation than in comparing means.

11.2. Because of a general lack of knowledge about the nature of effects, the random effects model is probably more common than the fixed effects one.

11.3. The experimenter does not test for homogeneity of variance unless he has reason to doubt this customary assumption for the analysis of variance.

11.4. If Hartley's test is significant when performed on the original data, a suitable transformation will result in nonsignificance when the test is performed on the transformed data.

11.5. The proper transformation should provide a more powerful F test than one based on the original data, which do not meet the conditions for an analysis of variance.

11.6. If in a scientific journal an analysis of variance is based on the additive model $y_{ij} = \pi + \theta_i + \delta_{ij}$, the reader has enough information to distinguish whether or not it was a fixed effects model.

11.7. When the model is $y_{ij} = \mu + \alpha_i + \epsilon_{ij}$, the same F test will be performed whether the α_i's are fixed or random.

11.8. Multiple comparison procedures such as Tukey's honestly significant differences are used to determine differences among fixed effects, but for random effects the investigator is more interested in whether there is variability among the effects than in making comparisons among them.

11.9. If the sample sizes are large, the experimenter should always check for normality prior to an analysis of variance.

11.10. Transformations can correct non-normality, unequal variances, and lack of independence between the ϵ_{ij}'s and the α_i's.

12

Other Analysis of Variance Designs

The one-way analysis of variance described in Chapters 10 and 11 is only one of many designs for an experiment. Many experiments have a more complex design than the one-way completely randomized design. The investigator may be using replications or subsamples. There may be a need to control extraneous factors, or there may be interest in more than one set of treatments. In this chapter, we illustrate several different designs. In each case we discuss when they should be used and how the analysis is carried out.

12.1. NESTED DESIGN

A *nested design* (or hierarchal design) is used for experiments in which there is interest in one set of treatments and the experimental units are measured more than once or are subsampled. For example, if three diets are being tested for their effect on blood cholesterol level and four volunteers are assigned at random to each diet (a total of 12 volunteers), the investigator might want to obtain two lab determinations of cholesterol level for each volunteer (24 determinations) because of variability in this lab procedure (Figure 12.1). In this example, there are repeated observations of the subjects.

If four dyes are being tests for colorfastness on cotton, each dye might be used on two bolts of material (a total of eight bolts) and then six swatches of material from each bolt selected at random (48 swatches) for the test. In this example the experimental units (bolts) are subsampled.

Other examples of nested designs:

1. Three drugs are each used at two different clinics (a total of six clinics) and are given to five patients at each clinic.

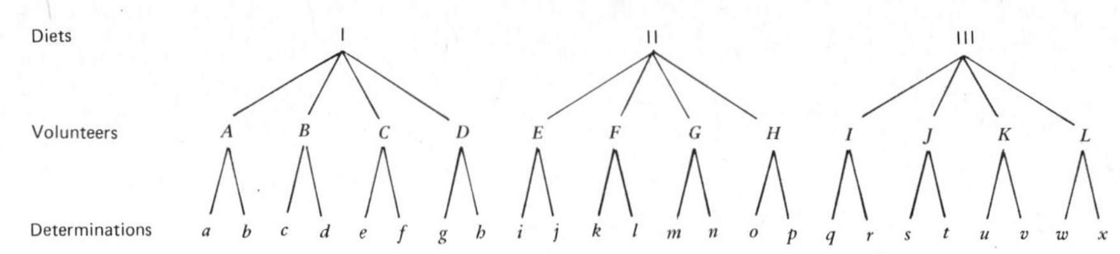

FIGURE 12.1. A nested design.

2. Ten roosters are mated to five hens each, and a random sample of six chicks from each hen is examined for a certain genetic characteristic.
3. Four fungicides are used on a certain type of tree. Each fungicide is applied to three trees, and 10 leaves are examined from each tree.
4. Each of three methods of teaching geometry is used by two teachers (six teachers are in the experiment), and a random sample of 10 students of each teacher is tested.

The additive model for these nested designs is

$$y_{ijk} = \mu + \alpha_i + \beta_{ij} + \epsilon_{ijk}$$

with

$$i = 1, \ldots, a$$

$$j = 1, \ldots, b$$

$$k = 1, \ldots, n$$

The terms in this model have the following meanings.

μ: A constant, the mean for all experiments of this type.

α_i: A constant for the ith treatment group, the effect of the ith treatment. If the treatments are fixed effects, $\sum_i \alpha_i = 0$; if the treatments are random effects, α_i is IND(0, σ_A^2).

β_{ij}: A random effect due to the ijth experimental unit; β_{ij} is IND(0,σ_B^2) for each i.

ϵ_{ijk}: A random effect due to the ijkth observation. It contains all uncontrolled variability; ϵ_{ijk} is IND(0,σ^2).

In the examples given above, all of the treatments are fixed effects except the roosters in example 2. The analysis of variance is computationally the same whether the treatments are fixed or random. We consider only cases in which the experimental units are random effects (if they are fixed the F test is different).

The analysis of variance is an extension of the one-way design. The main hypothesis to be tested is H_0: $\alpha_1 = \alpha_2 = \cdots = \alpha_a = 0$ for the FEM and H_0: $\sigma_A^2 = 0$ for the REM. A secondary hypothesis can be tested to determine if there is variability among the experimental units, H_0: $\sigma_B^2 = 0$.

Subscripts ijk are used in the following manner. The first subscript i refers to the treatment group. The second subscript j refers to the jth experimental unit within a treatment group. The third subscript k refers to the kth sub-sample or replicate within an experimental unit.

In the diet example at the beginning of this section, the diets are the treatments, so $i = 1, 2, 3$. The volunteers are the experimental units, so $j = 1, 2, 3, 4$. The lab determinations are replications, so $k = 1, 2$. Thus y_{241} is the cholesterol level from the first determination for the fourth person on Diet 2.

	Diet			
Volunteer	1	2	3	
1	y_{111}	y_{211}	y_{311}	
	y_{112}	y_{212}	y_{312}	
	$T_{11.}$	$T_{21.}$	$T_{31.}$	
2	y_{121}	y_{221}	y_{321}	
	y_{122}	y_{222}	y_{322}	
	$T_{12.}$	$T_{22.}$	$T_{32.}$	
3	y_{131}	y_{231}	y_{331}	
	y_{132}	y_{232}	y_{332}	
	$T_{13.}$	$T_{23.}$	$T_{33.}$	
4	y_{141}	y_{241}	y_{341}	
	y_{142}	y_{242}	y_{342}	
	$T_{14.}$	$T_{24.}$	$T_{34.}$	
	$T_{1..}$	$T_{2..}$	$T_{3..}$	$T_{...} = \sum_i T_{i..}$

There are four types of totals:

y_{ijk} = the individual observations, a total of one observation

$T_{ij.}$ = the subsample or replicate totals

$T_{i..}$ = the treatment group totals

$T_{...}$ = the grand total

These four types of totals lead to four uncorrected sums of squares, as shown:

Uncorrected Sums of Squares

Sum of Squares	Formula	Symbol	No. of Totals	Observations/ Total
Uncorrected total	$\sum_i \sum_j \sum_k y_{ijk}^2$	T	abn	1
Uncorrected treatment	$\sum_i (T_{i..}^2/bn)$	A	a	bn
Uncorrected experimental unit	$\sum_i \sum_j (T_{ij.}^2/n)$	B	ab	n
Correction factor	$T_{...}^2/abn$	CF	1	abn

The corrected sum of squares, as for the one-way ANOVA, are found by computational formulas in which the number of totals in the uncorrected sums of squares correspond to the degrees of freedom.

Corrected Sums of Squares

Sum of Squares	Symbol	df	Definition	Computational Formula
Total	SS_t	$abn - 1$	$\sum_i \sum_j \sum_k (y_{ijk} - \bar{y})^2$	$T - CF$
Among treatments	SS_a	$a - 1$	$bn \sum_i (\bar{y}_i - \bar{y})^2$	$A - CF$
Among units within treatments	SS_b	$a(b - 1)$	$n \sum_i \sum_j (\bar{y}_{ij} - \bar{y}_i)^2$	$B - A$
Among samples (or replicates) within units	SS_e	$ab(n - 1)$	$\sum_i \sum_j \sum_k (y_{ijk} - \bar{y}_{ij})^2$	$T - B$

In the definitions,

$$\bar{y} = T_{...}/abn \text{ is the overall experimental average}$$

$$\bar{y}_i = T_{i..}/bn \text{ is the } i\text{th treatment average}$$

$$\bar{y}_{ij} = T_{ij.}/n \text{ is the } ij\text{th experimental unit average}$$

Example 12.1. Nested Analysis of Variance

A taxicab company is going to choose among five models of cars for its fleet. The company has already determined that these five are comparable in initial cost and maintenance, and it wants to make a decision based on gas mileage. Ten cars are available for the experiment, two of each model. Each car is to be tested three times. Thus $a = 5$, $b = 2$, and $n = 3$.

	Model of Car					
Car	A	B	C	D	E	
	15.8	18.5	12.3	19.5	16.0	
1	15.6	18.0	13.0	17.5	15.7	
	16.0	18.4	12.7	19.1	16.1	
$T_{i1.}$	47.4	54.9	38.0	56.1	47.8	
	13.9	17.9	14.0	18.7	15.8	
2	14.2	18.1	13.1	19.0	15.6	
	13.5	17.4	13.5	18.8	16.3	
$T_{i2.}$	41.6	53.4	40.6	56.5	47.7	Total
$T_{i..}$	89.0	108.3	78.6	112.6	95.5	484.0
$\sum_j \sum_k y_{ijk}^2$	1326.10	1955.59	1031.44	2115.44	1520.39	7948.96

Uncorrected SS

$$T = \sum_i \sum_j \sum_k y_{ijk}^2 = 7948.96 \qquad B = \sum_i \sum_j T_{ij.}^2/n = 7944.95$$

$$A = \sum_i T_{i..}^2/bn = 7937.81 \qquad CF = T_{...}^2/abn = 7808.53$$

Source	df	SS	MS
Among models	$a - 1 = 4$	$SS_a = A - CF = 128.28$	$MS_a = SS_a/(a-1) = 32.32$
Among cars within models	$a(b-1) = 5$	$SS_b = B - A = 7.14$	$MS_b = SS_b/a(b-1) = 1.43$
Among trials within cars	$ab(n-1) = 20$	$SS_e = T - B = 4.01$	$MS_e = SS_e/ab(n-1) = 0.20$
Total	$abn - 1 = 29$	$SS_t = T - CF$	

In this design,

$$MS_a \text{ estimates } \sigma^2 + n\sigma_B^2 + bn \sum_i \alpha_i^2/(a-1)$$

$$MS_b \text{ estimates } \sigma^2 + n\sigma_B^2$$

$$MS_e \text{ estimates } \sigma^2$$

so the F tests take the following form.

Source	F	$F_{0.05}$	H_0
Among models	$MS_a/MS_b = 22.60$	5.192	$\alpha_1 = \alpha_2 = \cdots = \alpha_5 = 0$
Among cars	$MS_b/MS_e = 7.15$	2.711	$\sigma_B^2 = 0$

Thus, there is at least one significant difference among the average mileages for the models. A secondary conclusion is that there is significant variability among the different cars within models.

The term *expected mean square*, E(MS), is used to indicate the parameter being estimated by the mean square. These expected values will differ for treatments that are fixed or random (they are fixed in the car example). However, in both cases MS_b estimates everything in $E(MS_a)$ except for the term that is being tested in the null hypothesis, so the main F test has the form MS_a/MS_b.

Expected Mean Squares

Source	FEM (Treatments)	REM (Treatments)
Among treatments	$\sigma^2 + n\sigma_B^2 + bn\sum_i \alpha_i^2/(a-1)$	$\sigma^2 + n\sigma_B^2 + bn\sigma_A^2$
Among units within treatments	$\sigma^2 + n\sigma_B^2$	$\sigma^2 + n\sigma_B^2$
Among trials within units	σ^2	σ^2

If desired, multiple comparisons can be done following ANOVA to find specific differences among the treatment means. Only one modification is necessary: the standard error of the difference of two means is $\sqrt{2MS_b/bn}$ instead of $\sqrt{2MS_e/n}$. Estimation of parameters or linear combinations of parameters can also be carried out, again substituting MS_b for MS_e. The degrees of freedom are $a(b-1)$.

The procedure for ANOVA for a nested design is summarized as follows.

■ ■ ■

Procedure. Nested ANOVA for Equal Sample Sizes

Main Hypothesis

H_0: $\alpha_1 = \alpha_2 = \cdots = \alpha_a = 0$ or H_0: $\sigma_A^2 = 0$ against
H_a: *At least one inequality* or H_a: $\sigma_A^2 > 0$

Secondary Hypothesis

H_0: $\sigma_B^2 = 0$ against H_a: $\sigma_B^2 > 0$

Model:

$$y_{ijk} = \mu + \alpha_i + \beta_{ij} + \epsilon_{ijk}$$

$$i = 1, \ldots, a$$

$$j = 1, \ldots, b$$

$$k = 1, \ldots, n$$

Compute:

$$T = \sum_i \sum_j \sum_k y_{ijk}^2$$

$$A = \sum_i T_{i..}^2/bn$$

$$B = \sum_i \sum_j T_{ij.}^2/n$$

$$CF = T_{...}^2/abn$$

Source	df	SS	MS	F
Among treatments	$a - 1$	$SS_a = A - CF$	$MS_a = SS_a/(a - 1)$	MS_a/MS_b
Among units within treatments	$a(b - 1)$	$SS_b = B - A$	$MS_b = SS_b/a(b - 1)$	MS_b/MS_e
Among trials within units	$ab(n - 1)$	$SS_e = T - B$	$MS_e = SS_e/ab(n - 1)$	

Reject the main H_0 if $F = MS_a/MS_b \geq F_{\alpha, a-1, a(b-1)}$. Reject the secondary hypothesis if $F = MS_b/MS_e \geq F_{\alpha, a(b-1), ab(n-1)}$.

■ ■ ■

It is possible to analyze a nested design with unequal sample sizes. Modifications are necessary in the uncorrected sums of squares and the degrees of freedom.

EXERCISES

12.1.1. Ring-necked pheasants establish breeding colonies, each consisting of one male (cock), several hens per cock, and several chicks per hen. If adult males and females can be identified by wing band, a wildlife biologist can locate the nests of female pheasants in a hunting reserve, and he can collect eggs through random sampling in such a manner that they will represent the breeding colonies of five cocks, three hens per cock, and two eggs per hen. The eggs will be marked, incubated, and chicks weighed at 28 days of age.

a. Given that the linear model for this study is

$$y_{ijk} = \mu + \alpha_i + \beta_{ij} + \epsilon_{ijk}$$

i. What does α_i represent? Is it a fixed or a random effect?
ii. What does β_{ij} represent? Is it a fixed or a random effect?

b. Given the following computations

$$\sum_i T_{i..}^2/6 = 918.0 \qquad \sum_i \sum_j T_{ij.}^2/2 = 1833.0$$

$$\sum_i \sum_j \sum_k y_{ijk}^2 - \sum_i \sum_j T_{ij.}^2/n = 7.5 \qquad T_{...}^2/30 = 900$$

complete the analysis of variance and test for significance of variability due to males.

12.1.2. Soda crackers lose their crispness in damp climates unless they are packaged in containers that protect them from humidity. A bakery firm wishes to compare five methods of packaging (including a cardboard box control). Four boxes are selected at random from each method of packaging, assigned numbers, and placed in a chamber in which the humidity is maintained at 80% for 24 hours. The boxes are opened and three crackers are selected from each box at random to be measured for moisture content. The measurements on the 60 crackers are given in milligrams.

Control Boxes				Wax Paper Boxes				Metal Foil Boxes			
11	12	13	14	21	22	23	24	31	32	33	34
73	81	70	67	60	64	62	53	46	49	52	58
75	77	62	69	61	67	55	50	48	54	62	53
77	75	64	62	63	62	59	55	46	54	56	53
225	233	196	198	184	193	176	158	140	157	170	164

Plastic Boxes				Metal Foil and Plastic Boxes			
41	42	43	44	51	52	53	54
60	49	39	52	38	45	58	48
53	42	40	55	37	47	55	47
60	52	44	49	38	49	54	46
173	143	123	156	113	141	167	141

a. Give the linear model and the assumptions.
b. State the null hypothesis of greatest concern.
c. Given that $\sum_i \sum_j \sum_k y_{ijk}^2 = 193{,}661$, perform the ANOVA.
d. Are there significant differences among the methods of packaging?
e. Which method of packaging do you recommend?
f. Is there significant variability among boxes receiving the same method of packaging?

12.1.3. In the taxicab study in this section, Example 12.1, use Fisher's least significant difference to locate the pairs of means that are different. Which model or models would you recommend?

12.1.4. In the taxicab study of this section, Example 12.1, estimate μ, $\alpha_4 - \alpha_5$, and $\mu_4 - (\mu_1 + \mu_2 + \mu_3)/3$ with 95% confidence intervals.

12.2. RANDOMIZED COMPLETE BLOCK DESIGN

An experimenter uses a *randomized complete block* design if he is interested in one set of treatments and wants to control an extraneous source of variability. For example, a physiologist studying the effect of four different drugs A, B, C, and D on mice might feel that the responses will be influenced by the particular litter from which the mice came. He would not want this litter effect to interfere with the analysis of the drug effect. In order to remove this nuisance variability he can use litters as *blocks*, an extension of matched pairs. He chooses four mice at random from each litter, and each drug is assigned at random to one mouse from each litter (Figure 12.2). The design is called complete because each treatment appears in each block exactly once.

Other examples of a randomized complete block design:

1. Four varieties of corn are each planted on sections of five different farms (the farms are chosen at random and the sections assigned at random), and yields are measured. The farms are the blocks. This

LITTERS (*Blocks*)				
1	*B*	*A*	*D*	*C*
2	*A*	*C*	*B*	*D*
3	*C*	*D*	*A*	*B*

FIGURE 12.2. Four treatments assigned at random within three blocks.

design makes it possible to remove any differences in yield due to differences in fertilities.

2. Five dyes are each applied to portions of eight random strips of cloth from a bolt (the strips are chosen at random and the portions assigned at random to the dyes), and the dyes are tested for permanence. The strips are the blocks. This design makes it possible to remove any differences due to variability of the cloth.
3. Three social studies textbooks are used in three classes at each of four different schools (the assignment of textbook to a class is random), and average class performance is measured. Schools are the blocks.
4. Four formulas for sun protection are tested on the skin of five subjects. Each formula is applied to different randomly chosen portions of skin of each subject. The subjects are the blocks.
5. Six different bacteria to be treated with a drug are cultured in a medium which is prepared in four batches. Each type of bacterium is cultured once in a portion of each batch of medium. The batches are the blocks.

In all of these examples the investigator is primarily interested in the treatment effects (varieties of corn, dyes, textbooks, formulas, bacteria), and the blocking is done to avoid extraneous variability (from different fertilities on the farms, from differences in the cloth in different parts of the bolt, from differences in schools, from differences in skin types, from differences in batches of medium). If this extraneous variability is not removed, it will show up in the MS_e, making it difficult to detect treatment differences.

The additive model for a randomized complete block design is

$$y_{ij} = \mu + \alpha_i + \beta_j + \epsilon_{ij}$$

$$i = 1, \ldots, a$$

$$j = 1, \ldots, b$$

in which the terms have the following meanings.

μ: A constant, the overall mean of experiments of this type.

α_i: A constant for the ith treatment group, the deviation from the mean due to the ith treatment; $\sum_i \alpha_i = 0$ if the treatments are fixed effects, or α_i IND$(0,\sigma_A^2)$ if the treatments are random.

β_j: A constant for the jth block, the deviation from the mean caused by the jth block; $\sum_j \beta_j = 0$ if the blocks are fixed effects, or β_j IND$(0,\sigma_B^2)$ if they are random.

ϵ_{ij}: A random deviation associated with the ijth observation, containing all uncontrolled sources of variability; ϵ_{ij} IND$(0,\sigma^2)$.

Data for a randomized complete block design are arranged as follows, in which i designates the treatment and j the blocks.

		Treatments (i)				
		1	2	3	4	Totals
Blocks (j)	1	y_{11}	y_{21}	y_{31}	y_{41}	$T_{.1}$
	2	y_{12}	y_{22}	y_{32}	y_{42}	$T_{.2}$
	3	y_{13}	y_{23}	y_{33}	y_{43}	$T_{.3}$
Totals		$T_{1.}$	$T_{2.}$	$T_{3.}$	$T_{4.}$	$T_{..}$ = Grand total

Sometimes rows and columns are interchanged, but we will continue to use i for the treatment and j for the blocks even in that case. Treatment group totals are represented by $T_{i.}$, indicating that the summation was over j. Block totals are $T_{.j}$ and the grand total is $T_{..}$. The corresponding averages are $\bar{y}_{i.}$, $\bar{y}_{.j}$, and $\bar{y}_{..}$.

The uncorrected sums of squares, sums of squares, and ANOVA procedure are as follows. In a block design the error sum of squares is sometimes called the *residual sum of squares*.

Uncorrected Sums of Squares

Sum of Squares	Formula	Symbol	No. of Totals	Observations/ Total
Uncorrected total	$\sum_i \sum_j y_{ij}^2$	T	ab	1
Uncorrected treatment	$\sum_i T_{i.}^2/b$	A	a	b
Uncorrected block	$\sum_j T_{.j}^2/a$	B	b	a
Residual	$T_{..}^2/ab$	CF	1	ab

Corrected Sums of Squares

Sum of Squares	df	Symbol	Definition	Computational Formula
Total	$ab-1$	SS_t	$\sum_i \sum_j (y_{ij}-\bar{y}_{..})^2$	$T-CF$
Treatment	$a-1$	SS_a	$b\sum_i (\bar{y}_{i.}-\bar{y}_{..})^2$	$A-CF$
Block	$b-1$	SS_b	$a\sum_j (\bar{y}_{.j}-\bar{y}_{..})^2$	$B-CF$
Residual	$(a-1)(b-1)$	SS_e	$\sum_i \sum_j (y_{ij}-\bar{y}_{i.}-\bar{y}_{.j}+\bar{y}_{..})^2$	$T-A-B+CF$

As in the one-way design, the short computational formulas correspond to the degrees of freedom. For example, the residual degrees of freedom are $(a - 1)(b - 1) = ab - a - b + 1$, and the terms $T - A - B + CF$ contain ab, a, b, and 1 total respectively.

■ ■ ■

Procedure. Randomized Complete Block ANOVA

Main Hypothesis

H_0: $\alpha_1 = \alpha_2 = \cdots = \alpha_a = 0$ or H_0: $\sigma_A^2 = 0$

against

H_a: *At least one inequality* or H_a: $\sigma_A^2 > 0$

Model:

$$y_{ij} = \mu + \alpha_i + \beta_j + \epsilon_{ij}$$
$$i = 1, \ldots, a$$
$$j = 1, \ldots, b$$

Compute:

$$T = \sum_i \sum_j y_{ij}^2 \qquad B = \sum_j T_{.j}^2/a$$
$$A = \sum_i T_{i.}^2/\text{b} \qquad CF = T_{..}^2/ab$$

Source	df	SS	MS	F
Among treatments	$a-1$	$SS_a = A - CF$	$MS_a = SS_a/(a-1)$	MS_a/MS_e
Among blocks	$b-1$	$SS_b = B - CF$	$MS_b = SS_b/(b-1)$	MS_b/MS_e
Residual	$(a-1)(b-1)$	$SS_e = T - A - B + CF$	$MS_e = SS_e/(a-1)(b-1)$	
Total	$ab-1$	$SS_t = T - CF$		

■ ■ ■

It is also possible to test for a block difference, H_0: $\beta_1 = \beta_2 = \cdots = \beta_b = 0$ or H_0: $\sigma_B^2 = 0$. These hypotheses are tested by $F = MS_b/MS_e$ with the corresponding degrees of freedom. The form of the F test can be determined in each case by the expected mean squares. The denominator of the F test must estimate everything except the term being tested.

Expected Mean Squares for Randomized Complete Block Design

MS	E(MS) Fixed	E(MS) Random
MS_a	$\sigma^2 + b\sum_i \alpha_i^2/(a-1)$	$\sigma^2 + b\sigma_A^2$
MS_b	$\sigma^2 + a\sum_j \beta_j^2/(b-1)$	$\sigma^2 + a\sigma_B^2$
MS_e	σ^2	σ^2

Example 12.2. Randomized Complete Block ANOVA

A psychology experiment involving three treatments is planned with a randomized complete block design, the random subjects being the blocks. The three treatments are administered on three different days and the order in which each subject receives the treatment is random. There are four subjects and the random variable is the length of time required to complete a certain task.

		Treatments 1	2	3	$T_{.j}$
Subjects	1	4.7	9.4	6.3	20.4
	2	3.5	7.6	5.1	16.2
	3	0.1	5.3	1.8	7.2
	4	1.6	6.2	3.6	11.4
	$T_{i.}$	9.9	28.5	16.8	55.2 = $T_{..}$

$$a = 3$$

$$b = 4$$

$$\sum_i \sum_j y_{ij}^2 = 331.46$$

$$T = 331.460 \qquad B = 286.800$$

$$A = 298.125 \qquad CF = 253.920$$

Source	df	SS	MS	F	$F_{0.05}$
Among treatments	$a-1=2$	$A-CF=44.205$	22.102	290.8	5.143
Among blocks	$b-1=3$	$B-CF=32.880$	10.960	144.2	4.757
Residual	$(a-1)(b-1)=6$	$T-A-B+CF=0.455$	0.076		

Since the F statistic for treatments is significant, there is evidence of differences among the treatments.

Although the psychologist in the example above is not interested in block differences for their own sake, the fact that the F for blocks is significant shows that this design is appropriate for the experiment. The decision to use a block design must come before the experiment. The experimenter knows from previous experience that an extraneous source of variability is present and designs the experiment so that this effect can be removed and the statistical procedure can be more powerful.

It is not always advantageous to use a block design instead of a completely randomized design. In a block design, along with the reduction in the error sum of squares there is also a reduction in the associated degrees of freedom, so the critical F value is larger. Thus, if blocking is used when there really is no block effect, the reduction in the error sum of squares will not be sufficient to offset the reduction in power due to the loss of degrees of freedom in the denominator.

The power of the randomized complete block design will also be reduced if the treatment effect and block effect are not simply additive, as implied in the model:

$$y_{ij} = \mu + \alpha_i + \beta_j + \epsilon_{ij}$$

Additivity is not present if there is an interaction between treatments and blocks. An interaction is an additional boost or reduction due to the particular combination of a block and treatment. For example, in the psychologist's experiment, Subject 1 may be much faster than the average person under Treatment 1 but much slower than average under Treatment 2, whereas Subject 2 may be just the opposite. An absence of interactions means that although there are different reaction times for individuals, the general pattern is the same. If an interaction effect is present, there is no specific term for it in the block design. Since the variability due to the interaction will be in the total sum of squares and will not be removed by the treatment sum of squares or the block sum of squares, it will be left in the error sum of squares.

$$SS_t - SS_a - SS_b = SS_e$$

Thus the error sum of squares may contain not only variability due to sampling but also variability due to the interaction effect. (This is the reason for calling SS_e the *residual* sum of squares.) If an interaction is present the power of the test is reduced because of the inflated SS_e, which contributes to the denominator of the F statistic. If interactions are suspected, the randomized complete block design should not be used. The two-factor model described in Section 12.4 makes specific provision for an interaction effect.

A randomized complete block design with fixed treatment effects can be followed by multiple comparisons, one degree of freedom F tests, or estimation of the fixed effects. The MS_e is used in the standard error, and n must be replaced by a or b, whichever is appropriate in the formulas given in Chapter 10. Intraclass correlations can be done on the random effects. The total variability is $\sigma^2 + \sigma_A{}^2 + \sigma_B{}^2$.

Sometimes in carrying out a blocked experiment, an observation is missing for reasons extraneous to the experiment. For example, a plant dies because of an accident in the greenhouse, a subject leaves town or is ill and cannot complete the experiment (assuming the illness is not related to the treatment), or the data are lost or erased. One way to handle this situation is to remove the entire block that contains the missing value. The analysis is then carried out with $b - 1$ blocks.

Another approach is to estimate the missing value y_{ij} by

$$\hat{y}_{ij} = \frac{aT_{i.} + bT_{.j} - T_{..}}{(a - 1)(b - 1)}$$

and to decrease the residual degrees of freedom by one.

For example, in the psychology example in this section (Example 12.2), if y_{23} were missing it could be estimated as follows.

	Treatments			
Subjects	1	2	3	$T_{.j}$
1	4.7	9.4	6.3	20.4
2	3.5	7.6	5.1	16.2
3	0.1	___	1.8	1.9
4	1.6	6.2	3.6	11.4
$T_{i.}$	9.9	23.2	16.8	49.9 = $T_{..}$

$$\hat{y}_{23} = \frac{3(23.2) + 4(1.9) - 49.9}{(3 - 1)(4 - 1)} = 4.55$$

The residual degrees of freedom would be five.

If there are several missing values an iterative procedure may be used. For example, if there are three missing values a, b, and c, we guess values for b and c and then approximate a as above. Using the approximation of a and the original guess of c, b is approximated as above. Finally, c is approximated using the approximated values of a and b. The cycle is then repeated to obtain second approximations of each of the three values. Repetition of the cycle continues until there are no noticeable changes in the approximations. The total degrees of freedom and residual degrees of freedom are reduced by one for each missing value. For further details, see Cochran and Cox (1957).

EXERCISES

12.2.1. Four varieties of hybrid corn have been developed for resistance to the fungal infection known as smut. However, nothing is known about their potential for grain yield. Each hybrid is planted at each of five locations within the state, and the following yields are obtained.

	Location				
Hybrid	NW	NE	C	SE	SW
FR-11	62.3	64.0	64.3	65.0	66.4
BCM	63.3	62.7	66.2	66.8	64.5
DBC	60.8	64.3	65.2	62.2	65.1
RC-3	55.4	56.0	59.8	58.0	58.8

a. Give the linear model and the assumptions.
b. Perform the appropriate analysis of variance.
c. Are there differences in yield among the means of the hybrids?
d. Are there differences that can be attributed to location?
e. If a smut-resistant hybrid is used, which do you recommend?

12.2.2. In a study of reaction time under the influence of alcohol, age is thought to be another variable that could effect the time. A randomized complete block design is used, and reaction time is measured in seconds.

		Amount of Alcohol			
		None	1 oz.	2 oz.	$T_{.j}$
Age	20–39	0.42	0.47	0.65	1.54
	40–59	0.51	0.62	0.66	1.79
	60 or over	0.57	0.73	0.79	2.09
	$T_{i.}$	1.50	1.82	2.10	$5.42 = T_{..}$

$$\sum_i \sum_j y_{ij}^2 = 3.3818$$

a. Complete the ANOVA table.
b. Is there any difference in reaction time among the alcohol groups?
c. Use Student–Newman–Keuls' procedure to compare the alcohol means.
d. Is there a significant difference in reaction time due to age?

12.2.3. A large company is going to buy cars to be used by employees on business trips. Five models of cars are tested for mileage per gallon in five different randomly chosen cities. One car of each model is used, and the cities are tested in random order.

	CITY					
MODEL	1	2	3	4	5	TOTALS
A	15.83	17.56	21.11	20.48	26.04	101.02
B	14.80	16.22	21.30	20.84	19.27	92.43
C	17.43	19.54	17.67	22.58	19.86	97.08
D	16.60	16.34	17.01	15.82	16.57	82.34
E	21.24	21.29	20.34	19.43	25.05	107.35
TOTALS	85.90	90.95	97.43	99.15	106.79	480.22

a. What is the ANOVA model for this investigation?
b. Is the car effect random or fixed?
c. Is the city effect random or fixed?
d. What is the hypothesis of main interest to the investigator?
e. Complete the ANOVA.
f. Are there any differences in mileage among the cars?
g. Which mean separation procedure seems appropriate for this investigation? Why?
h. Use Fisher's least significant difference to find the best car or cars.
i. Is there significant variability due to cities?
j. What percentage of the total variability is due to the cities?

12.2.4. An experiment was conducted involving six schools and three teaching methods per school.

a. Identify the sources of variability represented by the sums of squares.

Source	No. of Squared Values	Observations/ Squared Value	Numerical Value
______	1	18	125
______	3	6	151
______	18	1	236
______	6	3	180

b. Complete the uncorrected sum of squares table and the ANOVA table.

c. Could Fisher's least significant difference be used to test for differences among teaching methods? Justify your answer.

12.2.5. Given the following analysis of variance:

Source	df	SS	MS
Treatments	3	150.0	50.0
Blocks	4	56.0	14.0
Residual	12	86.4	7.2

a. What are the values of a and b?

b. What is the numerical value of the standard error of a treatment average?

c. Use Duncan's procedure to compare the treatment means.

Treatment:	1	2	3	4
$\bar{y}_{i.}$:	6	9	12	13

12.2.6. a. Estimate the missing value in the block design.

b. Complete the ANOVA.

	Blocks		
Treatments	3	4	5
	1	2	2
	3	—	7
	5	8	2
	4	6	5

12.3. LATIN SQUARE DESIGN

Sometimes the investigator is aware of two causes of nuisance variability, and a blocked design is not adequate for the experiment. For example, in addition to a litter effect in a drug experiment on mice, there may also be a size-of-mouse effect. If there are no interactions present, and the experimenter is working with four drugs (A, B, C, D), four litters, and four sizes of mice, then a Latin square design may be used (Figure 12.3).

In a Latin square, each treatment appears exactly once in each row and column. This is a very economical design because it avoids the necessity of working with every combination possible. For example, in the mouse experiment, if all combinations of drug, litter, and size of mouse were used, 64 mice would be needed. In addition, litters of the proper number and with the needed assortment of sizes probably would not exist.

		Size 1	2	3	4
Litter	1	*A*	*B*	*C*	*D*
	2	*C*	*A*	*D*	*B*
	3	*B*	*D*	*A*	*C*
	4	*D*	*C*	*B*	*A*

FIGURE 12.3. A Latin square design.

The smallest Latin square that can be analyzed is 3 × 3. Squares larger than 9 × 9 are rarely used because of the difficulty of finding equal numbers of categories for the rows, columns, and treatments.

Standard Latin squares can be found in Fisher and Yates (1963). If more than one is available, the standard square should be selected by a random process, and the rows and columns should be randomized. For example, if

A	*B*	*C*
B	*C*	*A*
C	*A*	*B*

is the standard Latin square, two random sequences of the digits 1, 2, 3 are chosen, say (2, 1, 3) and (3, 1, 2). Then the columns are rearranged by the first sequence and the rows by the second (Figure 12.4).

Latin squares were originally used for agricultural experiments. Treatments were applied to a field in a Latin square design in order to randomize for any differences in fertility in different sections of the field. However, the design is very useful in other disciplines, and it is not necessary that the treatments be applied physically in a Latin square design. The mouse experiment which controls for litter and size is a typical nonagricultural application.

Column

1	2	3		2	1	3
A	*B*	*C*		*B*	*A*	*C*
B	*C*	*A*	$\xrightarrow{(2,1,3)}$	*C*	*B*	*A*
C	*A*	*B*		*A*	*C*	*B*

Row	1	*B*	*A*	*C*	$\xrightarrow{(3,1,2)}$	3	*A*	*C*	*B*
	2	*C*	*B*	*A*		1	*B*	*A*	*C*
	3	*A*	*C*	*B*		2	*C*	*B*	*A*

FIGURE 12.4. Randomizing columns and rows.

Other examples of a Latin square design:

1. Yield is measured for four varieties of wheat that were planted on four different farms and in four different corners of the farms, NE, NW, SE, and SW.
2. Miles per gallon are measured on six models of cars, using six brands of gasoline, each car using the brands in six different orders.
3. The strength of coated paper is measured for four different coatings, applied at four positions down the roll and four positions across the roll to control for variability in the strength of the uncoated paper.
4. A psychological experiment consists of six treatments given to six subjects in six different orders to control for learning.
5. Drug response is measured for three drugs, given in three dosages, and analyzed by three different lab technicians.
6. Time of assembly is measured for four products, four assemblers, and four positions in the assembly line.

The additive model for the Latin square design is

$$y_{ijk} = \mu + \alpha_i + \beta_j + \gamma_k + \epsilon_{ijk}$$

$$i = 1, \ldots, a$$

$$j = 1, \ldots, a$$

$$k = 1, \ldots, a$$

in which the terms have the following meaning.

μ: A constant, the overall mean for all experiments of this type.

α_i: A constant for the ith treatment; $\sum_i \alpha_i = 0$ if this effect is fixed or α_i IND$(0,\sigma_A^2)$ if it is random.

β_j: A constant for the jth first extraneous effect; $\sum_j \beta_j = 0$ if this effect is fixed or β_j IND$(0,\sigma_B^2)$ if it is random.

γ_k: A constant for the kth second extraneous effect; $\sum_k \gamma_k = 0$ if this effect is fixed or γ_k IND$(0,\sigma_C^2)$ if it is random.

ϵ_{ijk}: A random effect due to sampling; ϵ_{ijk} IND$(0,\sigma^2)$.

This model assumes that there are no interactions between the α_i's and β_j's, α_i's and γ_k's, and between β_j's and γ_k's.

Data for a Latin square design are arranged as in Figure 12.5, with the indicated notation. Treatments are indicated in parentheses within the cells. It does not matter which effect is placed in the rows, columns, or across the face of the table or which symbol, α_i, β_j, or γ_k, is assigned to a particular effect. The arrangement in Figure 12.5 is traditional because of the agricultural origins of this design, but other arrangements are common.

Averages are indicated by a notation corresponding to the totals, for example $\bar{y}_{.2.} = T_{.2.}/a$ and $\bar{y}_{...} = T_{...}/a^2$.

		β Effect 1	β Effect 2	β Effect 3	Totals
	1	y_{111} (1)	y_{221} (2)	y_{331} (3)	$T_{..1}$
γ Effect	2	y_{312} (3)	y_{122} (1)	y_{232} (2)	$T_{..2}$
	3	y_{213} (2)	y_{323} (3)	y_{133} (1)	$T_{..3}$
Totals		$T_{.1.}$	$T_{.2.}$	$T_{.3.}$	$T_{...}$ = Grand total

Treatment totals: $T_{1..}$, $T_{2..}$, $T_{3..}$

FIGURE 12.5. Notation for the Latin square design.

Uncorrected Sums of Squares

Sum of Squares	Symbol	Formula	No. of Totals	Observations/ Total
Uncorrected total	T	$\sum_j \sum_k y_{ijk}^2$	a^2	1
Uncorrected treatment	A	$\sum_i T_{i..}^2/a$	a	a
Uncorrected β effect	B	$\sum_j T_{.j.}^2/a$	a	a
Uncorrected γ effect	C	$\sum_k T_{..k}^2/a$	a	a
Correction factor	CF	$T_{...}^2/a^2$	1	a^2

Corrected Sums of Squares

Source	df	Symbol	Definition	Computational Formula
Total	a^2-1	SS_t	$\sum_i \sum_j (y_{ijk}-\bar{y}_{...})^2$	$T-CF$
Treatment	$a-1$	SS_a	$a\sum_i (\bar{y}_{i..}-\bar{y}_{...})^2$	$A-CF$
β effect	$a-1$	SS_b	$a\sum_j (\bar{y}_{.j.}-\bar{y}_{...})^2$	$B-CF$
γ effect	$a-1$	SS_c	$a\sum_k (\bar{y}_{..k}-\bar{y}_{...})^2$	$C-CF$
Residual	$(a-1)(a-2)$	SS_e	$\sum_i \sum_j \sum_k (y_{ijk}-\bar{y}_{i..} -\bar{y}_{.j.}-\bar{y}_{..k}+2\bar{y}_{...})^2$	$T-A-B-C+2CF$

Note that in the definition of SS_e not all the combinations of ijk exist. The missing terms can be thought of as having zero value.

■ ■ ■

Procedure. Latin Square ANOVA

Main Hypothesis

H_0: $\alpha_1 = \cdots = \alpha_a = 0$ or H_0: $\sigma_A^2 = 0$

Secondary Hypotheses

H_0: $\beta_i = \cdots = \beta_a = 0$ or H_0: $\sigma_B^2 = 0$
H_0: $\gamma_1 = \cdots = \gamma_a = 0$ or H_0: $\sigma_C^2 = 0$

Model:

$$y_{ijk} = \mu + \alpha_i + \beta_j + \gamma_k + \epsilon_{ijk}$$
$$i = 1, \ldots, a$$
$$j = 1, \ldots, a$$
$$k = 1, \ldots, a$$

Compute:

$$T = \sum_j \sum_k y_{ijk}^2 \qquad C = \sum_k T_{..k}^2/a$$
$$A = \sum_i T_{i..}^2/a \qquad CF = T_{...}^2/a^2$$
$$B = \sum_j T_{.j.}^2/a$$

Source	df	SS	MS	F
Among treatments	$a-1$	$SS_a = A - CF$	$MS_a = SS_a/(a-1)$	MS_a/MS_e
Among β effects	$a-1$	$SS_b = B - CF$	$MS_b = SS_b/(a-1)$	MS_b/MS_e
Among γ effects	$a-1$	$SS_c = C - CF$	$MS_c = SS_c/(a-1)$	MS_c/MS_e
Residual	$(a-1)(a-2)$	$SS_e = T - A - B - C + 2CF$	$MS_e = SS_e/(a-1)(a-2)$	
Total	a^2-1	$SS_t = T - CF$		

■ ■ ■

The F tests take the form given above because of the expectations of the mean squares.

Expected Mean Squares

MS	E(MS) Fixed	E(MS) Random
MS_a	$\sigma^2 + a\sum_i \alpha_i^2/(a-1)$	$\sigma^2 + a\sigma_A^2$
MS_b	$\sigma^2 + a\sum_j \beta_j^2/(a-1)$	$\sigma^2 + a\sigma_B^2$
MS_c	$\sigma^2 + a\sum_k \gamma_k^2/(a-1)$	$\sigma^2 + a\sigma_C^2$
MS_e	σ^2	σ^2

Example 12.3. Latin Square ANOVA

An audiologist is studying three different devices which help hearing in a certain type of deficiency. Three subjects with this type of hearing loss take hearing tests using each of the three devices. To control for learning, a Latin square design is used. Scores on the test are recorded. Devices are given in parentheses.

Subject (β_j)	Order of Test (γ_k): First	Second	Third	
1	74 (1)	57 (2)	50 (3)	$T_{1.} = 181$
2	6 (3)	94 (1)	78 (2)	$T_{2.} = 178$
3	40 (2)	29 (3)	112 (1)	$T_{3.} = 181$
	$T_{.1} = 120$	$T_{.2} = 180$	$T_{.3} = 240$	$T_{..} = 540$

Device Totals: $T_{1..} = 280$ $T_{2..} = 175$ $T_{3..} = 85$

The uncorrected sums of squares are:

$$T = 41{,}166 \quad B = 32{,}402 \quad CF = 32{,}400$$

$$A = 38{,}750 \quad C = 34{,}800$$

Source	df	SS	MS	F	H_0
Among devices	2	6350	3175	453.6	$\alpha_1 = \alpha_2 = \alpha_3 = 0$
Among subjects	2	2	1	0.1	$\sigma_B^2 = 0$
Among orders	2	2400	1200	171.4	$\gamma_1 = \gamma_2 = \gamma_3 = 0$
Residual	2	14	7		

Since $F_{0.01,2,2} = 99.000$, the audiologist concludes that there is a significant difference among the devices and there is a significant learning effect at the 0.01 level.

EXERCISES

12.3.1. A marketing expert for a publishing house wants to measure reader preference for five different covers of the same paperback novel. Five newsstands are selected at random and the novel is displayed at each newsstand for five weeks, one for each cover. One week is sufficient to determine sales potential because a new cover makes its impact immediately, followed by a pattern of diminishing returns. The number of sales are listed below with the cover given in parentheses.

	Week				
Newsstand	1	2	3	4	5
I	(D) 200	(C) 290	(A) 280	(E) 230	(B) 265
II	(C) 260	(B) 280	(E) 245	(A) 285	(D) 245
III	(A) 250	(D) 245	(C) 280	(B) 250	(E) 180
IV	(B) 260	(E) 190	(D) 230	(C) 205	(A) 200
V	(E) 340	(A) 335	(B) 265	(D) 270	(C) 230

a. Give the most logical null hypothesis with respect to covers.

b. Perform the ANOVA.

c. What should be concluded about covers?

d. Comment on the usefulness of the design employed.

12.3.2. A test is done on the miles per gallon for five models of cars using five brands of gasoline and tested in five different orders.

	Brand of Gasoline					
	A	B	C	D	E	Totals
I	15.8 (1)	15.9 (4)	17.9 (2)	17.3 (5)	13.3 (3)	80.2
II	18.1 (2)	17.5 (5)	18.5 (1)	18.5 (3)	16.3 (4)	88.9
III	18.5 (3)	18.2 (2)	17.9 (5)	17.1 (4)	19.2 (1)	90.9
IV	13.9 (4)	12.8 (1)	16.1 (3)	16.9 (2)	16.7 (5)	76.4
V	11.1 (5)	12.6 (3)	11.5 (4)	17.7 (1)	14.8 (2)	67.7
Totals	77.4	77.0	81.9	87.5	80.3	404.1

Order Totals:	(1)	(2)	(3)	(4)	(5)
	84.0	85.9	79.0	74.7	80.5

$$\sum_i \sum_j y_{ijk}^2 = 6665.35$$

a. What is the treatment of interest?
b. Why might the order of testing cause nuisance variability?
c. Carry out the ANOVA.
d. Test for differences among the models of cars.
e. Use Fisher's least significant difference to find the best car or cars.

12.3.3. The National Occupational Safety and Health Act was a comprehensive effort to improve industrial health and safety in this country. Part of this Act requires detailed reporting of industrial accidents. The data gained thereby can lead to the identification and elimination of unsafe practices in industry. With such a goal in mind, a safety engineer in a large chemical plant finds that the plant carries out five basic operations. Because he has to monitor each operation personally to record the number of unsafe incidents within a five-day work week, he decides to take a random sample of five weeks in order to have a Latin square design.

a. Give the additive model for the experiment, using subscripts i for weeks, j for days, and k for operations.
b. List the assumptions of this design and tell whether you feel it is appropriate in this case.
c. Given the following computations, complete the ANOVA.

$$\sum_i \sum_j y_{ijk}^2 = 10{,}990 \qquad \sum_i T_{i..}^2/5 = 2{,}750$$

$$T_{...}^2/25 = 2{,}250 \qquad \sum_j T_{.j.}^2/5 - T_{...}^2/25 = 710$$

$$\left(\sum_k T_{..k}^2/5 - T_{...}^2/25\right)/4 = 195$$

d. What hypothesis can be tested about the operations?
e. Are weeks random or fixed? Days? Operations?
f. What conclusions can the safety engineer draw from this analysis?

12.3.4. An apiarist conducts an experiment to determine the best method of insulating hives for winter survival of bee colonies. He has 16 hives and decides to expose four to each direction of the compass. He has colonies of four different origins and he compares four different insulating materials. He uses a design in which each combination of direction, colony, and material is assigned once and only once to the 16 hives.

a. What design is the apiarist using?
b. What special assumption is necessary for this analysis of variance design?
c. What is the null hypothesis for material effects?

d. What is the expected mean square for colonies?
e. What is the critical value at $\alpha = 0.05$ for a test of direction?
f. Complete the ANOVA table.

Source	df	SS	MS	F
Directions	3	105	35	___
Colonies	___	90	___	___
Materials	3	75	25	___
Residual	___	___	___	
Total	15	330		

12.3.5. Why is it impossible to analyze a 2×2 Latin square?

12.4. $a \times b$ FACTORIAL DESIGN

Often an investigator is interested in the combined effect of two types of treatments. For example, a study might be about weight loss for various diets combined with various levels of jogging per day (Figure 12.6).

This design differs from blocking in that neither of the treatments (diet or jogging) is considered extraneous to the experimental question. Subjects are assigned at random to each of the 12 combinations, and interest is in the combined effect as well as diet considered separately and jogging considered separately. This is an economical design since it accomplishes several things at once.

The sets of treatments are called *factors*, and the different treatments within the sets are called *levels*. If diet is factor A, it has $a = 4$ levels, and if jogging is factor B, it has $b = 3$ levels. (The levels need not be quantitative; the diets in this case have the same calories but different food-group proportions.) A design of this type is called a two-factor design or, more precisely, an *$a \times b$ factorial design*.† In this example, the design is 4×3 factorial. (In this text the first number, 4, refers to the number of levels of factor A. It could refer to either the number of rows or the number of columns in the diagram, depending upon how the diagram is specified.)

In a factorial design, the factors may be treatments in the strict sense or they may be certain classifications of existing populations. The following examples illustrate some of the many different types of study that follow this design.

1. In the jogging–diet example, both factors are treatments, the factor *diet* is qualitative, and the factor *jogging* is quantitative.

† Some statisticians prefer to call this a factorial experiment because combinations of treatments can be assigned in any kind of design.

		DIET			
		Normal	High Protein	High Fat	High Carbo-hydrate
	0 mi.				
JOGGING	1 mi.				
	2 mi.				

FIGURE 12.6. A two-factor design.

2. If change in blood sugar level is measured for various dosages of vitamin C combined with various dosages of aspirin, both factors—vitamin C and aspirin—are quantitative treatments.
3. If sales of a certain product are recorded in several standard metropolitan statistical areas and at several different types of chain stores, the factors—area and chain—are classifications and they are both qualitative.
4. If the lifetimes of tires made by different companies are measured on several different road surfaces, the factor *manufacturer* is a qualitative classification and the factor *roads* is a qualitative treatment.

In all cases, randomization is necessary. In the jogging–diet and vitamin-C–aspirin examples, subjects must be assigned at random to the combination of levels. In the sales example, stores must be chosen at random from the chain stores in the areas. In the tire example, tires from the companies are assigned at random to the type of road.

The tire example is not clearly distinct from a randomized complete block design; in fact, it can be thought of as a block design. However, if the investigator is interested in differences caused by various surfaces as well as differences in brand, and especially if he is interested in any interactions between road surface and brand, then it is a factorial design.

An interaction is an additional effect due to the particular combination of the two levels. For example, certain combinations of level of diet and level of jogging may produce a weight loss in excess of the sum of the effects of the two levels involved. Or a particular combination may produce less of weight loss than expected. In order to be able to analyze the data for possible interactions, the investigator must observe more than one subject at each combination of levels.

Geometrically, the absence of interactions yields parallel lines when the means of the response variable are graphed for the various combinations of levels of the factors. Interactions are indicated by deviations from parallelism; Figure 12.7 illustrates the effect of interactions in the blood sugar experiment.

In the jogging–diet study, $n = 2$ subjects are assigned to each combination

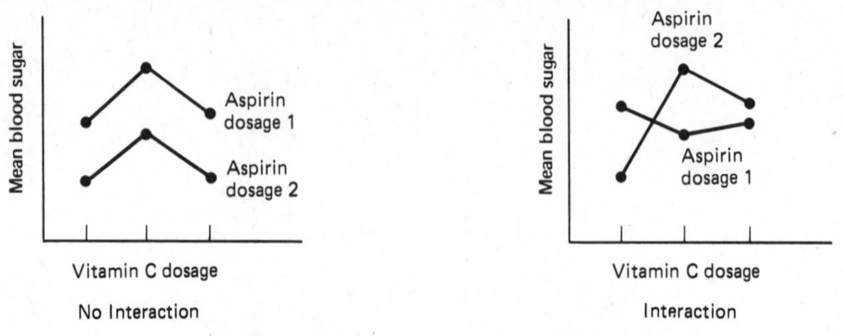

FIGURE 12.7. Effect of interactions on subclass means.

of levels, and the data are represented by the scheme and notation in Figure 12.8.

The model for an $a \times b$ factorial design is:

$$y_{ijk} = \mu + \alpha_i + \beta_j + \alpha\beta_{ij} + \epsilon_{ijk}$$

$$i = 1, \ldots, a$$

$$j = 1, \ldots, b$$

$$k = 1, \ldots, n$$

μ: The overall mean for all experiments of this type.

α_i: The effect of the ith level of factor A; the levels may be fixed or random.

β_j: The effect of the jth level of factor B; the levels may be fixed or random.

$\alpha\beta_{ij}$: The interaction effect between the ith level of factor A and the jth level of factor B. ($\alpha\beta$ is a single symbol and is not a product.)

ϵ_{ijk}: A random effect due to sampling; ϵ_{ijk} IND$(0,\sigma^2)$.

		FACTOR A (*Diet*)				
		N	HP	HF	HC	TOTALS
FACTOR B (*Jogging*)	0 mi.	y_{111} y_{112} $T_{11.}$	y_{211} y_{212} $T_{21.}$	y_{311} y_{312} $T_{31.}$	y_{411} y_{412} $T_{41.}$	$T_{.1.}$
	1 mi.	y_{121} y_{122} $T_{12.}$	y_{221} y_{222} $T_{22.}$	y_{321} y_{322} $T_{32.}$	y_{421} y_{422} $T_{42.}$	$T_{.2.}$
	2 mi.	y_{131} y_{132} $T_{13.}$	y_{231} y_{232} $T_{23.}$	y_{331} y_{332} $T_{33.}$	y_{431} y_{432} $T_{43.}$	$T_{.3.}$
TOTALS		$T_{1..}$	$T_{2..}$	$T_{3..}$	$T_{4..}$	$T_{...}$ = Grand total

FIGURE 12.8. Notation for a two factor design.

Uncorrected Sums of Squares

Sum of Squares	Symbol	Formula	No. of Totals	Observations/ Total
Uncorrected total	T	$\sum_i \sum_j \sum_k y_{ijk}^2$	abn	1
Uncorrected A factor	A	$\sum_i T_{i..}^2/bn$	a	bn
Uncorrected B factor	B	$\sum_j T_{.j.}^2/an$	b	an
Uncorrected subclass	S	$\sum_i \sum_j T_{ij.}^2/n$	ab	n
Correction factor	CF	$T_{...}^2/abn$	1	abn

Corrected Sums of Squares

Source	df	Symbol	Definition	Computational Formula
Total	$abn-1$	SS_t	$\sum_i \sum_j \sum_k (y_{ijk}-\bar{y}_{...})^2$	$T-CF$
Factor A	$a-1$	SS_a	$bn \sum_i (\bar{y}_{i..}-\bar{y}_{...})^2$	$A-CF$
Factor B	$b-1$	SS_b	$an \sum_j (\bar{y}_{.j.}-\bar{y}_{...})^2$	$B-CF$
$A \times B$	$(a-1)(b-1)$	SS_{ab}	$n \sum_i \sum_j (\bar{y}_{ij.}-\bar{y}_{i..}-\bar{y}_{.j.}+\bar{y}_{...})^2$	$S-A-B+CF$
Error	$ab(n-1)$	SS_e	$\sum_i \sum_j \sum_k (y_{ijk}-\bar{y}_{ij.})^2$	$T-S$

■ ■ ■

Procedure. $a \times b$ Factorial ANOVA

Hypotheses

H_0: $\alpha_1 = \cdots = \alpha_a = 0$ or H_0: $\sigma_A^2 = 0$

H_0: $\beta_1 = \cdots = \beta_b = 0$ or H_0: $\sigma_B^2 = 0$

H_0: $\alpha\beta_{11} = \alpha\beta_{12} = \cdots = \alpha\beta_{ab} = 0$ or H_0: $\sigma_{AB}^2 = 0$

Compute:

$$T = \sum_i \sum_j \sum_k y_{ijk}^2 \qquad S = \sum_i \sum_j T_{ij.}^2/n$$

$$A = \sum_i T_{i..}^2/bn \qquad CF = T_{...}^2/abn$$

$$B = \sum_j T_{.j.}^2/an$$

Source	df	SS	MS
Factor A	$a-1$	$SS_a = A - CF$	$MS_a = SS_a/(a-1)$
Factor B	$b-1$	$SS_b = B - CF$	$MS_b = SS_b/(b-1)$
$A \times B$	$(a-1)(b-1)$	$SS_{ab} = S - A - B + CF$	$MS_{ab} = SS_{ab}/(a-1)(b-1)$
Error	$ab(n-1)$	$SS_e = T - S$	$MS_e = SS_e/ab(n-1)$
Total	$abn-1$	$SS_t = T - CF$	

The appropriate F test depends upon whether factors A and B are fixed or random. The F test can be determined by the expected mean squares. The denominator of the F test must estimate everything that the numerator estimates except for the term being tested.

Expected Mean Squares

MS	A and B Fixed	F
A	$\sigma^2 + nb\sum_i \alpha_i^2/(a-1)$	MS_a/MS_e
B	$\sigma^2 + na\sum_j \beta_j^2/(b-1)$	MS_b/MS_e
$A \times B$	$\sigma^2 + n\sum_i\sum_j \alpha\beta_{ij}^2/(a-1)\,(b-1)$	MS_{ab}/MS_e

MS	A and B Random	F
A	$\sigma^2 + n\sigma_{AB}^2 + nb\sigma_A^2$	MS_a/MS_{ab}
B	$\sigma^2 + n\sigma_{AB}^2 + na\sigma_B^2$	MS_b/MS_{ab}
$A \times B$	$\sigma^2 + n\sigma_{AB}^2$	MS_{ab}/MS_e

MS	A Fixed, B Random	F
A	$\sigma^2 + n\sigma_{AB}^2 + nb\sum_i \alpha_i^2/(a-1)$	MS_a/MS_{ab}
B	$\sigma^2 + na\sigma_B^2$	MS_b/MS_e
$A \times B$	$\sigma^2 + n\sigma_{AB}^2$	MS_{ab}/MS_e

MS	A Random, B Fixed	F
A	$\sigma^2 + nb\sigma_A^2$	MS_a/MS_e
B	$\sigma^2 + n\sigma_{AB}^2 + na\sum_j \beta_j^2/(b-1)$	MS_b/MS_{ab}
$A \times B$	$\sigma^2 + n\sigma_{AB}^2$	MS_{ab}/MS_e

■ ■ ■

Example 12.4. ***a* × *b* Factorial ANOVA**

Because of energy shortages, oil companies are considering secondary and even tertiary recovery methods for obtaining more petroleum from exhausted oil wells. These methods attempt to free the oil from porous rock so that it can be pumped from the ground. To compare three such methods, an oil company takes a random sample of four exhausted oil fields and tries each method on two wells at each field. The results (in barrels of oil per day) are given below.

Method (*Factor A*)		Oil Field (*Factor B*) 1	2	3	4	Total
	Mechanical fracture	2, 1; 3	4, 2; 6	3, 1; 4	1, 1; 2	15
	Carbon dioxide	4, 5; 9	3, 3; 6	6, 7; 13	6, 5; 11	39
	Pressurized steam	6, 4; 10	8, 8; 16	7, 8; 15	5, 6; 11	52
	Total	22	28	32	24	106

$T = 596 \quad B = 478 \quad CF = 468.17$

$A = 556.25 \quad S = 587$

Methods are fixed because there are only three methods of interest. Fields are random, and whatever inference can be made from this experiment is to be extended to the entire population of exhausted oil fields from which this random sample was drawn.

Source	df	SS	MS	F	$F_{0.05}$
Methods	2	88.08	44.04	$MS_a/MS_{ab} = 12.62$	5.143
Fields	3	9.83	3.28	$MS_b/MS_e = 4.37$	3.490
$M \times F$	6	20.92	3.49	$MS_{ab}/MS_e = 4.65$	2.996
Error	12	9.00	0.75		

All three F values are significant. At least one method is superior to another in all the fields, but because of the significant interaction, the degree of superiority varies from field to field.

EXERCISES

12.4.1. Twenty-four men, each approximately 40 pounds overweight, are assigned to the 24 treatments that arise from four diets and three levels of jogging. Each man consumes the same number of calories per day, but the diets differ in their proportions of protein, fat, and carbohydrate.

		Diet				
		Normal	HP	HF	HC	TOTAL
Jogging	0 mi.	8.5 11.5 20.0	15.5 16.5 32.0	8.5 7.5 16.0	15.5 13.5 29.0	97.0
	1 mi.	14.0 16.0 30.0	20.0 23.0 43.0	13.0 11.0 24.0	21.0 18.0 39.0	136.0
	2 mi.	24.5 19.5 44.0	27.0 24.0 51.0	22.0 27.0 49.0	24.5 27.5 52.0	196.0
	TOTAL	94.0	126.0	89.0	120.0	429.0

a. Are the diets random or fixed?
b. Are the jogging levels random or fixed?
c. Carry out the ANOVA.
d. What hypotheses can be tested?
e. Are there significant differences related to the diets?
f. Are there significant differences related to jogging?
g. Are interactions present?
h. Which regime should be recommended for maximum weight loss?

12.4.2. The Council of Graduate Schools is an organization representing more than 700 U.S. institutions with graduate programs. Its member schools are used in a study of the difference in verbal Graduate Record Examination scores between males and females in mathematics graduate programs in the United States. Twelve institutions and six students of each sex are sampled in the study.

a. Are the effects due to sex of student random or fixed?
b. Are the effects due to institution of student random or fixed?
c. Complete the ANOVA table.

Source	df	MS	E(MS)	F
Institutions	—	132,250	______	______
Sexes	—	52,900	______	______
$I \times S$	—	26,450	______	______
Error	—	13,225	______	

d. Are any of the effects significant?

e. What is the final conclusion?

12.4.3. The State Road Commission decides to make a study of the soil erosion on hillsides that have been cut into in order to prepare roadbeds. A random sample is taken of native species of plants that can serve as ground cover. A random sample is selected among the affected hillsides around the state, and each species is planted on each hillside. After the plants are established, five observations on erosion are made on each plant and hillside combination.

a. Complete the following table.

Source	df	MS	E(MS)	F
Plant species	5	410	______	______
Hillsides	______	416	______	______
$P \times H$	20	80	______	______
Error	______	12	______	

b. Test the interaction variance for significance.

c. Which contributes more to the total variability, plant species or hillsides? Give numerical values to support your answer.

12.5. $a \times b \times c$ FACTORIAL DESIGN

The $a \times b$ factorial design can be generalized to three or more factors. In this section, we discuss the case of the $a \times b \times c$ factorial design, that is, the three-factor design.

The weight-loss problem of Exercise 12.4.1 becomes a three-factor design if we add an exercise program to the diet and jogging factors (Figure 12.9). Diet is factor A, and there are $a = 4$ levels. Amount of jogging is factor B,

		Factor C (*Exercise*)					
		No			Yes		
Factor B (*Jogging*)		0 mi.	1 mi.	2 mi.	0 mi.	1 mi.	2 mi.
Factor A (*Diet*)	N						
	HP						
	HF						
	C						

FIGURE 12.9. A three-factor design.

with $b = 3$ levels. Exercise is factor C, with $c = 2$ levels. Thus, this is a $4 \times 3 \times 2$ factorial design.

Some other examples of designs with three factors:

1. The amount of sales of a certain product at several different times of the year, both before and after an advertising campaign, using several different advertising media.
2. The achievement of foreign language classes taught by four different instructors using two different methods and involving three different workbooks.
3. The yield of a certain crop with various amounts of fertilizer, various amounts of water, and using various amounts of spacing between plants.
4. The quality of a certain product when inspected by three different inspectors using two different methods and at three different times of the day.

There is some resemblance between this design and a Latin square design. However, in the $a \times b \times c$ factorial design, it is not necessary that $a = b = c$; multiple observations are made at each combination of the three factors, and it is possible to test for interactions.

Each of the three factors may be fixed or random. The model may be entirely fixed, entirely random, or mixed with one or two random factors. In mixed models, it is not always possible to use the usual F test to test for the effect of each factor; in some cases, an exact F test does not exist.

We consider here only $a \times b \times c$ factorial designs in which the same number of subjects n are assigned at random to each combination of levels of the three factors. In the weight-loss problem, if $n = 2$ then the data are represented as in Figure 12.10.

The model for an $a \times b \times c$ factorial design is as follows:

$$y_{ijkl} = \mu + \alpha_i + \beta_j + \gamma_k + \alpha\beta_{ij} + \alpha\gamma_{ik} + \beta\gamma_{jk} + \alpha\beta\gamma_{ijk} + \epsilon_{ijkl}$$

$$i = 1, \ldots, a$$

$$j = 1, \ldots, b$$

$$k = 1, \ldots, c$$

$$l = 1, \ldots, n$$

μ: The overall mean for all experiments of this type.

α_i: The effect of the ith level of factor A; the levels may be fixed or random.

β_j: The effect of the jth level of factor B; the levels may be fixed or random.

		Factor C (*Exercise*)						
		No			Yes			
Factor B (*Jogging*)		0 mi.	1 mi.	2 mi.	0 mi.	1 mi.	2 mi.	A Totals
	N	y_{1111}	y_{1211}	y_{1311}	y_{1121}	y_{1221}	y_{1321}	
		y_{1112}	y_{1212}	y_{1312}	y_{1122}	y_{1222}	y_{1322}	
		$T_{111.}$	$T_{121.}$	$T_{131.}$	$T_{112.}$	$T_{122.}$	$T_{132.}$	$T_{1...}$
Factor A (*Diet*)	HP	y_{2111}	y_{2211}	y_{2311}	y_{2121}	y_{2221}	y_{2321}	
		y_{2112}	y_{2212}	y_{2312}	y_{2122}	y_{2222}	y_{2322}	
		$T_{211.}$	$T_{221.}$	$T_{231.}$	$T_{212.}$	$T_{222.}$	$T_{232.}$	$T_{2...}$
	HF	y_{3111}	y_{3211}	y_{3311}	y_{3121}	y_{3221}	y_{3321}	
		y_{3112}	y_{3212}	y_{3312}	y_{3122}	y_{3222}	y_{3322}	
		$T_{311.}$	$T_{321.}$	$T_{331.}$	$T_{312.}$	$T_{322.}$	$T_{332.}$	$T_{3...}$
	HC	y_{4111}	y_{4211}	y_{4311}	y_{4121}	y_{4221}	y_{4321}	
		y_{4112}	y_{4212}	y_{4312}	y_{4122}	y_{4222}	y_{4322}	
		$T_{411.}$	$T_{421.}$	$T_{431.}$	$T_{412.}$	$T_{422.}$	$T_{432.}$	$T_{4...}$
		C Totals		$T_{..1.}$			$T_{..2.}$	$T_{....}$ Grand Total

AC Totals

		Factor C	
		No	Yes
Factor A	N	$T_{1.1.}$	$T_{1.2.}$
	HP	$T_{2.1.}$	$T_{2.2.}$
	HF	$T_{3.1.}$	$T_{3.2.}$
	HC	$T_{4.1.}$	$T_{4.2.}$

AB Totals

		Factor B		
		0 mi.	1 mi.	2 mi.
Factor A	N	$T_{11..}$	$T_{12..}$	$T_{13..}$
	HP	$T_{21..}$	$T_{22..}$	$T_{23..}$
	HF	$T_{31..}$	$T_{32..}$	$T_{33..}$
	HC	$T_{41..}$	$T_{42..}$	$T_{43..}$

BC Totals

		Factor B		
		0 mi.	1 mi.	2 mi.
Factor C	No	$T_{.11.}$	$T_{.21.}$	$T_{.31.}$
	Yes	$T_{.12.}$	$T_{.22.}$	$T_{.32.}$
B Totals		$T_{.1..}$	$T_{.2..}$	$T_{.3..}$

FIGURE 12.10. Notation for a three-factor design.

γ_k: The effect of the kth level of factor C; the levels may be fixed or random.

$\alpha\beta_{ij}$: The interaction effect between the ith level of factor A and the jth level of factor B.

$\alpha\gamma_{ik}$: The interaction effect between the ith level of factor A and the kth level of factor C.

$\beta\gamma_{jk}$: The interaction effect between the jth level of factor B and the kth level of factor C.

$\alpha\beta\gamma_{ijk}$: The interaction effect among the ith level of factor A, the jth level of factor B, and the kth level of factor C.

ϵ_{ijkl}: A random effect due to sampling, ϵ_{ijkl} IND$(0,\sigma^2)$.

Uncorrected Sums of Squares

Sum of Squares	Symbol	Formula	No. of Totals	Observations/ Total
Uncorrected total	T	$\sum_i\sum_j\sum_k\sum_l y_{ijkl}^2$	$abcn$	1
Uncorrected subclass	S	$\sum_i\sum_j\sum_k T_{ijk.}^2/n$	abc	n
Uncorrected $B \times C$	BC	$\sum_j\sum_k T_{.jk.}^2/an$	bc	an
Uncorrected $A \times C$	AC	$\sum_i\sum_k T_{i.k.}^2/bn$	ac	bn
Uncorrected $A \times B$	AB	$\sum_i\sum_j T_{ij..}^2/cn$	ab	cn
Uncorrected C	C	$\sum_k T_{..k.}^2/abn$	c	abn
Uncorrected B	B	$\sum_j T_{.j..}^2/acn$	b	acn
Uncorrected A	A	$\sum_i T_{i...}^2/bcn$	a	bcn
Correction factor	CF	$T_{....}^2/abcn$	1	$abcn$

Corrected Sums of Squares

Source	df	Symbol	Definition	Computational Formula
Total	$abcn-1$	SS_t	$\sum_i\sum_j\sum_k\sum_l(y_{ijkl}-\bar{y}_{....})^2$	$T-CF$
A	$a-1$	SS_a	$bcn\sum_i(y_{i...}-\bar{y}_{....})^2$	$A-CF$
B	$b-1$	SS_b	$acn\sum_j(y_{.j..}-\bar{y}_{....})^2$	$B-CF$
C	$c-1$	SS_c	$abn\sum_k(y_{..k.}-\bar{y}_{....})^2$	$C-CF$
$A\times B$	$(a-1)(b-1)$	SS_{ab}	$cn\sum_i\sum_j(\bar{y}_{ij..}-\bar{y}_{i...}-\bar{y}_{.j..}+\bar{y}_{....})^2$	$AB-A-B+CF$

Corrected Sums of Squares (*Continued*)

Source	df	Symbol	Definition	Computational Formula
$A \times C$	$(a-1)(c-1)$	SS_{ac}	$bn\sum_i\sum_k(\bar{y}_{i.k.}-\bar{y}_{i...}-\bar{y}_{..k.}+\bar{y}_{....})^2$	$AC-A-C+CF$
$B \times C$	$(b-1)(c-1)$	SS_{bc}	$an\sum_j\sum_k(\bar{y}_{.jk.}-\bar{y}_{.j..}-\bar{y}_{..k.}+\bar{y}_{....})^2$	$BC-B-C+CF$
$A \times B \times C$	$(a-1)(b-1)(c-1)$	SS_{abc}	$n\sum_i\sum_j\sum_k(\bar{y}_{ijk.}-\bar{y}_{ij..}-\bar{y}_{i.k.}-\bar{y}_{.jk.}+\bar{y}_{i...}+\bar{y}_{.j..}+\bar{y}_{..k.}-\bar{y}_{....})^2$	$S-AB-AC-BC+A+B+C-CF$
Error	$abc(n-1)$	SS_e	$\sum_i\sum_j\sum_k\sum_l(y_{ijkl}-\bar{y}_{ijk.})^2$	$T-S$

■ ■ ■

Procedure. $a \times b \times c$ Factorial ANOVA

Hypotheses

H_0: $\alpha_1 = \cdots = \alpha_a$ or H_0: $\sigma_A^2 = 0$
H_0: $\beta_1 = \cdots = \beta_b$ or H_0: $\sigma_B^2 = 0$
H_0: $\gamma_1 = \cdots = \gamma_c$ or H_0: $\sigma_C^2 = 0$
H_0: $\alpha\beta_{11} = \cdots = \alpha\beta_{ab}$ or H_0: $\sigma_{AB}^2 = 0$
H_0: $\alpha\gamma_{11} = \cdots = \alpha\gamma_{ac}$ or H_0: $\sigma_{AC}^2 = 0$
H_0: $\beta\gamma_{11} = \cdots = \beta\gamma_{bc}$ or H_0: $\sigma_{BC}^2 = 0$
H_0: $\alpha\beta\gamma_{111} = \cdots = \alpha\beta\gamma_{abc}$ or H_0: $\sigma_{ABC}^2 = 0$

Compute:

$$T = \sum_i \sum_j \sum_k \sum_l y_{ijkl}^2$$

$$A = \sum_i T_{i...}^2/bcn$$

$$B = \sum_j T_{.j..}^2/acn$$

$$C = \sum_k T_{..k.}^2/abn$$

$$AB = \sum_i \sum_j T_{ij..}^2/cn$$

$$AC = \sum_i \sum_k T_{i.k.}^2/bn$$

$$BC = \sum_j \sum_k T_{.jk.}^2/an$$

$$S = \sum_i \sum_j \sum_k T_{ijk.}^2/n$$

$$CF = T_{....}^2/abcn$$

Source	df	SS	MS
A	$a-1$	$A-CF$	$SS_a/(a-1)$
B	$b-1$	$B-CF$	$SS_b/(b-1)$
C	$c-1$	$C-CF$	$SS_c/(c-1)$
$A \times B$	$(a-1)(b-1)$	$AB-A-B+CF$	$SS_{ab}/(a-1)(b-1)$
$A \times C$	$(a-1)(c-1)$	$AC-A-C+CF$	$SS_{ac}/(a-1)(c-1)$
$B \times C$	$(b-1)(c-1)$	$BC-B-C+CF$	$SS_{bc}/(b-1)(c-1)$
$A \times B \times C$	$(a-1)(b-1)(c-1)$	$S-AB-AC-BC$ $+A+B+C-CF$	$SS_{abc} \div$ $(a-1)(b-1)(c-1)$
Error	$abc(n-1)$	$T-S$	$SS_e/abc(n-1)$
Total	$abcn-1$	$T-CF$	

Expected mean squares vary depending upon whether the factors are fixed or random. Table 12.1 can be used to write the expected mean squares as sums of terms with the indicated coefficients. In the coefficients, $f(a) = 0$ if A is fixed and $f(a) = 1$ if A is random; similarly for $f(b)$ and $f(c)$. A term has the fixed form only if all factors in that term are fixed. F tests, if they exist, can be determined by the expected mean square. The denominator of the F test must estimate everything that the numerator estimates except for the term being tested.

■ ■ ■

Example 12.5. ***a* × *b* × *c* Factorial ANOVA**

The first attempts to freeze bull semen were made in the 1950s. It was found by accident that a solution of egg parts, glycerin, sodium citrate, and a buffer worked. The investigators wanted to know what effect different levels of buffer had on the viability of the semen. In addition, they wanted to try other antifreezes besides glycerin. They designed an $a \times b \times c$ factorial experiment which involved three levels of the buffer (a fixed effect), semen from two randomly chosen bulls (a random effect), and three randomly chosen antifreezes (a random effect). The design would enable them to test for interactions as well as separate effects.

TABLE 12.1. Expected Mean Squares for $a \times b \times c$ Factorial Design

Terms									
	Fixed	σ^2	$\frac{\sum_i \sum_j \sum_k (\alpha\beta\gamma_{ijk})^2}{(a-1)(b-1)(c-1)}$	$\frac{\sum_j \sum_k (\beta\gamma_{jk})^2}{(b-1)(c-1)}$	$\frac{\sum_i \sum_k (\alpha\gamma_{ik})^2}{(a-1)(c-1)}$	$\frac{\sum_i \sum_j (\alpha\beta_{ij})^2}{(a-1)(b-1)}$	$\frac{\sum_k \gamma_k^2}{(c-1)}$	$\frac{\sum_j \beta_j^2}{(b-1)}$	$\frac{\sum_i \alpha_i^2}{(a-1)}$
	Random	σ^2	σ_{ABC}^2	σ_{BC}^2	σ_{AC}^2	σ_{AB}^2	σ_C^2	σ_B^2	σ_A^2
Coefficients	MS_a	1	$nf(b)f(c)$	—	$nbf(c)$	$ncf(b)$	—	—	bcn
	MS_b	1	$nf(a)f(c)$	$naf(c)$	—	$ncf(a)$	—	acn	
	MS_c	1	$nf(a)f(b)$	$naf(b)$	$nbf(a)$	—	abn		
	MS_{ab}	1	$nf(c)$	—	—	nc			
	MS_{ac}	1	$nf(b)$	—	nb				
	MS_{bc}	1	$nf(a)$	na					
	MS_{abc}	1	n						
	MS_e	1							

The model is:

$$y_{ijkl} = \mu + \alpha_i + \beta_j + \gamma_k + \alpha\beta_{ij} + \alpha\gamma_{ik} + \beta\gamma_{jk} + \alpha\beta\gamma_{ijk} + \epsilon_{ijkl}$$

in which y_{ijkl} is a measure of viability, α_i is the buffer effect, β_j the bull effect, γ_k the antifreeze effect, and the other terms are the interactions.

FACTOR A, *Buffer* (*fixed*)

FACTOR C, *Antifreeze* (*random*)		A_1			A_2			A_3			
		C_1	C_2	C_3	C_1	C_2	C_3	C_1	C_2	C_3	B Totals
FACTOR B, *Bulls* (*random*)	B_1	4	1	9	8	18	11	8	7	16	
		2	3	4	5	10	9	3	11	12	
		6	0	8	1	13	6	10	6	9	
		5	5	3	7	11	5	12	6	14	
		17	9	24	21	52	31	33	30	51	268
	B_2	5	6	13	8	12	15	6	9	−2	
		1	2	8	3	12	7	12	15	3	
		7	4	9	12	9	10	4	4	5	
		−1	10	12	6	11	13	7	7	1	
		12	22	42	29	44	45	29	35	7	265
A Totals			126			222				185	533 Grand total

AC Totals

	C_1	C_2	C_3
A_1	29	31	66
A_2	50	96	76
A_3	62	65	58
C Totals	141	192	200

AB Totals

	B_1	B_2
A_1	50	76
A_2	104	118
A_3	114	71

BC Totals

	C_1	C_2	C_3
B_1	71	91	106
B_2	70	101	94

$a = 3, \quad b = 2, \quad c = 3, \quad n = 4$

$$T = \sum_i \sum_j \sum_k \sum_l y_{ijkl}^2 = 5269$$

$$A = \sum_i T_{i...}^2/bcn = \frac{126^2 + 222^2 + 185^2}{24} = 4141.04$$

$$B = \sum_j T_{.j..}^2/acn = \frac{268^2 + 265^2}{36} = 3945.81$$

$$C = \sum_k T_{..k.}^2/abn = \frac{141^2 + 192^2 + 200^2}{24} = 4031.04$$

$$AB = \sum_i \sum_j T_{ij..}^2/cn = \frac{50^2 + \cdots + 71^2}{12} = 4254.42$$

$$AC = \sum_i \sum_k T_{i.k.}^2/bn = \frac{29^2 + \cdots + 58^2}{8} = 4385.38$$

$$BC = \sum_j \sum_k T_{.jk.}^2/an = \frac{71^2 + \cdots + 94^2}{12} = 4041.25$$

$$S = \sum_i \sum_j \sum_k T_{ijk.}^2/n = \frac{17^2 + \cdots + 7^2}{4} = 4737.75$$

$$CF = T_{....}^2/abcn = \frac{533^2}{72} = 3945.68$$

Source	df	SS	MS
A	$a-1=2$	$A-CF=195.36$	97.68
B	$b-1=1$	$B-CF=0.13$	0.13
C	$c-1=2$	$C-CF=85.36$	42.68
$A\times B$	$(a-1)(b-1)=2$	$AB-A-B+CF=113.25$	56.63
$A\times C$	$(a-1)(c-1)=4$	$AC-A-C+CF=158.98$	39.75
$B\times C$	$(b-1)(c-1)=2$	$BC-B-C+CF=10.08$	5.04
$A\times B\times C$	$(a-1)(b-1)(c-1)=4$	$S-AB-AC-BC+A+B+C-CF=228.91$	57.23
Error	$abc(n-1)=54$	$T-S=531.25$	9.84
Total	$abcn-1=71$		

Mean Squares	E(MS)	F	Critical Value F
MS_a	$\sigma^2+4\sigma_{abc}^2+8\sigma_{ac}^2+12\sigma_{ab}^2+12\sum_i\alpha_i^2$	No appropriate F test	—
MS_b	$\sigma^2+12\sigma_{bc}^2+36\sigma_b^2$	$MS_b/MS_{bc}=0.03$	18.513
MS_c	$\sigma^2+12\sigma_{bc}^2+24\sigma_c^2$	$MS_c/MS_{bc}=8.47$	19.000
MS_{ab}	$\sigma^2+4\sigma_{abc}^2+12\sigma_{ab}^2$	$MS_{ab}/MS_{abc}=0.99$	6.944
MS_{ac}	$\sigma^2+4\sigma_{abc}^2+8\sigma_{ac}^2$	$MS_{ac}/MS_{abc}=0.69$	6.388
MS_{bc}	$\sigma^2+12\sigma_{bc}^2$	$MS_{bc}/MS_e=0.51$	3.170
MS_{abc}	$\sigma^2+4\sigma_{abc}^2$	$MS_{abc}/MS_e=5.82$	2.544
MS_e	σ^2		

The buffer effect cannot be tested in this design; however, the experimenter probably knows about this effect already and is more interested in the interactions. There are no differences between bulls and no differences among antifreezes. Although there are no significant interactions between two factors, there is a significant interaction among all three factors. This is the worst possible experimental result, since it indicates that different combinations of buffer and antifreeze give maximal viability for different bulls, and it will be difficult to find one combination that works best for all bulls.

EXERCISES

12.5.1. When land is in continuous production, it needs to be treated with a complete fertilizer, that is, one combining nitrogen (chemical symbol N), phosphorus (P), and potassium (K, from the Latin *kalium*). So, shortly after a new variety or hybrid is developed, an NPK factorial experiment is conducted in order to learn something about its response to fertilizers. Suppose there is developed a fescue grass hybrid which is resistant to white grubs, and it will be sold for use on lawns and golf courses. Before marketing, however, an NPK experiment is conducted so that fertilizer recommendations can be made. Forty-eight plots containing mature stands of grass are assigned at random to each of 24 different combinations of fertilizer, two plots to each combination. The fertilizer is applied and given time to have an effect. Each plot is mowed, and the clippings are dried and weighed to provide the data below:

		Potassium			
		0 cwt/acre		3 cwt/acre	
		Plot			
Nitrogen	Phosphorus	1	2	1	2
0 cwt	0 cwt	91	54	80	85
	3 cwt	56	72	62	90
	6 cwt	103	154	158	175
3 cwt	0 cwt	254	266	262	258
	3 cwt	173	252	238	317
	6 cwt	383	392	340	465
6 cwt	0 cwt	243	303	239	345
	3 cwt	238	303	287	252
	6 cwt	389	394	384	403
9 cwt	0 cwt	252	175	114	229
	3 cwt	263	281	205	241
	6 cwt	295	244	271	380

a. Give the linear model.
b. Which effects are fixed and which are random?
c. Compute a one-way ANOVA with the following sources of variation and degrees of freedom:

Source	df
Fertilizers	23
Within	24

d. From the sum of squares for fertilizers, break out the effects of N, P, and K and all of their interactions.
e. Give the expectations of mean squares for the three-factor ANOVA above.
f. Make F tests that are valid and draw conclusions.

12.5.2. In an effort to learn more about the shrinkage of cotton knit undershirts when washed and dried at military base laundries, the U.S. Army Quartermaster Corps takes a random sample of four brands of shirts from several hundred available for purchase. They further randomly sample enough shirts to have two from each brand to be washed at each of two water temperatures and dried at each of these temperatures. The results, measured by shrinkage of length (in centimeters), are given below.

	Cold Water Wash Drying Temperature			Hot Water Wash Drying Temperature		
Brand	210°F	218°F	226°F	210°F	218°F	226°F
A	1.9, 2.1	3.3, 3.7	7.5, 7.9	3.4, 3.6	8.0, 7.6	7.5, 7.7
B	2.2, 2.4	4.8, 5.0	9.8, 9.2	4.6, 4.4	9.3, 9.5	10.1, 9.7
C	2.8, 3.2	6.5, 6.6	13.2, 13.0	5.7, 6.3	12.9, 13.3	13.1, 13.3
D	3.1, 3.7	4.5, 4.8	10.8, 11.2	5.6, 5.0	10.9, 10.7	11.4, 11.7

a. Which effects are random and which are fixed?
b. Give the expectations of mean square.
c. Perform the ANOVA and make all valid F tests.
d. Draw conclusions about the washing and drying procedures that minimize shrinkage.

12.5.3. Holly trees are attractive and desirable for landscaping, but their propagation presents many problems. Individual trees are either male or female, so there is no production of seed through self-fertilization. Furthermore, once seed are produced, they lie in the ground for about two years before the germination and emergence of the seedlings that begin the next generation of trees. In an effort to find ways to speed up the process, a horticulturist takes a random sample of four male trees and another of four female trees and makes all possible cross-pollinations. When seeds are produced, he divides the seeds from each of the 16 crosses into two groups at random. The seeds in one group are used as a control, and those in the other are scarified because it is claimed this process frequently promotes germination. Seeds are then planted in individual pots. Three years later, two healthy seedlings are selected at random from each cross and treatment, and measured for height. The data (in inches) are recorded below.

Control

	F_1	F_2	F_3	F_4
M_1	4.6, 4.9	5.1, 6.1	4.4, 4.8	5.2, 6.3
M_2	8.6, 7.8	5.2, 5.4	3.4, 4.6	4.2, 3.8
M_3	8.7, 8.5	6.6, 7.4	2.0, 2.8	3.7, 4.3
M_4	7.6, 8.4	5.1, 5.4	5.3, 7.7	8.0, 7.5

Scarified

	F_1	F_2	F_3	F_4
M_1	5.3, 4.7	7.7, 8.5	5.3, 5.3	7.7, 6.5
M_2	7.3, 8.5	5.8, 5.4	7.7, 6.9	4.4, 4.6
M_3	6.6, 6.9	6.0, 7.0	8.0, 8.5	6.8, 7.2
M_4	6.9, 7.1	8.8, 8.2	8.9, 9.1	6.7, 7.3

a. Which effects are fixed and which are random?

b. Compute a two-way ANOVA with the following sources of variation and degrees of freedom:

Source	df
Crosses	15
Treatment	1
Crosses × treatment	15
Error	32

c. From the Crosses SS, break out the effects of Male Trees, Female Trees, and the Male × Female interaction.

d. From the Crosses × Treatment SS break out the effects of the three interactions Treatment × Male, Treatment × Female, and Treatment × Male × Female.

e. Give the expectations of mean squares for the three-factor ANOVA, and make all valid F tests.

f. Estimate the percentage to total variability in height due to Male Trees, Female Trees, and Male × Female.

g. What conclusions should be drawn from this study?

12.6. SPLIT-PLOT DESIGN

In this section we discuss a split-plot design that involves randomized complete blocks and two fixed factors. Many other variations of the split-plot design exist, and the reader should consult a reference such as Steel and Torrie (1960) or Cochran and Cox (1957) if one of these other variations is needed.

An example of a split-plot design that involves randomized complete blocks is a marketing experiment in which the investigator wants to study the effectiveness of different incentives used in buy-by-mail advertising for different types of products.

Four large cities are randomly selected for the experiment. From the city directories, 100 households are selected to receive mailings for each of three products (a total of 300 households in each city). The three products are ladies' hosiery, men's underwear, and household linens. Half of each group receives a mailing that offers an extra discount on an order placed within a short time, and the other half is offered a free pen-and-pencil set with each order (Figure 12.11). Total sales are recorded for each category.

This design differs from the $a \times b \times c$ factorial design discussed in Section 12.5, although the diagrams appear to be similar. Cities in this experiment are randomized complete blocks rather than a factor. The investigator is not interested in cities as such but is using them to control for extraneous variability caused by different locations. Within cities, three samples of 100 are assigned at random to the products, which are the *main-*

	CITIES (*blocks*)												
		A			B			C			D		
	PRODUCT (*main units*)	H	U	L	H	U	L	H	U	L	H	U	L
INCENTIVE (*subunits*)	DISCOUNT												
	PREMIUM												

FIGURE 12.11. A split-plot design.

unit treatment, or whole-unit treatment. Then, within these samples, half are assigned at random to each incentive group, the *subunit treatment*. The investigator's first interest is the incentive factor, but at the same time, he wishes to gather information about how the incentives work with different products.

Other examples of this split-plot design:

1. A study of vitamin C content of oranges grown in six different orchards (blocks), using four trees from each orchard which are each treated with a different spray (main-unit treatment), and using two oranges picked from each tree and stored at different temperatures (subunit treatment).
2. A study of yield of soybeans, using different types of seed with different fertilizer treatments. Farms are used for blocks, fertilizer is applied to large plots (whole units), and the different types of seed are planted on sections within the fertilizer plots (subunits).
3. A study of medications for reducing high blood pressure in males involving four different drugs (main-unit treatment), each assigned at random to three males from each of several ethnic groups (blocks), and within each medication group the drugs are administered once a day but at three different times of day (subunit treatment).
4. A study of the retention of historical facts in which students are blocked by schools, two techniques of teaching are used (main-unit treatment), and retention is measured on the same student after several different time periods (subunit treatment).

Here is a summary of the blocks, main-unit factor, and subunit factor for each of the examples above:

Example	Blocks	Treatment on Whole Units	Treatment on Subunits
Buy-by-mail	Cities	Products	Incentives
Vitamin C	Orchards	Sprays	Storage temperatures
Yield	Farms	Fertilizers	Seed types
Blood pressure	Ethnic groups	Drugs	Time of day
Retention	Schools	Techniques	Time periods

An example of the statistical analysis used for a split-plot design is helpful at this point.

Example 12.6. Split-Plot Design

A food scientist wishes to study the effects of tenderizer and length of cooking time on meat. Six beef carcasses are obtained at random from a meat packing plant. The right rib-eye muscle is excised from each carcass; from the midportion of each muscle, three rolled roasts are prepared, as nearly alike as possible. Each of the roasts is assigned at random to a tenderizing treatment: control, vinegar marinade, or papain marinade. After treatment, a coring device is used to make four cores of meat near the center of each roast. The cores, however, are left in place, and the three roasts from the same carcass are placed together in an oven preheated to 300°F and allowed to cook. After 30 minutes of roasting, one of the cores is taken at random from each roast, another randomly drawn set of three cores is taken after 36 minutes, a third set after 42 minutes, and the final set at 48 minutes. As each set is taken, the cores are allowed to cool to serving temperature and are then measured for tenderness using the Warner–Bratzler device, an instrument similar to a guillotine. The measurement is a number on the Warner–Bratzler scale. A large number indicates a tough piece of meat. The measurements from the six carcasses (blocks), three tenderizing treatments (on whole units), and four lengths of roasting time (on subunits) are the variables of analysis.

In this experiment, combinations of tenderizer and roasting time could not be assigned at random to the cores of meat; the nature of the experiment

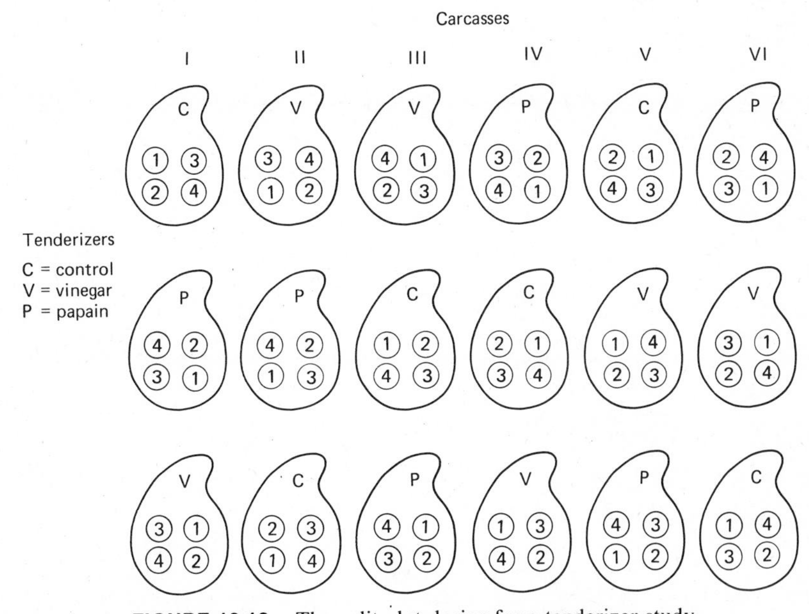

FIGURE 12.12. The split-plot design for a tenderizer study.

does not allow for that kind of assignment of treatment combinations. Instead, there were three distinct levels of randomization. Six carcasses were taken at random from a very large number of available carcasses. The right rib-eye muscle from each carcass (block) was divided into three roasts (whole units), to which three tenderizer treatments were assigned at random. Finally, four cores of meat (subunits) were taken to measure the interior tenderness of each roast at a specified time of cooking, but at the specified time a core was drawn randomly from each roast. The experiment can be visualized as in Figure 12.12. The random variable is tenderness (average of four determinations) of meat prepared with one of three tenderizers and roasted for one of four lengths of time.

Tenderizers (*Factor A*)		Roasting Time (*Factor B*)	Carcasses (*Factor C*) I	II	III	IV	V	VI
	Control	30	8.25	8.00	7.75	8.25	7.50	7.75
		36	7.50	7.00	6.75	6.25	6.75	6.25
		42	4.25	3.25	3.75	4.00	3.25	3.00
		48	3.50	3.25	3.25	3.50	3.50	3.00
			23.50	21.50	21.50	22.00	21.00	20.00
	Vinegar	30	7.25	7.00	6.75	6.75	6.50	6.25
		36	6.25	6.00	6.00	5.50	5.25	5.00
		42	3.50	3.50	4.00	3.50	3.25	3.25
		48	2.50	2.50	2.75	2.25	2.00	3.00
			19.50	19.00	19.50	18.00	17.00	17.50
	Papain	30	6.50	6.00	6.25	5.75	5.25	5.25
		36	4.50	4.75	5.00	4.50	4.50	4.25
		42	3.50	4.00	3.50	3.50	3.25	3.25
		48	3.50	3.75	3.75	3.25	3.00	3.25
			18.00	18.50	18.50	17.00	16.00	16.00
	TOTAL		61.00	59.00	59.50	57.00	54.00	53.50

Tenderizer	Roasting Time 30	36	42	48	Total
Control	47.50	40.50	21.50	20.00	129.5
Vinegar	40.50	34.00	21.00	15.00	110.5
Papain	35.00	27.50	21.00	20.50	104.0
TOTAL	123.00	102.00	63.50	55.50	344.0

Uncorrected Sum of Squares	Symbol	No. of Squared Values	Observations/ Squared Value	Calculations	Numerical Value
Total	T	$abc=72$	1	$(8.25)^2+(7.50)^2+\cdots+(3.25)^2$	1852.2500
Whole unit (roasts)	W	$ac=18$	$b=4$	$[(23.5)^2+\cdots+(16.00)^2]/4$	1663.0000
Factor A (tenderizers)	A	$a=3$	$bc=24$	$[(129.5)^2+(110.5)^2+(104.0)^2]/24$	1658.1875
Block (carcasses)	C	$c=6$	$ab=12$	$[(61.00)^2+\cdots+(53.50)^2]/12$	1647.4583
Factor B (roasting time)	B	$b=4$	$ac=18$	$[(123.00)^2+\cdots+(55.50)^2]/18$	1813.6388
$A\times B$ (tenderizer by time)	AB	$ab=12$	$c=6$	$[(47.50)^2+\cdots+(20.50)^2]/6$	1843.9167
Correction factor	CF	1	$abc=72$	$(344)^2/72$	1643.5556

The analysis can initially be approached as though the experiment involved nothing more than 18 roasts and four roasting times. One could then conduct a two-way ANOVA, which we call the preliminary analysis.

Preliminary Analysis

Source	df	SS
Roasts	$ac-1=17$	$W-CF=$ 19.4444
Roasting time	$b-1=3$	$B-CF=$ 170.0832
Residual	$(ac-1)(b-1)=51$	$T-W-B+CF=$ 19.1668

But the roasts (whole units) are not independent; some are associated because they came from the same carcass, and others because they received the same tenderizing treatment. Consequently, the variability due to these effects can be accounted for in the roast sum of squares in the following manner:

Source	df	SS
Roasts (whole units)	$ac - 1 = 17$	$W - CF = 19.4444$
Tenderizers	$a - 1 = 2$	$A - CF = 14.6319$
Carcasses	$c - 1 = 5$	$C - CF = 3.9027$
Whole-unit remainder	$(a - 1)(c - 1) = 10$	$W - A - B + CF = 0.9098$

Other variability in the preliminary analysis can be accounted for, and that is the variability due to interaction between tenderizer and roasting time ($A \times B$). This variability is, perforce, part of the residual sum of squares, so it should be computed and removed, as shown below.

Source	df	SS
Residual in preliminary analysis	$(ac - 1)(b - 1) = 51$	$T - W - B + CF = 19.1668$
$A \times B$	$(a - 1)(b - 1) = 6$	$AB - A - B + CF = 15.6460$
Subunit remainder	$a(b - 1)(c - 1) = 45$	$T - W - AB + A = 3.5208$

The complete ANOVA for the split-plot design can be obtained by putting together the sums of squares that have been broken out of the preliminary analysis. The final analysis is:

Source	df	SS	MS	F
WHOLE UNITS				
Tenderizers	$a - 1 = 2$	$A - CF = 14.6319$	7.3160	80.396*
Carcasses	$c - 1 = 5$	$C - CF = 3.9027$	0.7805	8.577*
Whole-unit remainder	$(a - 1)(c - 1) = 10$	$W - A - C + CF = 0.9098$	0.0910	
SUBUNITS				
Roasting time	$b - 1 = 3$	$B - CF = 170.0832$	56.6944	724.992*
Times × tenderizer	$(a - 1)(b - 1) = 6$	$AB - A - B + CF = 15.6460$	2.6077	33.347*
Subunit remainder	$a(b - 1)(c - 1) = 45$	$T - W - AB + A = 3.5208$	0.0782	

*Significant at $\alpha = .05$.

Not too surprisingly, the analysis results in claiming significance for all effects tested. This is largely due to the nature of the experiment. It has probably been known from the time of the cavemen that the longer meat is cooked, the more tender it becomes. Similarly, the benefits of marinating were discovered without benefit of statistical analysis. However, it is not uncommon in the split-plot design for the experimenter to know in advance of the experiment that the whole-unit treatments (tenderizers) and even subunit treatments (roasting times) are significant. The principal concern in the design is usually the interaction. Here, the food scientist wants to know about the best combinations of tenderizer and roasting time. Because the interaction term also proved to be significant, the food scientist will pay particular interest to a mean separation technique that allows for further examination of the interaction. This can be done with a two-way table of means:

Factor A	Factor B (Roasting Time)			
(Tenderizer)	30 min.	36 min.	42 min.	48 min.
Control	7.9167	6.7500	3.5833	3.3333
Vinegar	6.7500	5.6667	3.5000	2.5000
Papain	5.8333	4.5833	3.5000	3.4167

The Warner-Bratzler score employed here is a function of the force necessary to shear a piece of meat of a given size. Consequently, the greater the mean score, the less tender is the meat. The interactions can be best understood by comparing means of roasting times for the same tenderizer or, conversely, the means of tenderizers at the same roasting time. Multiple comparisons within a split-plot design differ from those in other designs (see Steel and Torrie 1960).

To compare means for roasting times (B means) with the same tenderizer (same A level), the least significant difference is:

$$t_{\alpha/2,a(b-1)(c-1)}\sqrt{\frac{2\text{MS}_{\text{subunit remainder}}}{c}} = 2.014\sqrt{\frac{2(0.0782)}{6}}$$

$$= 0.3254 \text{ at } \alpha = 0.05$$

To compare two tenderizer means (A means) at the same roasting time (same B level), an approximate test must be used because the two A means contain both A effects and AB interactions. The least significant difference is:

$$t_{\alpha/2*}\sqrt{\frac{2[(b-1)\text{MS}_{\text{subunit remainder}} + \text{MS}_{\text{whole unit remainder}}]}{cb}}$$

in which

$$t_{\alpha/2*} = \frac{(b-1)\text{MS}_{sr}\, t_{\alpha/2,a(b-1)(c-1)} + \text{MS}_{wr}\, t_{\alpha/2,(a-1)(c-1)}}{(b-1)\text{MS}_{sr} + \text{MS}_{wr}}$$

Thus,

$$t_{0.025^*} = \frac{3(0.0782)2.014 + (0.0910)2.228}{3(0.0782) + 0.0910}$$

$$= 2.074$$

and the least significant difference is:

$$2.074\sqrt{2[3(0.0782) + 0.0910]/24} = 0.1647$$

There is an indication that the use of a tenderizing marinade is important for those who prefer their roasts rare or medium rare, because the differences between all tenderizer treatments are significant for roasts cooked 30 or 36 minutes. The differences cease to be significant, however, for meat roasted for 42 minutes. Similarly, no matter what the precooking tenderizing treatment, the longer a roast is cooked, the more tender it will be. The significant difference between the vinegar marinade and the other two whole-plot treatments at 48 minutes could be an anomaly; there are $\binom{12}{2} = 66$ possible comparisons between means, so Type I errors could indeed be present. On the other hand, it could also represent a reproducible phenomenon which the food scientist might want to examine further. However, even if it is a real difference, the gain in tenderness may not merit the added broiling time if there is offsetting loss in meat texture, juiciness, or other components of palatability.

In general, a split-plot design may be arranged similar to Figure 12.13, in which three blocks, two whole-unit treatments, and four subunit treatments are used.

The model for this split-plot design, in which the whole-unit treatment is randomized within complete blocks, is:

$$y_{ijk} = \mu + \alpha_i + \beta_j + \alpha\beta_{ij} + \gamma_k + \delta_{ik} + \epsilon_{ijk}$$

$$i = 1, \ldots, a$$

$$j = 1, \ldots, b$$

$$k = 1, \ldots, c$$

The terms in this model have the following meanings.

μ: The overall mean for all experiments of this type.

α_i: The effect of the ith level of factor A, the whole unit treatment; a fixed effect, $\sum_i \alpha_i = 0$.

β_j: The effect of the jth level of factor B, the subunit treatment; a fixed effect, $\sum_j \beta_j = 0$.

Factor A (Whole-Unit treatment)		Factor B (Subunit Treatment)	Factor C (Blocks) C_1	C_2	C_3	
	A_1	B_1	y_{111}	y_{112}	y_{113}	
		B_2	y_{121}	y_{122}	y_{123}	
		B_3	y_{131}	y_{132}	y_{133}	
		B_4	y_{141}	y_{142}	y_{143}	
			$T_{1.1}$	$T_{1.2}$	$T_{1.3}$	$T_{1..}$
	A_2	B_1	y_{211}	y_{212}	y_{213}	
		B_2	y_{221}	y_{222}	y_{223}	
		B_3	y_{231}	y_{232}	y_{233}	
		B_4	y_{241}	y_{242}	y_{243}	
			$T_{2.1}$	$T_{2.2}$	$T_{2.3}$	$T_{2..}$
		C TOTALS	$T_{..1}$	$T_{..2}$	$T_{..3}$	$T_{...}$ GRAND TOTAL

	B_1	B_2	B_3	B_4
A_1	$T_{11.}$	$T_{12.}$	$T_{13.}$	$T_{14.}$
A_2	$T_{21.}$	$T_{22.}$	$T_{23.}$	$T_{24.}$
B TOTALS	$T_{.1.}$	$T_{.2.}$	$T_{.3.}$	$T_{.4.}$

FIGURE 12.13. Notation for a split-plot design.

$\alpha\beta_{ij}$: The interaction effect between the ith level of factor A and the jth level of factor B.

γ_k: The kth block effect; blocks are random.

δ_{ik}: The whole unit random component, $\delta_{ik}\text{IND}(0,\sigma_D^2)$.

ϵ_{ijk}: The subunit random component, $\epsilon_{ijk}\text{IND}(0,\sigma^2)$.

Uncorrected Sums of Squares

Sum of Squares	Symbol	Formula	No. of Totals	Observations/ Total
Uncorrected total	T	$\sum_i \sum_j \sum_k y_{ijk}^2$	abc	1
Uncorrected whole unit	W	$\sum_i \sum_k T_{i.k}^2/b$	ac	b
Uncorrected A factor	A	$\sum_i T_{i..}^2/bc$	a	bc
Uncorrected B factor	B	$\sum_j T_{.j.}^2/ac$	b	ac
Uncorrected block	C	$\sum_k T_{..k}^2/ab$	c	ab
Uncorrected $A \times B$	AB	$\sum_i \sum_j T_{ij.}^2/c$	ab	c
Correction factor	CF	$T_{...}^2/abc$	1	abc

■ ■ ■

Procedure. Split-Plot ANOVA with Randomized Complete Block (Factors *A* and *B* Fixed Effects)

Hypotheses:

H_0: $\alpha_1 = \cdots = \alpha_a = 0$ (no differences among secondary treatments)
H_0: $\beta_1 = \cdots = \beta_b = 0$ (no difference among main treatments)
H_0: $\alpha\beta_{11} = \cdots = \alpha\beta_{ab} = 0$ (no interactions)
H_0: $\sigma_C^2 = 0$ (no block effect)

Source	df	SS	E(MS)
Whole units			
Factor *A*	$a-1$	$A-CF$	$\sigma^2+b\sigma_D^2+cb\sum_i \alpha_i^2/(a-1)$
Blocks	$c-1$	$C-CF$	$\sigma^2+b\sigma_D^2+ab\sigma_C^2$
Whole-unit remainder	$(a-1)(c-1)$	$W-A-C+CF$	$\sigma^2+b\sigma_D^2$
Subunits			
Factor *B*	$b-1$	$B-CF$	$\sigma^2+ca\sum_j \beta_j^2/(b-1)$
$A \times B$	$(a-1)(b-1)$	$AB-A-B+CF$	$\sigma^2+c\sum_i \sum_j \alpha\beta_{ij}^2/(a-1)(b-1)$
Subunit remainder	$a(c-1)(b-1)$	$T-W-AB+A$	σ^2

Mean squares are found by dividing the sums of squares by the corresponding degrees of freedom. The appropriate *F* tests can be determined from the expected mean square. The standard errors needed for estimates and for multiple comparison are given in the following table:

Difference Between	Standard Error	df for t
Two overall *A* means	$\sqrt{\dfrac{2\text{MS}_{wr}}{bc}}$	$(a-1)(c-1)$
Two overall *B* means	$\sqrt{\dfrac{2\text{MS}_{sr}}{ac}}$	$a(b-1)(c-1)$
Two *B* means at the same *A* level	$\sqrt{\dfrac{2\text{MS}_{sr}}{c}}$	$a(b-1)(c-1)$
Two *A* means at the same *B* level or different *B* levels	$\sqrt{\dfrac{2[(b-1)\text{MS}_{sr}+\text{MS}_{wr}]}{bc}}$	use t_α^* below

$$t_\alpha^* = \frac{(b-1)\text{MS}_{sr}\, t_{\alpha,a(b-1)(c-1)} + \text{MS}_{wr}\, t_{\alpha,(a-1)(c-1)}}{(b-1)\text{MS}_{sr} + \text{MS}_{wr}}$$

It is appropriate to use a split-plot design if:

1. One of the treatments requires large quantities of material (such as the fertilizer in the yield example), and the whole units are used for this treatment.
2. An additional factor is to be incorporated into the experiment (such as the products in the buy-by-mail example). The main factor (incentives) is applied to the subunits and the additional factor to the whole units.
3. Larger differences are expected among the levels of one factor than among the levels of the other factor (as in the blood pressure example). The factor with the larger differences (drugs) is used for the whole units and the factor with small differences (time of day) for the subunits.
4. Greater precision is desired for comparisons among the levels of one factor than the other factor. The factor requiring the greater precision is used for the subunits.

Some split-plot designs could be laid out as an $a \times b \times c$ factorial design. For example, the achievements of foreign language classes taught by four different instructors using two different methods and three different workbooks is a $4 \times 2 \times 3$ factorial design if groups of students are assigned at random to each combination of teacher, method, and workbook. However, this could be planned as a split-plot design. If the students pick the teachers and each teacher is offering two classes, the teachers are the blocks. The classes are the whole units, and they are randomly assigned a method. Within classes, equal numbers of students (subunits) are randomly assigned to the three different workbooks.

The overall precision of the two experiments is probably the same. However, the split-plot design gives increased precision for subunit comparisons and a lower precision for whole-unit comparisons. Thus, if the experimenter wants to be able to detect differences among the workbooks, the split-plot design increases the probability of detecting these differences if they exist.

EXERCISE

12.6.1. Analyze the shrinkage data in Exercise 12.5.2 as if they arose from a split-plot design in which brands are the blocks, wash temperatures are applied to groups of shirts together (whole units), and the drying temperatures are randomly assigned within the whole units. Let the random variable be the average centimeters of shrinkage in length of the two shirts in each subgroup.

REVIEW EXERCISES

Decide whether each of the following is true or false. If a statement is false, explain why.

12.1. Whether an effect is fixed or random will determine the arithmetical procedure for computing the sums of squares.

12.2. If the additive model is $y_{ij} = \mu + \alpha_i + \epsilon_{ij}$, the appropriate ANOVA will produce two mean squares.

12.3. The model $y_{ij} = \mu + \alpha_i + \beta_j + \epsilon_{ij}$ identifies a nested classification of data.

12.4. The Latin square design is appropriate for pilot experiments in new areas of research, because it provides an economical design for measuring three different kinds of variability.

12.5. The model $y_{ijk} = \mu + \alpha_i + \beta_{ij} + \epsilon_{ijk}$ does not indicate whether the β is a block effect or a nested effect.

12.6. If an interaction exists in experimental data and no provision is made for it in the model and analysis, the interaction variability will be confounded with the estimate of random variation.

12.7. In a one-way analysis of variance, if the null hypothesis is true, the corrected total-SS and corrected treatment-group-SS divided by their respective degrees of freedom estimate the same value.

12.8. A randomized complete block design with four blocks and three treatments will have the same degrees of freedom for the F test of treatments as one consisting of three blocks and four treatments.

12.9. The chief advantage of the Latin square design is that it permits the analysis of main effects without any concern for interaction.

12.10. An investigator is not likely to design a 12×12 Latin square because of the problem of maintaining balance in so large an experiment.

12.11. Because the residual mean squares from a blocked design will have fewer degrees of freedom than the within mean square of a one-way analysis of the same data, one could obtain a poorer F test of treatments in a blocked design if the block effects are nonsignificant.

12.12. A mixed model means that certain effects are factorial and others are random.

12.13. When performing a randomized complete block ANOVA, the experimenter is usually as interested in finding differences among the blocks as among the treatments, so he uses some sort of multiple comparison technique on both sets of means.

12.14. Whether an effect is nested or factorial has no bearing on whether it is random or fixed.

12.15. In ANOVA, it may be possible to estimate a particular variance component, but still not be possible to have an exact test for significance.

12.16. In an experiment involving three effects in a factorial arrangement, if all three main effects are fixed, the interaction term drops out of the expectations of all mean squares.

12.17. The nested classification is a continued one-way classification of subgroups within the major groups.

12.18. Missing value techniques may be employed even when all observations in a row or column are missing.

12.19. To use a missing value technique does not cause the loss of one degree of freedom; the degree of freedom was lost when the observation was lost.

12.20. Because the experimenter is looking for an optimal combination of effects in a factorial experiment, such experiments usually fit the fixed effects model.

12.21. The assumption $\sigma_{ab}^2 = 0$ or $\alpha\beta_{ij} = 0$ for all i, j is necessary for both Latin square and split-plot design.

12.22. A nested effect may be either fixed or random.

12.23. A linear model may contain both factorial and nested effects.

12.24. A linear model may contain both random and fixed effects.

12.25. In ANOVA, if the row mean square is nonsignificant and the column mean square is also nonsignificant, it is unlikely that the row $\times$ column mean square will be significant.

12.26. Because the Latin square design does not permit a treatment to be found twice in the same row or column, it is impossible to randomize treatments in that design.

12.27. There are four types of interactions in an $a \times b \times c$ factorial design.

12.28. Data collected for an $a \times b \times c$ factorial design may be analyzed as a split-plot design.

12.29. A split-plot design cannot be used if the whole-unit treatment is a fixed effect.

12.30. Approximate tests must be used for some follow-up procedures after a split-plot analysis.

SELECTED READINGS

Bliss, C. I. (1967). *Statistics in Biology*, McGraw-Hill, New York.

Box, G. E. P. (1954). Some theorems on quadratic forms applied in the study of analysis of variance problems: II. Effect of inequality of variance and of correlation between errors in the two-way classification, *The Annals of Mathematical Statistics,* **25,** 484–498.

Brown, B. M. (1975). A short-cut test for outliers using residuals, *Biometrika,* **62,** 623–629.

Cochran, W. G., and G. M. Cox (1957). *Experimental Design*, Wiley, New York.

Daniel, C. (1960). Locating outliers in factorial experiments, *Technometrics,* **2,** 149–156.

——— (1978). Patterns in residuals in the two-way layout, *Technometrics,* **20,** 385–395.

Davies, O. L., Ed. (1956). *The Design and Analysis of Industrial Experiments*, 2nd Ed., Hafner, New York.

DeLury, D. B. (1946). The analysis of Latin squares when some observations are missing, *Journal of the American Statistical Association,* **41,** 370–389.

Fisher, R. A., and F. Yates (1963). *Statistical Tables for Biological, Agricultural, and Medical Research*, Hafner, New York.

Geisser, S. (1959). A method for testing treatment effects in the presence of learning, *Biometrics,* **15,** 389–395.

Glenn, W. A., and C. Y. Kramer (1958). Analysis of variance of a randomized block design with missing observations, *Applied Statistics,* **7,** 173–185.

Harter, H. L. (1970). Multiple comparison procedures for interactions, *The American Statistician,* **24** (Dec.), 30–32.

Kramer, C. Y., and S. Glass (1960). Analysis of variance of a Latin square design with missing observations, *Applied Statistics,* **9,** 43–50.

Monlezun, C. J. (1979). Two-dimensional plots for interpreting interactions in the three-factor analysis of variance model, *The American Statistician,* **33,** 63–69.

Schultz, E. F., Jr. (1955). Rules of thumb for determining expectations of mean squares in analysis of variance, *Biometrics,* **11,** 123–135.

Snedecor, G. W., and Cochran, W. G. (1973). *Statistical Methods*, Iowa State University Press.

Steel, R. G. D., and J. H. Torrie (1960). *Principles and Procedures of Statistics*, McGraw-Hill, New York.

Taylor, W. H., Jr., and H. G. Hilton (1981). A structure diagram symbolization for analysis of variance, *The American Statistician,* **35,** 85–93.

Wilk, M. B., and O. Kempthorne (1957). Nonadditives in a Latin square design, *Journal of the American Statistical Association,* **52,** 218–236.

Winer, B. J. (1971). *Statistical Principles in Experimental Design*, 2nd ed., McGraw-Hill, New York.

13

Analysis of Covariance

The analysis of covariance is a combination of regression analysis with an analysis of variance. Covariance is used when the response variable y, in addition to being affected by the treatments, is also linearly related to another variable x. In this chapter we discuss the analysis of covariance in which simple linear regression is combined with a one-way ANOVA. More complex designs exist but are beyond the scope of this book.

13.1. COMBINING REGRESSION WITH ANOVA

The analysis of covariance is useful in several types of research situations. For example, it can be used to:

1. Increase precision in an experiment.
2. Control for an extraneous variable in a survey.
3. Compare regressions within several groups.

Specific examples of these three types of application follow.

Increasing precision in an experiment is illustrated by the use of covariance analysis in a study of weight loss y under three different diets (the treatments). Ordinary ANOVA may fail to detect a significant difference among the treatment effects because the within-treatment-group variability is too large. Covariance sharpens the analysis of variance on y by utilizing a related variable x, called a *covariate*, or *concomitant variable*. Pounds lost, y, is linearly related to x, pounds overweight at the beginning of the experiment. By combining the regression of y on x with the analysis of variance on y, the within-treatment variability is reduced, making it more likely that treatment differences will be detected. Intuitively we can think of the analysis of covariance as removing that portion of the within-treatment variability which is accounted for by the regression. (Blocking by overweight classes could also be used to reduce within-group variability, but this cannot always be done since it requires equal numbers of subjects in each overweight class.)

Controlling for an extraneous variable in a survey is illustrated by a study of teachers' salaries y in three different school systems (treatment groups) in which the educational level in years attained by the teachers is an extraneous variable x. If y is linearly related to x, then the analysis of covariance can be used to adjust for differences in the educational attainment of the teachers. In this application, we can think of the analysis of covariance as transforming all of the data points (x_{ij}, y_{ij}) to $(\bar{x}_{..}, y_{ij}')$, a point on the vertical line at the overall average x value, by means of a translation parallel to the regression line (Figure 13.1).

Intuitively, this means that all the subjects are made average with respect to educational attainment, and then the corresponding adjusted y values are analyzed for significant differences due to school systems. Group averages are also transformed in this process; sometimes the adjusted y averages are further apart, sometimes they are closer together than the original averages (Figure 13.2). Because the regression lines are estimated from the data, the actual analysis is more complex than finding the lines, transforming all data points, and performing the ANOVA on the transformed points. However, the adjusted group averages can be found by this method.

In the third type of application of covariance, *comparing regressions within several groups*, the classifications (treatments) are not of primary concern, but rather the relationship of y to x within each classification is of

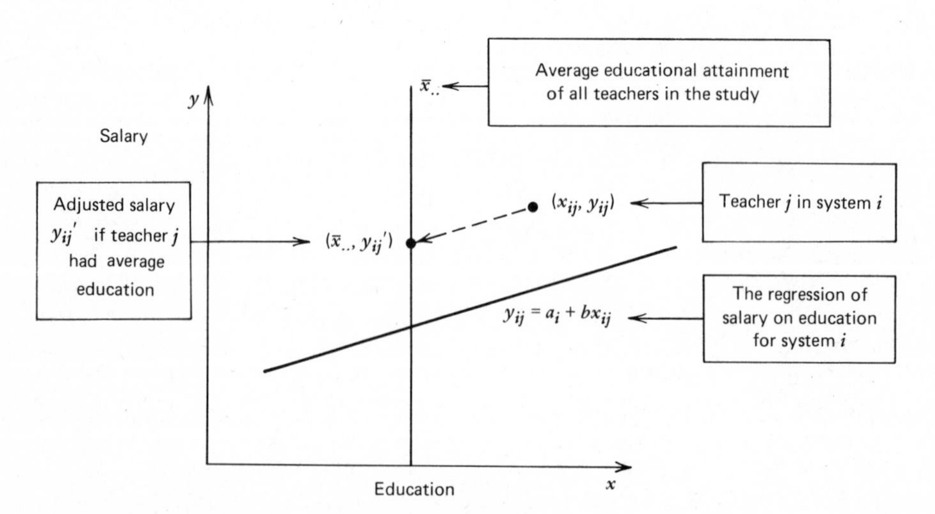

FIGURE 13.1. Adjusting observations by covariance analysis.

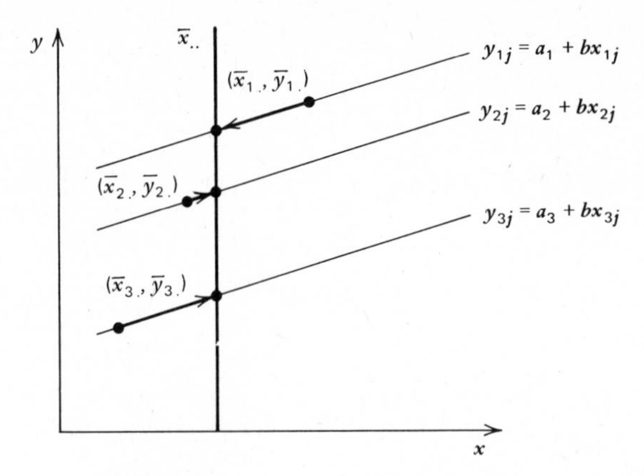

FIGURE 13.2. Adjusting group averages.

main interest. For example, it is known that high blood pressure is more common in some racial groups than in others. Data on the relationship of salt intake x and blood pressure y may be classified by racial groups and covariance used to determine whether the relationship between salt and blood pressure is the same for all the racial groups in the study.

The additive model for the analysis of covariance is

$$y_{ij} = \mu + \alpha_i + \beta(x_{ij} - \bar{x}_{..}) + \epsilon_{ij}$$

$$i = 1, \ldots, a$$

$$j = 1, \ldots, n_i$$

$$N = \sum_i n_i$$

The terms in this model have the following meanings.

μ: The true overall y mean for all studies of this type involving the specified treatments.

α_i: The deviation due to the ith treatment after allowance for the relationship of y to x; $\sum_i \alpha_i = 0$. (*Note:* α_i are the treatment effects and not the y intercepts.)

β: The true common slope of the a regression lines.

$\bar{x}_{..}$: The overall average of the covariate for the observations in the study.

ϵ_{ij}: A random effect for the jth element in the ith treatment group; ϵ_{ij} IND$(0,\sigma^2)$.

The model assumes that all of the regression lines have the same slope, that the variances about the regression lines are equal, and that the covariate x_{ij} is unaffected by the treatments, and it makes the usual assumptions for the

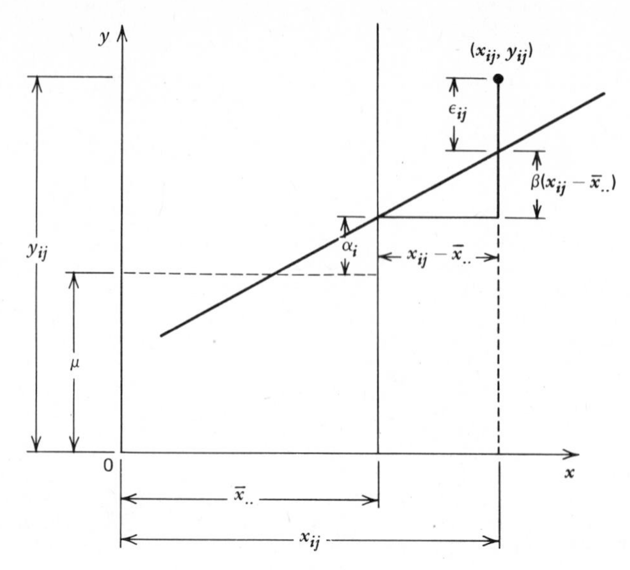

FIGURE 13.3. Terms in the covariance model.

analysis of variance. Figure 13.3 may be helpful in understanding the terms in the model.

In the study of teachers' salaries, μ is the true mean salary for all teachers in the three school systems. The fixed effect α_2 is the true deviation from the mean salary in the second school system after making allowance for the educational attainment of the teachers in that system. The common slope β is the change in salary per additional year of teacher's education. The average educational attainment for the teachers in the three samples is $\bar{x}_{..}$. The random effect ϵ_{32} is the deviation of the second teacher in the sample from the third school system from the regression line for the third system.

In an analysis of covariance, we are usually interested in testing for differences in the treatment effects

$$H_0: \quad \alpha_1 = \alpha_2 = \cdots = \alpha_a = 0 \qquad \text{against} \qquad H_a: \quad \textit{At least one inequality}$$

If the equality of slopes is of primary interest, it can also be tested within the covariance procedure. Since the equality of slopes is an assumption of the model, it is usually tested to verify that the proper model is being used (see Section 13.3).

EXERCISES

13.1.1. Samples of three varieties of wheat, A, B, and C, result in the following data for yield y in bushels per acre and rainfall x in inches.

A		B		C	
x	y	x	y	x	y
1	2	2	3	3	2
2	6	3	7	4	6
4	10	5	11	6	10
5	10	6	11	7	10

a. Draw the scatter plot for each variety on a common graph, keeping the varieties separate by using different colors or symbols.

b. Find the unadjusted group means $(\bar{x}_{i.}, \bar{y}_{i.})$ and add them to the graph.

c. Draw the vertical line at $x = \bar{x}_{..}$.

d. Estimate the regression equation for each variety and add these lines to the graph. (Note that the estimates of the slopes are the same.)

e. Compute $\bar{y}_{i.}'$ for each variety from the regression equations. Locate the adjusted means on the graph.

f. Will the analysis of covariance increase or decrease the differences among the variety averages? Does it change the rank order of the group averages?

13.1.2. The diagrams in Figure 13.4 show the unadjusted treatment averages and the regression lines for the treatment groups in experiments in which covariance is being considered as a method of analysis. In which case or cases can covariance be justified?

13.1.3. Match the following statistical symbols with the indicated distances on the graph in Figure 13.5.

(1) $y_{ij} - \mu$ (5) $y_{ij} - \hat{y}_{ij}$

(2) y_{ij} (6) $\bar{y}_{i.}'$

(3) μ (7) $\bar{y}_{i.}' - \mu$

(4) $\bar{y}_{i.}$

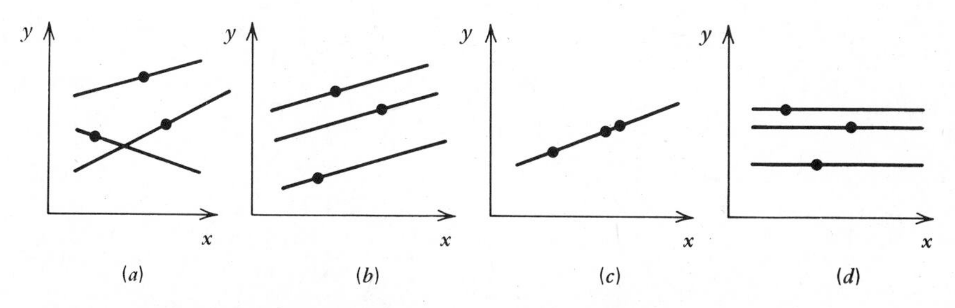

FIGURE 13.4. Regression lines and unadjusted treatment averages.

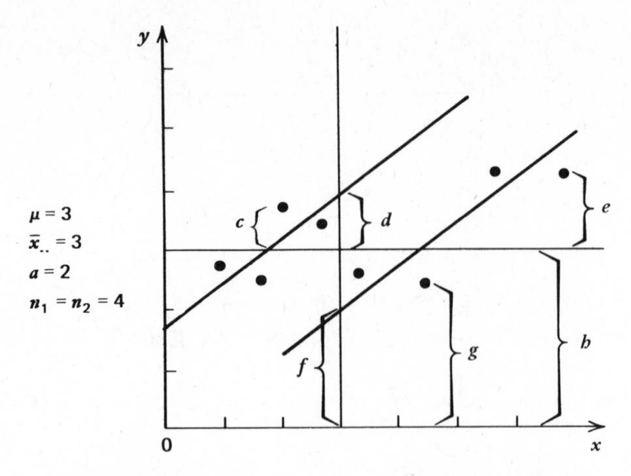

FIGURE 13.5. Distances used in covariance analysis.

13.2. ONE-WAY ANALYSIS OF COVARIANCE

Let the data for a one-way analysis of covariance with $a = 3$ treatments and $n_1 = n_2 = n_3 = 4$ observations per treatment group be arranged as follows.

Treatment

I		II		III		
x	y	x	y	x	y	
x_{11}	y_{11}	x_{21}	y_{21}	x_{31}	y_{31}	
x_{12}	y_{12}	x_{22}	y_{22}	x_{32}	y_{32}	
x_{13}	y_{13}	x_{23}	y_{23}	x_{33}	y_{33}	
x_{14}	y_{14}	x_{24}	y_{24}	x_{34}	y_{34}	Totals
$T_{1.(x)}$	$T_{1.(y)}$	$T_{2.(x)}$	$T_{2.(y)}$	$T_{3.(x)}$	$T_{3.(y)}$	$T_{..(x)}$ $T_{..(y)}$

Using a similar layout for a treatment groups and n_i observations per group, the general analysis of covariance procedure can be summarized as follows.

■ ■ ■

Procedure. Analysis of Covariance

Hypothesis

H_0: $\alpha_1 = \alpha_2 = \cdots = \alpha_a$ against H_a: *At least one inequality*

$$\text{Model: } y_{ij} = \mu + \alpha_i + \beta(x_{ij} - \bar{x}_{..}) + \epsilon_{ij}$$

$$i = 1, 2, \ldots, a$$

$$j = 1, 2, \ldots, n_i$$

$$N = \sum_i n_i$$

Uncorrected Sums of Squares and Products

x	xy	y
$T_{(x)} = \sum_i \sum_j x_{ij}^2$	$T_{(xy)} = \sum_i \sum_j x_{ij} y_{ij}$	$T_{(y)} = \sum_i \sum_j y_{ij}^2$
$A_{(x)} = \sum_i T_{i.(x)}^2/n_i$	$A_{(xy)} = \sum_i T_{i.(x)} T_{i.(y)}/n_i$	$A_{(y)} = \sum_i T_{i.(y)}^2/n_i$
$CF_{(x)} = T_{..(x)}^2/N$	$CF_{(xy)} = T_{..(x)} T_{..(y)}/N$	$CF_{(y)} = T_{..(y)}^2/N$

Corrected Sums of Squares and Products

Source	df	$SS_{(x)}$	SP	$SS_{(y)}$
Treatment	$a-1$	$SS_{a(x)} = A_{(x)} - CF_{(x)}$	$SP_a = A_{(xy)} - CF_{(xy)}$	$SS_{a(y)} = A_{(y)} - CF_{(y)}$
Error	$N-a$	$SS_{e(x)} = T_{(x)} - A_{(x)}$	$SP_e = T_{(xy)} - A_{(xy)}$	$SS_{e(y)} = T_{(y)} - A_{(y)}$
Total	$N-1$	$SS_{t(x)} = T_{(x)} - CF_{(x)}$	$SP_t = T_{(xy)} - CF_{(xy)}$	$SS_{t(y)} = T_{(y)} - CF_{(y)}$

Adjusted Sums of Squares

Source	df′	$SS_{(y)}'$	$MS_{(y)}'$
Treatment	$a-1$	$SS_{a(y)}' = SS_{t(y)}' - SS_{e(y)}'$	$MS_{a(y)}' = SS_{a(y)}'/(a-1)$
Error	$N-a-1$	$SS_{e(y)}' = SS_{e(y)} - SP_e^2/SS_{e(x)}$	$MS_{e(y)}' = SS_{e(y)}'/(N-a-1)$
Total	$N-2$	$SS_{t(y)}' = SS_{t(y)} - SP_t^2/SS_{t(x)}$	

Reject H_0 if $F = MS_{a(y)}'/MS_{e(y)}' > F_{\alpha,a-1,N-a-1}$ at the α level of significance.

The procedure is illustrated by the following example.

Example 13.1. One-Way Analysis of Covariance

An experiment is conducted involving three different advertising media, each used for five stores of a certain franchise. The fifteen stores are located in different but comparable cities, and the stores are randomly assigned to the three advertising media: radio, newspaper, television. All advertising takes place during the same time period. Profits y in thousands of dollars are recorded for this time period. Although all stores are of the same size, they employ different numbers of workers. Since additional employees may be related to profits, the number of employees is used as a concomitant variable x ($x = 1.5$ means the equivalent of one full-time and one half-time employee).

	Media I x	I y	II x	II y	III x	III y	
	1.0	26	2.0	21	3.0	10	$a = 3$
	1.5	24	2.5	13	3.5	6	$n_1 = n_2 = n_3 = 5$
	2.0	16	3.0	7	4.0	-1	$N = 15$
	2.5	7	3.5	3	4.5	-8	
	3.0	2	4.0	-4	5.0	-12	Totals
$T_{i.(x)}$	10.0		15.0		20.0		$T_{..(x)} = 45.0$
$T_{i.(y)}$		75		40		-5	$T_{..(y)} = 110.0$
$\sum_j x_{ij}^2$	22.5		47.5		82.5		152.5
$\sum_j x_{ij}y_{ij}$	117.5		90		-49.0		158.5
$\sum_j y_{ij}^2$		1561		684		345	2590.0

Uncorrected SS and SP

	x	xy	y
T	152.5	158.5	2590
A	$(10^2+15^2+20^2)/5 = 145$	$[10(75)+15(40)+20(-5)]/5 = 250$	$[75^2+40^2+(-5)^2]/5 = 1450$
CF	$45^2/15 = 135$	$45(110)/15 = 330$	$110^2/15 = 806.67$

Corrected SS and SP

Source	df	$SS_{(x)}$	SP	$SS_{(y)}$
Treatment	$3-1 = 2$	$145-135 = 10$	$250-330 = -80$	$1450-806.67 = 643.33$
Error	$15-3 = 12$	$152.5-145 = 7.5$	$158.5-250 = -91.5$	$2590-1450 = 1140.00$
Total	14	17.5	-171.5	1783.33

Adjusted SS

Source	df′	$SS_{(y)}'$	$MS_{(y)}'$	F
Treatment	2	$102.63-23.70 = 78.93$	39.465	18.313
Error	11	$1140-(-91.5)^2/7.5 = 23.70$	2.155	
Total		$1783.33-(-171.5)^2/17.5 = 102.63$		

Since $F_{0.05,2,11} = 3.982$, the null hypothesis is rejected. There is a significant difference among the media effects on average profits after adjusting for numbers of employees.

If the F test had been performed on the unadjusted y values, no significant difference would have been observed:

Source	df	$SS_{(y)}$	$MS_{(y)}$	F
Treatment	2	643.33	321.66	3.386
Error	12	1140.00	95.00	

and $F_{0.05,2,12} = 3.885$.

EXERCISES

13.2.1. A certain airplane part must withstand extremes of temperature. The part can be made from a number of metal alloys; the one to be chosen must have the greatest strength y for a given density x. An experiment is designed involving five alloys and five parts per alloy. In hopes of obtaining a lighter part, the density of each alloy is deliberately varied within a safe range. The data are analyzed by covariance procedures to yield the following information.

Source	df	$SS_{(x)}$	SP	$SS_{(y)}$
Alloys	4	200	300	2500
Error	20	300	1200	7500

a. What is the linear model?

b. What assumptions must be made in order to perform the analysis of covariance?

c. Complete the analysis of covariance.

13.2.2. Complete the analysis of covariance for the data given in Exercise 13.1.1.

13.3. TESTING THE ASSUMPTIONS FOR ANALYSIS OF COVARIANCE

In order for an analysis of covariance to be valid, we may need to verify that:

1. All the treatment groups have the same variance about their regression lines, $\sigma_1^2 = \sigma_2^2 = \cdots = \sigma_a^2$.
2. All the regression lines have the same slope, $\beta_1 = \beta_2 = \cdots = \beta_a = \beta$.
3. The common slope β is not equal to 0, that is, the regression lines are not horizontal.

In this section, we illustrate these tests using the advertising media study, Example 13.1.

We begin by estimating the individual regression lines for each treatment group.

	Media					
	I		II		III	
	x	y	x	y	x	y
S_{xx}	2.5		2.5		2.5	
S_{xy}	−32.5		−30.0		−29.0	
S_{yy}		436		364		340
$\bar{x}_{i.}$	2.0		3.0		4.0	
$\bar{y}_i$		15.0		8.0		−1.0
b_i	−13.0		−12.0		−11.6	
a_i	41.0		44.0		45.4	

In this table, the sums of squares and cross-products are computed as for simple linear regression. Thus, for medium I:

$$S_{xx} = \sum_j x_{1j}^2 - T_{1.(x)}^2/n_1 = 22.2 - (10)^2/5 = 2.5$$

$$S_{xy} = \sum_j x_{1j}y_{1j} - T_{1.(x)}T_{1.(y)}/n_1 = 117.5 - (100)(75)/5 = -32.5$$

$$S_{yy} = \sum_j y_{ij}^2 - T_{1.(y)}^2/n_1 = 1561 - (75)^2/5 = 436$$

and so on.

The slope and y intercept are also computed as in simple linear regression. For example, for medium I:

$$b_1 = \frac{S_{xy(\text{I})}}{S_{xx(\text{I})}} = \frac{-32.5}{2.5} = -13.0$$

and

$$a_1 = \bar{y}_{1.} - b_1\bar{x}_{1.} = 15.0 - (-13.0)2 = 41.0$$

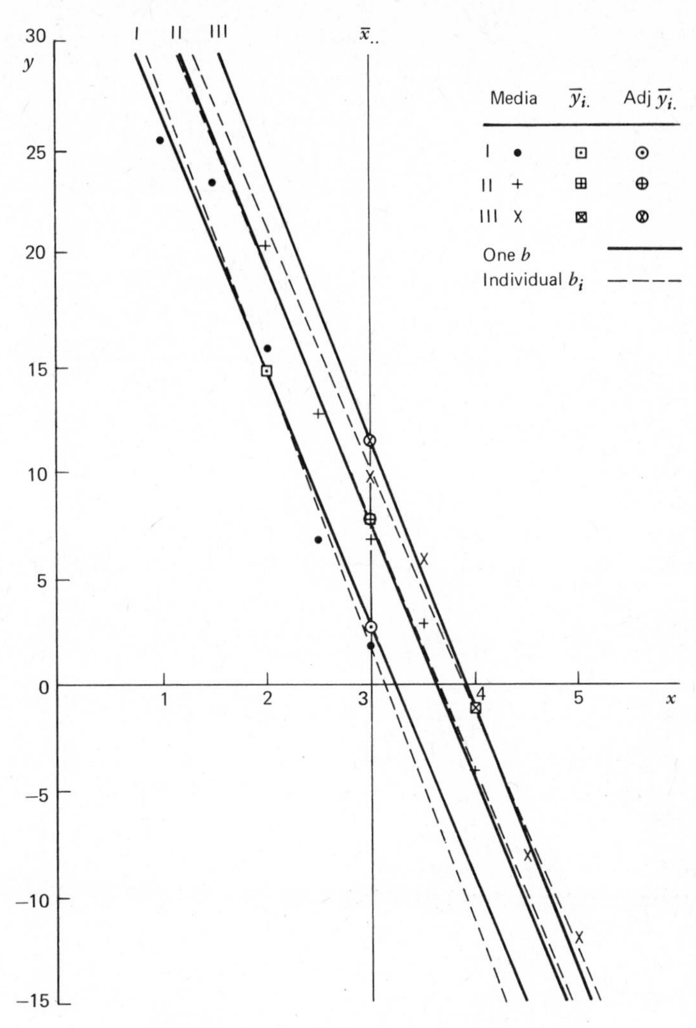

FIGURE 13.6. Media study, Example 13.1.

The data points and the individual regression lines are graphed in Figure 13.6.

To test for the equality of variances about the trend lines, we may use the F-max test or Bartlett's test (see Section 11.2). The variability about each line is computed, using

$$s_i^2 = \frac{\sum_j (y_{ij} - \hat{y}_{ij})^2}{n_i - 2} = \frac{S_{yy(i)} - S_{xy(i)}^2/S_{xx(i)}}{n_i - 2}$$

Using the sums of squares and cross-products above, we have:

Medium	df	$S_{yy(i)} - S_{xy(i)}^2/S_{xx(i)}$	s_i^2
I	$n_1 - 2 = 3$	13.5	4.50
II	$n_2 - 2 = 3$	4.0	1.33
III	$n_3 - 2 = 3$	3.6	1.20
		21.1	

$$F_{\max} = \frac{\text{largest } s_i^2}{\text{smallest } s_i^2} = \frac{4.50}{1.20} = 3.75$$

and $F_{\max\, \alpha, a, n-2} = F_{\max\, 0.05,3,3} = 27.8$ from Table A.15 in the Appendix with $a = 3$ groups, and $n_i - 2 = 3$ degrees of freedom for each estimated variance. Since $F_{\max}$ is not significant, we conclude that the variances are the same, and we proceed to test the other assumptions necessary for an analysis of covariance.

The equality of the slopes $\beta_1 = \beta_2 = \beta_3$ is tested by comparing the sum of squared deviations from the regression lines $\sum\sum_{ij}(y_{ij} - \hat{y}_{ij})^2$ when the lines are found two different ways. First, using the individual estimates of the slopes

$$b_1 = -13.0 \qquad b_2 = -12.0 \qquad b_3 = 11.6$$

and second, using a pooled estimate of the slope

$$b = \frac{\text{SP}_e}{\text{SS}_{e(x)}} = \frac{-91.5}{7.5} = -12.2$$

If the three separate estimates b_1, b_2, b_3 are all estimates of the same parameter, the difference between these two sums of squared deviations should not be significant.

The sum of squared deviations about the regression lines using b, the pooled estimate of the slope, is

$$D = \text{SS}_{e(y)}' = 23.70$$

and the sum of squared deviations using the individual estimates of the b_i's is

$$E = \sum_i [S_{yy(i)} - S_{xy(i)}^2/S_{xx(i)}]$$

$$= 21.1$$

The test can be summarized in the following table.

H_0: $\beta_1 = \beta_2 = \beta_3$ against H_a: *At least one inequality*

Source	df	SS	MS
About regression lines using one b	$N-a-1=11$	$D=SS_{e(y)}'=23.70$	
About regression lines using three b_i	$N-2a=\ 9$	$E=21.10$	2.34
Difference	$a-1=\ 2$	$D-E=\ 2.60$	1.30

$$F = \frac{(D - E)/(a - 1)}{E/(N - 2a)} = \frac{1.30}{2.34} = 0.56$$

Reject H_0 if $F > F_{0.05,2,9} = 4.256$, so there is no evidence of unequal slopes.

In order for the analysis of covariance to be a significant improvement over a simple one-way analysis of variance, the common slope β must not be zero.

$$H_0\colon\ \beta = 0 \qquad \text{against} \qquad H_a\colon\ \beta \neq 0$$

is tested by

$$F = \frac{SP_e^2/SS_{e(x)}}{MS_{e(y)}'} = \frac{(-91.5)^2/7.5}{2.155} = 518.00$$

with 1 and $N - a - 1$ degrees of freedom. Since $F_{0.05,1,11} = 4.844$, we reject $\beta = 0$, and it is appropriate to use an analysis of covariance.

There are times when the hypothesis H_0: $\beta = 0$ is not rejected, but covariance still provides a more powerful test of H_0: $\alpha_1 = \alpha_2 = \cdots = \alpha_a$ than would a comparable ANOVA of the y variable. If the experimenter has reason to suspect that a sizable portion of the variability in y is attributable to a covariate x, the experiment should be designed and data collected with covariance analysis in mind. The worst that can happen is the loss of one degree of freedom attributable to a nonsignificant b. But even with that loss, $MS_{e(y)}'$ may still be sufficiently smaller than $MS_{e(y)}$ for covariance analysis to be more powerful than ANOVA.

EXERCISES

13.3.1. In Exercise 13.2.1

a. What is the pooled estimate of the slope?

b. Test that the slope is not zero.

13.3.2. Given that $n_1 = n_2 = n_3 = 10$, y is yield of a certain crop, x is the amount of limestone added to the soil, and

Soil Type	S_{xx}	S_{xy}	S_{yy}
A	4500	4200	4300
B	5800	3600	2400
C	5100	5100	5300

a. Estimate the individual slopes for each type of soil.
b. Estimate the variances about the trend lines.
c. Test for homogeneity of variances.
d. Estimate the common slope.
e. Test that the three slopes are equal.

13.3.3. See Exercise 13.2.2. Was the analysis of covariance on the data from Exercise 13.1.1 justified?

13.3.4. Darwin's theory of evolution postulates that there is a struggle for existence and only the most fit survive. Using these two principles, experimental geneticists can quantify the relative fitness of different species by comparing their survival under some stressful conditions. Suppose a researcher wishes to compare the relative survival of three species of *Drosophila* under increasing levels of organic phosphorus insecticide. Four batches of medium are prepared and all batches are identical except for the level of insecticide they contain. One hundred eggs from each species are deposited on each preparation. The variables recorded for each container are: level of insecticide x in parts per million (ppm) and number of flies that survive to adulthood y. The researcher knows that the experiment may show either of two results: the mean number of survivors is not the same from species to species, or the effect of increasing the level of insecticide is not the same for all species.

a. Give the null and alternative hypotheses used to test each of these responses.
b. Which null hypothesis should be tested first?
c. Given the following data

Level of Insecticide (ppm)	Species		
	Drosophila melanogaster	*Drosophila pseudoobscura*	*Drosophila serrata*
0.0	91	89	87
0.3	71	77	43
0.6	23	12	22
0.9	5	2	8

i. Test the hypothesis that all species show that same response to increasing levels of insecticide in the medium.
ii. Should the researcher compute adjusted average survival for an average level of insecticide? Why?
iii. Draw a graph to show how each species responds to increased levels of insecticide.
iv. If the three species were competing for existence in an environment in which insecticide is accumulating, which species seems to have the best advantage, that is, the greatest relative fitness?

13.4. MULTIPLE COMPARISON PROCEDURES

If the analysis of covariance is justified and leads to a significant F test for differences among the adjusted averages, then we will want to follow this procedure with a test that compares the means, a *multiple comparison procedure*.

The adjustments must be performed, of course, before we can test the adjusted averages. Intuitively, the original group averages $(\bar{x}_{i.}, \bar{y}_{i.})$ are transformed along the regressions lines to the vertical line $x = \bar{x}_{..}$ (see Figure 13.6 in Section 13.3). Algebraically the transformed y averages can be found by the formula

$$\text{adj } \bar{y}_{i.} = \bar{y}_{i.} - b(\bar{x}_{i.} - \bar{x}_{..})$$

Thus, in the media experiment (Example 13.1):

$$\text{adj } \bar{y}_{1.} = 15.0 - (-12.2)(2.0 - 3.0) = 2.8$$

$$\text{adj } \bar{y}_{2.} = 8.0 - (-12.2)(3.0 - 3.0) = 8.0$$

$$\text{adj } \bar{y}_{3.} = -1.0 - (-12.2)(4.0 - 3.0) = 11.8$$

If desired, confidence intervals can be found for the adjusted means:

$$\text{CI}_{1-\alpha}\text{: adj } \bar{y}_{i.} \pm t_{\alpha/2, N-a-1}\sqrt{\text{MS}_{e(y)}'}\sqrt{\frac{1}{n_i} + \frac{(\bar{x}_{i.} - \bar{x}_{..})^2}{\text{SS}_{e(x)}}}$$

For example, for the adjusted mean of the third group, we have:

$$\text{CI}_{0.95}\text{: } 11.8 \pm t_{0.025,11}\sqrt{2.155}\sqrt{\frac{1}{5} + \frac{(4.0 - 3.0)^2}{7.5}}$$

$$11.8 \pm 1.87$$

If the treatment groups are the same size n, comparisons of two adjusted averages adj $\bar{y}_{i.}$ − adj $\bar{y}_{i'.}$ can be made with the significant difference at the α level being given by

$$t_{\alpha/2,N-a-1}\sqrt{\text{MS}_{e(y)}'}\sqrt{\frac{2}{n}+\frac{(\bar{x}_{i.}-\bar{x}_{i'.})^2}{\text{SS}_{e(x)}}}$$

In the media example, adj $\bar{y}_{2.}$ − adj $\bar{y}_{1.}$ = 8.0 − 2.8 = 5.2, and at the 0.05 level of significance the critical difference is:

$$2.201\sqrt{2.155}\sqrt{\frac{2}{5}+\frac{(3.0-2.0)^2}{7.5}}=2.36$$

Thus $\alpha_1 \neq \alpha_2$ after adjusting for the number of employees.

Similarly, adj $\bar{y}_{3.}$ − adj $\bar{y}_{2.}$ = 11.8 − 8.0 = 3.8, and the critical difference is:

$$2.201\sqrt{2.155}\sqrt{\frac{2}{5}+\frac{(4.0-3.0)^2}{7.5}}=2.36$$

Thus $\alpha_2 \neq \alpha_3$ after adjusting for the number of employees.

Finally, adj $\bar{y}_{3.}$ − adj $\bar{y}_{1.}$ = 11.8 − 2.8 = 9.0, and the critical difference is:

$$2.201\sqrt{2.155}\sqrt{\frac{2}{5}+\frac{(4.0-2.0)^2}{7.5}}=3.12$$

Thus $\alpha_1 \neq \alpha_3$ after adjusting for the covariate.

The final conclusion of the media study is that each of the media used has a different effect on profits, and medium III has the greatest positive effect. We would not have come to this conclusion if we had not adjusted for the number of employees—before the adjustment, medium III had the lowest of the group averages.

EXERCISES

13.4.1. Given the following information from a one-way analysis of covariance involving three groups and eight observations per group:

Source	$SS_{(x)}$	SP	$SS_{(y)}$
Groups	144	120	208
Within	175	140	132

Group	$\bar{x}_{i.}$	$\bar{y}_{i.}$
1	27	20
2	30	18
3	33	25

a. Graph the unadjusted group averages.

b. Find the estimate of the common slope, b.

c. Graph the trend lines using the common slope.

d. Find the adjusted y averages graphically.
e. Find the adjusted y averages algebraically.
f. Find the 95% confidence intervals on the adjusted means.
g. Test the adjusted group means for significant differences.

REVIEW EXERCISES

Decide whether each of the following statements is true or false. Correct each false statement.

13.1. The model $y_{ij} = \mu + \alpha_i + \beta_j + \epsilon_{ij}$ would apply to covariance analysis if $\bar{x}_{..} = 0$.

13.2. Covariance techniques require that unadjusted y_{ij} as well as adjusted y_{ij} have homogeneous variance.

13.3. The analysis of covariance techniques in this chapter are appropriate whether the α_i are fixed or random.

13.4. Analysis of covariance techniques are appropriate whether the x_{ij} are fixed or random.

13.5. Analysis of covariance techniques are appropriate even though H_0: $\beta_1 = \beta_2 = \beta_3$ is rejected.

13.6. Analysis of covariance techniques may be appropriate even though H_0: $\beta = 0$ is not rejected.

13.7. Analysis of covariance techniques are appropriate even though $\sum_j (y_{ij} - \hat{y}_{ij})^2/(n_i - 2)$ are significantly different from group to group.

13.8. Analysis of covariance techniques are appropriate even though $\sum_j (x_{ij} - \bar{x}_{i.})^2$ are significantly different from group to group.

13.9. The model for a one-way analysis of covariance is $y_{ij} = \mu + \alpha_i + \beta(x_{ij} - \bar{x}_{..}) + \epsilon_{ij}$.

13.10. For a valid analysis of covariance, both the x variable and the y variable must be normally distributed.

13.11. For a valid analysis of covariance, both the x variable and the y variable must be random.

13.12. Accepting the hypothesis that the common slope $\beta = 0$ means that there is no relationship between x and y.

13.13. When the hypothesis of parallel regression lines is rejected, it becomes meaningless to discuss differences among adjusted averages that have been based on a common slope.

13.14. Because $\sum_i \sum_j (y_{ij} - \hat{y}_{ij})^2 \leq \sum_i \sum_j (y_{ij} - \bar{y}_{i.})^2$, the adjusted within-group sum of squares can never be greater than the unadjusted within-group SS.

13.15. It is possible that an analysis of variance on the unadjusted group averages can yield a significant F test for treatments, but in a similar test after adjustment for the x variable by covariance techniques, group differences may be nonsignificant.

SELECTED READINGS

Cochran, W. G. (1957). Analysis of covariance: Its nature and uses, *Biometrics,* **13,** 261–281.

Cox, D. R. (1957). The use of a concomitant variable in selecting an experimental design, *Biometrika,* **44,** 150–158.

Gourlay, N. (1953). Covariance analysis and its application in psychological research, *British Journal of Statistical Psychology,* **6,** 25–34.

Huitema, B. E. (1980). *The Analysis of Covariance and Alternatives*, Wiley, New York.

Snedecor, G. W., and W. G. Cochran (1973). *Statistical Methods*, 6th ed., Iowa State University Press, Ames.

Winer, B. J. (1971). *Statistical Principles in Experimental Design*, 2nd ed., McGraw-Hill, New York.

14

Multiple Regression and Correlation

In Chapter 9 we discussed simple linear regression and correlation. We were interested in an independent variable x and a dependent variable y. In this chapter we generalize the discussion and speak of k independent variables $x_1, x_2, \ldots, x_k$ and a dependent variable y. For example, achievement in school y may be related to the student's IQ x_1, the class size x_2, the amount of homework x_3, and parents' educational level x_4. We also generalize simple linear regression another way by discussing curvilinear regression.

14.1. MATRICES

In simple linear regression, we assumed that x and y were linearly related, and we used the model $y = \alpha + \beta x + \epsilon$ to represent this relationship. One task related to this model was to estimate α and β in the regression equation. The estimates a and b were the solutions to two linear equations in a and b called the normal equations:

$$na + b \sum x = \sum y$$

$$a \sum x + b \sum x^2 = \sum xy$$

In a similar way, multiple regression involves using a model of the form

$$y = \alpha + \beta_1 x_1 + \beta_2 x_2 + \cdots + \beta_k x_k$$

and the estimates $a, b_1, \ldots, b_k$ of α and the β's are the solutions to several simultaneous linear equations. In this section we develop a computational technique that aids in solving systems of linear equations and also yields some information that is useful for inferences related to multiple regression.

In algebra, we learn how to solve two simultaneous equations in two unknowns when such solutions exist. For example, if we have two equations that express the relationship between chains, rods, and feet as

$$3 \text{ chains} + 2 \text{ rods} = 231 \text{ feet}$$

$$2 \text{ chains} + 4 \text{ rods} = 198 \text{ feet}$$

we can solve these equations to find out how many feet there are in a chain and how many feet there are in a rod. In symbols, we could write

$$3b_1 + 2b_2 = 231$$

$$2b_1 + 4b_2 = 198$$

in which b_1 is the number of feet in one chain and b_2 is the number of feet in one rod. To solve this system of equations, we may multiply or divide an equation by a nonzero constant and we may add or subtract a multiple of one equation from another. We repeat these operations until we get an equivalent system of equations of the form

$$1b_1 + 0b_2 = d_1$$

$$0b_1 + 1b_2 = d_2$$

from which we can read the solution $b_1 = d_1$ and $b_2 = d_2$.

For example, to solve the system above, we might use the following steps.

STEP 1. Divide the first equation by 3:

$$1.00b_1 + 0.67b_2 = 77.00$$
$$2.00b_1 + 4.00b_2 = 198.00$$

STEP 2. Multiply the first equation by 2 and subtract the product from the second equation:

$$1.00b_1 + 0.67b_2 = 77.00$$
$$0.00b_1 + 2.67b_2 = 44.00$$

STEP 3. Divide the second equation by 2.67:

$$1.00b_1 + 0.67b_2 = 77.00$$
$$0.00b_1 + 1.00b_2 = 16.50$$

STEP 4. Multiply the second equation by 0.67 and subtract the product from the first equation:

$$1.00b_1 + 0.00b_2 = 66.00$$
$$0.00b_1 + 1.00b_2 = 16.50$$

Thus is follows that $b_1 = 66.00$ and $b_2 = 16.50$; that is,

one chain is 66 feet

one rod is 16.5 feet

This same sequence of operations can be carried out by using a simple matrix approach. A matrix is a rectangular array of numbers. For example,

$$M = \begin{bmatrix} 3 & 2 \\ 2 & 4 \end{bmatrix}$$

is the matrix of coefficients of the original system of equations and

$$N = \begin{bmatrix} 231 \\ 198 \end{bmatrix}$$

is the matrix of constants. The solution above can be represented in the following manner. Begin with

$$[M|N] = \left[\begin{array}{cc|c} 3 & 2 & 231 \\ 2 & 4 & 198 \end{array}\right]$$

a matrix form made up of the matrix of coefficients on the left and the matrix of constants on the right.

STEP 1. Divide row one by 3:

$$\left[\begin{array}{cc|c} 1.00 & 0.67 & 77.00 \\ 2.000 & 4.00 & 198.00 \end{array}\right]$$

STEP 2. Multiply the first row by 2 and subtract the product from the second row:

$$\left[\begin{array}{cc|c} 1.00 & 0.67 & 77.00 \\ 0.00 & 2.67 & 44.00 \end{array}\right]$$

STEP 3. Divide the second row by 2.67:

$$\left[\begin{array}{cc|c} 1.00 & 0.67 & 77.00 \\ 0.00 & 1.00 & 16.50 \end{array}\right]$$

STEP 4. Multiply the second row by 0.67 and subtract the product from the first row:

$$\left[\begin{array}{cc|c} 1.00 & 0.00 & 66.00 \\ 0.00 & 1.00 & 16.50 \end{array}\right]$$

The original matrix form

$$[M|N] = \left[\begin{array}{cc|c} m_{11} & m_{12} & n_1 \\ m_{21} & m_{22} & n_2 \end{array}\right]$$

representing

$$m_{11}b_1 + m_{12}b_2 = n_1$$
$$m_{21}b_1 + m_{22}b_2 = n_2 .$$

has been transformed into

$$[I|B] = \left[\begin{array}{cc|c} 1 & 0 & b_1 \\ 0 & 1 & b_2 \end{array}\right]$$

in which

$$I = \begin{bmatrix} 1 & 0 \\ 0 & 1 \end{bmatrix}$$

is the identity matrix and

$$B = \begin{bmatrix} b_1 \\ b_2 \end{bmatrix}$$

is the matrix of solutions.

Although this matrix procedure gives the solutions desired, by itself it yields no useful additional information when applied to regression. To increase the usefulness of this procedure, we augment the beginning matrix form with the identity matrix

$$I = \begin{bmatrix} 1 & 0 \\ 0 & 1 \end{bmatrix}$$

as follows

$$[M|N|I] = \left[\begin{array}{cc|c|cc} m_{11} & m_{12} & n_1 & 1 & 0 \\ m_{21} & m_{22} & n_2 & 0 & 1 \end{array}\right]$$

If the same row operations are applied to this form, it is changed into

$$\left[\begin{array}{cc|c|cc} 1 & 0 & b_1 & p_{11} & p_{12} \\ 0 & 1 & b_2 & p_{21} & p_{22} \end{array}\right]$$

in which

$$P = \begin{bmatrix} p_{11} & p_{12} \\ p_{21} & p_{22} \end{bmatrix} = M^{-1}$$

is the inverse of the matrix of coefficients, that is

$$PM = \begin{bmatrix} p_{11} & p_{12} \\ p_{21} & p_{22} \end{bmatrix} \begin{bmatrix} m_{11} & m_{12} \\ m_{21} & m_{22} \end{bmatrix} = \begin{bmatrix} 1 & 0 \\ 0 & 1 \end{bmatrix} = I$$

(Inverses of this type are used for inferences related to regression.)

Using the system of equations above, with the same row operations, we get

$$[M|N|I] = \left[\begin{array}{cc|c|cc} 3 & 2 & 231 & 1 & 0 \\ 2 & 4 & 198 & 0 & 1 \end{array}\right]$$

divide row 1 by 3 ↓

$$\left[\begin{array}{cc|c|cc} 1.00 & 0.67 & 77.00 & 0.33 & 0.00 \\ 2.00 & 4.00 & 198.00 & 0.00 & 1.00 \end{array}\right]$$

subtract 2 times row 1 from row 2 ↓

$$\left[\begin{array}{cc|c|cc} 1.00 & 0.67 & 77.00 & 0.33 & 0.00 \\ 0.00 & 2.67 & 44.00 & -0.67 & 1.00 \end{array}\right]$$

divide row 2 by 2.67 ↓

$$\left[\begin{array}{cc|c|cc} 1.00 & 0.67 & 77.00 & 0.33 & 0.000 \\ 0.00 & 1.00 & 16.50 & -0.25 & 0.375 \end{array}\right]$$

subtract 0.67 times row 2 from row 1 ↓

$$\left[\begin{array}{cc|c|cc} 1.00 & 0.00 & 66.00 & 0.50 & -0.250 \\ 0.00 & 1.00 & 16.50 & -0.25 & 0.375 \end{array}\right] = [I \mid B \mid P]$$

The matrix on the right

$$P = \begin{bmatrix} 0.50 & -0.250 \\ -0.25 & 0.375 \end{bmatrix}$$

is the inverse M^{-1} of

$$M = \begin{bmatrix} 3 & 2 \\ 2 & 4 \end{bmatrix}$$

This can be verified by using the definition of matrix multiplication:

$$\begin{bmatrix} p_{11} & p_{12} \\ p_{21} & p_{22} \end{bmatrix} \begin{bmatrix} m_{11} & m_{12} \\ m_{21} & m_{22} \end{bmatrix} = \begin{bmatrix} p_{11}m_{11} + p_{12}m_{21} & p_{11}m_{12} + p_{12}m_{22} \\ p_{21}m_{11} + p_{22}m_{21} & p_{21}m_{12} + p_{22}m_{22} \end{bmatrix}$$

Thus

$$\begin{bmatrix} 0.50 & -0.250 \\ -0.25 & 0.375 \end{bmatrix} \begin{bmatrix} 3 & 2 \\ 2 & 4 \end{bmatrix}$$
$$= \begin{bmatrix} (0.50)3 + (-0.25\ )2 & (0.50)2 + (-.25\ )4 \\ (-0.25)3 + (0.375)2 & (-0.25)2 + (0.375)4 \end{bmatrix}$$
$$= \begin{bmatrix} 1 & 0 \\ 0 & 1 \end{bmatrix}$$

In general, if

$$M = \begin{bmatrix} m_{11} & m_{12} \\ m_{21} & m_{22} \end{bmatrix} \qquad M^{-1} = \begin{bmatrix} p_{11} & p_{12} \\ p_{21} & p_{22} \end{bmatrix}$$

$$I = \begin{bmatrix} 1 & 0 \\ 0 & 1 \end{bmatrix} \qquad N = \begin{bmatrix} n_1 \\ n_2 \end{bmatrix} \quad \text{and} \quad B = \begin{bmatrix} b_1 \\ b_2 \end{bmatrix}$$

then we have demonstrated that

$$M^{-1}M = I$$

and have solved $MB = N$ for B. It can be shown that $M^{-1}N = B$, so we can use this as a numerical check of our solutions.

$$M^{-1}N = \begin{bmatrix} 0.50 & -0.250 \\ -0.25 & 0.375 \end{bmatrix} \begin{bmatrix} 231 \\ 198 \end{bmatrix}$$
$$= \begin{bmatrix} (0.50)231 + (-0.25\ )198 \\ (-0.25)231 + (0.375)198 \end{bmatrix}$$
$$= \begin{bmatrix} 66.0 \\ 16.5 \end{bmatrix} = \begin{bmatrix} b_1 \\ b_2 \end{bmatrix} = B$$

Three or more simultaneous equations can be solved and the matrix of coefficients inverted by means of the same type of row operations. The permissible operations are:

1. Multiplication or division of every element in a row by a nonzero constant.
2. Addition or subtraction of any multiple of a row from the corresponding elements in another row.

The objective is to transform the matrix of coefficients, as

$$\begin{bmatrix} m_{11} & m_{12} & m_{13} \\ m_{21} & m_{22} & m_{23} \\ m_{31} & m_{32} & m_{33} \end{bmatrix}$$

into the corresponding identity matrix

$$\begin{bmatrix} 1 & 0 & 0 \\ 0 & 1 & 0 \\ 0 & 0 & 1 \end{bmatrix}$$

in a systematic way. This can be done by first transforming m_{11} into 1, then using this new row 1 to get zeros in the remainder of the first column. Then the value in the m_{22} position is transformed into 1 and the second row used to get zeros in the rest of the second column. This process is repeated for each diagonal element m_{ii}.

Example 14.1. Solution of Three Simultaneous Equations

Consider the following three equations:

$$\begin{aligned} 2 \text{ pints} + 4 \text{ ounces} + 2 \text{ drams} &= 1072.78 \text{ milliliters} \\ 4 \text{ pints} + 9 \text{ ounces} + 5 \text{ drams} &= 2178.14 \text{ milliliters} \\ 2 \text{ pints} + 5 \text{ ounces} + 7 \text{ drams} &= 1125.20 \text{ milliliters} \end{aligned}$$

In symbols, we could write

$$\begin{aligned} 2b_1 + 4b_2 + 2b_3 &= 1072.78 \\ 4b_1 + 9b_2 + 5b_3 &= 2178.14 \\ 2b_1 + 5b_2 + 7b_3 &= 1125.20 \end{aligned}$$

in which b_1, b_2, b_3 are the number of milliliters in one pint, one ounce, and one dram, respectively. Three matrices are used:

Matrix of Coefficients

$$M = \begin{bmatrix} 2 & 4 & 2 \\ 4 & 9 & 5 \\ 2 & 5 & 7 \end{bmatrix}$$

Matrix of Constants

$$N = \begin{bmatrix} 1072.78 \\ 2178.14 \\ 1125.20 \end{bmatrix}$$

Identity Matrix

$$I = \begin{bmatrix} 1 & 0 & 0 \\ 0 & 1 & 0 \\ 0 & 0 & 1 \end{bmatrix}$$

Combining these three matrices, we form the following, on which row operations are performed.

$$[M|N|I] = \left[\begin{array}{ccc|c|ccc} 2 & 4 & 2 & 1072.78 & 1 & 0 & 0 \\ 4 & 9 & 5 & 2178.14 & 0 & 1 & 0 \\ 2 & 5 & 7 & 1125.20 & 0 & 0 & 1 \end{array}\right]$$

divide row 1 by 2
↓

$$\left[\begin{array}{ccc|c|ccc} 1 & 2 & 1 & 536.39 & 0.5 & 0 & 0 \\ 4 & 9 & 5 & 2178.14 & 0.0 & 1 & 0 \\ 2 & 5 & 7 & 1125.20 & 0.0 & 0 & 1 \end{array}\right]$$

subtract 4 times row 1 from row 2
and subtract 2 times row 1 from row 3
↓

$$\left[\begin{array}{ccc|c|ccc} 1 & 2 & 1 & 536.39 & 0.5 & 0 & 0 \\ 0 & 1 & 1 & 32.58 & -2.0 & 1 & 0 \\ 0 & 1 & 5 & 52.42 & -1.0 & 0 & 1 \end{array}\right]$$

subtract 2 times row 2 from row 1
and subtract row 2 from row 3
↓

$$\left[\begin{array}{ccc|c|ccc} 0 & 0 & -1 & 471.23 & 4.5 & -2 & 0 \\ 0 & 1 & 1 & 23.58 & -2.0 & 1 & 0 \\ 0 & 0 & 4 & 19.84 & 1.0 & -1 & 1 \end{array}\right]$$

divide row 3 by 4
↓

$$\left[\begin{array}{ccc|c|ccc} 1 & 0 & -1 & 471.23 & 4.50 & -2.00 & 0.00 \\ 0 & 1 & 1 & 32.58 & -2.00 & 1.00 & 0.00 \\ 0 & 0 & 1 & 4.96 & 0.25 & -0.25 & 0.25 \end{array}\right]$$

add row 3 to row 1,
subtract row 3 from row 2
↓

$$\left[\begin{array}{ccc|c|ccc} 1 & 0 & 0 & 476.19 & 4.75 & -2.25 & 0.25 \\ 0 & 1 & 0 & 27.62 & -2.25 & 1.25 & -0.25 \\ 0 & 0 & 1 & 4.96 & 0.25 & -0.25 & 0.25 \end{array}\right] = [I|B|M^{-1}]$$

Thus

$$b_1 = 476.19, \qquad b_2 = 27.62, \qquad b_3 = 4.96$$

that is,

one pint equals 476.19 milliliters

one ounce equals 27.62 milliliters

one dram equals 4.96 milliliters

The identity matrix was incorporated into the initial matrix form in order to obtain the inverse M^{-1} of the coefficient matrix M. Although M^{-1} is not needed to solve the set of simultaneous equations showing the relationship between British and metric liquid measures, it is a common computational feature with which users of multiple regression need to become familiar. The inverse can be thought of as a memory of the computations that had to be performed in order to solve for the unknowns. This is demonstrated by the fact that $M^{-1}N = B$, that is

$$M^{-1}N = \begin{bmatrix} 4.75(1072.78) - 2.25(2178.14) + 0.25(1125.20) \\ -2.25(1072.78) + 1.25(2178.14) - 0.25(1125.20) \\ 0.25(1072.78) - 0.25(2178.14) + 0.25(1125.20) \end{bmatrix}$$

$$= \begin{bmatrix} 476.19 \\ 27.62 \\ 4.96 \end{bmatrix} = B$$

Because of this fact, the inverse can be useful even in cases of simultaneous equations in which no statistical inferences are made. Not only does it provide a check of the computations, but it can reduce computations if the matrix of constants is omitted from the initial form and then the inverse is used to solve for the unknowns.

The square matrices used in this section are a special type known as symmetric matrices; the symmetry is around the diagonal so that $m_{ij} = m_{ji}$. Not all simultaneous equations have a symmetric matrix of coefficients, but the equations that arise in multiple regression do. The inverse of a symmetric matrix is also symmetric (so we have another check on our computations).

The matrices in this section were easily inverted by the row operations. We used this procedure because it shows the relationship between the inverse and the original matrix of coefficients. There are many other ways of finding an inverse, but it is often difficult to understand what is happening in the computations, and there is the danger that the inverse may appear to be a cabalistic set of numbers with which statisticians perform magic. The procedure in this section, therefore, was used to show that the inverse is the computational consequence, on the identity matrix, of the procedures necessary to solve for the unknowns. It does not necessarily provide the easiest or best method of computing the inverse.

Most computational experts have a favorite algorithm for matrix inversion, but few contend that one procedure is best for all situations. Even computer programs are subject to inaccuracy in such situations as:

1. A large matrix with many elements.

2. A matrix that contains some elements that are very large numerically and others that are comparatively quite small.

3. A matrix in which many off-diagonal elements are near zero.

4. A matrix in which one row is very nearly a multiple of another.

Special methods may be needed to invert matrices of these types as well as some others.

Most researchers have access to computers with standard programs for multiple regression. They are spared the tedious computations associated with matrix inversion. There may be occasions, however, when the investigator chooses to perform the calculations. When this is the case, it may be helpful to examine the procedures presented in such books as Searle (1971).

EXERCISES

14.1.1. Multiply the following matrices:

a. $$\begin{bmatrix} 2 & 5 \\ 5 & 4 \end{bmatrix} \begin{bmatrix} 4 & 1 \\ 1 & 5 \end{bmatrix}$$

b. $$\begin{bmatrix} -2 & 4 \\ 7 & 3 \end{bmatrix} \begin{bmatrix} 9 & -3 \\ -3 & 1 \end{bmatrix}$$

c. $$\begin{bmatrix} 4 & -2 \\ -2 & 7 \end{bmatrix} \begin{bmatrix} 10 \\ 12 \end{bmatrix}$$

d. $$\begin{bmatrix} 6 & 7 & -1 \\ 7 & -2 & 5 \\ -1 & 5 & 4 \end{bmatrix} \begin{bmatrix} 3 \\ -2 \\ 5 \end{bmatrix}$$

14.1.2. a. Solve the following system of equations using row operations:

$$4b_1 + 3b_2 = 10$$
$$3b_1 + 5b_2 = 16$$

b. Find the inverse M^{-1} of the matrix of coefficients

$$M = \begin{bmatrix} 4 & 3 \\ 3 & 5 \end{bmatrix}$$

c. Show that $M^{-1}M = I$.

d. Show that $M^{-1}N = B$ where $N = \begin{bmatrix} 10 \\ 16 \end{bmatrix}$ and B is the matrix of solutions.

14.1.3. Find the inverse of $\begin{bmatrix} 1 & 2 \\ 2 & 5 \end{bmatrix}$

14.1.4. Simple linear regression can be approached using matrices. Using the example of employee training in Section 9.1,

Hours of Instruction	x:	1	2	3	4	5
Units per Hour	y:	5	4	6	8	7

find the estimates of the y intercept and slope as solutions of the systems of normal equations

$$na + b \sum x = \sum y$$
$$a \sum x + b \sum x^2 = \sum xy$$

Compare your answers with the results in Chapter 9.

14.2. PROCEDURE FOR MULTIPLE REGRESSION AND CORRELATION

In multiple regression, we ask if a dependent variable y is linearly related to two or more independent variables $x_1, \ldots, x_k$. If there is a relationship of the form

$$y = \alpha + \beta_1 x_1 + \beta_2 x_2 + \cdots + \beta_k x_k + \epsilon$$

then we want to estimate α and the β's (the partial regression coefficients) by a least-squares approach and use the least-squares surface

$$\hat{y} = a + b_1 x_2 + b_2 x_2 + \cdots + b_k x_k$$

to predict y.

If there are only two independent variables x_1, x_2, then we can visualize the least-squares procedure as fitting a plane to a set of n data points (x_1, x_2, y) in such a way that $\sum(y - \hat{y})^2$, the sum of the squared deviations of the actual y's from the predicted $\hat{y}$'s, is minimized (Figure 14.1). This is analogous to the fitting of the least-squares trend line in simple linear regression. (If there are more than two independent variables, then the least-

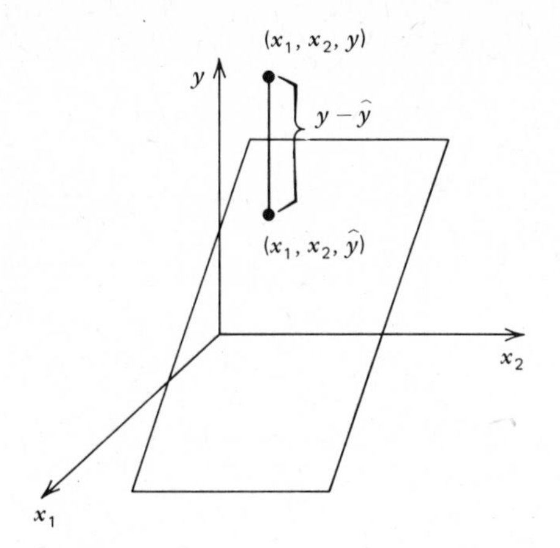

FIGURE 14.1. The least-squares plane.

squares procedure is fitting a hyperplane, the generalization of a plane in higher than three dimensions, to the data points.) It is possible to use the plane for prediction if the plane is not parallel to the x_1, x_2 plane.

To illustrate the least-squares procedure, imagine that a study is being done on a random sample of five migrant-worker families. The data include, among other variables, the highest grade in school completed by the mother x_1, the father x_2, and the oldest child y.

Family	x_1	x_2	y
1	10	10	11
2	11	9	12
3	8	11	12
4	9	12	12
5	12	8	8

We want to know if the educational attainment of either parent significantly influences the attainment of the oldest child. (A sample size of five is unrealistically small for such a study, but we use this sample fictional data to illustrate the procedure.)

We first compute

$$\sum x_1 = 50 \qquad \sum x_2 = 50 \qquad \sum y = 55 \qquad \sum x_1x_2 = 491$$

$$\sum x_1^2 = 510 \qquad \sum x_2^2 = 510 \qquad \sum y^2 = 617 \qquad \sum x_1y = 542$$

$$\bar{x}_1 = 10 \qquad \bar{x}_2 = 10 \qquad \bar{y} = 11 \qquad \sum x_2y = 558$$

and find

$$S_{11} = \sum(x_1 - \bar{x}_1)^2 = \sum x_1^2 - (\sum x_1)^2/n = 10$$

$$S_{12} = \sum(x_1 - \bar{x}_1)(x_2 - \bar{x}_2) = \sum x_1x_2 - (\sum x_1)(\sum x_2)/n = -9$$

$$S_{22} = \sum(x_2 - \bar{x}_2)^2 = \sum x_2^2 - (\sum x_2)^2/n = 10$$

$$S_{1y} = \sum(x_1 - \bar{x}_1)(y - \bar{y}) = \sum x_1 y - (\sum x_1)(\sum y)/n = -8$$

$$S_{2y} = \sum(x_2 - \bar{x}_2)(y - \bar{y}) = \sum x_2 y - (\sum x_2)(\sum y)/n = 8$$

Minimizing

$$\sum(y - \hat{y})^2 = \sum[y - (a + b_1x_1 + b_2x_2)]^2$$

leads to three equations

$$a = \bar{y} - b_1\bar{x}_1 - b_2\bar{x}_2$$

$$S_{11}b_1 + S_{12}b_2 = S_{1y}$$

$$S_{12}b_1 + S_{22}b_2 = S_{2y}$$

or

$$a = 11 - 10b_1 - 10b_2$$

$$10b_1 - 9b_2 = -8$$

$$-9b_1 + 10b_2 = 8$$

The last two equations can be easily solved using the matrix approach of Section 14.1, and then b_1 and b_2 can be used in the first equation to find a.

$$M = \begin{bmatrix} S_{11} & S_{12} \\ S_{12} & S_{22} \end{bmatrix} = \begin{bmatrix} 10 & -9 \\ -9 & 10 \end{bmatrix}$$

is called the matrix of sums of squares and cross-products. We augment M with

$$\begin{bmatrix} S_{1y} \\ S_{2y} \end{bmatrix} = \begin{bmatrix} -8 \\ 8 \end{bmatrix}$$

and the identity matrix

$$\begin{bmatrix} 1 & 0 \\ 0 & 1 \end{bmatrix}$$

$$\left[\begin{array}{cc|c|cc} 10 & -9 & -8 & 1 & 0 \\ -9 & 10 & 8 & 0 & 1 \end{array}\right]$$

divide row 1 by 10
↓

$$\left[\begin{array}{cc|c|cc} 1 & -0.9 & -0.8 & 0.1 & 0 \\ -9 & 10.0 & 8.0 & 0.0 & 1 \end{array}\right]$$

add 9 times row 1 to row 2
↓

$$\left[\begin{array}{cc|c|cc} 1 & -0.9 & -0.8 & 0.1 & 0 \\ 0 & 1.9 & 0.8 & 0.9 & 1 \end{array}\right]$$

divide row 2 by 1.9
↓

$$\left[\begin{array}{cc|c|cc} 1 & -0.9 & -0.800 & 0.100 & 0.000 \\ 0 & 1.0 & 0.421 & 0.474 & 0.526 \end{array}\right]$$

add 0.9 times row 2 to row 1
↓

$$\left[\begin{array}{cc|c|cc} 1 & 0 & -0.421 & 0.527 & 0.474 \\ 0 & 1 & 0.421 & 0.474 & 0.526 \end{array}\right]$$

Thus

$$b_1 = -0.421, \; b_2 = 0.421$$

and

$$a = 11 - 10(-0.421) - 10(0.421) = 11$$

The least-squares regression plane is

$$\hat{y} = 11 - 0.421x_1 + 0.421x_2$$

(Although we do not use the inverse of M in this example, we include its computation because it can be used to check the solutions and because we should develop the habit, for it will be needed in follow-up procedures.)

To determine whether the plane is parallel to the x_1, x_2 plane, we test H_0: $\beta_1 = \beta_2 = 0$ (parallel) against H_a: $\beta_1 \neq 0$ or $\beta_2 \neq 0$ (not parallel). As in simple linear regression, this test requires the variance of the data points from the regression plane

$$s^2 = \frac{\sum(y - \hat{y})^2}{n - k - 1}$$

in which n is the number of data points and k is the number of independent variables. We can compute this directly:

x_1	x_2	y	$11 - 0.421x_1 + 0.421x_2 = \hat{y}$	$y - \hat{y}$	$(y - \hat{y})^2$
10	10	11	$11 - 0.421(10) + 0.421(10) = 11.000$	0.000	0.000
11	9	12	$11 - 0.421(11) + 0.421(9) = 10.158$	1.842	3.393
8	11	12	$11 - 0.421(8) + 0.421(11) = 12.263$	−0.263	0.069
9	12	12	$11 - 0.421(9) + 0.421(12) = 12.263$	−0.263	0.069
12	8	8	$11 - 0.421(12) + 0.421(8) = 9.316$	−1.316	1.732
					5.263

$$s^2 = \frac{\sum(y - \hat{y})^2}{n - k - 1} = \frac{5.263}{5 - 2 - 1} = 2.632$$

Or we can use a more convenient computation procedure:

$$\sum(y - \hat{y})^2 = S_{yy} - b_1 S_{1y} - b_2 S_{2y}$$

in which

$$S_{yy} = \sum(y - \bar{y})^2 = \sum y^2 - \left(\sum y\right)^2/n = 12$$

Thus

$$s^2 = \frac{\sum(y - \hat{y})^2}{n - k - 1} = \frac{12 - (-0.421)(-8) - (0.421)8}{5 - 2 - 1}$$

$$= \frac{12 - 6.736}{2} = \frac{5.264}{2} = 2.632$$

(These two approaches sometimes give slightly different results because of rounding error.)

The test of H_0: $\beta_1 = \beta_2 = 0$ is an F test and can be set up in the form of an ANOVA table.

Source	df	SS	MS	F
Due regression	$k = 2$	$b_1S_{1y} + b_2S_{2y} = 6.736$	3.368	1.28
Deviations	$n-k-1 = 2$	$12 - 6.736 = 5.264$	2.632	
Total	$n-1 = 4$	$S_{yy} = 12.000$		

Since $F_{0.05,2,2} = 19.000$, the computed F of 1.28 is not significant and the plane is parallel to the x_1, x_2 plane, that is, H_0: $\beta_1 = \beta_2 = 0$ is accepted. Thus the plane does not explain a significant portion of the variability, and the educational attainment of the mother or father in migrant-worker families cannot be used to predict the educational attainment of the oldest child.

Instead of the F test for $\beta_1 = \beta_2 = 0$, the reliability of the regression equation is frequently measured by the *multiple correlation coefficient*. The multiple correlation coefficient, $R_{y\hat{y}}$ or R, can be thought of as the correlation between the observed y's and the $\hat{y}$'s predicted by the multiple regression equation. It can be computed in much the same way as the correlation coefficient was computed for bivariate data:

$$R = \frac{\sum(y - \bar{y})(\hat{y} - \bar{y})}{\sqrt{\sum(y - \bar{y})^2 \sum(\hat{y} - \bar{y})^2}}$$

Unlike the case of simple correlation, $0 \leq R \leq 1$.

The square of the multiple correlation coefficient R^2 can be interpreted as the proportion of the variability that has been accounted for by the regression equation. R^2 is between 0 and 1. If the equation fits the data well, R^2 is close to 1; if the linear model is a poor fit, R^2 will be close to 0.

The formula given above for R is usually cumbersome computationally; instead, R^2 can be computed directly using the following formula:

$$R^2 = \frac{\sum b_iS_{iy}}{S_{yy}}$$

Then R can be found by taking the positive square root of R^2.

As in the case of simple linear regression and correlation, different assumptions are used when deriving multiple regression and multiple correlation procedures. Multiple regression assumes that the x's are fixed and predetermined by the investigator, the relationship is linear, and the ϵ's are IND$(0,\sigma^2)$. Multiple correlation assumes that the x's are random and that $y, x_1, \ldots, x_k$ have a multivariate normal distribution.

As a result of these assumptions, all the procedures we discuss in this chapter may be applied to situations that fit the correlation model. If a research situation fits the regression model and has fixed x's, then correlation statistics as R^2 may be calculated, but inference should not be made from these statistics.

If the correlation model is being used, R^2 may be tested. To test the significance of the multiple correlation coefficient, we use hypotheses

$$H_0: \mathbf{P}^2 = 0 \quad \text{against} \quad H_a: \mathbf{P}^2 > 0$$

in which **P** (the upper case Greek letter rho) is the true population multiple correlation coefficient. The test statistic is

$$F = \frac{R^2/k}{(1 - R^2)/(n - k - 1)}$$

with $\nu_1 = k$ and $\nu_2 = n - k - 1$ degrees of freedom.

The following example illustrates these procedures.

Example 14.2. Multiple Regression and Correlation

In Midwest farming areas, the land is divided into quarter-acre tracts called sections, and frequently county roads separate the sections. This arrangement makes it easy for wildlife biologists to estimate a game bird population. They can select a random sample of sections planted in a certain crop such as corn, count the number of pheasant or other game species in those sections, and then multiply by the number of sections of corn in the area. The only difficulty with this procedure is counting the pheasant in a randomly chosen section. A number of biologists and volunteers are needed to walk through the section, flushing all pheasant as they go, while others have to be stationed along the perimeter to count the birds as they fly out. To curtail this labor-intensive task, simpler methods are employed to estimate the number of pheasant in a section. Then, if there is close agreement between the estimated number and the actual count, the counts can be discontinued.

Multiple regression procedures are commonly employed to construct the estimation equations used by wildlife biologists. Suppose they decide to use as the independent variables in a prediction equation the number of birds seen x_1 and heard x_2 in a section. A random sample of sections of corn is taken, and the wildlife biologists drive around the perimeter of each section at a constant speed and at the same time of day. They record the number of pheasant they see in the section, x_1, and the number of times they hear a cockerel crow while they are driving around the section, x_2. Then they use volunteers to flush all of the birds out of the section and obtain an accurate count y of the actual number of pheasant in the section.

Suppose that a random sample is viewed as a pilot study and contains only six sections from which the following information is produced:

Section	x_1	x_2	y
1	0	1	2
2	4	5	8
3	2	3	3
4	3	6	6
5	1	0	1
6	2	3	4

$$\begin{aligned} S_{11} &= 10 & n &= 6 \\ S_{22} &= 26 & \bar{x}_1 &= 2 \\ S_{yy} &= 34 & \bar{x}_2 &= 3 \\ S_{12} &= 14 & \bar{y} &= 4 \\ S_{1y} &= 17 \\ S_{2y} &= 27 \end{aligned}$$

The augmented matrix of sums of squares and cross-products is

$$\left[\begin{array}{cc|c|cc} 10 & 14 & 17 & 1 & 0 \\ 14 & 26 & 27 & 0 & 1 \end{array}\right]$$

and row operations transform it into

$$\left[\begin{array}{cc|c|cc} 1 & 0 & 1.0 & 0.40625 & -0.21875 \\ 0 & 1 & 0.5 & -0.21875 & 0.15625 \end{array}\right]$$

Thus the estimated partial regression coefficients are

$$b_1 = 1.0 \qquad \text{and} \qquad b_2 = 0.5$$

and

$$a = \bar{y} - b_1\bar{x}_1 - b_2\bar{x}_2 = 4.0 - 1.0(2) - 0.5(3) = 0.5$$

The prediction equation is

$$\hat{y} = 0.5 + 1.0x_1 + 0.5x_2$$

To test H_0: $\beta_1 = \beta_2 = 0$ against H_a: $\beta_1 \neq 0$ or $\beta_2 \neq 0$, the wildlife biologists compute

$$\begin{aligned} \sum(y - \hat{y})^2 &= S_{yy} - b_1S_{1y} - b_2S_{2y} \\ &= 34 - 1.0(17) - 0.5(27) \\ &= 34 - 30.5 \\ &= 3.5 \end{aligned}$$

and use the ANOVA format

Source	df	SS	MS	F
Due regression	2	30.5	15.2500	13.07
Deviations	3	3.5	1.1667	
Total	5	34.0		

Since $F_{0.05,2,3} = 9.552$, the null hypothesis is rejected, and the regression plane may be used for prediction.

Alternatively, the researchers could compute

$$R^2 = \frac{\sum b_i S_{iy}}{S_{yy}} = \frac{30.5}{34} = 0.8971$$

and since x_1 and x_2 are random, they could test H_0: $\mathbf{P}^2 = 0$ against H_a: $\mathbf{P}^2 > 0$ using

$$F = \frac{R^2/k}{(1 - R^2)/(n - k - 1)} = \frac{0.8971/2}{0.1029/3} = 13.08$$

and $F_{0.05,2,3} = 9.552$. The two tests give the same result (except for rounding error).

There is significant agreement between the observed values and their estimates obtained from the least-squares equation. The wildlife biologists can feel optimistic about being able to estimate the number of pheasant in a section of corn without having to recruit volunteers to drive the birds out for an exact count. However, they will very likely want to confirm their results by taking a larger sample. They can undertake the larger project with the encouragement of the significant relationship they found in the pilot study.

As in simple linear regression, R^2 is a summary measure that helps judge the fit of the linear model to the data. R^2 is the proportion of the variability that is explained by the model. However, a large R^2 does not necessarily mean that we have chosen the proper model for the data. It could be that the data do not fit the assumptions of the model. In particular, we assume for linear regression that y is a linear function of the x's and ϵ's, and that the ϵ's are independently distributed with a mean of zero and constant variance σ^2. As in Section 9.2, a simple and effective way of detecting model deficiencies is to examine the residuals

$$e = y - \hat{y}$$

and plot e against the predicted value $\hat{y}$. For a more complete discussion of residual analysis in the multiple regression case, see Chatterjee and Price (1977).

The procedure for fitting the multiple regression equation can be summarized as follows.

■ ■ ■

Procedure. Multiple Regression and Correlation

Given n data points consisting of a dependent variable y and k independent variables $x_1, \ldots, x_k$, the least-squares regression equation

$$\hat{y} = a + b_1x_1 + \cdots + b_kx_k$$

is found by computing the augmented matrix of sums of squares and cross-products

$$\left[\begin{array}{cccc|c|cccc}
S_{11} & S_{12} & \ldots & S_{1k} & S_{1y} & 1 & 0 & \ldots & 0 \\
S_{12} & S_{22} & \ldots & S_{2k} & & 0 & 1 & & \\
\vdots & & & \vdots & \vdots & \vdots & & & \vdots \\
 & & & & & & & \ddots & 0 \\
S_{1k} & S_{2k} & \ldots & S_{ky} & S_{ky} & 0 & \ldots & 0 & 1
\end{array}\right]$$

and transforming this by row operations to

$$\left[\begin{array}{cccc|c|cccc}
1 & 0 & \ldots & 0 & b_1 & p_{11} & p_{11} & \ldots & p_{1k} \\
0 & 1 & \ldots & 0 & & p_{21} & & & \vdots \\
 & & & & \vdots & & & & \\
\vdots & & & \vdots & & \vdots & & & \\
 & & \ddots & 0 & & & & & \\
0 & \ldots & 0 & 1 & b_k & p_{k1} & \ldots & & p_{kk}
\end{array}\right]$$

The estimates of the partial regression coefficients appear in the center of the transformed matrix and

$$a = \bar{y} - \sum b_i\bar{x}_i$$

H_0: $\beta_1 = \cdots = \beta_k = 0$ is tested by

$$F = \frac{\sum b_iS_{iy}/k}{(S_{yy} - \sum b_iS_{iy})/(n - k - 1)}$$

with $\nu_1 = k$ and $\nu_2 = n - k - 1$ degrees of freedom.

The proportion of the variability explained by the regression equation is

$$R^2 = \frac{\sum b_iS_{iy}}{S_{yy}}$$

For the multiple correlation model, R^2 is significant if the F test for H_0: $\beta_1 = \cdots = \beta_k = 0$ is significant. Equivalently,

$$F = \frac{R^2/k}{(1 - R^2)/(n - k - 1)}$$

may be calculated and tested with k and $n - k - 1$ degrees of freedom.

■ ■ ■

In practice, most research workers have computer programs available to them to do these computations. However, understanding how the regression equation is derived and tested can lead to a more intelligent use of such programs.

EXERCISES

14.2.1. Given the following sums of squares and cross-products for 27 data points (x_1, x_2, y)

$$\begin{array}{lll} S_{11} = 10 & S_{22} = 41 & S_{yy} = 50 \\ S_{1y} = \ \ 4 & S_{2y} = \ \ 2 & S_{12} = 20 \end{array}$$

a. Complete the augmented matrix of sums of squares and cross-products and the final matrix after row operations.

$$\overset{M}{\left[\begin{array}{cc:c:cc} \_\_ & \_\_ & \_\_ & 1 & 0 \\ \_\_ & \_\_ & \_\_ & 0 & 1 \end{array}\right]} \rightarrow \overset{M^{-1}}{\left[\begin{array}{cc:c:cc} \_\_ & \_\_ & \_\_ & 4.1 & -2.0 \\ \_\_ & \_\_ & \_\_ & -2.0 & 1.0 \end{array}\right]}$$

(It is not necessary to actually do the row operations.)

b. Complete the ANOVA table for multiple regression.

Source	df	SS	MS	F	$F_{0.05}$
Regression					
Deviations					

14.2.2. When the age y of a grazing animal is unknown, it can be estimated from the extent of tooth wear x_1 and the amount of gray hair x_2 on the animal's muzzle. In an effort to evaluate and refine this procedure, a random sample of horses of known ages is measured on indices developed to determine tooth wear and graying. The following information is derived:

Augmented Sum of Squares
and Cross-Products Matrix

$$\begin{bmatrix} 64.00 & -39.20 & -20.00 & 1 & 0 \\ -39.20 & 49.00 & 19.39 & 0 & 1 \end{bmatrix}$$

$$\rightarrow \begin{bmatrix} 1 & 0 & -0.1375 & 0.0306 & 0.0245 \\ 0 & 1 & 0.2856 & 0.2856 & 0.0400 \end{bmatrix}$$

a. Complete the ANOVA table.

Source	df	SS	MS
Regression	____	8.29	4.145
Deviations	____	16.71	1.671

b. What percentage of the variation in the horses' ages can be explained on the basis of tooth condition and graying?
c. If a multiple regression prediction equation is used, will it explain a significant portion of the variability of age? Why, or why not?
d. Do you think the prediction equation would be very useful in estimating the ages of horses when their ages are not known? Why, or why not?

14.2.3. In studies of the effect of acid rain on the biomass in freshwater lakes, biologists have found that biomass decreases as acid concentration increases. If the lakes have sources of phosphorus, however, biomass increases with an increase in the amount of phosphorus available. In an effort to make a more thorough study, researchers take water samples from 18 randomly selected lakes and measure the acidity x_1, available phosphorus x_2, and population density y of a certain species of algal plant. The following statistics are computed:

$$\begin{array}{lll} \bar{y} = 1400 & S_{yy} = 14{,}400 & S_{12} = 900 \\ \bar{x}_1 = 2100 & S_{11} = 1{,}600 & S_{1y} = -3000 \\ \bar{x}_2 = 760 & S_{22} = 3{,}600 & S_{2y} = 2100 \end{array}$$

$$\begin{array}{ll} p_{11} = 0.000727 & b_1 = -2.563 \\ p_{12} = -0.000182 & b_2 = 1.224 \\ p_{22} = 0.000323 & s^2 = 276 \end{array}$$

a. Compute R^2.
b. Test R^2 for significance.
c. Test $\beta_1 = \beta_2 = 0$.

d. What is the equation of the least-squares plane?
e. If acidity is increased one unit and phosphorus held fixed, what is the effect on population density?

14.3. INFERENCES RELATIVE TO MULTIPLE REGRESSION

In an analysis of variance, if the F test of means is significant, the investigator will do further tests to pinpoint the specific differences. Similarly, in regression, if H_0: $\beta_1 = \cdots = \beta_k = 0$ is rejected, the investigator will want to know which of the x variables contribute to this overall significance. As with ANOVA, different people prefer different methods. Some common follow-up procedures are:

1. Tests on the partial regression coefficients.
2. Tests on the standardized partial regression coefficients.
3. Tests on the partial correlation coefficients.
4. Confidence intervals on the partial regression coefficients.

To illustrate these different approaches, we return to the study of five migrant-worker families of Section 14.2. This time, however, let the data collected be as follows, in which x_1 is the educational attainment of the mother, x_2 is the educational attainment of the father, and y is the educational attainment of the oldest child.

x_1	x_2	y
9.8	7.0	6.6
11.4	8.0	8.2
9.0	9.0	9.0
8.2	10.0	10.4
10.6	11.0	11.8
49.0	45.0	46.0

$$S_{11} = 6.4 \quad S_{12} = -1.6 \quad S_{22} = 10.0 \quad S_{1y} = -1.9 \quad S_{2y} = 12.6 \quad S_{yy} = 16.0$$

$$n = 5 \quad \bar{x}_1 = 9.8 \quad \bar{x}_2 = 9.0 \quad \bar{y} = 9.2$$

The augmented matrix of sums of squares and cross-products is

$$\left[\begin{array}{cc|c|cc} 6.4 & -1.6 & -1.9 & 1 & 0 \\ -1.6 & 10.0 & 12.6 & 0 & 1 \end{array}\right]$$

and row operations transform it to

$$\left[\begin{array}{cc|c|cc} 1 & 0 & 0.01888 & 0.16276 & 0.02604 \\ 0 & 1 & 1.26303 & 0.02604 & 0.10417 \end{array}\right]$$

Thus

$$b_1 = 0.01888 \text{ and } b_2 = 1.26302$$

We find

$$a = \bar{y} - \sum b_i\bar{x}_i = 9.2 - 0.01888(9.8) - 1.26302(9.0) = -2.35220$$

and the least-squares equation is

$$\hat{y} = -2.35220 + 0.01888x_1 + 1.26302x_2$$

Testing H_0: $\beta_1 = \beta_2 = 0$ against H_a: $\beta_1 \neq 0$ or $\beta_2 \neq 0$, we find:

Source	df	SS	MS	F	$F_{0.05,2,2}$
Regression	$k = 2$	$\sum b_i S_{iy} = 15.878$	7.939	130.148	19.000
Deviations	$n-k-1 = 2$	0.122	0.061		
Total	$n-1 = 4$	$S_{yy} = 16.000$			

The test is significant and the regression plane can be used for prediction. Now we should like to know if this is due to x_1, x_2, or both variables.

The estimated partial regression coefficients b_1 and b_2 can be interpreted as partial slopes. The coefficient $b_1 = 0.01888$ tells us that when x_1 is increased by one unit and x_2 is held constant, y is increased by 0.01888 units. Similarly for b_2. However, we must be cautious about directly comparing b_1 and b_2. If x_1 and x_2 are completely unrelated to each other, then the partial regression coefficients would be the same as if x_1 and x_2 were regressed one at a time on y. However, the x's are usually interrelated, and although we can hold an x fixed in the statistical sense, it may not be possible to do that in the real world.

To judge the contribution of each x to the model, we test the partial regression coefficients separately.

H_0: $\beta_1 = 0$ against H_a: $\beta_1 \neq 0$ is tested with

$$t = \frac{b_1 - \beta_{1_0}}{\sqrt{p_{11}s^2}} = \frac{0.01888 - 0}{\sqrt{0.16276(0.061)}} = 0.189$$

with $\nu = n - k - 1 = 2$ degrees of freedom and in which β_{1_0} is the value of β_{1_0} specified in the null hypothesis (β_{1_0} could be some value other than zero), and p_{11} is the value in the first row and column of M^{-1}, the inverse of the matrix of sums of squares and cross-products found in the process of solving for b_1 and b_2.

Similarly, H_0: $\beta_2 = 0$ against H_a: $\beta_2 \neq 0$ is tested with

$$t = \frac{b_2 - \beta_{2_0}}{\sqrt{p_{22}s^2}} = \frac{1.26302 - 0}{\sqrt{0.10417(0.061)}} = 15.844$$

In both cases, the null hypothesis is rejected if $|t| \geq t_{0.025,2} = 4.303$. We accept $\beta_1 = 0$ and reject $\beta_2 = 0$. We conclude that the educational attainment of the mother cannot be used to predict the educational attainment of the oldest offspring, but that the educational attainment of the father can be used.

If desired, equivalent F tests can be used to test $\beta_i = 0$.

$$F = \frac{b_i^2}{p_{ii}s^2} = \frac{b_i S_{iy}}{s^2}$$

with $\nu_1 = 1$, $\nu_2 = n - k - 1$.

It should be remembered that b_1 and b_2 are not pure numbers but contain units of measurement. In this example, b_1 is the years of education of offspring per year of education for the mother, and b_2 is the years of education of offspring per year of education for the father. Sometimes this natural scale of measurement does not have a particularly meaningful interpretation, and the two units of measurement are so different that any comparison is meaningless. In such cases, it is useful to express the estimated partial regression coefficients in standardized form.

The *standardized partial regression coefficients* are symbolized by b_i' and

$$b_i' = b_i\sqrt{S_{ii}/S_{yy}}$$

In the example,

$$b_1' = b_1\sqrt{S_{11}/S_{yy}} = 0.01888\sqrt{6.4/16.0} = 0.0119$$

$$b_2' = b_2\sqrt{S_{22}/S_{yy}} = 1.26302\sqrt{10.0/16.0} = 0.9985$$

The standardized coefficients are unit-free and express the partial slopes in terms of standard deviations. If the educational achievement of the father is increased by one standard deviation and the mother's educational achievement is held fixed, then the educational achievement of the child is increased by 0.9985 standard deviations.

The standardized partial regression coefficients can be tested directly.

$$H_0:\ \beta_1' = 0 \qquad H_0:\ \beta_2' = 0$$

and

$$H_a:\ \beta_1' \neq 0 \qquad H_a:\ \beta_2' \neq 0$$

are tested with

$$\begin{aligned} t &= \frac{b_1' - 0}{\sqrt{p_{11}s^2 S_{11}/S_{yy}}} \\ &= \frac{0.0119}{\sqrt{(0.16276)(0.061)(6.4)/16.0}} \\ &= 0.189 \end{aligned}$$

and

$$t = \frac{b_2' - 0}{\sqrt{p_{22}s^2S_{22}/S_{yy}}}$$

$$= \frac{0.9985}{\sqrt{(0.10417)(0.061)(10.0)/16.0}}$$

$$= 15.844$$

Note that there are the same t values as for the tests of $\beta_1 = 0$ and $\beta_2 = 0$. The tests for the standardized partial regression coefficients are equivalent to the corresponding tests for the unstandardized coefficients.

Another unit-free approach which may be used if the correlation model is being considered is *partial correlation coefficients*. We first compute all simple correlation coefficients for the sample:

$$r_{y1} = \frac{S_{1y}}{\sqrt{S_{11}S_{yy}}} = \frac{-1.9}{\sqrt{6.4(16.0)}} = -0.188$$

$$r_{y2} = \frac{S_{2y}}{\sqrt{S_{22}S_{yy}}} = \frac{12.6}{\sqrt{10(16.0)}} = 0.996$$

$$r_{12} = \frac{S_{12}}{\sqrt{S_{11}S_{22}}} = \frac{-16}{\sqrt{6.4(10)}} = -0.200$$

Since the father's education is probably related to the mother's education, we ask what the correlation between the mother's education and the education of the offspring would be if we removed the effect of the father's education. This is expressed by the estimated partial correlation coefficient $r_{y1\cdot 2}$.

To derive the partial correlation coefficient, the effect of x_2 is removed from y and x_1 by regressing y on x_2 and x_1 on x_2. This leads to least-squares equations of the form

$$\hat{y} = a + bx_2 \qquad \text{and} \qquad \hat{x}_1 = a' + b'x_2$$

The part of y that is unexplained by x_2 is $y - \hat{y}$, and $x_1 - \hat{x}_1$ is the part of x_1 unexplained by x_2. The relationship between the unexplained portions can be measured by the simple correlation between $y - \hat{y}$ and $x_1 - \hat{x}_1$. Fortunately, it is not necessary to carry out all of the steps, since the result can be expressed in a form involving the simple correlations between y, x_1, and x_2.

$$r_{y1\cdot 2} = \frac{r_{y1} - r_{y2}r_{12}}{\sqrt{(1 - r_{y2}^2)(1 - r_{12}^2)}}$$

$$= \frac{-0.188 - (0.996)(-0.200)}{\sqrt{(1 - 0.992)(1 - 0.040)}}$$

$$= 0.128$$

The estimated correlation between the father's education and the education of the offspring when the effect of the mother's education is removed is

$$r_{y2\cdot 1} = \frac{r_{y2} - r_{y1}r_{12}}{\sqrt{(1 - r_{y1}^2)(1 - r_{12}^2)}}$$
$$= \frac{0.996 - (-0.188)(-0.200)}{\sqrt{(1 - 0.035)(1 - 0.040)}}$$
$$= 0.996$$

The tests for significance of $r_{y1\cdot 2}$ and $r_{y2\cdot 1}$ are t tests.

$$H_0: \ \rho_{yi\cdot j} = 0 \qquad \text{against} \qquad H_a: \ \rho_{yi\cdot j} \neq 0$$

is tested with

$$t = \frac{r_{yi.j}\sqrt{n - 3}}{\sqrt{1 - r_{yi\cdot j}^2}}$$

with $n - 3$ degrees of freedom. These tests are only for $\rho_{yi\cdot j} = 0$ and are equivalent to testing $\beta_i = 0$.

The follow-up tests to determine the contribution of each x_i to the regression equation are now summarized.

■ ■ ■

Procedure. Follow-Up Tests to Multiple Regression

Partial Regression Coefficients b_i

$$H_0: \ \beta_i = \beta_{i_0} \qquad \text{against} \qquad H_a: \ \beta_i \neq \beta_{i_0}$$

is tested with

$$t = \frac{b_i - \beta_{i_0}}{\sqrt{p_{ii}s^2}}$$

with $\nu = n - k - 1$ in which p_{ii} is the ith diagonal element of M^{-1}, the inverse of the matrix of sums of squares and cross-products, and s^2 is the mean squared deviation from the regression surface. The test of H_0: $\beta_i = 0$ against H_a: $\beta_i \neq 0$ can be carried out using

$$F = \frac{b_i^2}{p_{ii}s^2}$$

with $\nu_1 = 1$ and $\nu_2 = n - k - 1$; it is equivalent to the above t test.

Standardized Partial Regression Coefficients b_i'

$$H_0: \ \beta_i' = \beta_{i_0}' \qquad \text{against} \qquad H_a: \ \beta_i' \neq \beta_{i_0}'$$

is tested with

$$t = \frac{b_i' - \beta_{i_0}'}{\sqrt{p_{ii}s^2 S_{ii}/S_{yy}}}$$

with $\nu = n - k - 1$. This test is equivalent to the t test for β_i.

Partial Correlation Coefficients (for $k = 2$)

$$r_{yi \cdot j} = \frac{r_{yi} - r_{yj}r_{ij}}{\sqrt{(1 - r_{yj}^2)(1 - r_{ij}^2)}}$$

in which

$$r_{yi} = \frac{S_{iy}}{\sqrt{S_{ii}S_{yy}}}$$

$$r_{yj} = \frac{S_{yj}}{\sqrt{S_{jj}S_{yy}}}$$

$$r_{ij} = \frac{S_{ij}}{\sqrt{S_{ii}S_{jj}}}$$

$$H_0: \ \rho_{yi \cdot j} = 0 \qquad \text{against} \qquad H_a: \ \rho_{yi \cdot j} \neq 0$$

is tested with

$$t = \frac{r_{yi \cdot j}\sqrt{n - 3}}{\sqrt{1 - r_{yi \cdot j}^2}}$$

with $\nu = n - 3$.

■ ■ ■

In addition to tests of hypotheses, several types of estimates are possible within a regression analysis. For example, confidence intervals for β_i are found using

$$\text{CI}_{1-\alpha}: \quad b_i \pm t_{\alpha/2, n-k-1}\sqrt{p_{ii}s^2}$$

Confidence intervals for the difference between two partial regression coefficients $\beta_i - \beta_j$ are

$$\text{CI}_{1-\alpha}: \quad b_i - b_j \pm t_{\alpha/2, n-k-1}\sqrt{(p_{ii} + p_{jj} - 2p_{ij})s^2}$$

The term $-2p_{ij}$ in the standard error is due to the possible relationship of x_i and x_j.

It is also possible to find confidence intervals for the estimates obtained using the fitted equation

$$\hat{y} = a + b_i x_i + \cdots + b_k x_k$$

Given the particular values $x_1^*, x_2^*, \ldots, x_k^*$ the point estimate of $E(y|x_1 = x_1^*, \ldots, x_k = x_k^*)$ is $\hat{y}$ and the confidence interval is

$$\mathrm{CI}_{1-\alpha}: \quad \hat{y} \pm t_{\alpha/2, n-k-1}\sqrt{s^2[1/n + \sum_i \sum_j p_{ij}(x_i^* - \bar{x}_i)(x_j^* - \bar{x}_j)]}$$

For example, if we wanted to estimate the mean educational attainment of the oldest child in migrant-worker families in which the mother had nine years of education and the father had ten years of education, then

$$\hat{y} = -2.35220 + 0.01888(9) + 1.26302(10) = 10.4$$

and the 95% confidence interval on $E(y|x_1 = 9, x_2 = 10)$ is

$$\mathrm{CI}_{0.95}: \quad \hat{y} \pm t_{0.025,2}\sqrt{\begin{aligned}&s^2[1/n + p_{11}(x_1^* - \bar{x}_1)^2\\ &\quad + 2p_{12}(x_1^* - \bar{x}_1)(x_2^* - \bar{x}_2)\\ &\quad + p_{22}(x_2^* - \bar{x}_2)^2]\end{aligned}}$$

$$10.4 \pm 4.305\sqrt{\begin{aligned}&0.061[1/5 + 0.16276(9 - 9.8)^2\\ &\quad + 2(0.02604)(9 - 9.8)(10 - 9.0)\\ &\quad + 0.10417(10 - 9.0)^2]\end{aligned}}$$

$$10.4 \pm 0.15$$

$$10.25 \leq E(y|x_1 = 9, x_2 = 10) \leq 10.55$$

Note that if $k = 1$, this reduces to a form equivalent to that used in simple linear regression.

If an individual y is predicted, the point estimate is $\hat{y}$ and the prediction interval is

$$\mathrm{PI}_{1-\alpha}: \quad \hat{y} \pm t_{\alpha/2, n-k-1}\sqrt{s^2[1 + 1/n + \sum_i \sum_j p_{ij}(x_i^* - \bar{x}_i)(x_j^* - \bar{x}_j)]}$$

Because the complexity of these standard errors increases with additional independent variables, we do not want to include in the model variables that provide little or no additional information. In the next section, we show how to simplify the prediction equation by eliminating x variables that contribute little to the reliability of the prediction.

EXERCISES

14.3.1. In Exercise 14.2.2 on grazing animals:

a. Compute the partial correlation coefficients.

b. Is there a linear relationship between graying and age when tooth condition is held constant?

14.3.2. In Exercise 14.2.3 on lake biomass:

a. Compute all possible simple correlations between the variables.

b. Compute the standardized partial regression coefficients.

c. Test $\beta_1 = 0$ and $\beta_2 = 0$ separately, and interpret the results.

d. Estimate the mean population density of the algae in a lake with an acidity measurement of 2000 and a phosphorus measurement of 860. Place a 95% confidence interval on this estimate.

14.3.3. Using the education example of this section, show that

$$R^2 = \frac{r_{1y}^2 + r_{2y}^2 - 2r_{1y}r_{2y}r_{12}}{1 - r_{12}^2}$$

14.3.4. Using the education example of this section, show that the F statistic to test H_0: $\beta_1 = \beta_2 = 0$ can be computed from the multiple correlation coefficient

$$F = \frac{R^2/k}{(1 - R^2)/(n - k - 1)}$$

14.4. MODEL FITTING

The object of model fitting is to obtain the simplest model that at the same time adequately fits the data, so that it can be useful for prediction purposes. The investigator may be considering several possible independent variables that could be included in the model; however, they may be time-consuming or expensive to obtain, they may increase the complexity of the standard errors of the estimates (as noted in the previous section), or they may add little accuracy to the estimates.

Two approaches are commonly used to select a subset of the possible independent variables:

1. Backward elimination, a step-down procedure in which the investigator begins with a full model, including all possible independent variables, and the x variables are eliminated one by one as it is determined that they contribute little to the model.
2. Stepwise regression, an addition procedure in which the model is built by adding one independent variable at a time and measuring its contribution to the model.

We discuss the elimination approach in some detail.

The basic idea behind the elimination procedure can be illustrated with the migrant-worker family example of the previous section. The investigator begins with the full model containing x_1 and x_2 and examines the t tests for the b_i (or b_i' or $r_{yi \cdot j}$) to see which contributes least to the accuracy of the prediction.

Coefficient	t Value for H_0: $\beta_i = 0$
$b_1 = 0.01888$	0.189
$b_2 = 1.26302$	15.844

The x_1 variable has the lower t values, and it is deleted. Multiple regression (in this case simple regression) is performed again without x_1. The new estimate of β_2 is

$$b_2^* = S_{2y}/S_{22} = 12.6/10.0 = 1.26$$

The difference in the sum of squares due to regression shows how much (or little) accuracy in prediction is lost by eliminating the x_1 variable. This comparison can be done in tabular form.

Source	df	SS	MS	F
Full regression model	$k=2$	$\text{SSR} = \sum b_i S_{iy} = 15.878$		
x_2 alone	1	$\text{SS2} = b_2^* S_{2y} = 15.876$	15.876	
x_1 after x_2	1	$\text{SSR} - \text{SS2} = 0.002$	0.002	0.03
Deviations	$n-k-1=2$	$\text{SST} - \text{SSR} = 0.122$	0.061	
Total	$n-1=4$	$\text{SST} = S_{yy} = 16.000$		

Since $F_{0.05,1,2} = 18.513$ and the F value for the addition of x_1 to the model after x_2 is only 0.03, x_1 adds no significant predictive efficiency to a model already containing x_2. In order to keep the prediction equation as simple as possible, the experimenter may delete x_1 without losing significant accuracy in prediction.

The example just used involved a very small sample $n = 5$. This was merely for purposes of illustration to keep the computations simple. Some statisticians recommend for multiple regression a sample size of 100, or 20 times the number of independent variables, whichever is larger. This may be larger than necessary in some cases. It depends on the width of the confidence intervals that is desired for the estimates.

We now look at the elimination procedure applied to a slightly more complex investigation.

Example 14.3. Multiple Regression Analysis and Model Building

The phantom midge, genus *Chaoborus*, resembles the mosquito in appearance but not in bloodsucking behavior. Swarms of adult chaoborids are a familiar sight along the shoreline of lakes and other bodies of fresh water, but a great portion of the life cycle is spent in the water in the larval stage. The larva burrows into the sediment at the bottom of a lake or pond and remains there during the daylight hours. At night it migrates vertically toward the surface to feed on the fauna in the plankton layer. The larva is itself prey for larger animals and consequently has an important role in the food chain of freshwater fish.

Man-made lakes and other water impoundments seem to create good habitats for chaoborids, so much so that they can become a nuisance. They seem to be little affected by the brackish nature of such water; the reduced oxygen content may even be favorable for an increase in population density. The steep banks and greater depths of man-made impoundments also seem to be favorable to the genus.

In order to learn more about the contribution of various environmental factors to the habitat of *Chaoborus* larvae, a team of biologists made a study of a recreational lake that was created by damming a small stream. The lake has a surface area of approximately 20 hectares, and in order to obtain random samples from it, a grid was superimposed on a map of the lake and 14 random sampling points were taken on the grid. By means of surveying equipment, these sampling points were located on the lake. The following variables were measured at each sampling point:

x_1: The depth of the lake at the sampling point, measured to the nearest decimeter (recorded in meters).

x_2: The brackishness (conductivity) of the water, measured from a sample taken at the bottom (recorded in mhos per decimeter).

x_3: The dissolved oxygen (milligrams per liter) in the water sampled from the lake bottom.

y: The number of *Chaoborus* larvae collected in a grab sample of the sediment at the sampling point. The sampling device collected sediment from an area of approximately 225 cm^2 of lake bottom.

The data obtained were:

Sample	x_1	x_2	x_3	y
1	8.4	8.0	1.0	35
2	2.0	6.5	8.5	10
3	3.5	6.2	6.5	9
4	10.4	5.0	1.5	30
5	6.5	6.5	7.5	20
6	6.2	7.3	4.5	23
7	12.4	6.4	4.0	28
8	7.0	6.0	10.0	8
9	5.8	6.1	3.0	29
10	3.0	5.4	11.0	4

Sample	x_1	x_2	x_3	y
11	6.0	7.3	4.5	18
12	5.5	6.6	5.5	14
13	9.0	6.5	2.5	32
14	1.1	5.8	7.0	6
Total	86.8	89.6	77.0	266
Mean	6.2	6.4	5.5	19.0

$$\begin{aligned} \sum x_1^2 &= 669.52 \\ (\sum x_1)^2/n &= \underline{538.16} \\ S_{11} &= 131.36 \end{aligned} \qquad \begin{aligned} \sum x_1 y &= 1993.10 \\ (\sum x_1)(\sum y)/n &= \underline{1649.20} \\ S_{1y} &= 343.90 \end{aligned}$$

$$\begin{aligned} \sum x_2^2 &= 581.30 \\ (\sum x_2)^2/n &= \underline{573.44} \\ S_{22} &= 7.86 \end{aligned} \qquad \begin{aligned} \sum x_2 y &= 1741.00 \\ (\sum x_2)(\sum y)/n &= \underline{1702.40} \\ S_{2y} &= 38.60 \end{aligned}$$

$$\begin{aligned} \sum x_3^2 &= 546.00 \\ (\sum x_3)^2/n &= \underline{423.50} \\ S_{33} &= 122.50 \end{aligned} \qquad \begin{aligned} \sum x_1 x_3 &= 397.85 \\ (\sum x_1)(\sum x_3)/n &= \underline{477.40} \\ S_{13} &= -79.55 \end{aligned}$$

$$\begin{aligned} \sum y^2 &= 6520.00 \\ (\sum y)^2/n &= \underline{5054.00} \\ S_{yy} &= 1466.00 \end{aligned} \qquad \begin{aligned} \sum x_2 x_3 &= 481.95 \\ (\sum x_2)(\sum x_3)/n &= \underline{492.80} \\ S_{23} &= -10.85 \end{aligned}$$

$$\begin{aligned} \sum x_1 x_2 &= 559.33 \\ (\sum x_1)(\sum x_2)/n &= \underline{555.52} \\ S_{12} &= 3.81 \end{aligned} \qquad \begin{aligned} \sum x_3 y &= 1080.00 \\ (\sum x_3)(\sum y)/n &= \underline{1463.00} \\ S_{3y} &= -383.00 \end{aligned}$$

The simultaneous equations to be solved are:

$$\begin{aligned} 131.36b_1 + 3.81b_2 - 79.55b_3 &= 343.90 \\ 3.81b_1 + 7.86b_2 - 10.85b_3 &= 38.60 \\ -79.55b_1 - 10.85b_2 + 122.50b_3 &= -383.00 \end{aligned}$$

and the augmented matrix upon which row operations will be performed is $[M|N|I]$ or

$$\begin{array}{l} \text{Row 1:} \\ \\ \text{Row 2:} \\ \\ \text{Row 3:} \end{array} \left[\begin{array}{ccc|c|ccc} 131.36 & 3.81 & -79.55 & 343.90 & 1 & 0 & 0 \\ \\ 3.81 & 7.86 & -10.85 & 38.60 & 0 & 1 & 0 \\ \\ -79.55 & -10.85 & 122.50 & -383.00 & 0 & 0 & 1 \end{array}\right]$$

So that a practical example can be examined in detail, the individual steps are given with six decimal places used for accuracy.

Row 1a = Row 1/131.36:

$$1 \quad 0.029004 \quad -0.605588 \qquad 2.617996 \qquad 0.007613 \quad 0 \quad 0$$

Row 1a as above, Row 2a = Row 2 − 3.81(Row 1a),
Row 3a = Row 3 + 79.55(Row 1a)

$$\begin{array}{l} \text{Row } 1a: \\ \\ \text{Row } 2a: \\ \\ \text{Row } 3a: \end{array} \left[\begin{array}{rrr|r|rrr} 1 & 0.029004 & -0.605588 & 2.617996 & 0.007613 & 0 & 0 \\ \\ 0 & 7.749494 & -8.542711 & 28.625434 & -0.029004 & 1 & 0 \\ \\ 0 & -8.542711 & 74.325500 & -174.738391 & 0.605588 & 0 & 1 \end{array}\right]$$

Because of carrying so many decimal places, it may be helpful to list the three multiples a_{ij} that are used in the next set of row operations:

$$a_{12} = 0.029004 \qquad a_{22} = 7.749494 \qquad a_{32} = -8.542711$$

Row 2b = Row 2a/a_{22}:

$$0 \quad 1 \quad -1.102357 \qquad 3.693846 \quad -0.003743 \qquad 0.129041 \quad 0$$

Row 1a − a_{12}(Row 2b), Row 2b as above,
Row 3b = Row 3a − a_{32}(Row 2b)

$$\begin{array}{l} \text{Row } 1b: \\ \\ \text{Row } 2b: \\ \\ \text{Row } 3b: \end{array} \left[\begin{array}{rrr|r|rrr} 1 & 0 & -0.573615 & 2.510859 & 0.007721 & -0.003743 & 0 \\ \\ 0 & 1 & -1.102357 & 3.693846 & -0.003743 & 0.129041 & 0 \\ \\ 0 & 0 & 64.908379 & -143.182936 & 0.573614 & 1.102357 & 1 \end{array}\right]$$

The a_{ij} values that are now used in the third, and final, set of row operations are:

$$a_{13} = -0.573615 \qquad a_{23} = -1.102357 \qquad a_{33} = 64.908791$$

and Row 3c is

Row 3c = Row 3b/a_{33}:

$$0 \quad 0 \quad 1 \quad -2.205924 \quad 0.008837 \quad 0.016983 \qquad 0.015406$$

Row 1c = Row 1b − a_{13}(Row 3c), Row 2c = Row 1b − a_{23}Row 3c,
Row 3c as above

$$\begin{array}{l} \text{Row } 1c: \\ \\ \text{Row } 2c: \\ \\ \text{Row } 3c: \end{array} \left[\begin{array}{rrr|r|rrr} 1 & 0 & 0 & 1.245509 & 0.012790 & 0.005999 & 0.008837 \\ \\ 0 & 1 & 0 & 1.262129 & 0.005999 & 0.147762 & 0.016983 \\ \\ 0 & 0 & 1 & -2.205924 & 0.008837 & 0.016983 & 0.015406 \end{array}\right]$$

The arithmetic can be partially checked by remembering that the b_i can also be computed by $M^{-1}N$:

$$b_1 = 0.012790(343.90) + 0.005999(38.60) + 0.008837(-383.0)$$
$$= 1.245471 \text{ vs. } 1.245509$$
$$b_2 = 0.005999(343.90) + 0.147762(38.60) + 0.016983(-383.0)$$
$$= 1.262180 \text{ vs. } 1.262129$$
$$b_3 = 0.008837(343.90) + 0.016983(38.60) + 0.015406(-383.0)$$
$$= -2.205910 \text{ vs. } -2.205924$$

The discrepancies are $+0.000062$, $+0.000053$, and -0.000014, which are well under the units of measurement of the original data and near enough to zero to be attributable to the rounding error that would be expected for six-decimal-place accuracy. Therefore the calculations are accepted as accurate to at least three decimal places, and the estimated partial regression coefficients can be reported as:

$$b_1 = 1.246 \qquad b_2 = 1.262 \qquad b_3 = -2.206$$

The experimenters would want to know whether H_0: $\beta_1 = \beta_2 = \beta_3 = 0$ can be rejected, for they choose these variables with the expectation that they will explain a significant portion of the variability in population density of the larvae. Thus the variability due to the regression equation is computed by $\sum b_i S_{iy}$:

$$\begin{aligned} b_1 S_{2y} &= 1.246(343.90) = 428.4994 \\ b_2 S_{2y} &= 1.262(38.60) = 48.7132 \\ b_3 S_{3y} &= -2.206(-383.00) = 844.8980 \\ \text{SS due to regression} &= 1322.1106 \end{aligned}$$

ANOVA procedures can be used to test the significance of the variability due to regression.

Source	df	SS	MS	F	$F_{0.05,3,10}$
Regression	3	SSR = 1322.11	MSR = 440.70	30.63	3.708
Deviations	10	SSE = 143.89	s^2 = 14.39		
Total		S_{yy} = 1466			

The null hypothesis is rejected; the prediction equation is significant, and a measure of the reliability can be obtained in $R^2 = \text{SSR}/S_{yy} = 1322/1466 = 0.902$. It can be seen that slightly more than 90% of the variability in population density can be accounted for by the depth, conductivity, and available oxygen of a given site in the lake.

The next step in the analysis is to determine the relative importance of each of the x variables in explaining variability in population density. Each partial regression coefficient can be tested by a t test with $n - k - 1 = 10$ degrees of freedom, or because t^2 is F with 1 and 10 degrees of freedom, a very useful F test can be employed.

Hypothesis	$t = b_i/\sqrt{p_{ii}s^2}$	$F = (b_i^2/p_{ii})/s^2$
$\beta_1 = 0$	$1.246/\sqrt{(0.012790)14.39} = 2.904$	$121.39/14.39 = 8.435$
$\beta_2 = 0$	$1.262/\sqrt{(0.147762)14.39} = 0.866$	$10.78/14.39 = 0.749$
$\beta_3 = 0$	$-2.206/\sqrt{(0.015406)14.39} = -4.684$	$315.88/14.39 = 21.951$
Critical value	$t_{0.05,10} = 2.228$	$F_{0.05,1,10} = 4.965$

From the preceding tests, it can be seen that x_2 (conductivity, or brackishness), adds no significantly predictive ability to a multiple regression equation that already contains x_1 (depth) and x_3 (oxygen). Thus it can be dropped from the equation. If this is done, however, new b_1 and b_3 values will have to be computed which do not take into account the covariability between x_2 and x_1 and that between x_2 and x_3. One could compute the new b_i values (symbolized here as b_i*) by solving a set of simultaneous equations based on the sums of squares and cross-products that had been previously computed:

$$\begin{aligned} 131.36b_1^* - 79.55b_3^* &= 343.90 \\ -79.55b_1^* + 122.50b_3^* &= -383.00 \end{aligned}$$

This is not necessary, however, because matrix algebra can be used to show that

$$\begin{aligned} b_1^* &= b_1 - p_{12}b_2/p_{22} \\ &= 1.246 - (0.005999)(1.262)/0.147762 \\ &= 1.246 - 0.051 \\ &= 1.195 \end{aligned}$$

and

$$\begin{aligned} b_3^* &= b_3 - p_{32}b_2/p_{22} \\ &= -2.206 - (0.016983)(1.262)/0.147762 \\ &= -2.206 - 0.145 \\ &= -2.351 \end{aligned}$$

The elements of the new inverse can be similarly found by

$$p_{11}^* = p_{11} - p_{12}^2/p_{22}$$
$$= 0.012790 - (0.005999)^2/0.147762 = 0.012546$$
$$p_{33}^* = p_{33} - p_{32}^2/p_{22}$$
$$= 0.015406 - (0.016983)^2/0.147762 = 0.013454$$
$$p_{13}^* = p_{13} - p_{12}p_{32}/p_{22}$$
$$= 0.008837 - (0.005999)(0.016983)/0.147762 = 0.008148$$

The sum of squares for deviations from regression will be greater than it was when all three x variables were employed in the equation, but this is to be expected from the fact that the least-squares techniques are employed. The new value, $\sum(y - \hat{y}^*)^2$, can be computed by

$$SS_{yy} - \sum b_i^* S_{iy} = 1466 - 1.195(343.90) - 2.3511(383.00)$$
$$= 1466 - 1311.3935$$
$$= 154.6065$$

or by the equation

$$SSE^* = \sum(y - \hat{y}^*)^2 = \sum(y - \hat{y})^2 + b_2^2/p_{22}$$
$$= 143.8894 + (1.262)^2/0.147762$$
$$= 143.8894 + 10.7784$$
$$= 154.6678$$

(The discrepancy is rounding error.) Note that b_2^2/p_{22} was the numerator of the F test used to test H_0: $\beta_2 = 0$. It gives the additional sum of squares explained by x_2 in an equation that already combines x_1 and x_3. If this is not a significant portion of the variability, x_2 does not improve the usefulness of the equation and this variable can be deleted, but SSE* is increased by b_2^2/p_{22}. Similarly, b_1^2/p_{11} is the additional variability explained by x_1 in an equation that already contains x_2 and x_3, and b_3^2/p_{33} is the additional amount explained by x_3 when it is included in an equation containing the other two. Thus the F tests show that depth at the sampling point x_1 and oxygen content x_3 explain significant portions of the variability in larval density. If statistical significance is the only criterion, they will both be kept in the equation, but we may also want to see if a simpler equation (one with fewer x values) may still suffice. So the process can be continued.

Model	df	SS	MS	F	R^2
Full model (x_1, x_2, x_3)	3	$\sum b_i S_{iy} = 1322.11$	440.67	30.58	0.902
Deviations from full model	10	SSE = 143.89	14.39		
$SS(x_2 \mid x_1, x_3)$	1	$b_2^2/p_{22} = 10.78$	10.78	0.75	
Two-variable model (x_1, x_3)	2	$\sum b_i^* S_{iy} = 1311.33$	655.66	46.63	0.894
Deviations from two-variable model	11	SSE* = 154.67	14.06		
$SS(x_1 \mid x_3)$	1	$b_1^{*2}/p_{11}^* = 113.82$	113.82	8.10	
Single-variable model (x_3)	1	$S_{3y}^2/S_{33} = 1197.46$	1197.46	53.51	0.817
Deviations from single-variable model	12	$S_{yy} - S_{3y}^2/S_{33} = 268.54$	22.38		

Conclusions. Oxygen x_3 is the most important variable in the study in accounting for the variability in population density. Knowledge of this variable reduces the variability in larval density by 81.7% (this is r^2 expressed as a percentage). The second most important variable is depth x_1, and knowledge of these two variables accounts for 89.4% of the variability. The additional variability accounted for by depth is 0.894 − 0.817 = 7.7%. This is a significant portion of the total variability, but it is relatively small. If it were a difficult measurement to obtain (which it is not), it might be justifiably expedient to exclude it because of its relatively small contribution. When oxygen content of the water and depth have already been taken into account, brackishness (conductivity) of the water has no significant linear relationship with population density, at least not in the ranges of that variable x_2 included in the study. Thus it is not an important variable in study of population density in the lake under consideration.

The signs of the partial regression coefficients are important. The positive relationship between x_1 and y indicates that density increases with the depth of the lake when oxygen content is held constant, whereas there is a negative association between larval count and oxygen content of the water when depth is held constant. Prediction equations based on x_1 and x_3 would be valid, but only for the one lake studied. Thus they are not very important to this study and are not demonstrated here. The important ecological information obtained from this lake is the knowledge that oxygen and depth explain a great deal of the variability in population density. These are variables that should be obtained in future studies in other lakes. Also, if the biologists decide to conduct experiments to regulate chaoborid larval population density, they have already identified oxygen and depth as two variables that can be used as treatment effects in a factorial experiment.

The backward elimination procedure can be summarized as follows.

■ ■ ■

Procedure. Backward Elimination for Model Fitting

STEP 1. Compute the regression equation containing all of the potential x variables. Test for significance. If the equation is not significant, stop; there is no appropriate model with these x's. If the equation is significant, proceed to Step 2.

STEP 2. Test the contribution of each x_i as if it were the last to enter the equation. This can be done by using a t test for H_0: $\beta_i = 0$ as $t = b_i/\sqrt{p_{ii}s^2}$ or by using the equivalent F test, $F = (b_i^2/p_{ii})/s^2$.

STEP 3. Pick the smallest t (or F) from these tests. If the smallest t (or F) is significant at a preselected α level, keep all of the x variables in the equation. If the smallest t (or F) is not significant, remove the x associated with it and recompute the regression equation. If x_j is being removed, then the new partial regression coefficients are

$$b_i^* = b_i - p_{ij}b_j/p_{jj}$$

and the new elements of the inverse of the matrix of coefficients are

$$p_{ii}^* = p_{ii} - p_{ij}^2/p_{jj}$$

and

$$p_{ik}^* = p_{ik} - p_{ij}p_{kj}/p_{jj}$$

STEP 4. Repeat Steps 2 and 3 with the reduced set of x variables.

■ ■ ■

Another popular procedure called *stepwise regression* involves building up the model by the addition of independent variables. This procedure sometimes produces a smaller model than the elimination procedure. This happens because it measures the contribution of each variable at every step, and sometimes an x that is already in the model is removed later when the presence of other variables makes its contribution redundant.

A computer with a statistical package is usually used for the computations, so we give only an outline here. A fuller discussion can be found in Draper and Smith (1981).

Stepwise Regression

STEP 1. Compute all simple correlations between y and the possible independent variables x, that is, $r_{y1}, r_{y2}, \ldots, r_{yk}$. Select the x with the largest simple correlation coefficient, say, x_j. Compute the regression $\hat{y} = a + b_j x_j$ and test H_0: $\beta_j = 0$. If this test is not significant, stop; there is no appropriate model using these x's. If it is significant, proceed to Step 2.

STEP 2. Compute the partial correlation coefficients with the effect of j removed, that is, $r_{y1\cdot j}, r_{y2\cdot j}, \ldots, r_{yk\cdot j}$. Select the x with the largest partial correlation coefficient, say, x_i. Compute $\hat{y} = a^* + b_i^* x_i + b_j^* x_j$. Test this model for significance. If it is not significant, use $\hat{y} = a + b_j x_j$. If it is significant, proceed to Step 3.

STEP 3. Test x_i and x_j as if each were the last to enter the equation. If both are significant, proceed to Step 4. If x_i is not significant, leave it out and use $\hat{y} = a + b_j x_j$ as the model. If x_i is significant but not x_j, remove x_j, put it back in the pool of possible independent variables, and proceed to Step 4.

STEP 4. Compute the partial correlations between y and each x that is not in the equation with the effects of the included x's removed. Select x with the largest partial correlation coefficient. Add this x to the equation, and test the enlarged equation for significance. If it is not significant, remove this last x and use the previous equation. If it is significant, proceed to Step 5.

STEP 5. Check all of the x's as if they were the last to enter the equation. If any are not significant, remove them and return them to the pool of possible independent variables. Compute the equation for the reduced set of x's. Repeat Step 4 with this reduced set of included x's. If all are significant, repeat Step 4 with this full set of included x's.

STEP 6. Repeat Steps 4 and 5 until the x with the largest partial correlation coefficient does not make a significant contribution or until all x's are included.

EXERCISES

14.4.1. Using the example of the *Chaoborus* larvae in this section:

a. Show numerically that the contribution of x_2 to a model already containing x_1 and x_3 can be tested by using

$$F = \frac{(R_{\text{full}}^2 - R_{\text{reduced}}^2)/(k - k')}{(1 - R_{\text{full}}^2)/(n - k - 1)}$$

with $\nu_1 = k - k'$ and $\nu_2 = n - k - 1$ in which R_{full}^2 is the multiple correlation coefficient for the full model containing x_1, x_2, x_3; R_{reduced}^2 is for the reduced model before the addition of x_2 (that is, for the model containing x_1, x_3); k is the number of independent variables in the full model, and k' is the number of independent variables in the reduced model.

b. Use the same approach to test the significance of the addition of x_1 to a model that already contains x_3.

14.5. LOGARITHMIC TRANSFORMATIONS

There is a tendency among those who use linear regression techniques to drop the term "linear" when they speak and write about the relationship between variables x and y. Also, most researchers wisely seek the simplest solution first and test for a linear association before looking for a more complex relationship between the variables. Thus there is the danger of implying that all relationships are linear or that least-squares techniques are not appropriate for nonlinear relationships.

The problem in testing for more complex relationships is knowing what sort of relationship we should test. If the relationship is not linear, there are an infinite number of other possible relationships in which y is a function of x. In this section and the next one, we examine some functions of x that are curves rather than straight lines. We assume as before that there will be deviations from the trend line, that these deviations are normally distributed, and that the deviations have the same variance for all x values.

We look at two techniques for nonlinear functions: logarithmic transformations and polynomial regression. Log transformations are discussed in this section and polynomial regression in Section 14.6.

If there is a log-linearizable relationship between x and y, then we can obtain a straight line by transforming x to logs, y to logs, or both x and y to logs. Each of these procedures rectifies (straightens out) a different sort of relationship. The three types of relationship are shown in Figure 14.2 along with the type of logarithmic transformation to be used.

The type of logarithmic transformation to use may be determined in several ways. The nature of the two variables may indicate it, such as the exponential growth rate of single-cell organisms, or investment strategy when earnings are reinvested. Sometimes there may be an absolute upper or lower bound to the y variable, and this asymptotic value is approached experimentally. Frequently, the research literature in the area reveals that earlier experimenters have successfully used a logarithmic transformation, and one can anticipate that such a procedure will serve again. Finally, the experimenter may choose to plot the data points on semilog graph paper or on log-log graph paper to see whether a certain transformation appears to work. It is worth remembering, however, that the experimental α level is

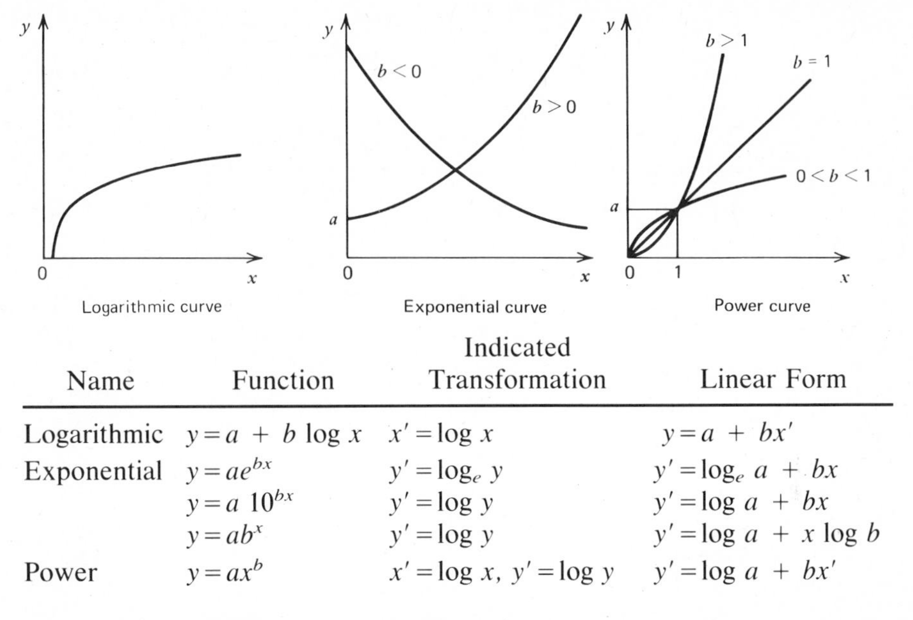

Name	Function	Indicated Transformation	Linear Form
Logarithmic	$y = a + b \log x$	$x' = \log x$	$y = a + bx'$
Exponential	$y = ae^{bx}$	$y' = \log_e y$	$y' = \log_e a + bx$
	$y = a\, 10^{bx}$	$y' = \log y$	$y' = \log a + bx$
	$y = ab^x$	$y' = \log y$	$y' = \log a + x \log b$
Power	$y = ax^b$	$x' = \log x,\ y' = \log y$	$y' = \log a + bx'$

FIGURE 14.2. Log-linearizable functions ($a > 0$, $x > 0$).

affected when one uses a "try it and see how it works" approach to data analysis. If one has a truly independent set of x and y variables, it may still be possible to find a seemingly significant relationship if enough different transformations are tried and the best fit is chosen for statistical analysis.

Example 14.4. Log *x* Transformation

Research workers in nuclear medicine have been interested in establishing cytogenetic dose-response relationships for various levels of radioactivity. Early work depended on evaluating cytogenetic lesions in tissue cultures of lymphocytes from individuals accidentally exposed to nuclear radiation and from those undergoing radiation chemotherapy. Now, procedures are available that make it possible to establish dose-response curves for human lymphocytes that are exposed *in vitro* (outside the body). Blood can be drawn from healthy individuals and the white cells collected, exposed to the appropriate dose, and placed in tissue-culture solution. Cell division is arrested at a stage when the chromosomes are clearly distinguishable and can be examined for radiation damage.

In the biological sciences associated with medicine, the logarithmic transformation of dosage is so common that consulting statisticians almost anticipate using it. Thus, when data are obtained, the statistician has it plotted on graph paper that has horizontal rulings on an arithmetic scale (to plot the y variable) and vertical ruling on a logarithmic scale (for the x variable). If his suspicions about log dose response are confirmed, he will proceed with the sort of analysis demonstrated below. (Specific activity, dosage, is measured in nano-Curies per milliliter, nCi/ml.)

Specific Activity	Log of Activity x	Dicentric Chromosomes y
40	1.6021	4
40	1.6021	5
80	1.9031	9
80	1.9031	6
160	2.2041	14
160	2.2041	19
320	2.5051	35
320	2.5051	32
Total	16.4288	124

$$\begin{aligned} \sum xy &= 283.5424 \\ (\sum x)(\sum y)/n &= 254.6464 \\ S_{xy} &= 28.8960 \end{aligned} \qquad \begin{aligned} \sum y^2 &= 2964 \\ (\sum y)^2/n &= 1922 \\ S_{yy} &= 1042 \end{aligned}$$

$$\begin{aligned} \sum x^2 &= 34.6442 \\ (\sum x)^2/n &= 33.7382 \\ S_{xx} &= 0.9060 \end{aligned} \qquad b = \frac{28.896}{0.906} = 31.8940$$

The variance from the trend line is obtained in the same manner as it was for simple regression

$$s^2 = \frac{\sum(y - \hat{y})^2}{n - 2} = \frac{S_{yy} - S_{xy}{}^2/S_{xx}}{n - 2}$$

$$= \frac{1042 - (28.896)^2/0.906}{6}$$

$$= 20.0650$$

and the test of significance for H_0: $\beta = 0$ against H_a: $\beta > 0$ is

$$t = \frac{b - 0}{\sqrt{s^2/S_{xx}}} = \frac{31.894}{\sqrt{20.065/0.906}}$$

$$= \frac{31.894}{4.706}$$

$$= 6.777$$

When compared with $t_{05,6} = 1.943$, the trend is found to be significant. The coefficient of determination is

$$r^2 = \frac{S_{xy}^2/S_{xx}}{S_{yy}} = \frac{921.61}{1042} = 0.884$$

which is a relatively large value, indicating a reasonably good fit that could be useful in predicting the chromosomal transmutations that result from specific levels of radioactivity. Additional studies would be necessary to determine the association between *in vivo* (within the body) chromosomal changes and those obtained by this procedure. However, the experimenter should feel encouraged by this experiment, for it indicates a useful technique in the study of genetic damage caused by exposure to radioactive substances.

Similar techniques can be used for exponential relationships. An example in which y is transformed to log y follows.

Example 14.5. Log y Transformation

The use of insecticides is a benefit but also a source of concern to the fruit industry. Insecticides protect the fruit from insect damage, but they are also toxic compounds that can be ingested by human beings. There are federally set tolerances on the amount of insecticides that fresh fruit and fruit pulp can contain, and fruit is carefully washed to meet those tolerances. Consequently, fruit processors are eager to gain as much information as they can about the deposition of insecticides and how they can be removed.

Insecticides are applied topically by spraying the fruit trees, so if the skin of the fruit has not been broken, all of the insecticide lies on the surface. Consequently, the larger the fruit, the more insecticide is deposited on it. To study the relationship between the size of peaches and the amount of insecticide sprayed on them, a horticulturist sprayed an orchard according to USDA recommendations and after the fruit was harvested, took a random sample of 10 peaches and measured their diameter x. He then washed each peach with a constant volume of detergent solution and made a chemical determination of the amount of insecticide i in the solution after cleaning. Because he expected the amount of insecticide i to be an exponential function of diameter, $i = a10^{bx}$, he transformed the measurements on the i variable to common logarithms.

Peach	Diameter x	Insecticide i	Log $i = y$
1	6.0 cm	0.5 ppm	−0.3010†
2	7.0	6.4	0.8062

† log(0.5) = log(5) − log(10)

Peach	Diameter x	Insecticide i	Log $i = y$
3	6.6	1.0	0.0000
4	5.8	0.2	−0.6990
5	6.8	5.5	0.7407
6	7.4	14.2	1.1523
7	7.2	8.2	0.9138
8	5.4	0.1	−1.0000
9	5.6	0.3	−0.5229
10	6.2	0.6	−0.2218
Total	64.0		0.8680

$$\sum x^2 = 414.0 \qquad \sum xy = 10.2209 \qquad \sum y^2 = 5.2628$$
$$(\sum x)^2/n = 409.6 \qquad (\sum x)(\sum y)/n = 5.5552 \qquad (\sum y)^2/n = 0.0753$$
$$S_{xx} = 4.4 \qquad S_{xy} = 4.6657 \qquad S_{yy} = 5.1875$$

To show the consistency between simple regression and multiple regression techniques, we can state the single equation analog to the simultaneous equations that must be solved in multiple regression

$$S_{xx}b = S_{xy}$$

and we can set up the augmented matrix

$$[4.4 \;\vdots\; 4.6657 \;\vdots\; 1]$$

and invert it by dividing by $S_{xx} = 4.4$ to obtain

$$[1 \;\vdots\; 1.0604 \;\vdots\; 0.2273]$$

We now have the values that are familiar from both simple and multiple regression

$$b = S_{xy}/S_{xx} = 1.0604$$

$$p_{11} = 1/S_{xx} = 0.2273$$

$$\sum(y - \hat{y})^2 = S_{yy} - bS_{xy} = S_{yy} - S_{xy}{}^2/S_{xx}$$

$$= 5.1875 - (4.6657)^2/4.4 = 0.24001$$

$$s^2 = \sum(y - \hat{y})^2/(n - k - 1) = 0.24001/8 = 0.03$$

$$s^2p_{11} = s^2/S_{xx} = 0.03(0.2273) = 0.0068$$

The test of significance for the relationship H_0: $\beta = 0$ against H_a: $\beta > 0$ is

$$t = \frac{1.0604}{\sqrt{.0068}} = \frac{1.0604}{0.0825} = 12.85, \qquad t_{.05,8} = 1.860$$

The coefficient of determination is

$$r^2 = \frac{bs_{xy}}{S_{yy}} = \frac{4.9475}{5.1875} = 0.9537$$

This indicates that the diameter of a peach in this orchard can be used as a very reliable indicator of the amount of insecticide that has been deposited on its surface. This may have some bearing on the thoroughness with which different sized peaches should be washed prior to marketing.

A similar technique can be used for exponential functions of the form $y = ae^{bx}$. In this case, $\log_e y$ is used for the transformation. If desired, common logarithms base 10 may be used and then converted to natural logarithms by the relationship

$$\log_e y = 2.303 \log y$$

If the function is of the form $y = ax^b$, then it can be linearized by transforming the variables to $\log y$ and $\log x$.

Consulting statisticians are frequently asked by economists to assist in the analysis of data that involve the regression of $\log y$ on $\log x$. The economists refer to the equations that are obtained as Cobb-Douglas functions. In other fields of research, there are also associations of the form

$$y = ax^b$$

but in economics they have been used with sufficient frequency to have gained a special designation. An example of their use would be a situation in which y is a measure of production in a certain industry and x is a measure of labor. Thus an economist could take a random sample of, say, bottling plants, gain access to their records, and find the regression of log(cases of soda) on log(man-hours). With this procedure, it is not uncommon to see multiple regression techniques employed as well. Thus the function becomes

$$y = ax_1^{b_1}x_2^{b_2}$$

Such a study might involve log(production) as a function of log(labor) and log(capital invested).

Having already demonstrated $\log x$ regression and $\log y$ regression, it seems unnecessary to give a numerical example of this procedure as well. However, it might be worthwhile to review the assumptions that are made in regression analysis. Irrespective of the units on the x and y axes, for the diagram, it is assumed that:

1. The relationship is linear for the units of x and y used.
2. y has a normal distribution.
3. y has the same variance throughout the range of x in the study.

Thus, if y is measured on the log scale, it implies that the log of the original units of measurement—log(cases of soda) in the Cobb-Douglas example—has a normal distribution with the same variance from the trend line irrespective of the number of workers involved. If the researcher is uncertain whether these assumptions should be made, then preliminary data should be obtained and used to investigate their distribution under the transformation. The arithmetic can be performed and numerical values obtained whether or not the assumptions hold true, but probability statements and inference are meaningless if the assumptions are not valid.

EXERCISES

14.5.1. Dicentric chromosomes result from the fusion of parts of two shattered chromosomes to form a single large chromosome. When dicentric chromosomes are formed, there are other chromosome fragments which are not reassembled and are eventually lost from the karyotype (chromosome composition). In the example demonstrating curvilinear regression rectified by log x, the dicentric chromosomes are used as the y variable; suppose that in the same experiment, chromosome fragments are also counted and the following results obtained.

Specific Activity	x:	40	80	160	320
Fragments	y:	10, 12	14, 20	22, 34	42, 70

a. Complete the regression of y on log x, and test it for significance.

b. In studies of this sort, the variances sometimes increase proportionally. Is there cause for concern about that possibility in these data? What might the experimenters do to determine whether or not variances are homogeneous irrespective of dosage?

c. Compute r and r^2.

d. Compute the expected number of chromosome fragments for 100 nCi/ml specific activity and place a 95% CI on the estimate.

14.5.2. In the log y transformation example in this section concerning insecticide residue, estimate a in the function $i = a10^{bx}$.

14.5.3. In order to study the efficiency of microwave cooking in sterilizing meat, a food scientist takes a random sample of eight sausage links, and by means of a hypodermic needle she inoculates each with the same volume of a nutrient broth containing a heavy suspension of salmonella. She then cooks each link for a different length of time in a microwave oven set for a constant temperature. The contents of the sausages are then mixed with an agar solution and poured into petri dishes. The dishes are placed in an incubator. After 18 hours of incubation, the number of salmonella colonies per dish are counted. The results are:

Minutes Cooked in Microwave x:	0	2	4	6	8	10	12	14	16
Number of Salmonella Colonies y:	740	410	210	100	45	25	10	6	4

a. Graph the data. What type of function seems to model the relationship of y to x?
b. Make a $\log_e$ transformation on the y variable, graph the transformed data, and compute:
   i. The regression coefficient.
   ii. The correlation coefficient.
   iii. The coefficient of determination.
c. In testing a hypothesis about the slope of the regression line,
   i. Why would the food scientist use a one-sided alternative?
   ii. Would she reject the null hypothesis for $\alpha = 0.05$?
d. Based on the results of this study,
   i. What is the expected number of colonies to develop in sausage cooked 15 minutes?
   ii. Place a 95% CI for the value estimated above.
   iii. How long should sausage be cooked in the microwave oven in order to produce an expected salmonella survival of zero?

14.5.4. A learning model used in experimental psychology is $T_i = ab^i$, in which T_i is the time it takes to perform a task on the ith occasion. Since $\log T_i = \log a + i \log b$, this relationship is log-linearizable. An experiment is performed which is believed to follow this model:

i:	1	2	3	4	5	6	7
T_i (min):	27	17	11	7	5	3	2

Estimate a and b.

14.6. POLYNOMIAL REGRESSION

Multiple regression procedures can be used to analyze for polynomial regression. A number of geometrical curves involve selected powers of x. For example, the quadratic curve (parabola) can be written

$$y = a + b_1x + b_2x^2$$

and the cubic curve can be written

$$y = a + b_1x + b_2x^2 + b_3x^3$$

In general, there are as many maximum or minimum points (extrema) on the curve as one less than the highest power of x in the model (Figure 14.3). It is possible to discuss quartic, quintic, and even more complex curves, but most experimenters find it difficult to explain curves with more than two maximum or minimum points. Thus we discuss only the quadratic and cubic curves.

A quadratic curve is utilized by agronomists when they study the effect of fertilizer. Agronomists know that there is a diminishing return from the use of more than a certain amount of fertilizer. In soil that is deficient in nitrogen, yield of crop increases with additional applications of nitrogen fertilizer, but it is possible to apply more nitrogen than the crop can use. In fact, too much fertilizer can damage and even kill the crop. Thus it is important to identify the range of safe application and to exclude applications beyond the maximum, the point of diminishing return.

To find the maximum point, agronomists set up experimental plots and use fertilizers in a series of applications. This series should extend through the supposed safe range and even into the range that is thought to be dangerous. The data can then be analyzed for a quadratic trend. A specific example follows.

Example 14.6. Quadratic Regression

The Jerusalem artichoke, *Helianthus tuberosus*, resembles the sunflower, but as its scientific name implies, it produces tubers. The polysaccharide

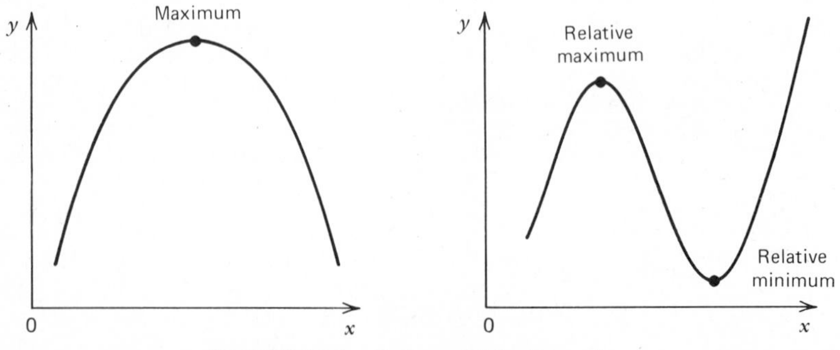

FIGURE 14.3. Polynomial funtions of x.

stored in the *Helianthus* tubers is inulin, which cannot be converted into sugars as can the starch stored in many tubers and roots. But it can be fermented to produce alcohol. The plant has the added advantage of being able to grow on relatively poor soil; consequently, it does not compete for the farmland used to grow beets, cane, corn, sorghum, and other sources of sugar and carbohydrates. Thus the Jerusalem artichoke has potential as a source of the polysaccharides needed to produce alcohol for use in industry, transportation, and beverages. However, the plant has been grown mainly as a flower, a curiosity, or a cover plant, and little is known about its culture as a cash crop. To gain some of this information, an agronomist plants Jerusalem artichoke on 12 hillside plots and randomly assigns three hillsides to each of four fertilizer regimes (0, 4, 8, and 12 hundredweight per acre). Yield, measured in hundredweight inulin per acre, is given below.

Fertilizer x	Yield y
0	35.0
0	38.7
0	33.1
4	42.6
4	40.5
4	43.8
8	41.0
8	42.1
8	36.9
12	36.1
12	40.8
12	37.4

$$\sum x = 72 \qquad \sum y = 468.0$$
$$\sum x^2 = 672 \qquad \sum xy = 2839.2$$
$$\sum x^3 = 6912 \qquad \sum x^2 y = 26{,}169.6$$
$$\sum x^4 = 75{,}264 \qquad \sum y^2 = 18{,}373.38$$
$$n = 12$$

$$\begin{aligned}
S_{11} &= \sum x^2 - (\sum x)^2/n = 240 \\
S_{12} &= \sum xx^2 - (\sum x)(\sum x^2)/n \\
&= \sum x^3 - (\sum x)(\sum x^2)/n = 2880 \\
S_{22} &= \sum x^2x^2 - (\sum x^2)^2/n \\
&= \sum x^4 - (\sum x^2)^2/n = 37{,}632 \\
S_{1y} &= \sum xy - (\sum x)(\sum y)/n = 31.2 \\
S_{2y} &= \sum x^2y - (\sum x^2)(\sum y)/n = -38.4 \\
S_{yy} &= \sum y^2 - (\sum y)^2/n = 121.38
\end{aligned}$$

In order to find b_1 and b_2 in

$$y = a + b_1x + b_2x^2$$

by least squares, it is necessary to solve the following simultaneous equations:

$$S_{11}b_1 + S_{12}b_2 = S_{1y}$$

$$S_{12}b_1 + S_{22}b_2 = S_{2y}$$

in which the sums of squares and cross-products are the same as for multiple regression, with $x = x_1$ and $x^2 = x_2$. Thus the equations to be solved are

$$240b_1 + 2{,}880b_2 = 31.2$$

$$2{,}880b_1 + 37{,}632b_2 = -38.4$$

and the augmented matrix to be inverted is

$$\left[\begin{array}{cc|c|cc} 240 & 2{,}880 & 31.2 & 1 & 0 \\ 2{,}880 & 37{,}632 & -38.4 & 0 & 1 \end{array}\right]$$

Row operations transform the matrix to

$$\left[\begin{array}{cc|c|cc} 1 & 0 & 1.74250000 & 0.0510417 & -0.0039062 \\ 0 & 1 & -0.1343750 & -0.0039062 & 0.0003255 \end{array}\right]$$

The intercept is obtained from $a = \bar{y} - b_1\bar{x} - b_2\bar{x}^2$ or $a = 39.0 - 1.7425(6) + 0.1344(56) = 36.0714$, and yield can be expressed as a function of fertilizer in the quadratic equation

$$\hat{y} = 36.0714 + 1.7425x - 0.1344x^2$$

Before any inference is drawn from this equation, the relationship must be tested for significance

$$\begin{aligned} \text{SSR} &= b_1S_{1y} + b_2S_{2y} \\ &= 1.7425(31.2) - 0.1344(-38.4) \\ &= 59.53 \end{aligned}$$

and

$$\begin{aligned} \sum(y - \hat{y})^2 &= S_{yy} - \text{SSR} \\ & 121.38 - 59.53 \\ &= 61.85 \end{aligned}$$

and the test of significance is

Source	df	SS	MS	F	$F_{0.05;2,9}$
Regression	2	59.53	29.765	4.331	4.256
Deviations	9	61.85	6.872		

The sum of squares due to regression can be separated into two parts by a stepwise regression procedure similar to that demonstrated previously.

Model	df	SS	MS	F
Full model (x and x^2)	2	$\sum b_iS_{iy} = 59.53$		
SS(x^2 after x)	1	$b_2^2/p_{22} = 55.474$	55.474	$55.474/6.872 = 8.072$

Model	df	SS	MS	F
SS(x alone)	1	$S_{1y}^2/S_{11} =$ 4.056	4.056	4.056/[(121.38 − 4.056)/10] = 0.346
Deviations from full model	9	61.85	6.872	

$$F_{0.05,1,9} = 5.117 \quad \text{and} \quad F_{0.05,1,10} = 4.965$$

The x^2 term is a significant addition to a model already containing x's; furthermore, the simple linear regression of y on x is nonsignificant. These results are not surprising, for the plant scientist had intended to apply fertilizer rates that would extend beyond the point of diminishing return. The response curve can be better described by a parabola than a straight line, hence the x^2 term is important to the model.

The maximum, or point of diminishing return, can be found by setting the first derivative of y with respect to x equal to zero. Thus the maximum y is at

$$x_m = \frac{-b_1}{2b_2} = \frac{-1.7425}{2(-0.1344)} = 6.48$$

as illustrated in Figure 14.4.

We have seen how polynomial regression can be used to fit a parabola, or quadratic curve. A cubic curve can be fitted by using x, x^2, and x^3 and $k = 3$ multiple regression procedures. A third extremum in the regression line can be obtained by using x, x^2, x^3, and x^4 and fitting a quartic curve.

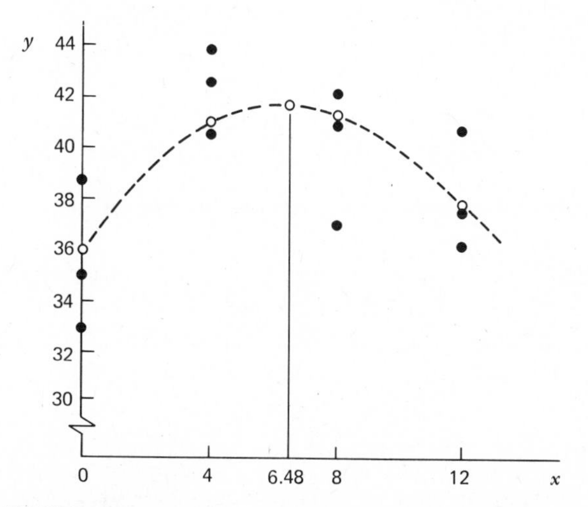

FIGURE 14.4. The maximum of the model in Example 14.6.

Polynomial regression is an extremely useful technique, but as with the other statistical techniques we have discussed, there are also limitations, cautions, and assumptions to be considered before drawing inference from these procedures. Here are some of the things the research worker should consider before using polynomial regression:

1. Not all curves with a single extremum are parabolas, and similarly polynomial curves may not provide the best fit for more complex curves. The polynomial curves have symmetrical features which make them unsuitable for fitting data that follow a nonsymmetrical trend. It is always useful to gather preliminary data, plot it, and then discuss with a statistician or mathematician what function may provide the best fit of y.
2. The number of different values of x is more important than the number of data points in polynomial regression. In the example where inulin yield was fitted to fertilizer applications, there were 12 data points but only $m = 4$ different values of x. The best possible fit (the maximum R^2) is obtained when $k = m - 1$, so it is a waste of time and effort to try to fit a very complex polynomial curve to data for which there are only a few different x values.
3. In polynomial regression, $x_k = x^k$, or as we saw in the Jerusalem artichoke example, $x_2 = x^2$. Because of this, if the x's are greater than 1, S_{22} will be larger than S_{11}, and if we use $x_3 = x^3$ and $x_4 = x^4$, then S_{33} and S_{44} will be still larger. A great disparity in the size of the S_{ii} makes it difficult to invert the sum of squares and cross-products matrix accurately.
4. As always, it is necessary to make the assumption that the deviations from the trend line are normally distributed, with the same variance all along the segment of the line for which inference will be made.

A technique called *orthogonal polynomials* further addresses some of the concerns given here and shows how good experimental design can permit easy tests of significance for higher order polynomial regression. We conclude this chapter with a discussion of orthogonal polynomials.

It might first be useful to review Section 10.4 on orthogonal contrasts, since very similar techniques are demonstrated here. If the x's are equally spaced, and there are a constant number of observations n at each x, then one can use tabulated orthogonal polynomials to determine which kind of polynomial curve best fits the data. This is usually done in conjunction with an analysis of variance in which each value of x is considered an experimental group.

The procedure can be demonstrated with the data obtained from the Jerusalem artichoke experiment, for the x's are equally spaced, that is, there is a four-hundredweight interval between adjacent levels of fertilizer, and there are $n = 3$ yields obtained for each level of fertilizer. The data can be grouped for an analysis of variance as follows:

	0 cwt	4 cwt	8 cwt	12 cwt
	35.0	42.6	41.0	36.1
	38.7	40.5	42.1	40.8
	33.1	43.8	36.9	37.4
$\sum y_i = T_i$	106.8	126.9	120.0	114.3

T = 18,373.38 (Uncorrected total sum of squares)

A = 18,324.78 (Uncorrected group sum of squares)

CF = 18,252.000

Source	df	SS	MS	F	$F_{0.05;3,8}$
Levels	3	72.78	24.26	4.02	4.066
Error	8	48.60	6.03		
Total		121.38			

The coefficients to be used for computing the contributions of x, x^2, and x^3 to the model can be obtained from Table A.18 (see Appendix) for $a = 4$ levels. These are used to compute the three sums of squares which partition the sum of squares for levels as follows.

Degree Polynomial	Level:	0	4	8	12			
	T_i:	106.8	126.9	120.0	114.3	$\sum a_i T_i$	$\sum a_i^2$	$(\sum a_i T_i)^2/n\sum a_i^2$
Linear	a_{Li}:	−3	−1	+1	+3	15.6	20	4.056
Quadratic	a_{Qi}:	+1	−1	−1	+1	−25.8	4	55.470
Cubic	a_{Ci}:	−1	+3	−3	+1	28.2	20	13.254
								72.780

The ANOVA table can be expanded to take into account these three orthogonal sums of squares, each with 1 degree of freedom.

Source	df	SS		MS	F	$F_{0.05;1,8}$
Levels	3	72.78				
Linear	1		4.056	4.056	0.673	5.318
Quadratic	1		55.470	55.470	9.199	5.318
Cubic	1		13.254	13.254	2.198	5.318
Error	8	48.60		6.03		

When we compare the three sums of squares computed here with the calculations performed in the Jerusalem artichoke yield analysis, we see the relationships:

Orthogonal Coefficients	Multiple Regression	SS
Linear	SS(x alone)	4.056
Quadratic	SS(x^2 after x)	55.470
Cubic	Not computed	13.254

The x^2 term is a significant addition to the model, and the quadratic curve provides the best fit. The sum of squares for cubic obtained from the polynomial coefficients yields the additional information that had we included x^3 as a third independent variable in the multiple regression equation, we would have found that SS(x^3 after x and x^2) is 13.254 and nonsignificant. Because x^3 was not considered in the earlier multiple regression analysis, there this sum of squares was combined with the within sum of squares to give

$$\text{SS}_e = 13.254 + 48.60 = 61.854$$

The statistical procedures discussed here are targeted for those who are or will be research workers. For the person who is interested in research, it is never very satisfying to be dismissed with an answer which smacks of "well, that's the way it is, so don't worry about it." The polynomial coefficients are given in Table A.18 in the Appendix for various levels of a, and they can be used as we have used them here, provided the a levels are equally spaced and n is the same at all levels. Still, the research worker is one who follows the Aristotelian approach of seeking truth through experience, and we consequently anticipate some curiosity about how the coefficients are computed. To satisfy those who want to experience the computations, the a_i values for linear, quadratic, and cubic coefficients will be derived for $a = 4$.

Let us examine again the data from the Jerusalem artichoke experiment.

	Level x	$(x-\bar{x})$	$(x-\bar{x})^2$	$a_{Li}=(x-\bar{x})/2$	a_{Li}^2	T_i	$(x-\bar{x})T_i$	$a_{Li}T_i$
	0	−6	36	−3	9	106.8	−640.8	−320.4
	4	−2	4	−1	1	126.9	−253.8	−126.9
	8	2	4	1	1	120.0	240.8	120.0
	12	6	36	3	9	114.3	685.8	342.9
Sum	24	0	80	0	20	468.0	31.8	15.6
Mean =	6							

The value $\sum(x - \bar{x})T_i$ is S_{1y}, as computed in the multiple regression analysis, and because there are $n = 3$ observations at each value of x, $S_{11} = n \sum(x - \bar{x})^2 = 3(80) = 240$. We have already seen that $S_{1y}{}^2/S_{11} = [\sum(x - \bar{x})T_i]^2/n \sum(x - \bar{x})^2 = (31.2)^2/3(80) = 4.056$, which is SS($x$ alone) in the multiple regression analysis. Because each of the $(x - \bar{x})$ is a multiple of 2, and because $(x - \bar{x})^2$ terms appear in both numerator and denominator, the arithmetic can be simplified by dividing by 2 to yield the a_{Li} found in the table and used in our analysis.

Quadratic terms can be obtained by squaring the a_{Li} and, if necessary, adjusting the a_{Li}^2 terms by covariance techniques so that the quadratic coefficients are uncorrelated to the linear coefficients.

Level	a_{Li}	a_{Li}^2	$a_{Li}^2 - \text{mean}$	$(a_{Li}^2 - \text{mean})/4 = a_{Qi}$	a_{Qi}^2
0	−3	9	4	1	1
4	−1	1	−4	−1	1
8	1	1	−4	−1	1
12	3	9	4	1	1
	Mean = 5				

When we examine the correlation between a_{Li} and a_{Qi}, we find $\sum(a_{Li}a_{Qi}) = -3(1) - 1(-1) + 1(-1) + 3(1) = 0$; these two sets of coefficients are orthogonal and no adjustments are necessary.

The cubic terms are somewhat more difficult to obtain, but we again start with a_{Li} and raise it to the power of interest, set the sum of zero, and test for correlation with the other coefficients:

Level	a_{Li}	a_{Li}^2	$a_{Li}^3 = a_{Ci'}$	$a_{Li}a_{Ci'}$	a_{Qi}	$a_{Qi}a_{Ci'}$
0	−3	9	−27	81	1	−27
4	−1	1	−1	1	−1	1
8	1	1	1	1	−1	−1
12	3	9	27	81	1	27
		20	0	164		0

The a_{Qi} and $a_{Ci'}$ are uncorrelated and hence orthogonal, but the a_{Li} and $a_{Ci'}$ are highly correlated and hence the $a_{Ci'}$ must be adjusted by the regression techniques employed in covariance. The regression of a_{Ci} on a_{Li} is 164/20 = 8.2, and the adjusted values are obtained as follows.

a_{Li}	$a_{Ci'}$	$a_{Ci'} - 8.2a_{Li}$	$a_{Ci} = (a_{Ci} - 8.2a_{Li})/2.4$	$a_{Li}a_{Ci}$	$a_{Qi}a_{Ci}$
−3	−27	−27 − 8.2(−3) = −2.4	−1	3	−1
−1	−1	−1 − 8.2(−1) = 7.2	3	−3	−3
1	1	1 − 8.2(1) = −7.2	−3	−3	3
3	27	27 − 8.2(3) = 2.4	1	3	1
			0	0	0

From the fourth column, we see that the a_{Ci} sum to zero, and in columns 5 and 6 we see that they are uncorrelated with the linear and quadratic coefficients. Hence they meet the conditions for orthogonality.

EXERCISES

14.6.1. An experiment similar to that studying the yield of inulin in Jerusalem artichoke is performed with sugar beets. The yield is measured in cwt of sugar.

x:	0	4	8	12
y:	34.5, 37.9, 31.4	39.2, 39.8, 43.4	45.1, 40.3, 43.0	43.2, 38.8, 43.4

a. Find the numerical values of b_1 and b_2 and test them for significance.

b. Does a quadratic curve fit the data significantly ($\alpha = .05$) better than a straight line? On what computations do you base your answers?

c. Find the maximum response of yield as a function of fertilizer.

d. The x values are deliberately kept the same in this problem as they were in the numerical example. This is to provide a computational guide for those who chose to invert the sum of squares and cross-products matrix. How can one use the results of the numerical example to perform the analysis without having to invert the matrix?

14.6.2. Low-temperature biologists are studying the effect of temperature on the germination of seed from cold-resistant trees. Seeds of Korean ash (*Fraxinus chinensis*) are collected and kept in dry storage for eight months, when 14 groups of 100 seeds each are established through random sampling. For seven groups, the pericarp (a plant ovary part which serves as a seed covering) is picked away from the seeds; for the other seven groups, it is left intact. Each group of seeds is placed in a separate flat containing vermiculite, and two flats, one of each kind of seed treatment, are assigned at random to each of seven temperature chambers. The numbers of seed germinating for each temperature and seed treatment are given below.

	Temperature (°C)						
Seed Treatment	5	10	15	20	25	30	35
With pericarp	4	5	9	31	58	75	77
Without pericarp	3	4	9	18	36	65	96

Computational hint: Because the settings of the temperature chambers are in multiples of 5, these observations can be easily coded by dividing by 5. This simplifies the arithmetic when powers of x are employed.

a. Test for curvilinear regression by using x, x^2, and x^3 in multiple regression:
   i. For germination of seed with pericarp.
   ii. For germination of seed with the pericarp removed.
b. The simple linear trend of germination on temperature (uncoded data) is 2.914 seed/degree. The regression coefficient for the coded data is $b_1 = 14.571$. What is the effect on the simple linear regression coefficient of dividing the x values by 5? What is the effect on the other coefficients if multiple regression is performed?
c. In order to determine how complex the model must be to explain Korean ash germination under different conditions, give the percentage of variability explained by each model below.
   i. Germination y as a simple linear function of temperature x: (1) For seed with pericarp. (2) For seed without pericarp.
   ii. Germination as a quadratic function of temperature: (1) For seed with pericarp. (2) For seed without.
   ii. Germination as a cubic function of temperature: (1) For seed with pericarp. (2) For seed without.
d. Is there evidence that the relationship to temperature is significantly different for the two seed treatments?
e. Based on the information gained here, could the biologists properly use the techniques covered in the section on covariance to adjust similar data for different temperatures in order to compare the two seed treatments at a common temperature?

14.6.3. The yield of sugar from beets has been studied, and there is interest in determining the response of yield to the amounts of fertilizer applied. The data are:

	Fertilizer (cwt)			
	0	4	8	12
	34.5	39.2	45.1	43.2
	37.9	39.8	40.3	38.8
	31.4	43.4	43.0	43.4
Total	103.8	122.4	128.4	125.4

a. Perform the ANOVA and test for differences among levels of fertilizer.

b. Test for linear, quadratic, and cubic trends.
c. Is there evidence that the range of applications of fertilizer encompasses the point of diminishing return?
d. These data were analyzed in the sugar beet experiment of Exercise 14.6.1, using x and x^2 as independent variables in multiple regression. Compare results from the two techniques.

14.6.4. A parasitologist conducts a study in which a vermifuge is incorporated into the diet of young swine infected with roundworms. The intent of the experiment is to determine the point at which the additive adversely affects the weight gain of the animals. The experiment is designed for five levels of vermifuge (0, 25, 50, 75, and 100 ppm) and ten pigs per level; however, a miscalculation in computing the level of 50 ppm additive renders those data unsuitable for analysis. The remaining results are:

Level of Additive (ppm):	0	25	75	100
Total gain (T_i):	149	306	251	64

Source	df	SS	MS
Levels	3	3,480.3	1160.10
Error	36	4,041.0	112.25

a. Show that the orthogonal polynomials for this set of levels are

Linear:	-2	-1	$+1$	$+2$
Quadratic:	$+1$	-1	-1	$+1$
Cubic:	-1	$+2$	-2	$+1$

b. Show how these coefficients meet the conditions of orthogonality.
c. Break out the three individual degrees of freedom and draw conclusions about the relationship between gain and additive.

REVIEW EXERCISES

Decide whether each of the following is true or false. If a statement is false, explain why.

14.1. Multiple regression techniques require that all x variables have the same variance.

14.2. If surface area of an animal seems to be a function of its weight raised to a power, a logarithmic transformation on the area is indicated before a regression analysis.

14.3. All F tests of coefficients in a multiple regression analysis have one and $n - k$ degrees of freedom associated with them.
14.4. The experimenter may be as interested in determining which variables are nonsignificant as in determining those which are related to the dependent variable.
14.5. The test of significance of the multiple regression coefficient R is against a one-sided alternative.
14.6. If the test of the simple correlation between y and x_1 is significant, then H_0: $\rho_{y1\cdot2} = 0$ is rejected.
14.7. In a regression of y on x_1 and x_2, it is possible to use the least-squares plane for prediction if it is perpendicular to the y axis.
14.8. The total variability in y can be split into two nonoverlapping parts: the portion explained by regression and the unexplained portion.
14.9. The multiple correlation coefficient R is never negative.
14.10. Multiple regression and multiple correlation analysis require the same assumptions.
14.11. It is possible for R^2 to equal 0.90 and the regression equation may be the wrong model for the data.
14.12. The partial regression coefficients are unit-free.
14.13. The partial regression coefficients are always in the same units.
14.14. Standardized partial regression coefficients are unit-free.
14.15. Partial correlation coefficients can never be negative.
14.16. Backward elimination and stepwise regression always lead to the same model.
14.17. The log transformations are used to simplify the computations involved in regression.
14.18. Polynomial regression is multiple regression with $x_i = x^i$.
14.19. The model $\hat{y} = \alpha + \beta_1 x + \beta_2 x^2$ will always fit a data set better than $\hat{y} = \alpha + \beta_1 x$ because it contains a term with a higher power of x.
14.20. If quadratic regression leads to the equation $\hat{y} = 3 + 50x - 5x^2$, then the greatest value of $\hat{y}$ is at $x = -5$.

SELECTED READINGS

Anderson, R. L., D. M. Allen, and F. B. Cady (1972). Selection of predictor variables in linear multiple regression, in *Statistical Papers in Honor of George W. Snedecor*, edited by T. A. Bancroft, Iowa State University Press, Ames.

Andrews, D. F. (1971). A note on the selection of data transformations, *Biometrika*, **58**, 249–254.

Bradley, R. A., and S. S. Srivastava (1979). Correlation in polynomial regression, *The American Statistician*, **33**, 11–14.

Brogden, H. E. (1946). On the interpretation of the correlation coefficient as a measure of predictive efficiency, *The Journal of Educational Psychology*, **37**, 65–76.

Chatterjee, S., and B. Price (1977). *Regression Analysis by Example*, Wiley, New York.

Cochran, W. G. (1970). Some effects of errors of measurement on multiple correlation, *Journal of the American Statistical Association*, **65**, 22–34.

Cramer, E. M. (1972). Significance tests and tests of models in multiple regression, *The American Statistician*, **26** (Oct.), 26–30.

Crocker, D. C. (1972). Some interpretations of the multiple correlation coefficient, *The American Statistician*, **26** (April), 31–33.

Draper, N., and H. Smith (1981). *Applied Regression Analysis*, 2nd ed., Wiley, New York.

Ellenberg, J. H. (1976). Testing for a single outlier from a general linear regression, *Biometrics*, **32**, 637–645.

Furnival, G. M. (1971). All possible regressions with less computation, *Technometrics*, **13**, 403–408.

Garside, M. J. (1965). The best subset in multiple regression analysis, *Applied Statistics*, 14, 196–200.

Gorman, J. W., and R. J. Toman (1966). Selection of variables for fitting equations to data, *Technometrics*, **8**, 27–51.

Heren, D. A. (1968). A note on log-linear regression, *Journal of the American Statistical Association*, **63**, 1034–1038.

Hill, R. C., G. G. Judge, and T. B. Fomby (1978). On testing the adequacy of a regression model, *Technometrics*, **20**, 491–494.

——— (1980). Is the regression equation adequate?—A reply, *Technometrics*, **22**, 127–128.

Hocking, R. R., and R. N. Leslie (1967). Selection of the best sub-set in regression analysis, *Technometrics*, **9**, 531–540.

Kramer, K. H. (1963). Tables for constructing confidence limits on the multiple correlation coefficient, *Journal of the American Statistical Association*, **58**, 1082–1085.

Lindley, D. V. (1968). The choice of variables in multiple regression, *Journal of the Royal Statistical Society*, Series B, **30**, 31–66.

Mullet, G. M. (1972). A graphical illustration of simple (total) and partial regression, *The American Statistician*, **26** (Dec.), 25–27.

Robson, D. S. (1959). A simple method for constructing orthogonal polynomials when the independent variable is unequally spaced, *Biometrics*, **15**, 187–191.

Searle, S. R. (1971). *Linear Models*, Wiley, New York.

Suich, R., and G. C. Derringer (1977). Is the regression equation adequate?–One criterion, *Technometrics*, **19**, 213–216.

——— (1980). Is the regression equation adequate–A further note, *Technometrics*, **22**, 125–126.

Weisberg, S. (1980). *Applied Linear Regression*, Wiley, New York.

Weiss, N. S. (1970). A graphical representation of the relationships between multiple regression and multiple correlation, *The American Statistician*, **24** (April), 25–29.

Wilkie, D. (1965). Complete set of leading coefficients, $\lambda(r,n)$, for orthogonal polynomials up to $n = 26$, *Technometrics*, **7**, 644–648.

Appendix of Useful Tables

Table A.1. 2500 Random Digits. These computer-produced pseudo-random digits may be read in any direction: vertical, up or down; horizontal, left-to-right or right-to-left; or along any diagonal, up or down. Single digits or groups of any size may be read; the five-digit groupings are only for ease of reading and should be ignored when reading the table. Care should be taken not to use the same portion of the table repeatedly, especially for the same experiment. This can be accomplished by using a random start (see Section 2.1) or by starting at one corner of the table and striking out the digits as they are used so that each portion of the table is used only once.

TABLE A.1. 2500 Random Digits

	1–5	6–10	11–15	16–20	21–25	26–30	31–35	36–40	41–45	46–50
1	38742	24201	25580	18631	30563	11548	08022	62261	74563	54597
2	01448	28091	45285	81470	09829	49377	88809	59780	46891	29447
3	34768	23715	37836	17206	26527	21554	62118	78918	30845	78748
4	89533	67552	74970	68065	50599	85529	20588	59726	84051	44388
5	74163	13487	64602	07271	03530	88954	66174	68319	25323	05476
6	92837	06594	01664	43011	27981	81256	75467	28245	29149	70357
7	69008	55983	22496	55337	74159	11283	13316	27479	63079	34060
8	92404	00156	38141	06269	51599	11371	24120	88150	99649	54740
9	45369	68854	67952	06245	32056	67900	84670	50098	29179	47904
10	16929	17418	70611	53752	39997	53621	67393	24891	53738	77251
11	95400	57951	64492	52389	86037	52586	42206	74681	82599	24606
12	36981	75140	26771	67681	54042	26121	70479	50295	43593	08220
13	37705	05124	60924	24374	99850	12414	13982	83219	26396	93876
14	67830	54660	89150	92919	90913	49560	49845	98239	78807	87479
15	32789	25115	44030	86301	61900	17173	34870	37043	40625	17954
16	60127	17491	59011	37625	03435	77178	08520	49910	34898	34345
17	17115	42174	81592	04300	68875	30353	48630	86132	55173	05788
18	27760	36661	85617	06242	09725	10642	44142	29625	49415	98360
19	04494	95805	16053	37126	54750	12617	09310	94021	38471	57427
20	34753	89545	33847	78318	41551	18705	64107	18200	56834	74584
21	63319	12471	56242	06344	94606	89207	26550	93261	17931	79259
22	98802	54600	92170	51425	74130	10301	08763	56046	00093	03793
23	82661	67501	01368	91079	54810	68160	11860	84288	27053	00917
24	99251	10088	48345	72786	81066	54353	17546	31595	77246	40514
25	72756	52088	29291	46169	14636	26380	35201	07490	28845	02341
26	96723	05193	38941	33288	13923	46860	12385	94973	43259	85010
27	96169	16158	24345	78561	46611	66869	17678	38209	24023	56259
28	96678	41518	88402	17882	79991	00083	29337	39994	06328	06476
29	97329	58496	55229	90839	93840	67032	77411	57137	06172	11036
30	38143	94319	58015	71878	42332	28120	80481	41745	68085	88776
31	83510	94405	93811	02145	74541	29582	24535	21485	54519	93320
32	98898	39140	50371	20646	07782	63276	66375	88305	77405	74749
33	04406	76609	46544	55985	72507	98678	48840	16601	44598	50487
34	55997	34203	29784	12914	37942	86041	48431	11784	28492	28049
35	95911	19810	65733	05412	18498	79393	37322	75911	92047	61599
36	67151	13303	12466	08918	27140	22886	61210	67131	52278	95829
37	59368	23548	60681	09171	18170	62627	48209	62135	44727	12937
38	75670	78997	76059	83474	15744	71892	52740	22930	92624	93036
39	94444	45866	42304	85506	26762	24841	47226	34746	90302	70785
40	73516	82157	24805	75928	02150	84557	12930	63123	11922	76960
41	89059	45446	56541	62549	21737	78963	30917	37046	81184	83397
42	94958	71785	47469	29362	91492	80902	80586	66162	74551	87221
43	21739	80710	61346	04257	09821	17188	80855	76589	36971	41982
44	93859	78783	46343	03715	12473	48553	02762	45114	75502	42382
45	14263	52552	17964	20078	82454	35167	35631	81815	18879	93676
46	22894	01894	47934	54594	43739	51301	22511	39456	51031	58121
47	29316	85620	09294	67074	77403	82789	22212	52358	69310	57604
48	31889	40095	98007	15605	93206	86857	29784	63937	83545	50407
49	60096	11744	74086	65948	37934	35941	25731	30787	68848	14320
50	42450	70020	43245	05233	21149	85898	73527	55648	65388	55211

TABLE A.2. Factorials

$n!$	n
1	1
2	2
6	3
24	4
120	5
720	6
5,040	7
40,320	8
362,880	9
3,628,800	10
39,916,800	11
479,001,600	12
6,227,020,800	13
87,178,291,200	14
1,307,674,368,000	15
20,922,789,888,000	16
355,687,428,096,000	17
6,402,373,705,728,000	18
121,645,100,408,832,000	19
2,432,902,008,176,640,000	20
51,090,942,171,709,440,000	21
1,124,000,727,777,607,680,000	22
25,852,016,738,884,976,640,000	23
620,448,401,733,239,439,360,000	24
15,511,210,043,330,985,984,000,000	25

$n! = 1 \cdot 2 \cdot 3 \cdot \ldots \cdot n$. $0! = 1$ by definition.

TABLE A.3. Binomial Coefficients $\binom{n}{y}$

n \ y	0	1	2	3	4	5	6	7	8	9	10	11	12
1	1	1											
2	1	2	1										
3	1	3	3	1									
4	1	4	6	4	1								
5	1	5	10	10	5	1							
6	1	6	15	20	15	6	1						
7	1	7	21	35	35	21	7	1					
8	1	8	28	56	70	56	28	8	1				
9	1	9	36	84	126	126	84	36	9	1			
10	1	10	45	120	210	252	210	120	45	10	1		
11	1	11	55	165	330	462	462	330	165	55	11	1	
12	1	12	66	220	495	792	924	792	495	220	66	12	1
13	1	13	78	286	715	1287	1716	1716	1287	715	286	78	13
14	1	14	91	364	1001	2002	3003	3432	3003	2002	1001	364	91
15	1	15	105	455	1365	3003	5005	6435	6435	5005	3003	1365	455
16	1	16	120	560	1820	4368	8008	11440	12870	11440	8008	4368	1820
17	1	17	136	680	2380	6188	12376	19448	24310	24310	19448	12376	6188
18	1	18	153	816	3060	8568	18564	31824	43758	48620	43758	31824	18564
19	1	19	171	969	3876	11628	27132	50388	75582	92378	92378	75582	50388
20	1	20	190	1140	4845	15504	38760	77520	125970	167960	184756	167960	125970
21	1	21	210	1330	5985	20349	54264	116280	203490	293930	352716	352716	293930
22	1	22	231	1540	7315	26334	74613	170544	319770	497420	646646	705432	646646
23	1	23	253	1771	8855	33649	100947	245157	490314	817190	1144066	1352078	1352078
24	1	24	276	2024	10626	42504	134596	346104	735471	1307504	1961256	2496144	2704156
25	1	25	300	2300	12650	53130	177100	480700	1081575	2042975	3268760	4457400	5200300

$\binom{n}{y} = \frac{n!}{y!(n-y)!}$. Use $\binom{n}{y} = \binom{n}{n-y}$ for $y > 12$.

TABLE A.4a. Binomial Distributions, n = 20

y \ π	.05	.10	.15	.20	.25	.30	.35	.40	.45
0	.358	.122	.039	.012	.003	.001	.000	.000	.000
1	.377	.270	.137	.058	.021	.007	.002	.000	.000
2	.189	.285	.229	.137	.067	.028	.010	.003	.001
3	.060	.190	.243	.205	.134	.072	.032	.012	.004
4	.013	.090	.182	.218	.190	.130	.074	.035	.014
5	.002	.032	.103	.175	.202	.179	.127	.075	.036
6	.000	.009	.045	.109	.169	.192	.171	.124	.075
7	.000	.002	.016	.055	.112	.164	.184	.166	.122
8	.000	.000	.005	.022	.061	.114	.161	.180	.162
9	.000	.000	.001	.007	.027	.065	.116	.160	.177
10	.000	.000	.000	.002	.010	.031	.069	.117	.159
11	.000	.000	.000	.000	.003	.012	.034	.071	.119
12	.000	.000	.000	.000	.001	.004	.014	.035	.073
13	.000	.000	.000	.000	.000	.001	.004	.015	.037
14	.000	.000	.000	.000	.000	.000	.001	.005	.015
15	.000	.000	.000	.000	.000	.000	.000	.001	.005
16	.000	.000	.000	.000	.000	.000	.000	.000	.001
17	.000	.000	.000	.000	.000	.000	.000	.000	.000
18	.000	.000	.000	.000	.000	.000	.000	.000	.000
19	.000	.000	.000	.000	.000	.000	.000	.000	.000
20	.000	.000	.000	.000	.000	.000	.000	.000	.000

Step boundaries give approximate 90% confidence intervals for π. (See Section 3.3.)

TABLE A.4a.—*Continued*

.50	.55	.60	.65	.70	.75	.80	.85	.90	.95 π \ y	
.000	.000	.000	.000	.000	.000	.000	.000	.000	.000	0
.000	.000	.000	.000	.000	.000	.000	.000	.000	.000	1
.000	.000	.000	.000	.000	.000	.000	.000	.000	.000	2
.001	.000	.000	.000	.000	.000	.000	.000	.000	.000	3
.005	.001	.000	.000	.000	.000	.000	.000	.000	.000	4
.015	.005	.001	.000	.000	.000	.000	.000	.000	.000	5
.037	.015	.005	.001	.000	.000	.000	.000	.000	.000	6
.074	.037	.015	.004	.001	.000	.000	.000	.000	.000	7
.120	.073	.035	.014	.004	.001	.000	.000	.000	.000	8
.160	.119	.071	.034	.012	.003	.000	.000	.000	.000	9
.176	.159	.117	.069	.031	.010	.002	.000	.000	.000	10
.160	.177	.160	.116	.065	.027	.007	.001	.000	.000	11
.120	.162	.180	.161	.114	.061	.022	.005	.000	.000	12
.074	.122	.166	.184	.164	.112	.055	.016	.002	.000	13
.037	.075	.124	.171	.192	.169	.109	.045	.009	.000	14
.015	.036	.075	.127	.179	.202	.175	.103	.032	.002	15
.005	.014	.035	.074	.130	.190	.218	.182	.090	.013	16
.001	.004	.012	.032	.072	.134	.205	.243	.190	.060	17
.000	.001	.003	.010	.028	.067	.137	.229	.285	.189	18
.000	.000	.000	.002	.007	.021	.058	.137	.270	.377	19
.000	.000	.000	.000	.001	.003	.012	.039	.122	.358	20

TABLE A.4b. Binomial Distributions, n = 25

y \ π	.05	.10	.15	.20	.25	.30	.35	.40	.45
0	.277	.072	.017	.004	.001	.000	.000	.000	.000
1	.365	.199	.076	.024	.006	.001	.000	.000	.000
2	.231	.266	.161	.071	.025	.007	.002	.000	.000
3	.093	.226	.217	.136	.064	.024	.008	.002	.000
4	.027	.138	.211	.187	.118	.057	.022	.007	.002
5	.006	.065	.156	.196	.165	.103	.051	.020	.006
6	.001	.024	.092	.163	.183	.147	.091	.044	.017
7	.000	.007	.044	.111	.165	.171	.133	.080	.038
8	.000	.002	.017	.062	.124	.165	.161	.120	.070
9	.000	.000	.006	.029	.078	.134	.163	.151	.108
10	.000	.000	.002	.012	.042	.092	.141	.161	.142
11	.000	.000	.000	.004	.019	.054	.103	.147	.158
12	.000	.000	.000	.001	.007	.027	.065	.114	.151
13	.000	.000	.000	.000	.002	.011	.035	.076	.124
14	.000	.000	.000	.000	.001	.004	.016	.043	.087
15	.000	.000	.000	.000	.000	.001	.006	.021	.052
16	.000	.000	.000	.000	.000	.000	.002	.009	.027
17	.000	.000	.000	.000	.000	.000	.001	.003	.012
18	.000	.000	.000	.000	.000	.000	.000	.001	.004
19	.000	.000	.000	.000	.000	.000	.000	.000	.001
20	.000	.000	.000	.000	.000	.000	.000	.000	.000
21	.000	.000	.000	.000	.000	.000	.000	.000	.000
22	.000	.000	.000	.000	.000	.000	.000	.000	.000
23	.000	.000	.000	.000	.000	.000	.000	.000	.000
24	.000	.000	.000	.000	.000	.000	.000	.000	.000
25	.000	.000	.000	.000	.000	.000	.000	.000	.000

Step boundaries give approximate 90% confidence intervals for π. (See Section 3.3.)

TABLE A.4b—*Continued*

.50	.55	.60	.65	.70	.75	.80	.85	.90	.95 π \ y	
.000	.000	.000	.000	.000	.000	.000	.000	.000	.000	0
.000	.000	.000	.000	.000	.000	.000	.000	.000	.000	1
.000	.000	.000	.000	.000	.000	.000	.000	.000	.000	2
.000	.000	.000	.000	.000	.000	.000	.000	.000	.000	3
.000	.000	.000	.000	.000	.000	.000	.000	.000	.000	4
.002	.000	.000	.000	.000	.000	.000	.000	.000	.000	5
.005	.001	.000	.000	.000	.000	.000	.000	.000	.000	6
.014	.004	.001	.000	.000	.000	.000	.000	.000	.000	7
.032	.012	.003	.001	.000	.000	.000	.000	.000	.000	8
.061	.027	.009	.002	.000	.000	.000	.000	.000	.000	9
.097	.052	.021	.006	.001	.000	.000	.000	.000	.000	10
.133	.087	.043	.016	.004	.001	.000	.000	.000	.000	11
.155	.124	.076	.035	.011	.002	.000	.000	.000	.000	12
.155	.151	.114	.065	.027	.007	.001	.000	.000	.000	13
.133	.158	.147	.103	.054	.019	.004	.000	.000	.000	14
.097	.142	.161	.141	.092	.042	.012	.002	.000	.000	15
.061	.108	.151	.163	.134	.078	.029	.006	.000	.000	16
.032	.070	.120	.161	.165	.124	.062	.017	.002	.000	17
.014	.038	.080	.133	.171	.165	.111	.044	.007	.000	18
.005	.017	.044	.091	.147	.183	.163	.092	.024	.001	19
.002	.006	.020	.051	.103	.165	.196	.156	.065	.006	20
.000	.002	.007	.022	.057	.118	.187	.211	.138	.027	21
.000	.000	.002	.008	.024	.064	.136	.217	.226	.093	22
.000	.000	.000	.002	.007	.025	.071	.161	.266	.231	23
.000	.000	.000	.000	.001	.006	.024	.076	.199	.365	24
.000	.000	.000	.000	.000	.001	.004	.017	.072	.277	25

Tables A.5a through A.5e Confidence Intervals on the Binomial parameter π. Each of the following tables gives central confidence intervals at the $\alpha = 0.10$, $\alpha = 0.05$, and $\alpha = 0.01$ levels. (L = lower confidence limit; U = upper confidence limit.) For sample sizes $n = 25$ and $n = 50$ (Tables A.5a and A.5b), if y cases of the outcome of interest occur in the sample, $CI_{1-\alpha}$: $L \leq \pi \leq U$ is found by referring to row y and reading L and U under the appropriate α level.

Example: If $\alpha = 0.10$, $n = 50$, and $y = 31$,

$$CI_{0.90}: 0.494 \leq \pi \leq 0.735$$

For $n = 100$, the procedure is the same except that if $y > 50$, row $100 - y$ must be used to find the confidence interval and $L = 1 - U$ (of row $100 - y$) and $U = 1 - L$ (of row $100 - y$).

Example: If $\alpha = 0.01$, $y = 75$, and $n = 100$, then $100 - y = 25$ and

$$CI_{0.99}: 1 - 0.337 \leq \pi \leq 1 - 0.148$$

and

$$CI_{0.99}: 0.663 \leq \pi \leq 0.852$$

For $n = 250$ and $n = 500$ (Tables A.5d and A.5e), the confidence interval is found using y/n.
Example: If $\alpha = 0.05$, $y = 100$, and $n = 250$, then $y/n = 100/250 = 0.40$ and

$$CI_{0.95}: 0.339 \leq \pi \leq 0.464$$

If $y/n > 0.50$, L is $1 - U$ (of row $1 - y/n$) and $U = 1 - L$ (of row $1 - y/n$).

Linear interpolation can be used with these tables for sample sizes intermediate to the ones given in the tables. Linear interpolation can also be used if y/n is intermediate to those values listed in Tables A.5d and A.5e.

The confidence intervals in these tables were derived with the use of the formulas given on page 960 of the *Handbook of Mathematical Functions With Formulas, Graphs, and Mathematical Tables*, edited by M. Abramowitz and I. A. Stegun, U.S. Department of Commerce, National Bureau of Standards, Applied Mathematics Series 55, 1964.

TABLE A.5a. Confidence Intervals on the Binomial Parameter π

sample size n = 25

	α = 0.10		α = 0.05		α = 0.01	
y	L	U	L	U	L	U
0	0.000	0.113	0.000	0.137	0.000	0.191
1	0.002	0.176	0.001	0.204	0.000	0.261
2	0.014	0.231	0.010	0.260	0.004	0.321
3	0.034	0.282	0.025	0.312	0.014	0.374
4	0.057	0.330	0.045	0.361	0.028	0.424
5	0.082	0.375	0.068	0.407	0.046	0.470
6	0.110	0.420	0.094	0.451	0.066	0.514
7	0.139	0.462	0.121	0.494	0.089	0.556
8	0.170	0.504	0.150	0.535	0.113	0.596
9	0.202	0.544	0.180	0.575	0.140	0.633
10	0.236	0.583	0.211	0.613	0.167	0.670
11	0.270	0.621	0.244	0.651	0.198	0.705
12	0.305	0.659	0.278	0.687	0.228	0.740
13	0.341	0.695	0.313	0.722	0.260	0.772
14	0.379	0.730	0.349	0.756	0.295	0.802
15	0.417	0.764	0.387	0.789	0.330	0.833
16	0.456	0.798	0.425	0.820	0.367	0.860
17	0.496	0.830	0.465	0.850	0.404	0.887
18	0.538	0.861	0.506	0.879	0.444	0.911
19	0.580	0.890	0.549	0.906	0.486	0.934
20	0.625	0.918	0.593	0.932	0.530	0.954
21	0.670	0.943	0.639	0.955	0.576	0.972
22	0.718	0.966	0.688	0.975	0.626	0.986
23	0.769	0.986	0.740	0.990	0.679	0.996
24	0.824	0.998	0.796	0.999	0.739	1.000
25	0.887	1.000	0.863	1.000	0.809	1.000

TABLE A.5b. Confidence Intervals on the Binomial Parameter π

sample size n = 50

	α = 0.10		α = 0.05		α = 0.01	
y	L	U	L	U	L	U
0	0.000	0.058	0.000	0.071	0.000	0.101
1	0.001	0.091	0.001	0.107	0.000	0.140
2	0.007	0.121	0.005	0.137	0.002	0.172
3	0.017	0.148	0.013	0.165	0.007	0.203
4	0.028	0.174	0.022	0.192	0.014	0.231
5	0.040	0.199	0.033	0.218	0.022	0.258
6	0.054	0.223	0.045	0.243	0.032	0.284
7	0.068	0.247	0.058	0.267	0.043	0.309
8	0.082	0.270	0.072	0.291	0.054	0.333
9	0.097	0.293	0.086	0.314	0.066	0.358
10	0.113	0.316	0.100	0.337	0.078	0.380
11	0.129	0.338	0.115	0.360	0.092	0.403
12	0.145	0.360	0.131	0.382	0.106	0.426
13	0.161	0.381	0.146	0.403	0.120	0.447
14	0.178	0.403	0.162	0.425	0.134	0.469
15	0.195	0.424	0.179	0.446	0.149	0.490
16	0.212	0.445	0.195	0.467	0.164	0.511
17	0.230	0.465	0.212	0.488	0.180	0.531
18	0.247	0.486	0.229	0.508	0.196	0.552
19	0.265	0.506	0.246	0.528	0.212	0.571
20	0.283	0.526	0.264	0.548	0.229	0.591
21	0.301	0.546	0.282	0.568	0.246	0.610
22	0.320	0.566	0.300	0.587	0.262	0.629
23	0.339	0.585	0.318	0.607	0.280	0.648
24	0.357	0.605	0.337	0.626	0.298	0.666
25	0.376	0.624	0.355	0.645	0.315	0.685
26	0.395	0.643	0.374	0.663	0.334	0.702
27	0.415	0.661	0.393	0.682	0.352	0.720
28	0.434	0.680	0.413	0.700	0.371	0.738
29	0.454	0.699	0.432	0.718	0.390	0.754
30	0.474	0.717	0.452	0.736	0.409	0.771
31	0.494	0.735	0.472	0.754	0.429	0.788
32	0.514	0.753	0.492	0.771	0.448	0.804
33	0.535	0.770	0.512	0.788	0.469	0.820
34	0.555	0.788	0.533	0.805	0.489	0.836
35	0.576	0.805	0.554	0.821	0.510	0.851
36	0.597	0.822	0.575	0.838	0.531	0.866
37	0.619	0.839	0.597	0.854	0.553	0.880
38	0.640	0.855	0.618	0.869	0.574	0.894
39	0.662	0.871	0.640	0.885	0.597	0.908
40	0.684	0.887	0.663	0.900	0.620	0.922
41	0.707	0.903	0.686	0.914	0.642	0.934
42	0.730	0.918	0.709	0.928	0.667	0.946
43	0.753	0.932	0.733	0.942	0.691	0.957
44	0.777	0.946	0.757	0.955	0.716	0.968
45	0.801	0.960	0.782	0.967	0.742	0.978
46	0.826	0.972	0.808	0.978	0.769	0.986
47	0.852	0.983	0.835	0.987	0.797	0.993
48	0.879	0.993	0.863	0.995	0.828	0.998
49	0.909	0.999	0.893	0.999	0.860	1.000
50	0.942	1.000	0.929	1.000	0.899	1.000

TABLE A.5c. Confidence Intervals on the Binomial Parameter π

sample size n = 100

	α = 0.10		α = 0.05		α = 0.01	
y	L	U	L	U	L	U
0	0.000	0.029	0.000	0.036	0.000	0.052
1	0.001	0.047	0.000	0.054	0.000	0.072
2	0.004	0.062	0.002	0.070	0.001	0.089
3	0.008	0.076	0.006	0.085	0.003	0.106
4	0.014	0.089	0.011	0.099	0.007	0.121
5	0.020	0.102	0.016	0.113	0.011	0.135
6	0.026	0.115	0.022	0.126	0.016	0.149
7	0.033	0.127	0.029	0.139	0.021	0.163
8	0.040	0.140	0.035	0.152	0.026	0.176
9	0.048	0.152	0.042	0.164	0.032	0.189
10	0.055	0.164	0.049	0.176	0.038	0.202
11	0.063	0.175	0.056	0.188	0.044	0.215
12	0.071	0.187	0.064	0.200	0.051	0.227
13	0.079	0.199	0.071	0.212	0.058	0.239
14	0.087	0.210	0.079	0.224	0.064	0.251
15	0.095	0.222	0.086	0.235	0.072	0.263
16	0.103	0.233	0.094	0.247	0.079	0.275
17	0.111	0.244	0.102	0.258	0.086	0.287
18	0.120	0.255	0.110	0.269	0.093	0.298
19	0.128	0.266	0.118	0.281	0.101	0.310
20	0.137	0.277	0.127	0.292	0.108	0.321
21	0.145	0.288	0.135	0.303	0.116	0.332
22	0.154	0.299	0.143	0.314	0.124	0.344
23	0.163	0.310	0.152	0.325	0.132	0.355
24	0.171	0.321	0.160	0.336	0.140	0.366
25	0.180	0.331	0.169	0.347	0.148	0.377
26	0.189	0.342	0.177	0.357	0.156	0.388
27	0.198	0.353	0.186	0.368	0.164	0.398
28	0.207	0.363	0.195	0.379	0.172	0.409
29	0.216	0.374	0.204	0.389	0.181	0.420
30	0.225	0.384	0.212	0.400	0.189	0.431
31	0.234	0.395	0.221	0.410	0.198	0.441
32	0.243	0.405	0.230	0.421	0.206	0.452
33	0.252	0.415	0.239	0.431	0.215	0.462
34	0.261	0.426	0.248	0.442	0.223	0.473
35	0.271	0.436	0.257	0.452	0.232	0.483
36	0.280	0.446	0.266	0.462	0.240	0.493
37	0.289	0.457	0.276	0.472	0.250	0.503
38	0.299	0.467	0.285	0.483	0.259	0.514
39	0.308	0.477	0.294	0.493	0.267	0.523
40	0.318	0.487	0.303	0.503	0.276	0.533
41	0.327	0.497	0.313	0.513	0.286	0.544
42	0.336	0.507	0.322	0.523	0.294	0.554
43	0.346	0.517	0.331	0.533	0.303	0.563
44	0.356	0.527	0.341	0.543	0.313	0.573
45	0.365	0.537	0.350	0.553	0.322	0.583
46	0.375	0.547	0.360	0.563	0.331	0.593
47	0.384	0.557	0.369	0.572	0.341	0.603
48	0.394	0.567	0.379	0.582	0.350	0.612
49	0.404	0.577	0.389	0.592	0.359	0.622
50	0.414	0.586	0.398	0.602	0.369	0.631

TABLE A.5d. Confidence Intervals on the Binomial Parameter π

sample size $n = 250$

	$\alpha = 0.10$		$\alpha = 0.05$		$\alpha = 0.01$	
y/n	L	U	L	U	L	U
0.00	0.000	0.012	0.000	0.015	0.000	0.021
0.02	0.008	0.042	0.007	0.046	0.004	0.056
0.04	0.022	0.067	0.019	0.072	0.015	0.084
0.06	0.037	0.091	0.034	0.097	0.028	0.110
0.08	0.054	0.114	0.050	0.121	0.042	0.134
0.10	0.070	0.137	0.066	0.144	0.057	0.159
0.12	0.088	0.159	0.082	0.167	0.073	0.182
0.14	0.105	0.181	0.099	0.189	0.089	0.205
0.16	0.123	0.203	0.117	0.211	0.105	0.228
0.18	0.141	0.225	0.134	0.233	0.122	0.250
0.20	0.159	0.246	0.152	0.255	0.139	0.273
0.22	0.178	0.267	0.170	0.277	0.156	0.295
0.24	0.196	0.289	0.188	0.298	0.174	0.316
0.26	0.215	0.310	0.207	0.319	0.192	0.338
0.28	0.233	0.331	0.225	0.340	0.210	0.359
0.30	0.252	0.351	0.244	0.361	0.228	0.380
0.32	0.271	0.372	0.263	0.382	0.246	0.401
0.34	0.290	0.393	0.281	0.402	0.264	0.422
0.36	0.310	0.413	0.300	0.423	0.283	0.442
0.38	0.329	0.433	0.320	0.443	0.302	0.463
0.40	0.348	0.454	0.339	0.464	0.321	0.483
0.42	0.368	0.474	0.358	0.484	0.340	0.503
0.44	0.387	0.494	0.377	0.504	0.359	0.523
0.46	0.407	0.514	0.397	0.524	0.378	0.543
0.48	0.426	0.534	0.417	0.544	0.398	0.563
0.50	0.446	0.554	0.436	0.564	0.417	0.583

TABLE A.5e. Confidence Intervals on the Binomial Parameter π

sample size n = 500

	α = 0.10		α = 0.05		α = 0.01	
y/n	L	U	L	U	L	U
0.00	0.000	0.006	0.000	0.007	0.000	0.011
0.01	0.004	0.021	0.003	0.023	0.002	0.028
0.02	0.011	0.034	0.010	0.036	0.007	0.042
0.03	0.019	0.046	0.017	0.049	0.014	0.056
0.04	0.027	0.058	0.025	0.061	0.021	0.068
0.05	0.035	0.069	0.033	0.073	0.028	0.081
0.06	0.044	0.081	0.041	0.085	0.036	0.093
0.07	0.052	0.092	0.049	0.096	0.044	0.105
0.08	0.061	0.103	0.058	0.107	0.052	0.116
0.09	0.070	0.114	0.066	0.119	0.060	0.128
0.10	0.079	0.125	0.075	0.130	0.068	0.139
0.11	0.088	0.136	0.084	0.141	0.077	0.151
0.12	0.097	0.147	0.093	0.152	0.085	0.162
0.13	0.106	0.157	0.102	0.163	0.094	0.173
0.14	0.115	0.168	0.111	0.174	0.103	0.184
0.15	0.124	0.179	0.120	0.184	0.111	0.196
0.16	0.134	0.189	0.129	0.195	0.120	0.207
0.17	0.143	0.200	0.138	0.206	0.129	0.217
0.18	0.152	0.211	0.147	0.217	0.138	0.228
0.19	0.162	0.221	0.157	0.227	0.147	0.239
0.20	0.171	0.232	0.166	0.238	0.156	0.250
0.21	0.180	0.242	0.175	0.248	0.165	0.261
0.22	0.190	0.253	0.184	0.259	0.174	0.271
0.23	0.199	0.263	0.194	0.269	0.183	0.282
0.24	0.209	0.274	0.203	0.280	0.192	0.292
0.25	0.218	0.284	0.213	0.290	0.202	0.303
0.26	0.228	0.294	0.222	0.301	0.211	0.314
0.27	0.237	0.305	0.232	0.311	0.220	0.324
0.28	0.247	0.315	0.241	0.322	0.230	0.335
0.29	0.257	0.325	0.251	0.332	0.239	0.345
0.30	0.266	0.336	0.260	0.342	0.248	0.356
0.31	0.276	0.346	0.270	0.353	0.258	0.366
0.32	0.286	0.356	0.279	0.363	0.267	0.376
0.33	0.295	0.366	0.289	0.373	0.277	0.387
0.34	0.305	0.376	0.299	0.383	0.286	0.397
0.35	0.315	0.387	0.308	0.394	0.296	0.407
0.36	0.324	0.397	0.318	0.404	0.305	0.417
0.37	0.334	0.407	0.328	0.414	0.315	0.428
0.38	0.344	0.417	0.337	0.424	0.325	0.438
0.39	0.354	0.427	0.347	0.434	0.334	0.448
0.40	0.363	0.437	0.357	0.444	0.344	0.458
0.41	0.373	0.448	0.367	0.455	0.353	0.468
0.42	0.383	0.458	0.376	0.465	0.363	0.478
0.43	0.393	0.468	0.386	0.475	0.373	0.489
0.44	0.403	0.478	0.396	0.485	0.383	0.498
0.45	0.413	0.488	0.406	0.495	0.392	0.509
0.46	0.423	0.498	0.416	0.505	0.402	0.519
0.47	0.432	0.508	0.426	0.515	0.412	0.529
0.48	0.442	0.518	0.435	0.525	0.422	0.539
0.49	0.452	0.528	0.445	0.535	0.432	0.548
0.50	0.462	0.538	0.455	0.545	0.442	0.558

TABLE A.6. Values of $e^{-\lambda}$

λ	.00	.05	.10	.15	.20
0.00	1.000000	.951229	.904837	.860708	.818731
1.00	.367879	.349938	.332871	.316637	.301194
2.00	.135335	.128735	.122456	.116484	.110803
3.00	.049787	.047359	.045049	.042852	.040762
4.00	.018316	.017422	.016573	.015764	.014996
5.00	.006738	.006409	.006097	.005799	.005517
6.00	.002479	.002358	.002243	.002133	.002029
7.00	.000912	.000867	.000825	.000785	.000747
8.00	.000335	.000319	.000304	.000289	.000275
9.00	.000123	.000117	.000112	.000106	.000101
10.00	.000045	.000043	.000041	.000039	.000037
11.00	.000017	.000016	.000015	.000014	.000014
12.00	.000006	.000006	.000006	.000005	.000005
13.00	.000002	.000002	.000002	.000002	.000002
14.00	.000001	.000001	.000001	.000001	.000001

λ	.25	.30	.35	.40	.45
0.00	.778801	.740818	.704688	.670320	.637628
1.00	.286505	.272532	.259240	.246597	.234570
2.00	.105399	.100259	.095369	.090718	.086294
3.00	.038774	.036883	.035084	.033373	.031746
4.00	.014264	.013569	.012907	.012277	.011679
5.00	.005248	.004992	.004748	.004517	.004296
6.00	.001930	.001836	.001747	.001662	.001581
7.00	.000710	.000676	.000643	.000611	.000581
8.00	.000261	.000249	.000236	.000225	.000214
9.00	.000096	.000091	.000087	.000083	.000079
10.00	.000035	.000034	.000032	.000030	.000029
11.00	.000013	.000012	.000012	.000011	.000011
12.00	.000005	.000005	.000004	.000004	.000004
13.00	.000002	.000002	.000002	.000002	.000001
14.00	.000001	.000001	.000001	.000001	.000001

TABLE A.6.—*Continued*

λ	.50	.55	.60	.65	.70
0.00	.606531	.576950	.548812	.522046	.496585
1.00	.223130	.212248	.201897	.192050	.182684
2.00	.082085	.078082	.074274	.070651	.067206
3.00	.030197	.028725	.027324	.025991	.024724
4.00	.011109	.010567	.010052	.009562	.009095
5.00	.004087	.003887	.003698	.003518	.003346
6.00	.001503	.001430	.001360	.001294	.001231
7.00	.000553	.000526	.000500	.000476	.000453
8.00	.000203	.000194	.000184	.000175	.000167
9.00	.000075	.000071	.000068	.000064	.000061
10.00	.000028	.000026	.000025	.000024	.000023
11.00	.000010	.000010	.000009	.000009	.000008
12.00	.000004	.000004	.000003	.000003	.000003
13.00	.000001	.000001	.000001	.000001	.000001
14.00	.000001	.000000	.000000	.000000	.000000

λ	.75	.80	.85	.90	.95
0.00	.472367	.449329	.427415	.406570	.386741
1.00	.173774	.165299	.157237	.149569	.142274
2.00	.063928	.060810	.057844	.055023	.052340
3.00	.023518	.022371	.021280	.020242	.019255
4.00	.008652	.008230	.007828	.007447	.007083
5.00	.003183	.003028	.002880	.002739	.002606
6.00	.001171	.001114	.001059	.001008	.000959
7.00	.000431	.000410	.000390	.000371	.000353
8.00	.000158	.000151	.000143	.000136	.000130
9.00	.000058	.000055	.000053	.000050	.000048
10.00	.000021	.000020	.000019	.000018	.000018
11.00	.000008	.000008	.000007	.000007	.000006
12.00	.000003	.000003	.000003	.000002	.000002
13.00	.000001	.000001	.000001	.000001	.000001
14.00	.000000	.000000	.000000	.000000	.000000

TABLE A.7. **Poisson Distributions**

y \ λ	0.05	0.10	0.20	0.30	0.40	0.50	0.60
0	.9512	.9048	.8187	.7408	.6703	.6065	.5488
1	.0476	.0905	.1637	.2222	.2681	.3033	.3293
2	.0012	.0045	.0164	.0333	.0536	.0758	.0988
3	.0000	.0002	.0011	.0033	.0072	.0126	.0198
4	.0000	.0000	.0001	.0003	.0007	.0016	.0030
5	.0000	.0000	.0000	.0000	.0001	.0002	.0004
6	.0000	.0000	.0000	.0000	.0000	.0000	.0000

y \ λ	0.70	0.80	0.90	1.00	1.20	1.40	1.60
0	.4966	.4493	.4066	.3679	.3012	.2466	.2019
1	.3476	.3595	.3659	.3679	.3614	.3452	.3230
2	.1217	.1438	.1647	.1839	.2169	.2417	.2584
3	.0284	.0383	.0494	.0613	.0867	.1128	.1378
4	.0050	.0077	.0111	.0153	.0260	.0395	.0551
5	.0007	.0012	.0020	.0031	.0062	.0111	.0176
6	.0001	.0002	.0003	.0005	.0012	.0026	.0047
7	.0000	.0000	.0000	.0001	.0002	.0005	.0011
8	.0000	.0000	.0000	.0000	.0000	.0001	.0002
9	.0000	.0000	.0000	.0000	.0000	.0000	.0000

Step boundaries give approximate 80% confidence intervals for λ. (See Section 4.3.)

TABLE A.7.—*Continued*

$y \backslash \lambda$	1.80	2.00	2.20	2.40	2.60	2.80	3.00
0	.1653	.1353	.1108	.0907	.0743	.0608	.0498
1	.2975	.2707	.2438	.2177	.1931	.1703	.1494
2	.2678	.2707	.2681	.2613	.2510	.2384	.2240
3	.1607	.1804	.1966	.2090	.2176	.2225	.2240
4	.0723	.0902	.1082	.1254	.1414	.1557	.1680
5	.0260	.0361	.0476	.0602	.0735	.0872	.1008
6	.0078	.0120	.0174	.0241	.0319	.0407	.0504
7	.0020	.0034	.0055	.0083	.0118	.0163	.0216
8	.0005	.0009	.0015	.0025	.0038	.0057	.0081
9	.0001	.0002	.0004	.0007	.0011	.0018	.0027
10	.0000	.0000	.0001	.0002	.0003	.0005	.0008
11	.0000	.0000	.0000	.0000	.0001	.0001	.0002
12	.0000	.0000	.0000	.0000	.0000	.0000	.0001
13	.0000	.0000	.0000	.0000	.0000	.0000	.0000

$y \backslash \lambda$	3.50	4.00	4.50	5.00	5.50	6.00	6.50
0	.0302	.0183	.0111	.0067	.0041	.0025	.0015
1	.1057	.0733	.0500	.0337	.0225	.0149	.0098
2	.1850	.1465	.1125	.0842	.0618	.0446	.0318
3	.2158	.1954	.1687	.1404	.1133	.0892	.0688
4	.1888	.1954	.1898	.1755	.1558	.1339	.1118
5	.1322	.1563	.1708	.1755	.1714	.1606	.1454
6	.0771	.1042	.1281	.1462	.1571	.1606	.1575
7	.0385	.0595	.0824	.1044	.1234	.1377	.1462
8	.0169	.0298	.0463	.0653	.0849	.1033	.1188
9	.0066	.0132	.0232	.0363	.0519	.0688	.0858
10	.0023	.0053	.0104	.0181	.0285	.0413	.0558
11	.0007	.0019	.0043	.0082	.0143	.0225	.0330
12	.0002	.0006	.0016	.0034	.0065	.0113	.0179
13	.0001	.0002	.0006	.0013	.0028	.0052	.0089
14	.0000	.0001	.0002	.0005	.0011	.0022	.0041
15	.0000	.0000	.0001	.0002	.0004	.0009	.0018
16	.0000	.0000	.0000	.0000	.0001	.0003	.0007
17	.0000	.0000	.0000	.0000	.0000	.0001	.0003
18	.0000	.0000	.0000	.0000	.0000	.0000	.0001
19	.0000	.0000	.0000	.0000	.0000	.0000	.0000

TABLE A.7.—*Continued*

y \ λ	7.00	8.00	9.00	10.00	11.00	12.00	13.00
0	.0009	.0003	.0001	.0000	.0000	.0000	.0000
1	.0064	.0027	.0011	.0005	.0002	.0001	.0000
2	.0223	.0107	.0050	.0023	.0010	.0004	.0002
3	.0521	.0286	.0150	.0076	.0037	.0018	.0008
4	.0912	.0573	.0337	.0189	.0102	.0053	.0027
5	.1277	.0916	.0607	.0378	.0224	.0127	.0070
6	.1490	.1221	.0911	.0631	.0411	.0255	.0152
7	.1490	.1396	.1171	.0901	.0646	.0437	.0281
8	.1304	.1396	.1318	.1126	.0888	.0655	.0457
9	.1014	.1241	.1318	.1251	.1085	.0874	.0661
10	.0710	.0993	.1186	.1251	.1194	.1048	.0859
11	.0452	.0722	.0970	.1137	.1194	.1144	.1015
12	.0263	.0481	.0728	.0948	.1094	.1144	.1099
13	.0142	.0296	.0504	.0729	.0926	.1056	.1099
14	.0071	.0169	.0324	.0521	.0728	.0905	.1021
15	.0033	.0090	.0194	.0347	.0534	.0724	.0885
16	.0014	.0045	.0109	.0217	.0367	.0543	.0719
17	.0006	.0021	.0058	.0128	.0237	.0383	.0550
18	.0002	.0009	.0029	.0071	.0145	.0255	.0397
19	.0001	.0004	.0014	.0037	.0084	.0161	.0272
20	.0000	.0002	.0006	.0019	.0046	.0097	.0177
21	.0000	.0001	.0003	.0009	.0024	.0055	.0109
22	.0000	.0000	.0001	.0004	.0012	.0030	.0065
23	.0000	.0000	.0000	.0002	.0006	.0016	.0037
24	.0000	.0000	.0000	.0001	.0003	.0008	.0020
25	.0000	.0000	.0000	.0000	.0001	.0004	.0010
26	.0000	.0000	.0000	.0000	.0000	.0002	.0005
27	.0000	.0000	.0000	.0000	.0000	.0001	.0002
28	.0000	.0000	.0000	.0000	.0000	.0000	.0001
29	.0000	.0000	.0000	.0000	.0000	.0000	.0001
30	.0000	.0000	.0000	.0000	.0000	.0000	.0000

Table A.8. Critical Chi-Square Values.

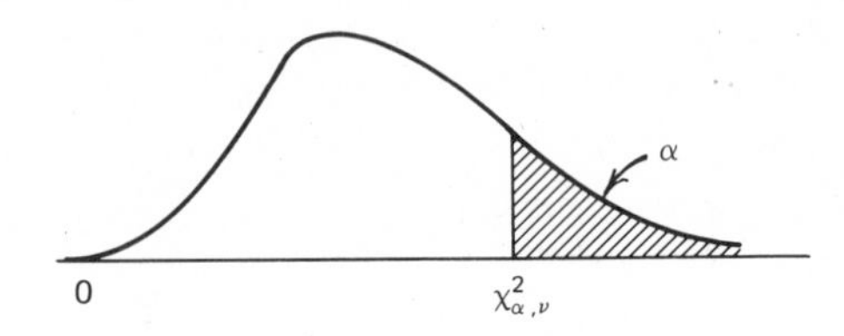

$$P(\chi^2 > \chi^2_{\alpha,\nu}) = P(\chi^2 > \text{tabular value}) = \alpha$$

Examples:

1. $P(\chi^2 > \chi^2_{0.025,5}) = P(\chi^2 > 12.833) = 0.025$
2. $P(\chi^2 > \chi^2_{0.995,10}) = P(\chi^2 > 2.156) = 0.995$

TABLE A.8. Critical Chi-Square Values

ν \ α	0.995	0.990	0.975	0.950	0.050	0.025	0.010	0.005
1	0.000	0.000	0.001	0.004	3.841	5.024	6.635	7.879
2	0.010	0.020	0.051	0.103	5.991	7.378	9.210	10.597
3	0.072	0.115	0.216	0.352	7.815	9.348	11.345	12.838
4	0.207	0.297	0.484	0.711	9.488	11.143	13.277	14.860
5	0.412	0.554	0.831	1.145	11.070	12.833	15.086	16.750
6	0.676	0.872	1.237	1.635	12.592	14.449	16.812	18.548
7	0.989	1.239	1.690	2.167	14.067	16.013	18.475	20.278
8	1.344	1.646	2.180	2.733	15.507	17.535	20.090	21.955
9	1.735	2.088	2.700	3.325	16.919	19.023	21.666	23.589
10	2.156	2.558	3.247	3.940	18.307	20.483	23.209	25.188
11	2.603	3.053	3.816	4.575	19.675	21.920	24.725	26.757
12	3.074	3.571	4.404	5.226	21.026	23.337	26.217	28.300
13	3.565	4.107	5.009	5.892	22.362	24.736	27.688	29.819
14	4.075	4.660	5.629	6.571	23.685	26.119	29.141	31.319
15	4.601	5.229	6.262	7.261	24.996	27.488	30.578	32.801
16	5.142	5.812	6.908	7.962	26.296	28.845	32.000	34.267
17	5.697	6.408	7.564	8.672	27.587	30.191	33.409	35.718
18	6.265	7.015	8.231	9.390	28.869	31.526	34.805	37.156
19	6.844	7.633	8.907	10.117	30.144	32.852	36.191	38.582
20	7.434	8.260	9.591	10.851	31.410	34.170	37.566	39.997
21	8.034	8.897	10.283	11.591	32.671	35.479	38.932	41.401
22	8.643	9.542	10.982	12.338	33.924	36.781	40.289	42.796
23	9.260	10.196	11.689	13.091	35.172	38.076	41.638	44.181
24	9.886	10.856	12.401	13.848	36.415	39.364	42.980	45.559
25	10.520	11.524	13.120	14.611	37.652	40.646	44.314	46.928
26	11.160	12.198	13.844	15.379	38.885	41.923	45.642	48.290
27	11.808	12.879	14.573	16.151	40.113	43.195	46.963	49.645
28	12.461	13.565	15.308	16.928	41.337	44.461	48.278	50.993
29	13.121	14.256	16.047	17.708	42.557	45.722	49.588	52.336
30	13.787	14.953	16.791	18.493	43.773	46.979	50.892	53.672
32	15.134	16.362	18.291	20.072	46.194	49.480	53.486	56.328
34	16.501	17.789	19.806	21.664	48.602	51.966	56.061	58.964
36	17.887	19.233	21.336	23.269	50.998	54.437	58.619	61.581
38	19.289	20.691	22.878	24.884	53.384	56.896	61.162	64.181
40	20.707	22.164	24.433	26.509	55.758	59.342	63.691	66.766
42	22.138	23.650	25.999	28.144	58.124	61.777	66.206	69.336
44	23.584	25.148	27.575	29.787	60.481	64.201	68.710	71.893
46	25.041	26.657	29.160	31.439	62.830	66.617	71.201	74.437
48	26.511	28.177	30.755	33.098	65.171	69.023	73.683	76.969
50	27.991	29.707	32.357	34.764	67.505	71.420	76.154	79.490
60	35.534	37.485	40.482	43.188	79.082	83.298	88.379	91.952
70	43.275	45.442	48.758	51.739	90.531	95.023	100.425	104.215
80	51.172	53.540	57.153	60.391	101.879	106.629	112.329	116.321
90	59.196	61.754	65.647	69.126	113.145	118.136	124.116	128.299
100	67.328	70.065	74.222	77.929	124.342	129.561	135.807	140.169

Table A.9. The Standard Normal Distribution.

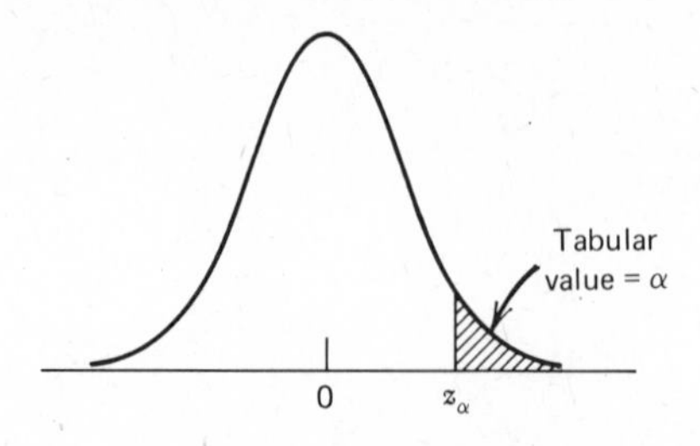

Values α in the body of the table are the probability that z is greater than the positive value z_α given in the margins.

Example:

$$P(z > 1.54) = 0.062$$

or

$$z_{0.062} = 1.54$$

For negative z values, the probability of a greater value can be found using the symmetry of the distribution.

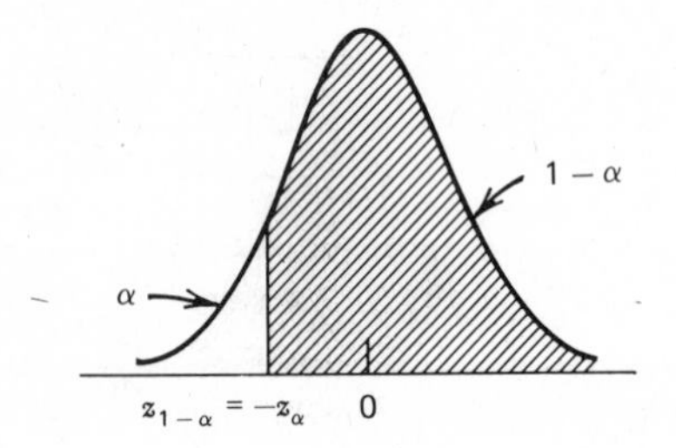

$$P(z > -z_\alpha) = 1 - \alpha = P(z > z_{1-\alpha})$$

Example:

$$P(z > -1.54) = 1 - 0.062 = 0.938$$

or

$$z_{0.938} = -1.54$$

TABLE A.9. The Standard Normal Distribution

z	.00	.01	.02	.03	.04	.05	.06	.07	.08	.09
0.00	.500	.496	.492	.488	.484	.480	.476	.472	.468	.464
0.10	.460	.456	.452	.448	.444	.440	.436	.433	.429	.425
0.20	.421	.417	.413	.409	.405	.401	.397	.394	.390	.386
0.30	.382	.378	.374	.371	.367	.363	.359	.356	.352	.348
0.40	.345	.341	.337	.334	.330	.326	.323	.319	.316	.312
0.50	.309	.305	.302	.298	.295	.291	.288	.284	.281	.278
0.60	.274	.271	.268	.264	.261	.258	.255	.251	.248	.245
0.70	.242	.239	.236	.233	.230	.227	.224	.221	.218	.215
0.80	.212	.209	.206	.203	.200	.198	.195	.192	.189	.187
0.90	.184	.181	.179	.176	.174	.171	.169	.166	.164	.161
1.00	.159	.156	.154	.152	.149	.147	.145	.142	.140	.138
1.10	.136	.133	.131	.129	.127	.125	.123	.121	.119	.117
1.20	.115	.113	.111	.109	.107	.106	.104	.102	.100	.099
1.30	.097	.095	.093	.092	.090	.089	.087	.085	.084	.082
1.40	.081	.079	.078	.076	.075	.074	.072	.071	.069	.068
1.50	.067	.066	.064	.063	.062	.061	.059	.058	.057	.056
1.60	.055	.054	.053	.052	.051	.049	.048	.047	.046	.046
1.70	.045	.044	.043	.042	.041	.040	.039	.038	.038	.037
1.80	.036	.035	.034	.034	.033	.032	.031	.031	.030	.029
1.90	.029	.028	.027	.027	.026	.026	.025	.024	.024	.023
2.00	.023	.022	.022	.021	.021	.020	.020	.019	.019	.018
2.10	.018	.017	.017	.017	.016	.016	.015	.015	.015	.014
2.20	.014	.014	.013	.013	.013	.012	.012	.012	.011	.011
2.30	.011	.010	.010	.010	.010	.009	.009	.009	.009	.008
2.40	.008	.008	.008	.008	.007	.007	.007	.007	.007	.006
2.50	.006	.006	.006	.006	.006	.005	.005	.005	.005	.005
2.60	.005	.005	.004	.004	.004	.004	.004	.004	.004	.004
2.70	.003	.003	.003	.003	.003	.003	.003	.003	.003	.003
2.80	.003	.002	.002	.002	.002	.002	.002	.002	.002	.002
2.90	.002	.002	.002	.002	.002	.002	.002	.001	.001	.001
3.00	.001	.001	.001	.001	.001	.001	.001	.001	.001	.001

Table A.10. Critical t Values.

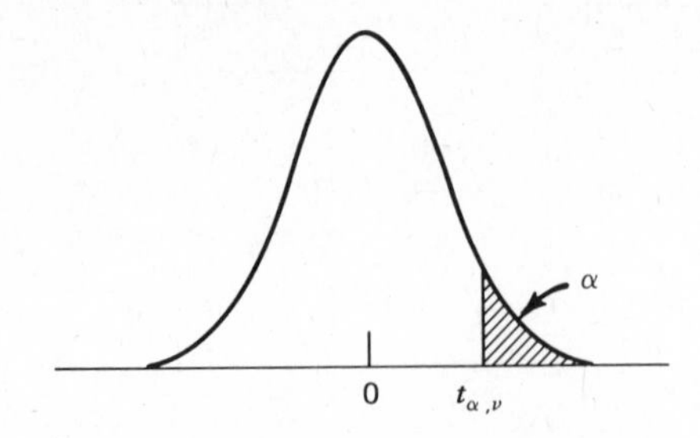

$$P(t > t_{\alpha,\nu}) = P(t > \text{tabular value}) = \alpha$$

Example:

$$P(t > t_{0.05,10}) = P(t > 1.812) = 0.05$$

Symmetry is used to find negative t values.

Example:

$$t_{0.95,10} = -t_{0.05,10} = -1.812$$

The last row of the t table gives critical z values, that is,

$$t_{\alpha,\infty} = z_{\alpha}$$

TABLE A.10. Critical *t* Values

$\nu \backslash \alpha$	0.100	0.050	0.025	0.010	0.005
1	3.078	6.314	12.706	31.821	63.657
2	1.886	2.920	4.303	6.965	9.925
3	1.638	2.353	3.182	4.541	5.841
4	1.533	2.132	2.776	3.747	4.604
5	1.476	2.015	2.571	3.365	4.032
6	1.440	1.943	2.447	3.143	3.707
7	1.415	1.895	2.365	2.998	3.499
8	1.397	1.860	2.306	2.896	3.355
9	1.383	1.833	2.262	2.821	3.250
10	1.372	1.812	2.228	2.764	3.169
11	1.363	1.796	2.201	2.718	3.106
12	1.356	1.782	2.179	2.681	3.055
13	1.350	1.771	2.160	2.650	3.012
14	1.345	1.761	2.145	2.624	2.977
15	1.341	1.753	2.131	2.602	2.947
16	1.337	1.746	2.120	2.583	2.921
17	1.333	1.740	2.110	2.567	2.898
18	1.330	1.734	2.101	2.552	2.878
19	1.328	1.729	2.093	2.539	2.861
20	1.325	1.725	2.086	2.528	2.845
21	1.323	1.721	2.080	2.518	2.831
22	1.321	1.717	2.074	2.508	2.819
23	1.319	1.714	2.069	2.500	2.807
24	1.318	1.711	2.064	2.492	2.797
25	1.316	1.708	2.060	2.485	2.787
26	1.315	1.706	2.056	2.479	2.779
27	1.314	1.703	2.052	2.473	2.771
28	1.313	1.701	2.048	2.467	2.763
29	1.311	1.699	2.045	2.462	2.756
30	1.310	1.697	2.042	2.457	2.750
40	1.303	1.684	2.021	2.423	2.704
60	1.296	1.671	2.000	2.390	2.660
120	1.289	1.658	1.980	2.358	2.617
INF	1.282	1.645	1.960	2.326	2.576

Tables A.11a Through A.11x. Critical F Values.

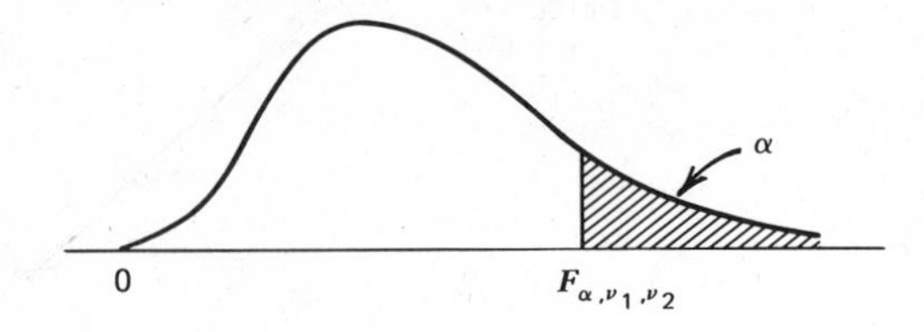

ν_1 = numerator degrees of freedom

ν_2 = denominator degrees of freedom

$$P(F > F_{\alpha,\nu_1,\nu_2}) = \alpha$$

Example:

$$F_{0.025,2,4} = 10.649$$

For lower critical F values, use the relationship

$$F_{1-\alpha,\nu_1,\nu_2} = 1/F_{\alpha,\nu_2,\nu_1}$$

Example:

$$F_{0.995,10,8} = 1/F_{0.005,8,10} = 1/6.116 = 0.1635$$

Table for a Given Pair of Degrees of Freedom

		Numerator Degrees of Freedom					
		1–5	6–10	11–15	16–20	21–25	26–30
Denominator	1–10	A.11a	A.11b	A.11c	A.11d	A.11e	A.11f
Degrees of	11–20	A.11g	A.11h	A.11i	A.11j	A.11k	A.11l
Freedom	21–30	A.11m	A.11n	A.11o	A.11p	A.11q	A.11r
	40–200	A.11s	A.11t	A.11u	A.11v	A.11w	A.11x

TABLE A.11a. Critical F Values

Denominator ν	α	Numerator ν 1	2	3	4	5
1	0.050	161.448	199.500	215.707	224.583	230.162
	0.025	647.790	799.500	864.163	899.583	921.848
	0.010	4052.194	4999.506	5403.355	5624.584	5763.660
	0.005	16210.873	19999.499	21614.726	22499.596	23055.762
	0.001	405293.184	499996.121	540378.670	562498.442	576406.763
2	0.050	18.513	19.000	19.164	19.247	19.296
	0.025	38.506	39.000	39.165	39.248	39.298
	0.010	98.503	99.000	99.166	99.249	99.299
	0.005	198.501	199.000	199.166	199.250	199.300
	0.001	998.505	998.991	999.168	999.257	999.302
3	0.050	10.128	9.552	9.277	9.117	9.013
	0.025	17.443	16.044	15.439	15.101	14.885
	0.010	34.116	30.817	29.457	28.710	28.237
	0.005	55.552	49.799	47.467	46.195	45.392
	0.001	167.030	148.501	141.109	137.099	134.581
4	0.050	7.709	6.944	6.591	6.388	6.256
	0.025	12.218	10.649	9.979	9.605	9.364
	0.010	21.198	18.000	16.694	15.977	15.522
	0.005	31.333	26.284	24.259	23.155	22.456
	0.001	74.137	61.245	56.177	53.436	51.711
5	0.050	6.608	5.786	5.409	5.192	5.050
	0.025	10.007	8.434	7.764	7.388	7.146
	0.010	16.258	13.274	12.060	11.392	10.967
	0.005	22.785	18.314	16.530	15.556	14.940
	0.001	47.181	37.122	33.203	31.085	29.753
6	0.050	5.987	5.143	4.757	4.534	4.387
	0.025	8.813	7.260	6.599	6.227	5.988
	0.010	13.745	10.925	9.780	9.148	8.746
	0.005	18.635	14.544	12.917	12.028	11.464
	0.001	35.508	27.000	23.703	21.924	20.803
7	0.050	5.591	4.737	4.347	4.120	3.972
	0.025	8.073	6.542	5.890	5.523	5.285
	0.010	12.246	9.547	8.451	7.847	7.460
	0.005	16.236	12.404	10.882	10.050	9.522
	0.001	29.245	21.689	18.772	17.198	16.206
8	0.050	5.318	4.459	4.066	3.838	3.687
	0.025	7.571	6.059	5.416	5.053	4.817
	0.010	11.259	8.649	7.591	7.006	6.632
	0.005	14.688	11.042	9.596	8.805	8.302
	0.001	25.415	18.494	15.829	14.392	13.485
9	0.050	5.117	4.256	3.863	3.633	3.482
	0.025	7.209	5.715	5.078	4.718	4.484
	0.010	10.561	8.022	6.992	6.422	6.057
	0.005	13.614	10.107	8.717	7.956	7.471
	0.001	22.857	16.387	13.902	12.560	11.714
10	0.050	4.965	4.103	3.708	3.478	3.326
	0.025	6.937	5.456	4.826	4.468	4.236
	0.010	10.044	7.559	6.552	5.994	5.636
	0.005	12.826	9.427	8.081	7.343	6.872
	0.001	21.040	14.905	12.553	11.283	10.481

TABLE A.11b. Critical F Values

Denominator ν	α	Numerator ν 6	7	8	9	10
1	0.050	233.986	236.768	238.883	240.543	241.882
	0.025	937.110	948.218	956.656	963.285	968.627
	0.010	5858.981	5928.349	5981.073	6022.471	6055.850
	0.005	23437.141	23714.565	23925.451	24091.033	24224.533
	0.001	585927.903	592864.102	598136.821	602279.789	605630.027
2	0.050	19.330	19.353	19.371	19.385	19.396
	0.025	39.331	39.355	39.373	39.387	39.398
	0.010	99.333	99.356	99.374	99.388	99.399
	0.005	199.333	199.357	199.375	199.388	199.399
	0.001	999.329	999.360	999.376	999.387	999.409
3	0.050	8.941	8.887	8.845	8.812	8.786
	0.025	14.735	14.624	14.540	14.473	14.419
	0.010	27.911	27.672	27.489	27.345	27.229
	0.005	44.838	44.434	44.126	43.882	43.686
	0.001	132.848	131.584	130.619	129.860	129.247
4	0.050	6.163	6.094	6.041	5.999	5.964
	0.025	9.197	9.074	8.980	8.905	8.844
	0.010	15.207	14.976	14.799	14.659	14.546
	0.005	21.975	21.622	21.352	21.139	20.967
	0.001	50.525	49.658	48.996	48.474	48.053
5	0.050	4.950	4.876	4.818	4.772	4.735
	0.025	6.978	6.853	6.757	6.681	6.619
	0.010	10.672	10.456	10.289	10.158	10.051
	0.005	14.513	14.200	13.961	13.772	13.618
	0.001	28.834	28.163	27.649	27.245	26.916
6	0.050	4.284	4.207	4.147	4.099	4.060
	0.025	5.820	5.695	5.600	5.523	5.461
	0.010	8.466	8.260	8.102	7.976	7.874
	0.005	11.073	10.786	10.566	10.391	10.250
	0.001	20.030	19.463	19.030	18.688	18.411
7	0.050	3.866	3.787	3.726	3.677	3.637
	0.025	5.119	4.995	4.899	4.823	4.761
	0.010	7.191	6.993	6.840	6.719	6.620
	0.005	9.155	8.885	8.678	8.514	8.380
	0.001	15.521	15.019	14.634	14.330	14.083
8	0.050	3.581	3.500	3.438	3.388	3.347
	0.025	4.652	4.529	4.433	4.357	4.295
	0.010	6.371	6.178	6.029	5.911	5.814
	0.005	7.952	7.694	7.496	7.339	7.211
	0.001	12.858	12.398	12.046	11.767	11.540
9	0.050	3.374	3.293	3.230	3.179	3.137
	0.025	4.320	4.197	4.102	4.026	3.964
	0.010	5.802	5.613	5.467	5.351	5.257
	0.005	7.134	6.885	6.693	6.541	6.417
	0.001	11.128	10.698	10.368	10.107	9.894
10	0.050	3.217	3.135	3.072	3.020	2.978
	0.025	4.072	3.950	3.855	3.779	3.717
	0.010	5.386	5.200	5.057	4.942	4.849
	0.005	6.545	6.302	6.116	5.968	5.847
	0.001	9.926	9.517	9.204	8.956	8.754

TABLE A.11c. Critical *F* Values

Denominator		Numerator ν				
ν	α	11	12	13	14	15
1	0.050	242.984	243.906	244.690	245.364	245.950
	0.025	973.025	976.709	979.837	982.527	984.866
	0.010	6083.321	6106.329	6125.853	6142.674	6157.294
	0.005	24334.361	24426.333	24504.525	24571.721	24630.203
	0.001	608357.024	610674.243	612614.192	614311.903	615752.317
2	0.050	19.405	19.413	19.419	19.424	19.429
	0.025	39.407	39.415	39.421	39.426	39.431
	0.010	99.408	99.416	99.422	99.428	99.432
	0.005	199.408	199.416	199.423	199.428	199.433
	0.001	999.412	999.421	999.422	999.437	999.426
3	0.050	8.763	8.745	8.729	8.715	8.703
	0.025	14.374	14.337	14.304	14.277	14.253
	0.010	27.133	27.052	26.983	26.924	26.872
	0.005	43.524	43.387	43.272	43.172	43.085
	0.001	128.742	128.317	127.957	127.645	127.376
4	0.050	5.936	5.912	5.891	5.873	5.858
	0.025	8.794	8.751	8.715	8.684	8.657
	0.010	14.452	14.374	14.307	14.249	14.198
	0.005	20.824	20.705	20.603	20.515	20.438
	0.001	47.704	47.412	47.163	46.948	46.761
5	0.050	4.704	4.678	4.655	4.636	4.619
	0.025	6.568	6.525	6.488	6.456	6.428
	0.010	9.963	9.888	9.825	9.770	9.722
	0.005	13.491	13.384	13.293	13.215	13.146
	0.001	26.646	26.418	26.224	26.057	25.911
6	0.050	4.027	4.000	3.976	3.956	3.938
	0.025	5.410	5.366	5.329	5.297	5.269
	0.010	7.790	7.718	7.657	7.605	7.559
	0.005	10.133	10.034	9.950	9.877	9.814
	0.001	18.182	17.989	17.824	17.682	17.559
7	0.050	3.603	3.575	3.550	3.529	3.511
	0.025	4.709	4.666	4.628	4.596	4.568
	0.010	6.538	6.469	6.410	6.359	6.314
	0.005	8.270	8.176	8.097	8.028	7.968
	0.001	13.879	13.707	13.561	13.434	13.324
8	0.050	3.313	3.284	3.259	3.237	3.218
	0.025	4.243	4.200	4.162	4.130	4.101
	0.010	5.734	5.667	5.609	5.559	5.515
	0.005	7.104	7.015	6.938	6.872	6.814
	0.001	11.352	11.194	11.060	10.943	10.841
9	0.050	3.102	3.073	3.048	3.025	3.006
	0.025	3.912	3.868	3.831	3.798	3.769
	0.010	5.178	5.111	5.055	5.005	4.962
	0.005	6.314	6.227	6.153	6.089	6.032
	0.001	9.718	9.570	9.443	9.334	9.238
10	0.050	2.943	2.913	2.887	2.865	2.845
	0.025	3.665	3.621	3.583	3.550	3.522
	0.010	4.772	4.706	4.650	4.601	4.558
	0.005	5.746	5.661	5.589	5.526	5.471
	0.001	8.586	8.445	8.324	8.220	8.129

TABLE A.11d. Critical *F* Values

Denominator ν	α	Numerator ν 16	17	18	19	20
1	0.050	246.464	246.918	247.323	247.686	248.013
	0.025	986.919	988.733	990.350	991.797	993.102
	0.010	6170.090	6181.436	6191.527	6200.577	6208.737
	0.005	24681.450	24726.829	24767.214	24803.335	24835.957
	0.001	617053.889	618188.763	619195.633	620086.602	620918.989
2	0.050	19.433	19.437	19.440	19.443	19.446
	0.025	39.435	39.439	39.442	39.445	39.448
	0.010	99.437	99.440	99.444	99.447	99.449
	0.005	199.437	199.441	199.444	199.447	199.449
	0.001	999.428	999.436	999.440	999.441	999.443
3	0.050	8.692	8.683	8.675	8.667	8.660
	0.025	14.232	14.213	14.196	14.181	14.167
	0.010	26.827	26.787	26.751	26.719	26.690
	0.005	43.008	42.941	42.880	42.826	42.778
	0.001	127.136	126.927	126.738	126.572	126.418
4	0.050	5.844	5.832	5.821	5.811	5.803
	0.025	8.633	8.611	8.592	8.575	8.560
	0.010	14.154	14.115	14.080	14.048	14.020
	0.005	20.371	20.311	20.258	20.210	20.167
	0.001	46.597	46.451	46.322	46.205	46.100
5	0.050	4.604	4.590	4.579	4.568	4.558
	0.025	6.403	6.381	6.362	6.344	6.329
	0.010	9.680	9.643	9.610	9.580	9.553
	0.005	13.086	13.033	12.985	12.942	12.903
	0.001	25.783	25.669	25.568	25.477	25.395
6	0.050	3.922	3.908	3.896	3.884	3.874
	0.025	5.244	5.222	5.202	5.184	5.168
	0.010	7.519	7.483	7.451	7.422	7.396
	0.005	9.758	9.709	9.664	9.625	9.589
	0.001	17.450	17.353	17.267	17.190	17.120
7	0.050	3.494	3.480	3.467	3.455	3.445
	0.025	4.543	4.521	4.501	4.483	4.467
	0.010	6.275	6.240	6.209	6.181	6.155
	0.005	7.915	7.868	7.826	7.788	7.754
	0.001	13.227	13.140	13.063	12.994	12.932
8	0.050	3.202	3.187	3.173	3.161	3.150
	0.025	4.076	4.054	4.034	4.016	3.999
	0.010	5.477	5.442	5.412	5.384	5.359
	0.005	6.763	6.718	6.678	6.641	6.608
	0.001	10.752	10.672	10.601	10.537	10.480
9	0.050	2.989	2.974	2.960	2.948	2.936
	0.025	3.744	3.722	3.701	3.683	3.667
	0.010	4.924	4.890	4.860	4.833	4.808
	0.005	5.983	5.939	5.899	5.864	5.832
	0.001	9.154	9.079	9.012	8.952	8.898
10	0.050	2.828	2.812	2.798	2.785	2.774
	0.025	3.496	3.474	3.453	3.435	3.419
	0.010	4.520	4.487	4.457	4.430	4.405
	0.005	5.422	5.379	5.340	5.305	5.274
	0.001	8.048	7.977	7.913	7.856	7.804

TABLE A.11e. Critical *F* Values

Denominator ν	α	Numerator ν 21	22	23	24	25
1	0.050	248.309	248.579	248.826	249.052	249.260
	0.025	994.286	995.363	996.346	997.249	998.081
	0.010	6216.126	6222.855	6228.993	6234.629	6239.826
	0.005	24865.611	24892.464	24916.926	24939.664	24960.416
	0.001	621653.353	622320.075	622924.674	623495.668	624013.102
2	0.050	19.448	19.450	19.452	19.454	19.456
	0.025	39.450	39.452	39.454	39.456	39.458
	0.010	99.452	99.454	99.456	99.457	99.459
	0.005	199.452	199.454	199.456	199.458	199.460
	0.001	999.452	999.452	999.456	999.456	999.460
3	0.050	8.654	8.648	8.643	8.639	8.634
	0.025	14.155	14.144	14.134	14.124	14.115
	0.010	26.664	26.640	26.618	26.598	26.579
	0.005	42.733	42.693	42.656	42.622	42.591
	0.001	126.281	126.155	126.041	125.935	125.840
4	0.050	5.795	5.787	5.781	5.774	5.769
	0.025	8.546	8.533	8.522	8.511	8.501
	0.010	13.994	13.970	13.949	13.929	13.911
	0.005	20.128	20.093	20.060	20.030	20.002
	0.001	46.005	45.918	45.839	45.766	45.699
5	0.050	4.549	4.541	4.534	4.527	4.521
	0.025	6.314	6.301	6.289	6.278	6.268
	0.010	9.528	9.506	9.485	9.466	9.449
	0.005	12.868	12.836	12.807	12.780	12.755
	0.001	25.320	25.252	25.190	25.133	25.080
6	0.050	3.865	3.856	3.849	3.841	3.835
	0.025	5.154	5.141	5.128	5.117	5.107
	0.010	7.372	7.351	7.331	7.313	7.296
	0.005	9.556	9.526	9.499	9.474	9.451
	0.001	17.057	16.999	16.946	16.897	16.853
7	0.050	3.435	3.426	3.418	3.410	3.404
	0.025	4.452	4.439	4.426	4.415	4.405
	0.010	6.132	6.111	6.092	6.074	6.058
	0.005	7.723	7.695	7.669	7.645	7.623
	0.001	12.875	12.823	12.776	12.732	12.692
8	0.050	3.140	3.131	3.123	3.115	3.108
	0.025	3.985	3.971	3.959	3.947	3.937
	0.010	5.336	5.316	5.297	5.279	5.263
	0.005	6.578	6.551	6.526	6.503	6.482
	0.001	10.427	10.379	10.336	10.295	10.258
9	0.050	2.926	2.917	2.908	2.900	2.893
	0.025	3.652	3.638	3.626	3.614	3.604
	0.010	4.786	4.765	4.746	4.729	4.713
	0.005	5.803	5.776	5.752	5.729	5.708
	0.001	8.848	8.803	8.762	8.724	8.689
10	0.050	2.764	2.754	2.745	2.737	2.730
	0.025	3.403	3.390	3.377	3.365	3.355
	0.010	4.383	4.363	4.344	4.327	4.311
	0.005	5.245	5.219	5.195	5.173	5.153
	0.001	7.757	7.713	7.674	7.638	7.604

TABLE A.11f. Critical *F* Values

Denominator ν	α	Numerator ν 26	27	28	29	30
1	0.050	249.453	249.631	249.797	249.951	250.095
	0.025	998.849	999.561	1000.222	1000.839	1001.414
	0.010	6244.624	6249.061	6253.195	6257.053	6260.644
	0.005	24979.489	24997.314	25013.859	25029.224	25043.644
	0.001	624504.229	624947.959	625346.713	625750.603	626089.462
2	0.050	19.457	19.459	19.460	19.461	19.462
	0.025	39.459	39.461	39.462	39.463	39.465
	0.010	99.461	99.462	99.464	99.465	99.466
	0.005	199.461	199.462	199.464	199.465	199.466
	0.001	999.456	999.462	999.464	999.466	999.474
3	0.050	8.630	8.626	8.623	8.620	8.617
	0.025	14.107	14.100	14.093	14.087	14.081
	0.010	26.562	26.546	26.531	26.517	26.505
	0.005	42.562	42.536	42.511	42.487	42.466
	0.001	125.749	125.666	125.587	125.517	125.448
4	0.050	5.763	5.759	5.754	5.750	5.746
	0.025	8.492	8.483	8.475	8.468	8.461
	0.010	13.894	13.878	13.864	13.850	13.838
	0.005	19.977	19.953	19.931	19.911	19.891
	0.001	45.637	45.579	45.525	45.475	45.428
5	0.050	4.515	4.510	4.505	4.500	4.496
	0.025	6.258	6.250	6.242	6.234	6.227
	0.010	9.433	9.418	9.404	9.391	9.379
	0.005	12.732	12.711	12.691	12.673	12.656
	0.001	25.032	24.987	24.944	24.906	24.869
6	0.050	3.829	3.823	3.818	3.813	3.808
	0.025	5.097	5.088	5.080	5.072	5.065
	0.010	7.280	7.266	7.253	7.240	7.229
	0.005	9.430	9.410	9.392	9.374	9.358
	0.001	16.811	16.773	16.737	16.703	16.672
7	0.050	3.397	3.391	3.386	3.381	3.376
	0.025	4.395	4.386	4.378	4.370	4.362
	0.010	6.043	6.029	6.016	6.003	5.992
	0.005	7.603	7.584	7.566	7.550	7.534
	0.001	12.655	12.620	12.588	12.558	12.530
8	0.050	3.102	3.095	3.090	3.084	3.079
	0.025	3.927	3.918	3.909	3.901	3.894
	0.010	5.248	5.234	5.221	5.209	5.198
	0.005	6.462	6.444	6.427	6.411	6.396
	0.001	10.224	10.192	10.162	10.135	10.109
9	0.050	2.886	2.880	2.874	2.869	2.864
	0.025	3.594	3.584	3.576	3.568	3.560
	0.010	4.698	4.685	4.672	4.660	4.649
	0.005	5.689	5.671	5.655	5.639	5.625
	0.001	8.656	8.626	8.598	8.572	8.548
10	0.050	2.723	2.716	2.710	2.705	2.700
	0.025	3.345	3.335	3.327	3.319	3.311
	0.010	4.296	4.283	4.270	4.258	4.247
	0.005	5.134	5.116	5.100	5.085	5.071
	0.001	7.573	7.544	7.517	7.492	7.469

TABLE A.11g. Critical F Values

Denominator ν	α	Numerator ν: 1	2	3	4	5
11	0.050	4.844	3.982	3.587	3.357	3.204
	0.025	6.724	5.256	4.630	4.275	4.044
	0.010	9.646	7.206	6.217	5.668	5.316
	0.005	12.226	8.912	7.600	6.881	6.422
	0.001	19.687	13.812	11.561	10.346	9.578
12	0.050	4.747	3.885	3.490	3.259	3.106
	0.025	6.554	5.096	4.474	4.121	3.891
	0.010	9.330	6.927	5.953	5.412	5.064
	0.005	11.754	8.510	7.226	6.521	6.071
	0.001	18.643	12.974	10.804	9.633	8.892
13	0.050	4.667	3.806	3.411	3.179	3.025
	0.025	6.414	4.965	4.347	3.996	3.767
	0.010	9.074	6.701	5.739	5.205	4.862
	0.005	11.374	8.186	6.926	6.233	5.791
	0.001	17.815	12.313	10.209	9.073	8.354
14	0.050	4.600	3.739	3.344	3.112	2.958
	0.025	6.298	4.857	4.242	3.892	3.663
	0.010	8.862	6.515	5.564	5.035	4.695
	0.005	11.060	7.922	6.680	5.998	5.562
	0.001	17.143	11.779	9.729	8.622	7.922
15	0.050	4.543	3.682	3.287	3.056	2.901
	0.025	6.200	4.765	4.153	3.804	3.576
	0.010	8.683	6.359	5.417	4.893	4.556
	0.005	10.798	7.701	6.476	5.803	5.372
	0.001	16.587	11.339	9.335	8.253	7.567
16	0.050	4.494	3.634	3.239	3.007	2.852
	0.025	6.115	4.687	4.077	3.729	3.502
	0.010	8.531	6.226	5.292	4.773	4.437
	0.005	10.575	7.514	6.303	5.638	5.212
	0.001	16.120	10.971	9.006	7.944	7.272
17	0.050	4.451	3.592	3.197	2.965	2.810
	0.025	6.042	4.619	4.011	3.665	3.438
	0.010	8.400	6.112	5.185	4.669	4.336
	0.005	10.384	7.354	6.156	5.497	5.075
	0.001	15.722	10.658	8.727	7.683	7.022
18	0.050	4.414	3.555	3.160	2.928	2.773
	0.025	5.978	4.560	3.954	3.608	3.382
	0.010	8.285	6.013	5.092	4.579	4.248
	0.005	10.218	7.215	6.028	5.375	4.956
	0.001	15.379	10.390	8.487	7.459	6.808
19	0.050	4.381	3.522	3.127	2.895	2.740
	0.025	5.922	4.508	3.903	3.559	3.333
	0.010	8.185	5.926	5.010	4.500	4.171
	0.005	10.073	7.093	5.916	5.268	4.853
	0.001	15.081	10.157	8.280	7.265	6.622
20	0.050	4.351	3.493	3.098	2.866	2.711
	0.025	5.871	4.461	3.859	3.515	3.289
	0.010	8.096	5.849	4.938	4.431	4.103
	0.005	9.944	6.986	5.818	5.174	4.762
	0.001	14.819	9.953	8.098	7.096	6.461

TABLE A.11h. Critical F Values

Denominator ν	α	Numerator ν 6	7	8	9	10
11	0.050	3.095	3.012	2.948	2.896	2.854
	0.025	3.881	3.759	3.664	3.588	3.526
	0.010	5.069	4.886	4.744	4.632	4.539
	0.005	6.102	5.865	5.682	5.537	5.418
	0.001	9.047	8.655	8.355	8.116	7.922
12	0.050	2.996	2.913	2.849	2.796	2.753
	0.025	3.728	3.607	3.512	3.436	3.374
	0.010	4.821	4.640	4.499	4.388	4.296
	0.005	5.757	5.525	5.345	5.202	5.085
	0.001	8.379	8.001	7.710	7.480	7.292
13	0.050	2.915	2.832	2.767	2.714	2.671
	0.025	3.604	3.483	3.388	3.312	3.250
	0.010	4.620	4.441	4.302	4.191	4.100
	0.005	5.482	5.253	5.076	4.935	4.820
	0.001	7.856	7.489	7.206	6.982	6.799
14	0.050	2.848	2.764	2.699	2.646	2.602
	0.025	3.501	3.380	3.285	3.209	3.147
	0.010	4.456	4.278	4.140	4.030	3.939
	0.005	5.257	5.031	4.857	4.717	4.603
	0.001	7.436	7.077	6.802	6.583	6.404
15	0.050	2.790	2.707	2.641	2.588	2.544
	0.025	3.415	3.293	3.199	3.123	3.060
	0.010	4.318	4.142	4.004	3.895	3.805
	0.005	5.071	4.847	4.674	4.536	4.424
	0.001	7.092	6.741	6.471	6.256	6.081
16	0.050	2.741	2.657	2.591	2.538	2.494
	0.025	3.341	3.219	3.125	3.049	2.986
	0.010	4.202	4.026	3.890	3.780	3.691
	0.005	4.913	4.692	4.521	4.384	4.272
	0.001	6.805	6.460	6.195	5.984	5.812
17	0.050	2.699	2.614	2.548	2.494	2.450
	0.025	3.277	3.156	3.061	2.985	2.922
	0.010	4.102	3.927	3.791	3.682	3.593
	0.005	4.779	4.559	4.389	4.254	4.142
	0.001	6.562	6.223	5.962	5.754	5.584
18	0.050	2.661	2.577	2.510	2.456	2.412
	0.025	3.221	3.100	3.005	2.929	2.866
	0.010	4.015	3.841	3.705	3.597	3.508
	0.005	4.663	4.445	4.276	4.141	4.030
	0.001	6.355	6.021	5.763	5.558	5.390
19	0.050	2.628	2.544	2.477	2.423	2.378
	0.025	3.172	3.051	2.956	2.880	2.817
	0.010	3.939	3.765	3.631	3.523	3.434
	0.005	4.561	4.345	4.177	4.043	3.933
	0.001	6.175	5.845	5.590	5.388	5.222
20	0.050	2.599	2.514	2.447	2.393	2.348
	0.025	3.128	3.007	2.913	2.837	2.774
	0.010	3.871	3.699	3.564	3.457	3.368
	0.005	4.472	4.257	4.090	3.956	3.847
	0.001	6.019	5.692	5.440	5.239	5.075

TABLE A.11i. Critical F Values

Denominator ν	α	Numerator ν 11	12	13	14	15
11	0.050	2.818	2.788	2.761	2.739	2.719
	0.025	3.474	3.430	3.392	3.359	3.330
	0.010	4.462	4.397	4.342	4.293	4.251
	0.005	5.320	5.236	5.165	5.103	5.049
	0.001	7.761	7.626	7.509	7.409	7.321
12	0.050	2.717	2.687	2.660	2.637	2.617
	0.025	3.321	3.277	3.239	3.206	3.177
	0.010	4.220	4.155	4.100	4.052	4.010
	0.005	4.988	4.906	4.836	4.775	4.721
	0.001	7.136	7.005	6.892	6.794	6.709
13	0.050	2.635	2.604	2.577	2.554	2.533
	0.025	3.197	3.153	3.115	3.082	3.053
	0.010	4.025	3.960	3.905	3.857	3.815
	0.005	4.724	4.643	4.573	4.513	4.460
	0.001	6.647	6.519	6.409	6.314	6.231
14	0.050	2.565	2.534	2.507	2.484	2.463
	0.025	3.095	3.050	3.012	2.979	2.949
	0.010	3.864	3.800	3.745	3.698	3.656
	0.005	4.508	4.428	4.359	4.299	4.247
	0.001	6.256	6.130	6.023	5.930	5.848
15	0.050	2.507	2.475	2.448	2.424	2.403
	0.025	3.008	2.963	2.925	2.891	2.862
	0.010	3.730	3.666	3.612	3.564	3.522
	0.005	4.329	4.250	4.181	4.122	4.070
	0.001	5.935	5.812	5.707	5.615	5.535
16	0.050	2.456	2.425	2.397	2.373	2.352
	0.025	2.934	2.889	2.851	2.817	2.788
	0.010	3.616	3.553	3.498	3.451	3.409
	0.005	4.179	4.099	4.031	3.972	3.920
	0.001	5.668	5.547	5.443	5.353	5.274
17	0.050	2.413	2.381	2.353	2.329	2.308
	0.025	2.870	2.825	2.786	2.753	2.723
	0.010	3.519	3.455	3.401	3.353	3.312
	0.005	4.050	3.971	3.903	3.844	3.793
	0.001	5.443	5.324	5.221	5.132	5.054
18	0.050	2.374	2.342	2.314	2.290	2.269
	0.025	2.814	2.769	2.730	2.696	2.667
	0.010	3.434	3.371	3.316	3.269	3.227
	0.005	3.938	3.860	3.793	3.734	3.683
	0.001	5.250	5.132	5.031	4.943	4.866
19	0.050	2.340	2.308	2.280	2.256	2.234
	0.025	2.765	2.720	2.681	2.647	2.617
	0.010	3.360	3.297	3.242	3.195	3.153
	0.005	3.841	3.763	3.696	3.638	3.587
	0.001	5.084	4.967	4.867	4.780	4.704
20	0.050	2.310	2.278	2.250	2.225	2.203
	0.025	2.721	2.676	2.637	2.603	2.573
	0.010	3.294	3.231	3.177	3.130	3.088
	0.005	3.756	3.678	3.611	3.553	3.502
	0.001	4.939	4.823	4.724	4.637	4.562

TABLE A.11j. Critical F Values

Denominator ν	α	Numerator ν 16	17	18	19	20
11	0.050	2.701	2.685	2.671	2.658	2.646
	0.025	3.304	3.282	3.261	3.243	3.226
	0.010	4.213	4.180	4.150	4.123	4.099
	0.005	5.001	4.959	4.921	4.886	4.855
	0.001	7.244	7.175	7.113	7.058	7.008
12	0.050	2.599	2.583	2.568	2.555	2.544
	0.025	3.152	3.129	3.108	3.090	3.073
	0.010	3.972	3.939	3.909	3.883	3.858
	0.005	4.674	4.632	4.595	4.561	4.530
	0.001	6.634	6.567	6.507	6.454	6.405
13	0.050	2.515	2.499	2.484	2.471	2.459
	0.025	3.027	3.004	2.983	2.965	2.948
	0.010	3.778	3.745	3.716	3.689	3.665
	0.005	4.413	4.372	4.334	4.301	4.270
	0.001	6.158	6.093	6.034	5.982	5.934
14	0.050	2.445	2.428	2.413	2.400	2.388
	0.025	2.923	2.900	2.879	2.861	2.844
	0.010	3.619	3.586	3.556	3.529	3.505
	0.005	4.200	4.159	4.122	4.089	4.059
	0.001	5.776	5.712	5.655	5.604	5.557
15	0.050	2.385	2.368	2.353	2.340	2.328
	0.025	2.836	2.813	2.792	2.773	2.756
	0.010	3.485	3.452	3.423	3.396	3.372
	0.005	4.024	3.983	3.946	3.913	3.883
	0.001	5.464	5.402	5.345	5.294	5.248
16	0.050	2.333	2.317	2.302	2.288	2.276
	0.025	2.761	2.738	2.717	2.698	2.681
	0.010	3.372	3.339	3.310	3.283	3.259
	0.005	3.875	3.834	3.797	3.764	3.734
	0.001	5.205	5.143	5.087	5.037	4.992
17	0.050	2.289	2.272	2.257	2.243	2.230
	0.025	2.697	2.673	2.652	2.633	2.616
	0.010	3.275	3.242	3.212	3.186	3.162
	0.005	3.747	3.707	3.670	3.637	3.607
	0.001	4.986	4.924	4.869	4.820	4.775
18	0.050	2.250	2.233	2.217	2.203	2.191
	0.025	2.640	2.617	2.596	2.576	2.559
	0.010	3.190	3.158	3.128	3.101	3.077
	0.005	3.637	3.597	3.560	3.527	3.498
	0.001	4.798	4.738	4.683	4.634	4.590
19	0.050	2.215	2.198	2.182	2.168	2.155
	0.025	2.591	2.567	2.546	2.526	2.509
	0.010	3.116	3.084	3.054	3.027	3.003
	0.005	3.541	3.501	3.465	3.432	3.402
	0.001	4.636	4.576	4.522	4.474	4.430
20	0.050	2.184	2.167	2.151	2.137	2.124
	0.025	2.547	2.523	2.501	2.482	2.464
	0.010	3.051	3.018	2.989	2.962	2.938
	0.005	3.457	3.416	3.380	3.347	3.318
	0.001	4.495	4.435	4.382	4.334	4.290

TABLE A. 11k. Critical *F* Values

Denominator		Numerator ν				
ν	α	21	22	23	24	25
11	0.050	2.636	2.626	2.617	2.609	2.601
	0.025	3.211	3.197	3.184	3.173	3.162
	0.010	4.077	4.057	4.038	4.021	4.005
	0.005	4.827	4.801	4.778	4.756	4.736
	0.001	6.962	6.920	6.882	6.847	6.815
12	0.050	2.533	2.523	2.514	2.505	2.498
	0.025	3.057	3.043	3.031	3.019	3.008
	0.010	3.836	3.816	3.798	3.780	3.765
	0.005	4.502	4.476	4.453	4.431	4.412
	0.001	6.361	6.320	6.283	6.249	6.217
13	0.050	2.448	2.438	2.429	2.420	2.412
	0.025	2.932	2.918	2.905	2.893	2.882
	0.010	3.643	3.622	3.604	3.587	3.571
	0.005	4.243	4.217	4.194	4.173	4.153
	0.001	5.891	5.851	5.815	5.781	5.751
14	0.050	2.377	2.367	2.357	2.349	2.341
	0.025	2.828	2.814	2.801	2.789	2.778
	0.010	3.483	3.463	3.444	3.427	3.412
	0.005	4.031	4.006	3.983	3.961	3.942
	0.001	5.514	5.475	5.440	5.407	5.377
15	0.050	2.316	2.306	2.297	2.288	2.280
	0.025	2.740	2.726	2.713	2.701	2.689
	0.010	3.350	3.330	3.311	3.294	3.278
	0.005	3.855	3.830	3.807	3.786	3.766
	0.001	5.207	5.168	5.133	5.101	5.071
16	0.050	2.264	2.254	2.244	2.235	2.227
	0.025	2.665	2.651	2.637	2.625	2.614
	0.010	3.237	3.216	3.198	3.181	3.165
	0.005	3.707	3.682	3.659	3.638	3.618
	0.001	4.951	4.913	4.878	4.846	4.817
17	0.050	2.219	2.208	2.199	2.190	2.181
	0.025	2.600	2.585	2.572	2.560	2.548
	0.010	3.139	3.119	3.101	3.084	3.068
	0.005	3.580	3.555	3.532	3.511	3.492
	0.001	4.734	4.697	4.663	4.631	4.602
18	0.050	2.179	2.168	2.159	2.150	2.141
	0.025	2.543	2.529	2.515	2.503	2.491
	0.010	3.055	3.035	3.016	2.999	2.983
	0.005	3.471	3.446	3.423	3.402	3.382
	0.001	4.549	4.512	4.478	4.447	4.418
19	0.050	2.144	2.133	2.123	2.114	2.106
	0.025	2.493	2.478	2.465	2.452	2.441
	0.010	2.981	2.961	2.942	2.925	2.909
	0.005	3.375	3.350	3.327	3.306	3.287
	0.001	4.390	4.353	4.319	4.288	4.259
20	0.050	2.112	2.102	2.092	2.082	2.074
	0.025	2.448	2.434	2.420	2.408	2.396
	0.010	2.916	2.895	2.877	2.859	2.843
	0.005	3.291	3.266	3.243	3.222	3.203
	0.001	4.250	4.214	4.180	4.149	4.121

TABLE A.11I. Critical *F* Values

Denominator ν	α	Numerator ν 26	27	28	29	30
11	0.050	2.594	2.588	2.582	2.576	2.570
	0.025	3.152	3.142	3.133	3.125	3.118
	0.010	3.990	3.977	3.964	3.952	3.941
	0.005	4.717	4.700	4.684	4.668	4.654
	0.001	6.785	6.757	6.731	6.707	6.684
12	0.050	2.491	2.484	2.478	2.472	2.466
	0.025	2.998	2.988	2.979	2.971	2.963
	0.010	3.750	3.736	3.724	3.712	3.701
	0.005	4.393	4.376	4.360	4.345	4.331
	0.001	6.188	6.161	6.136	6.112	6.090
13	0.050	2.405	2.398	2.392	2.386	2.380
	0.025	2.872	2.862	2.853	2.845	2.837
	0.010	3.556	3.543	3.530	3.518	3.507
	0.005	4.134	4.117	4.101	4.087	4.073
	0.001	5.722	5.695	5.671	5.647	5.626
14	0.050	2.333	2.326	2.320	2.314	2.308
	0.025	2.767	2.758	2.749	2.740	2.732
	0.010	3.397	3.383	3.371	3.359	3.348
	0.005	3.923	3.906	3.891	3.876	3.862
	0.001	5.349	5.323	5.298	5.275	5.254
15	0.050	2.272	2.265	2.259	2.253	2.247
	0.025	2.679	2.669	2.660	2.652	2.644
	0.010	3.264	3.250	3.237	3.225	3.214
	0.005	3.748	3.731	3.715	3.701	3.687
	0.001	5.043	5.018	4.994	4.971	4.950
16	0.050	2.220	2.212	2.206	2.200	2.194
	0.025	2.603	2.594	2.584	2.576	2.568
	0.010	3.150	3.137	3.124	3.112	3.101
	0.005	3.600	3.583	3.567	3.553	3.539
	0.001	4.789	4.764	4.740	4.718	4.697
17	0.050	2.174	2.167	2.160	2.154	2.148
	0.025	2.538	2.528	2.519	2.510	2.502
	0.010	3.053	3.039	3.026	3.014	3.003
	0.005	3.473	3.457	3.441	3.426	3.412
	0.001	4.575	4.550	4.526	4.504	4.484
18	0.050	2.134	2.126	2.119	2.113	2.107
	0.025	2.481	2.471	2.461	2.453	2.445
	0.010	2.968	2.955	2.942	2.930	2.919
	0.005	3.364	3.347	3.332	3.317	3.303
	0.001	4.391	4.366	4.343	4.321	4.301
19	0.050	2.098	2.090	2.084	2.077	2.071
	0.025	2.430	2.420	2.411	2.402	2.394
	0.010	2.894	2.880	2.868	2.855	2.844
	0.005	3.269	3.252	3.236	3.221	3.208
	0.001	4.233	4.208	4.185	4.163	4.143
20	0.050	2.066	2.059	2.052	2.045	2.039
	0.025	2.385	2.375	2.366	2.357	2.349
	0.010	2.829	2.815	2.802	2.790	2.778
	0.005	3.184	3.168	3.152	3.137	3.123
	0.001	4.094	4.070	4.047	4.025	4.005

TABLE A.11m. Critical F Values

Denominator ν	α	Numerator ν: 1	2	3	4	5
21	0.050	4.325	3.467	3.072	2.840	2.685
	0.025	5.827	4.420	3.819	3.475	3.250
	0.010	8.017	5.780	4.874	4.369	4.042
	0.005	9.830	6.891	5.730	5.091	4.681
	0.001	14.587	9.772	7.938	6.947	6.318
22	0.050	4.301	3.443	3.049	2.817	2.661
	0.025	5.786	4.383	3.783	3.440	3.215
	0.010	7.945	5.719	4.817	4.313	3.988
	0.005	9.727	6.806	5.652	5.017	4.609
	0.001	14.380	9.612	7.796	6.814	6.191
23	0.050	4.279	3.422	3.028	2.796	2.640
	0.025	5.750	4.349	3.750	3.408	3.183
	0.010	7.881	5.664	4.765	4.264	3.939
	0.005	9.635	6.730	5.582	4.950	4.544
	0.001	14.195	9.469	7.669	6.696	6.078
24	0.050	4.260	3.403	3.009	2.776	2.621
	0.025	5.717	4.319	3.721	3.379	3.155
	0.010	7.823	5.614	4.718	4.218	3.895
	0.005	9.551	6.661	5.519	4.890	4.486
	0.001	14.028	9.339	7.554	6.589	5.977
25	0.050	4.242	3.385	2.991	2.759	2.603
	0.025	5.686	4.291	3.694	3.353	3.129
	0.010	7.770	5.568	4.675	4.177	3.855
	0.005	9.475	6.598	5.462	4.835	4.433
	0.001	13.877	9.223	7.451	6.493	5.885
26	0.050	4.225	3.369	2.975	2.743	2.587
	0.025	5.659	4.265	3.670	3.329	3.105
	0.010	7.721	5.526	4.637	4.140	3.818
	0.005	9.406	6.541	5.409	4.785	4.384
	0.001	13.739	9.116	7.357	6.406	5.802
27	0.050	4.210	3.354	2.960	2.728	2.572
	0.025	5.633	4.242	3.647	3.307	3.083
	0.010	7.677	5.488	4.601	4.106	3.785
	0.005	9.342	6.489	5.361	4.740	4.340
	0.001	13.613	9.019	7.272	6.326	5.726
28	0.050	4.196	3.340	2.947	2.714	2.558
	0.025	5.610	4.221	3.626	3.286	3.063
	0.010	7.636	5.453	4.568	4.074	3.754
	0.005	9.284	6.440	5.317	4.698	4.300
	0.001	13.498	8.931	7.193	6.253	5.656
29	0.050	4.183	3.328	2.934	2.701	2.545
	0.025	5.588	4.201	3.607	3.267	3.044
	0.010	7.598	5.420	4.538	4.045	3.725
	0.005	9.230	6.396	5.276	4.659	4.262
	0.001	13.391	8.849	7.121	6.186	5.593
30	0.050	4.171	3.316	2.922	2.690	2.534
	0.025	5.568	4.182	3.589	3.250	3.026
	0.010	7.562	5.390	4.510	4.018	3.699
	0.005	9.180	6.355	5.239	4.623	4.228
	0.001	13.293	8.773	7.054	6.125	5.534

TABLE A.11n. Critical *F* Values

Denominator		Numerator ν				
ν	α	6	7	8	9	10
21	0.050	2.573	2.488	2.420	2.366	2.321
	0.025	3.090	2.969	2.874	2.798	2.735
	0.010	3.812	3.640	3.506	3.398	3.310
	0.005	4.393	4.179	4.013	3.880	3.771
	0.001	5.881	5.557	5.308	5.109	4.946
22	0.050	2.549	2.464	2.397	2.342	2.297
	0.025	3.055	2.934	2.839	2.763	2.700
	0.010	3.758	3.587	3.453	3.346	3.258
	0.005	4.322	4.109	3.944	3.812	3.703
	0.001	5.758	5.438	5.190	4.993	4.832
23	0.050	2.528	2.442	2.375	2.320	2.275
	0.025	3.023	2.902	2.808	2.731	2.668
	0.010	3.710	3.539	3.406	3.299	3.211
	0.005	4.259	4.047	3.882	3.750	3.642
	0.001	5.649	5.331	5.085	4.890	4.730
24	0.050	2.508	2.423	2.355	2.300	2.255
	0.025	2.995	2.874	2.779	2.703	2.640
	0.010	3.667	3.496	3.363	3.256	3.168
	0.005	4.202	3.991	3.826	3.695	3.587
	0.001	5.550	5.235	4.991	4.797	4.638
25	0.050	2.490	2.405	2.337	2.282	2.236
	0.025	2.969	2.848	2.753	2.677	2.613
	0.010	3.627	3.457	3.324	3.217	3.129
	0.005	4.150	3.939	3.776	3.645	3.537
	0.001	5.462	5.148	4.906	4.713	4.555
26	0.050	2.474	2.388	2.321	2.265	2.220
	0.025	2.945	2.824	2.729	2.653	2.590
	0.010	3.591	3.421	3.288	3.182	3.094
	0.005	4.103	3.893	3.730	3.599	3.492
	0.001	5.381	5.070	4.829	4.637	4.480
27	0.050	2.459	2.373	2.305	2.250	2.204
	0.025	2.923	2.802	2.707	2.631	2.568
	0.010	3.558	3.388	3.256	3.149	3.062
	0.005	4.059	3.850	3.687	3.557	3.450
	0.001	5.308	4.998	4.759	4.568	4.412
28	0.050	2.445	2.359	2.291	2.236	2.190
	0.025	2.903	2.782	2.687	2.611	2.547
	0.010	3.528	3.358	3.226	3.120	3.032
	0.005	4.020	3.811	3.649	3.519	3.412
	0.001	5.241	4.933	4.695	4.505	4.349
29	0.050	2.432	2.346	2.278	2.223	2.177
	0.025	2.884	2.763	2.669	2.592	2.529
	0.010	3.499	3.330	3.198	3.092	3.005
	0.005	3.983	3.775	3.613	3.483	3.377
	0.001	5.179	4.873	4.636	4.447	4.292
30	0.050	2.421	2.334	2.266	2.211	2.165
	0.025	2.867	2.746	2.651	2.575	2.511
	0.010	3.473	3.304	3.173	3.067	2.979
	0.005	3.949	3.742	3.580	3.450	3.344
	0.001	5.122	4.817	4.581	4.393	4.239

TABLE A.11o. Critical F Values

Denominator ν	α	Numerator ν 11	12	13	14	15
21	0.050	2.283	2.250	2.222	2.197	2.176
	0.025	2.682	2.637	2.598	2.564	2.534
	0.010	3.236	3.173	3.119	3.072	3.030
	0.005	3.680	3.602	3.536	3.478	3.427
	0.001	4.811	4.696	4.597	4.512	4.437
22	0.050	2.259	2.226	2.198	2.173	2.151
	0.025	2.647	2.602	2.563	2.528	2.498
	0.010	3.184	3.121	3.067	3.019	2.978
	0.005	3.612	3.535	3.469	3.411	3.360
	0.001	4.697	4.583	4.486	4.401	4.326
23	0.050	2.236	2.204	2.175	2.150	2.128
	0.025	2.615	2.570	2.531	2.497	2.466
	0.010	3.137	3.074	3.020	2.973	2.931
	0.005	3.551	3.475	3.408	3.351	3.300
	0.001	4.596	4.483	4.386	4.301	4.227
24	0.050	2.216	2.183	2.155	2.130	2.108
	0.025	2.586	2.541	2.502	2.468	2.437
	0.010	3.094	3.032	2.977	2.930	2.889
	0.005	3.497	3.420	3.354	3.296	3.246
	0.001	4.505	4.393	4.296	4.212	4.139
25	0.050	2.198	2.165	2.136	2.111	2.089
	0.025	2.560	2.515	2.476	2.441	2.411
	0.010	3.056	2.993	2.939	2.892	2.850
	0.005	3.447	3.370	3.304	3.247	3.196
	0.001	4.423	4.312	4.216	4.132	4.059
26	0.050	2.181	2.148	2.119	2.094	2.072
	0.025	2.536	2.491	2.451	2.417	2.387
	0.010	3.021	2.958	2.904	2.857	2.815
	0.005	3.402	3.325	3.259	3.202	3.151
	0.001	4.349	4.238	4.142	4.059	3.986
27	0.050	2.166	2.132	2.103	2.078	2.056
	0.025	2.514	2.469	2.429	2.395	2.364
	0.010	2.988	2.926	2.871	2.824	2.783
	0.005	3.360	3.284	3.218	3.161	3.110
	0.001	4.281	4.171	4.075	3.993	3.920
28	0.050	2.151	2.118	2.089	2.064	2.041
	0.025	2.494	2.448	2.409	2.374	2.344
	0.010	2.959	2.896	2.842	2.795	2.753
	0.005	3.322	3.246	3.180	3.123	3.073
	0.001	4.219	4.109	4.014	3.932	3.859
29	0.050	2.138	2.104	2.075	2.050	2.027
	0.025	2.475	2.430	2.390	2.355	2.325
	0.010	2.931	2.868	2.814	2.767	2.726
	0.005	3.287	3.211	3.145	3.088	3.038
	0.001	4.162	4.053	3.958	3.876	3.804
30	0.050	2.126	2.092	2.063	2.037	2.015
	0.025	2.458	2.412	2.372	2.338	2.307
	0.010	2.906	2.843	2.789	2.742	2.700
	0.005	3.255	3.179	3.113	3.056	3.006
	0.001	4.110	4.001	3.907	3.825	3.753

TABLE A.11p. Critical F Values

Denominator ν	α	Numerator ν 16	17	18	19	20
21	0.050	2.156	2.139	2.123	2.109	2.096
	0.025	2.507	2.483	2.462	2.442	2.425
	0.010	2.993	2.960	2.931	2.904	2.880
	0.005	3.382	3.342	3.305	3.273	3.243
	0.001	4.371	4.311	4.258	4.210	4.167
22	0.050	2.131	2.114	2.098	2.084	2.071
	0.025	2.472	2.448	2.426	2.407	2.389
	0.010	2.941	2.908	2.879	2.852	2.827
	0.005	3.315	3.275	3.239	3.206	3.176
	0.001	4.260	4.201	4.149	4.101	4.058
23	0.050	2.109	2.091	2.075	2.061	2.048
	0.025	2.440	2.416	2.394	2.374	2.357
	0.010	2.894	2.861	2.832	2.805	2.781
	0.005	3.255	3.215	3.179	3.146	3.116
	0.001	4.162	4.103	4.051	4.004	3.961
24	0.050	2.088	2.070	2.054	2.040	2.027
	0.025	2.411	2.386	2.365	2.345	2.327
	0.010	2.852	2.819	2.789	2.762	2.738
	0.005	3.201	3.161	3.125	3.092	3.062
	0.001	4.074	4.015	3.963	3.916	3.873
25	0.050	2.069	2.051	2.035	2.021	2.007
	0.025	2.384	2.360	2.338	2.318	2.300
	0.010	2.813	2.780	2.751	2.724	2.699
	0.005	3.151	3.111	3.075	3.043	3.013
	0.001	3.994	3.936	3.884	3.837	3.794
26	0.050	2.052	2.034	2.018	2.003	1.990
	0.025	2.360	2.335	2.314	2.294	2.276
	0.010	2.778	2.745	2.715	2.688	2.664
	0.005	3.107	3.067	3.031	2.998	2.968
	0.001	3.921	3.864	3.812	3.765	3.723
27	0.050	2.036	2.018	2.002	1.987	1.974
	0.025	2.337	2.313	2.291	2.271	2.253
	0.010	2.746	2.713	2.683	2.656	2.632
	0.005	3.066	3.026	2.990	2.957	2.928
	0.001	3.856	3.798	3.747	3.700	3.658
28	0.050	2.021	2.003	1.987	1.972	1.959
	0.025	2.317	2.292	2.270	2.251	2.232
	0.010	2.716	2.683	2.653	2.626	2.602
	0.005	3.028	2.988	2.952	2.919	2.890
	0.001	3.795	3.738	3.687	3.640	3.598
29	0.050	2.007	1.989	1.973	1.958	1.945
	0.025	2.298	2.273	2.251	2.231	2.213
	0.010	2.689	2.656	2.626	2.599	2.574
	0.005	2.993	2.953	2.917	2.885	2.855
	0.001	3.740	3.683	3.632	3.585	3.543
30	0.050	1.995	1.976	1.960	1.945	1.932
	0.025	2.280	2.255	2.233	2.213	2.195
	0.010	2.663	2.630	2.600	2.573	2.549
	0.005	2.961	2.921	2.885	2.853	2.823
	0.001	3.689	3.632	3.581	3.535	3.493

TABLE A.11q. Critical *F* Values

Denominator ν	α	Numerator ν 21	22	23	24	25
21	0.050	2.084	2.073	2.063	2.054	2.045
	0.025	2.409	2.394	2.380	2.368	2.356
	0.010	2.857	2.837	2.818	2.801	2.785
	0.005	3.216	3.191	3.168	3.147	3.128
	0.001	4.127	4.091	4.058	4.027	3.999
22	0.050	2.059	2.048	2.038	2.028	2.020
	0.025	2.373	2.358	2.344	2.331	2.320
	0.010	2.805	2.785	2.766	2.749	2.733
	0.005	3.149	3.125	3.102	3.081	3.061
	0.001	4.019	3.983	3.949	3.919	3.891
23	0.050	2.036	2.025	2.014	2.005	1.996
	0.025	2.340	2.325	2.312	2.299	2.287
	0.010	2.758	2.738	2.719	2.702	2.686
	0.005	3.089	3.065	3.042	3.021	3.001
	0.001	3.921	3.886	3.853	3.822	3.794
24	0.050	2.015	2.003	1.993	1.984	1.975
	0.025	2.311	2.296	2.282	2.269	2.257
	0.010	2.716	2.695	2.676	2.659	2.643
	0.005	3.035	3.011	2.988	2.967	2.947
	0.001	3.834	3.799	3.766	3.735	3.707
25	0.050	1.995	1.984	1.974	1.964	1.955
	0.025	2.284	2.269	2.255	2.242	2.230
	0.010	2.677	2.657	2.638	2.620	2.604
	0.005	2.986	2.961	2.939	2.918	2.898
	0.001	3.756	3.720	3.687	3.657	3.629
26	0.050	1.978	1.966	1.956	1.946	1.938
	0.025	2.259	2.244	2.230	2.217	2.205
	0.010	2.642	2.621	2.602	2.585	2.569
	0.005	2.941	2.917	2.894	2.873	2.853
	0.001	3.684	3.649	3.616	3.586	3.558
27	0.050	1.961	1.950	1.940	1.930	1.921
	0.025	2.237	2.222	2.208	2.195	2.183
	0.010	2.609	2.589	2.570	2.552	2.536
	0.005	2.900	2.876	2.853	2.832	2.812
	0.001	3.619	3.584	3.551	3.521	3.493
28	0.050	1.946	1.935	1.924	1.915	1.906
	0.025	2.216	2.201	2.187	2.174	2.161
	0.010	2.579	2.559	2.540	2.522	2.506
	0.005	2.863	2.838	2.815	2.794	2.775
	0.001	3.560	3.524	3.492	3.462	3.434
29	0.050	1.932	1.921	1.910	1.901	1.891
	0.025	2.196	2.181	2.167	2.154	2.142
	0.010	2.552	2.531	2.512	2.495	2.478
	0.005	2.828	2.803	2.780	2.759	2.740
	0.001	3.505	3.470	3.437	3.407	3.380
30	0.050	1.919	1.908	1.897	1.887	1.878
	0.025	2.178	2.163	2.149	2.136	2.124
	0.010	2.526	2.506	2.487	2.469	2.453
	0.005	2.796	2.771	2.748	2.727	2.708
	0.001	3.454	3.419	3.387	3.357	3.330

TABLE A.11r. Critical *F* Values

Denominator ν	α	Numerator ν 26	27	28	29	30
21	0.050	2.037	2.030	2.023	2.016	2.010
	0.025	2.345	2.335	2.325	2.317	2.308
	0.010	2.770	2.756	2.743	2.731	2.720
	0.005	3.110	3.093	3.077	3.063	3.049
	0.001	3.972	3.948	3.925	3.904	3.884
22	0.050	2.012	2.004	1.997	1.990	1.984
	0.025	2.309	2.299	2.289	2.280	2.272
	0.010	2.718	2.704	2.691	2.679	2.667
	0.005	3.043	3.026	3.011	2.996	2.982
	0.001	3.864	3.840	3.817	3.796	3.776
23	0.050	1.988	1.981	1.973	1.967	1.961
	0.025	2.276	2.266	2.256	2.247	2.239
	0.010	2.671	2.657	2.644	2.632	2.620
	0.005	2.983	2.966	2.951	2.936	2.922
	0.001	3.768	3.744	3.721	3.700	3.680
24	0.050	1.967	1.959	1.952	1.945	1.939
	0.025	2.246	2.236	2.226	2.217	2.209
	0.010	2.628	2.614	2.601	2.589	2.577
	0.005	2.929	2.912	2.897	2.882	2.868
	0.001	3.681	3.657	3.634	3.613	3.593
25	0.050	1.947	1.939	1.932	1.926	1.919
	0.025	2.219	2.209	2.199	2.190	2.182
	0.010	2.589	2.575	2.562	2.550	2.538
	0.005	2.880	2.863	2.847	2.833	2.819
	0.001	3.603	3.579	3.556	3.535	3.515
26	0.050	1.929	1.921	1.914	1.907	1.901
	0.025	2.194	2.184	2.174	2.165	2.157
	0.010	2.554	2.540	2.526	2.514	2.503
	0.005	2.835	2.818	2.802	2.788	2.774
	0.001	3.532	3.508	3.486	3.464	3.445
27	0.050	1.913	1.905	1.898	1.891	1.884
	0.025	2.171	2.161	2.151	2.142	2.133
	0.010	2.521	2.507	2.494	2.481	2.470
	0.005	2.794	2.777	2.761	2.747	2.733
	0.001	3.467	3.443	3.421	3.400	3.380
28	0.050	1.897	1.889	1.882	1.875	1.869
	0.025	2.150	2.140	2.130	2.121	2.112
	0.010	2.491	2.477	2.464	2.451	2.440
	0.005	2.756	2.739	2.724	2.709	2.695
	0.001	3.408	3.384	3.362	3.341	3.321
29	0.050	1.883	1.875	1.868	1.861	1.854
	0.025	2.131	2.120	2.110	2.101	2.092
	0.010	2.463	2.449	2.436	2.423	2.412
	0.005	2.722	2.705	2.689	2.674	2.660
	0.001	3.354	3.330	3.308	3.287	3.267
30	0.050	1.870	1.862	1.854	1.847	1.841
	0.025	2.112	2.102	2.092	2.083	2.074
	0.010	2.437	2.423	2.410	2.398	2.386
	0.005	2.689	2.672	2.657	2.642	2.628
	0.001	3.304	3.280	3.258	3.237	3.217

TABLE A.11s. Critical *F* Values

Denominator ν	α	Numerator ν 1	2	3	4	5
40	0.050	4.085	3.232	2.839	2.606	2.449
	0.025	5.424	4.051	3.463	3.126	2.904
	0.010	7.314	5.179	4.313	3.828	3.514
	0.005	8.828	6.066	4.976	4.374	3.986
	0.001	12.609	8.251	6.595	5.698	5.128
45	0.050	4.057	3.204	2.812	2.579	2.422
	0.025	5.377	4.009	3.422	3.086	2.864
	0.010	7.234	5.110	4.249	3.767	3.454
	0.005	8.715	5.974	4.892	4.294	3.909
	0.001	12.392	8.086	6.450	5.564	5.001
50	0.050	4.034	3.183	2.790	2.557	2.400
	0.025	5.340	3.975	3.390	3.054	2.833
	0.010	7.171	5.057	4.199	3.720	3.408
	0.005	8.626	5.902	4.826	4.232	3.849
	0.001	12.222	7.956	6.336	5.459	4.901
60	0.050	4.001	3.150	2.758	2.525	2.368
	0.025	5.286	3.925	3.343	3.008	2.786
	0.010	7.077	4.977	4.126	3.649	3.339
	0.005	8.495	5.795	4.729	4.140	3.760
	0.001	11.973	7.768	6.171	5.307	4.757
70	0.050	3.978	3.128	2.736	2.503	2.346
	0.025	5.247	3.890	3.309	2.975	2.754
	0.010	7.011	4.922	4.074	3.600	3.291
	0.005	8.403	5.720	4.661	4.076	3.698
	0.001	11.799	7.637	6.057	5.201	4.656
80	0.050	3.960	3.111	2.719	2.486	2.329
	0.025	5.218	3.864	3.284	2.950	2.730
	0.010	6.963	4.881	4.036	3.563	3.255
	0.005	8.335	5.665	4.611	4.029	3.652
	0.001	11.671	7.540	5.972	5.123	4.582
90	0.050	3.947	3.098	2.706	2.473	2.316
	0.025	5.196	3.844	3.265	2.932	2.711
	0.010	6.925	4.849	4.007	3.535	3.228
	0.005	8.282	5.623	4.573	3.992	3.617
	0.001	11.573	7.466	5.908	5.064	4.526
100	0.050	3.936	3.087	2.696	2.463	2.305
	0.025	5.179	3.828	3.250	2.917	2.696
	0.010	6.895	4.824	3.984	3.513	3.206
	0.005	8.241	5.589	4.542	3.963	3.589
	0.001	11.495	7.408	5.857	5.017	4.482
150	0.050	3.904	3.056	2.665	2.432	2.274
	0.025	5.126	3.781	3.204	2.872	2.652
	0.010	6.807	4.749	3.915	3.447	3.142
	0.005	8.118	5.490	4.453	3.878	3.508
	0.001	11.267	7.236	5.707	4.879	4.351
200	0.050	3.888	3.041	2.650	2.417	2.259
	0.025	5.100	3.758	3.182	2.850	2.630
	0.010	6.763	4.713	3.881	3.414	3.110
	0.005	8.057	5.441	4.408	3.837	3.467
	0.001	11.155	7.152	5.634	4.812	4.287

TABLE A.11t. Critical F Values

Denominator ν	α	Numerator ν 6	7	8	9	10
40	0.050	2.336	2.249	2.180	2.124	2.077
	0.025	2.744	2.624	2.529	2.452	2.388
	0.010	3.291	3.124	2.993	2.888	2.801
	0.005	3.713	3.509	3.350	3.222	3.117
	0.001	4.731	4.436	4.207	4.024	3.874
45	0.050	2.308	2.221	2.152	2.096	2.049
	0.025	2.705	2.584	2.489	2.412	2.348
	0.010	3.232	3.066	2.935	2.830	2.743
	0.005	3.638	3.435	3.276	3.149	3.044
	0.001	4.608	4.316	4.090	3.909	3.760
50	0.050	2.286	2.199	2.130	2.073	2.026
	0.025	2.674	2.553	2.458	2.381	2.317
	0.010	3.186	3.020	2.890	2.785	2.698
	0.005	3.579	3.376	3.219	3.092	2.988
	0.001	4.512	4.222	3.998	3.818	3.671
60	0.050	2.254	2.167	2.097	2.040	1.993
	0.025	2.627	2.507	2.412	2.334	2.270
	0.010	3.119	2.953	2.823	2.718	2.632
	0.005	3.492	3.291	3.134	3.008	2.904
	0.001	4.372	4.086	3.865	3.687	3.541
70	0.050	2.231	2.143	2.074	2.017	1.969
	0.025	2.595	2.474	2.379	2.302	2.237
	0.010	3.071	2.906	2.777	2.672	2.585
	0.005	3.431	3.232	3.075	2.950	2.846
	0.001	4.275	3.992	3.773	3.596	3.452
80	0.050	2.214	2.126	2.056	1.999	1.951
	0.025	2.571	2.450	2.355	2.277	2.213
	0.010	3.036	2.871	2.742	2.637	2.551
	0.005	3.387	3.188	3.032	2.907	2.803
	0.001	4.204	3.923	3.705	3.530	3.386
90	0.050	2.201	2.113	2.043	1.986	1.938
	0.025	2.552	2.432	2.336	2.259	2.194
	0.010	3.009	2.845	2.715	2.611	2.524
	0.005	3.352	3.154	2.999	2.873	2.770
	0.001	4.150	3.870	3.653	3.479	3.336
100	0.050	2.191	2.103	2.032	1.975	1.927
	0.025	2.537	2.417	2.321	2.244	2.179
	0.010	2.988	2.823	2.694	2.590	2.503
	0.005	3.325	3.127	2.972	2.847	2.744
	0.001	4.107	3.829	3.612	3.439	3.296
150	0.050	2.160	2.071	2.001	1.943	1.894
	0.025	2.494	2.373	2.278	2.200	2.135
	0.010	2.924	2.761	2.632	2.528	2.441
	0.005	3.245	3.048	2.894	2.770	2.667
	0.001	3.981	3.706	3.493	3.321	3.179
200	0.050	2.144	2.056	1.985	1.927	1.878
	0.025	2.472	2.351	2.256	2.178	2.113
	0.010	2.893	2.730	2.601	2.497	2.411
	0.005	3.206	3.010	2.856	2.732	2.629
	0.001	3.920	3.647	3.434	3.264	3.123

TABLE A.11u. Critical *F* Values

Denominator ν	α	Numerator ν 11	12	13	14	15
40	0.050	2.038	2.003	1.974	1.948	1.924
	0.025	2.334	2.288	2.248	2.213	2.182
	0.010	2.727	2.665	2.611	2.563	2.522
	0.005	3.028	2.953	2.888	2.831	2.781
	0.001	3.749	3.642	3.551	3.471	3.400
45	0.050	2.009	1.974	1.945	1.918	1.895
	0.025	2.294	2.248	2.208	2.172	2.141
	0.010	2.670	2.608	2.553	2.506	2.464
	0.005	2.956	2.881	2.816	2.759	2.709
	0.001	3.636	3.530	3.439	3.360	3.290
50	0.050	1.986	1.952	1.921	1.895	1.871
	0.025	2.263	2.216	2.176	2.140	2.109
	0.010	2.625	2.562	2.508	2.461	2.419
	0.005	2.900	2.825	2.760	2.703	2.653
	0.001	3.548	3.443	3.352	3.273	3.204
60	0.050	1.952	1.917	1.887	1.860	1.836
	0.025	2.216	2.169	2.129	2.093	2.061
	0.010	2.559	2.496	2.442	2.394	2.352
	0.005	2.817	2.742	2.677	2.620	2.570
	0.001	3.419	3.315	3.226	3.147	3.078
70	0.050	1.928	1.893	1.863	1.836	1.812
	0.025	2.183	2.136	2.095	2.059	2.028
	0.010	2.512	2.450	2.395	2.348	2.306
	0.005	2.759	2.684	2.619	2.563	2.513
	0.001	3.330	3.227	3.138	3.060	2.991
80	0.050	1.910	1.875	1.845	1.817	1.793
	0.025	2.158	2.111	2.071	2.035	2.003
	0.010	2.478	2.415	2.361	2.313	2.271
	0.005	2.716	2.641	2.577	2.520	2.470
	0.001	3.265	3.162	3.074	2.996	2.927
90	0.050	1.897	1.861	1.830	1.803	1.779
	0.025	2.140	2.092	2.051	2.015	1.983
	0.010	2.451	2.389	2.334	2.286	2.244
	0.005	2.683	2.608	2.544	2.487	2.437
	0.001	3.215	3.113	3.024	2.947	2.879
100	0.050	1.886	1.850	1.819	1.792	1.768
	0.025	2.124	2.077	2.036	2.000	1.968
	0.010	2.430	2.368	2.313	2.265	2.223
	0.005	2.657	2.583	2.518	2.461	2.411
	0.001	3.176	3.074	2.986	2.908	2.840
150	0.050	1.853	1.817	1.786	1.758	1.734
	0.025	2.080	2.032	1.991	1.955	1.922
	0.010	2.368	2.305	2.251	2.203	2.160
	0.005	2.580	2.506	2.441	2.385	2.335
	0.001	3.061	2.959	2.872	2.795	2.727
200	0.050	1.837	1.801	1.769	1.742	1.717
	0.025	2.058	2.010	1.969	1.932	1.900
	0.010	2.338	2.275	2.220	2.172	2.129
	0.005	2.543	2.468	2.404	2.347	2.297
	0.001	3.005	2.904	2.816	2.740	2.672

TABLE A.11v. Critical F Values

Denominator ν	α	Numerator ν 16	17	18	19	20
40	0.050	1.904	1.885	1.868	1.853	1.839
	0.025	2.154	2.129	2.107	2.086	2.068
	0.010	2.484	2.451	2.421	2.394	2.369
	0.005	2.737	2.697	2.661	2.628	2.598
	0.001	3.338	3.282	3.232	3.186	3.145
45	0.050	1.874	1.855	1.838	1.823	1.808
	0.025	2.113	2.088	2.066	2.045	2.026
	0.010	2.427	2.393	2.363	2.336	2.311
	0.005	2.665	2.625	2.589	2.556	2.527
	0.001	3.228	3.172	3.122	3.077	3.036
50	0.050	1.850	1.831	1.814	1.798	1.784
	0.025	2.081	2.056	2.033	2.012	1.993
	0.010	2.382	2.348	2.318	2.290	2.265
	0.005	2.609	2.569	2.533	2.500	2.470
	0.001	3.142	3.086	3.037	2.992	2.951
60	0.050	1.815	1.796	1.778	1.763	1.748
	0.025	2.033	2.008	1.985	1.964	1.944
	0.010	2.315	2.281	2.251	2.223	2.198
	0.005	2.526	2.486	2.450	2.417	2.387
	0.001	3.017	2.962	2.912	2.867	2.827
70	0.050	1.790	1.771	1.753	1.737	1.722
	0.025	1.999	1.974	1.950	1.929	1.910
	0.010	2.268	2.234	2.204	2.176	2.150
	0.005	2.468	2.428	2.392	2.359	2.329
	0.001	2.930	2.875	2.826	2.781	2.741
80	0.050	1.772	1.752	1.734	1.718	1.703
	0.025	1.974	1.948	1.925	1.904	1.884
	0.010	2.233	2.199	2.169	2.141	2.115
	0.005	2.425	2.385	2.349	2.316	2.286
	0.001	2.867	2.812	2.763	2.718	2.677
90	0.050	1.757	1.737	1.720	1.703	1.688
	0.025	1.955	1.929	1.905	1.884	1.864
	0.010	2.206	2.172	2.142	2.114	2.088
	0.005	2.393	2.353	2.316	2.283	2.253
	0.001	2.818	2.763	2.714	2.670	2.629
100	0.050	1.746	1.726	1.708	1.691	1.676
	0.025	1.939	1.913	1.890	1.868	1.849
	0.010	2.185	2.151	2.120	2.092	2.067
	0.005	2.367	2.326	2.290	2.257	2.227
	0.001	2.780	2.725	2.676	2.632	2.591
150	0.050	1.711	1.691	1.673	1.656	1.641
	0.025	1.893	1.867	1.843	1.821	1.801
	0.010	2.122	2.088	2.057	2.029	2.003
	0.005	2.290	2.250	2.213	2.180	2.150
	0.001	2.667	2.613	2.564	2.519	2.479
200	0.050	1.694	1.674	1.656	1.639	1.623
	0.025	1.870	1.844	1.820	1.798	1.778
	0.010	2.091	2.057	2.026	1.997	1.971
	0.005	2.252	2.212	2.175	2.142	2.112
	0.001	2.612	2.558	2.509	2.465	2.424

TABLE A.11w. Critical *F* Values

Denominator ν	α	Numerator ν 21	22	23	24	25
40	0.050	1.826	1.814	1.803	1.793	1.783
	0.025	2.051	2.035	2.020	2.007	1.994
	0.010	2.346	2.325	2.306	2.288	2.271
	0.005	2.571	2.546	2.523	2.502	2.482
	0.001	3.107	3.073	3.041	3.011	2.984
45	0.050	1.795	1.783	1.772	1.762	1.752
	0.025	2.009	1.993	1.978	1.965	1.952
	0.010	2.288	2.267	2.248	2.230	2.213
	0.005	2.499	2.474	2.451	2.430	2.410
	0.001	2.998	2.964	2.932	2.902	2.875
50	0.050	1.771	1.759	1.748	1.737	1.727
	0.025	1.976	1.960	1.945	1.931	1.919
	0.010	2.242	2.221	2.202	2.183	2.167
	0.005	2.443	2.418	2.395	2.373	2.353
	0.001	2.913	2.879	2.847	2.817	2.790
60	0.050	1.735	1.722	1.711	1.700	1.690
	0.025	1.927	1.911	1.896	1.882	1.869
	0.010	2.175	2.153	2.134	2.115	2.098
	0.005	2.360	2.335	2.311	2.290	2.270
	0.001	2.789	2.755	2.723	2.694	2.667
70	0.050	1.709	1.696	1.685	1.674	1.664
	0.025	1.892	1.876	1.861	1.847	1.833
	0.010	2.127	2.106	2.086	2.067	2.050
	0.005	2.302	2.276	2.253	2.231	2.211
	0.001	2.703	2.669	2.637	2.608	2.581
80	0.050	1.689	1.677	1.665	1.654	1.644
	0.025	1.866	1.850	1.835	1.820	1.807
	0.010	2.092	2.070	2.050	2.032	2.015
	0.005	2.259	2.233	2.210	2.188	2.168
	0.001	2.640	2.606	2.574	2.545	2.518
90	0.050	1.675	1.662	1.650	1.639	1.629
	0.025	1.846	1.830	1.814	1.800	1.787
	0.010	2.065	2.043	2.023	2.004	1.987
	0.005	2.226	2.200	2.177	2.155	2.134
	0.001	2.592	2.558	2.526	2.497	2.469
100	0.050	1.663	1.650	1.638	1.627	1.616
	0.025	1.830	1.814	1.798	1.784	1.770
	0.010	2.043	2.021	2.001	1.983	1.965
	0.005	2.199	2.174	2.150	2.128	2.108
	0.001	2.554	2.519	2.488	2.458	2.431
150	0.050	1.627	1.614	1.602	1.590	1.580
	0.025	1.783	1.766	1.750	1.736	1.722
	0.010	1.979	1.957	1.937	1.918	1.900
	0.005	2.122	2.096	2.072	2.050	2.030
	0.001	2.442	2.407	2.376	2.346	2.319
200	0.050	1.609	1.596	1.583	1.572	1.561
	0.025	1.759	1.742	1.726	1.712	1.698
	0.010	1.947	1.925	1.905	1.886	1.868
	0.005	2.084	2.058	2.034	2.012	1.991
	0.001	2.387	2.353	2.321	2.292	2.264

TABLE A.11x. Critical *F* Values

Denominator ν	α	Numerator ν 26	27	28	29	30
40	0.050	1.775	1.766	1.759	1.751	1.744
	0.025	1.983	1.972	1.962	1.952	1.943
	0.010	2.256	2.241	2.228	2.215	2.203
	0.005	2.464	2.447	2.431	2.416	2.401
	0.001	2.958	2.935	2.912	2.892	2.872
45	0.050	1.743	1.735	1.727	1.720	1.713
	0.025	1.940	1.929	1.919	1.909	1.900
	0.010	2.197	2.183	2.169	2.156	2.144
	0.005	2.392	2.374	2.358	2.343	2.329
	0.001	2.850	2.826	2.804	2.783	2.763
50	0.050	1.718	1.710	1.702	1.694	1.687
	0.025	1.907	1.895	1.885	1.875	1.866
	0.010	2.151	2.136	2.123	2.110	2.098
	0.005	2.335	2.317	2.301	2.286	2.272
	0.001	2.765	2.741	2.719	2.698	2.679
60	0.050	1.681	1.672	1.664	1.656	1.649
	0.025	1.857	1.845	1.835	1.825	1.815
	0.010	2.083	2.068	2.054	2.041	2.028
	0.005	2.251	2.234	2.217	2.202	2.187
	0.001	2.641	2.617	2.595	2.574	2.555
70	0.050	1.654	1.646	1.637	1.629	1.622
	0.025	1.821	1.810	1.799	1.789	1.779
	0.010	2.034	2.019	2.005	1.992	1.980
	0.005	2.192	2.175	2.158	2.143	2.128
	0.001	2.555	2.532	2.509	2.489	2.469
80	0.050	1.634	1.626	1.617	1.609	1.602
	0.025	1.795	1.783	1.772	1.762	1.752
	0.010	1.999	1.983	1.969	1.956	1.944
	0.005	2.149	2.131	2.115	2.099	2.084
	0.001	2.492	2.468	2.446	2.425	2.406
90	0.050	1.619	1.610	1.601	1.593	1.586
	0.025	1.774	1.763	1.752	1.741	1.731
	0.010	1.971	1.956	1.942	1.928	1.916
	0.005	2.115	2.098	2.081	2.065	2.051
	0.001	2.444	2.420	2.398	2.377	2.357
100	0.050	1.607	1.598	1.589	1.581	1.573
	0.025	1.758	1.746	1.735	1.725	1.715
	0.010	1.949	1.934	1.919	1.906	1.893
	0.005	2.089	2.071	2.054	2.039	2.024
	0.001	2.406	2.382	2.360	2.339	2.319
150	0.050	1.570	1.560	1.552	1.543	1.535
	0.025	1.709	1.697	1.686	1.675	1.665
	0.010	1.884	1.868	1.854	1.840	1.827
	0.005	2.010	1.992	1.975	1.959	1.944
	0.001	2.293	2.270	2.247	2.226	2.206
200	0.050	1.551	1.542	1.533	1.524	1.516
	0.025	1.685	1.673	1.661	1.650	1.640
	0.010	1.851	1.836	1.821	1.807	1.794
	0.005	1.972	1.953	1.936	1.920	1.905
	0.001	2.239	2.215	2.192	2.171	2.151

TABLE A.12a. Fisher's z Transformation

r	0.00	0.01	0.02	0.03	0.04	0.05	0.06	0.07	0.08	0.09
0.0	0.000	0.010	0.020	0.030	0.040	0.050	0.060	0.070	0.080	0.090
0.1	0.100	0.110	0.121	0.131	0.141	0.151	0.161	0.172	0.182	0.192
0.2	0.203	0.213	0.224	0.234	0.245	0.255	0.266	0.277	0.288	0.299
0.3	0.310	0.321	0.332	0.343	0.354	0.365	0.377	0.388	0.400	0.412
0.4	0.424	0.436	0.448	0.460	0.472	0.485	0.497	0.510	0.523	0.536
0.5	0.549	0.563	0.576	0.590	0.604	0.618	0.633	0.648	0.662	0.678
0.6	0.693	0.709	0.725	0.741	0.758	0.775	0.793	0.811	0.829	0.848
0.7	0.867	0.887	0.908	0.929	0.950	0.973	0.996	1.020	1.045	1.071
0.8	1.099	1.127	1.157	1.188	1.221	1.256	1.293	1.333	1.376	1.422

r	0.000	0.001	0.002	0.003	0.004	0.005	0.006	0.007	0.008	0.009
0.90	1.472	1.478	1.483	1.488	1.494	1.499	1.505	1.510	1.516	1.522
0.91	1.528	1.533	1.539	1.545	1.551	1.557	1.564	1.570	1.576	1.583
0.92	1.589	1.596	1.602	1.609	1.616	1.623	1.630	1.637	1.644	1.651
0.93	1.658	1.666	1.673	1.681	1.689	1.697	1.705	1.713	1.721	1.730
0.94	1.738	1.747	1.756	1.764	1.774	1.783	1.792	1.802	1.812	1.822
0.95	1.832	1.842	1.853	1.863	1.874	1.886	1.897	1.909	1.921	1.933
0.96	1.946	1.959	1.972	1.986	2.000	2.014	2.029	2.044	2.060	2.076
0.97	2.092	2.109	2.127	2.146	2.165	2.185	2.205	2.227	2.249	2.273
0.98	2.298	2.323	2.351	2.380	2.410	2.443	2.477	2.515	2.555	2.599
0.99	2.647	2.700	2.759	2.826	2.903	2.994	3.106	3.250	3.453	3.800

Tabular value is $\mathbf{z}_r = \log_e\sqrt{(1 + r)/(1 - r)}$. For example, $\mathbf{z}_{0.35} = \log_e\sqrt{(1 + 0.35)/(1 - 0.35)} = 0.365$.

TABLE A.12b. Inverse of Fisher's z Transformation

z	.00	.01	.02	.03	.04	.05	.06	.07	.08	.09
0.0	.000	.010	.020	.030	.040	.050	.060	.070	.080	.090
0.1	.100	.110	.119	.129	.139	.149	.159	.168	.178	.188
0.2	.197	.207	.217	.226	.235	.245	.254	.264	.273	.282
0.3	.291	.300	.310	.319	.327	.336	.345	.354	.363	.371
0.4	.380	.388	.397	.405	.414	.422	.430	.438	.446	.454
0.5	.462	.470	.478	.485	.493	.501	.508	.515	.523	.530
0.6	.537	.544	.551	.558	.565	.572	.578	.585	.592	.598
0.7	.604	.611	.617	.623	.629	.635	.641	.647	.653	.658
0.8	.664	.670	.675	.680	.686	.691	.696	.701	.706	.711
0.9	.716	.721	.726	.731	.735	.740	.744	.749	.753	.757
1.0	.762	.766	.770	.774	.778	.782	.786	.789	.793	.797
1.1	.800	.804	.808	.811	.814	.818	.821	.824	.827	.831
1.2	.834	.837	.840	.843	.845	.848	.851	.854	.856	.859
1.3	.862	.864	.867	.869	.872	.874	.876	.879	.881	.883
1.4	.885	.887	.890	.892	.894	.896	.898	.900	.901	.903
1.5	.905	.907	.909	.910	.912	.914	.915	.917	.919	.920
1.6	.922	.923	.925	.926	.927	.929	.930	.932	.933	.934
1.7	.935	.937	.938	.939	.940	.941	.943	.944	.945	.946
1.8	.947	.948	.949	.950	.951	.952	.953	.954	.954	.955
1.9	.956	.957	.958	.959	.960	.960	.961	.962	.963	.963
2.0	.964	.965	.965	.966	.967	.967	.968	.969	.969	.970
2.1	.970	.971	.972	.972	.973	.973	.974	.974	.975	.975
2.2	.976	.976	.977	.977	.978	.978	.978	.979	.979	.980
2.3	.980	.980	.981	.981	.982	.982	.982	.983	.983	.983
2.4	.984	.984	.984	.985	.985	.985	.986	.986	.986	.986
2.5	.987	.987	.987	.987	.988	.988	.988	.988	.989	.989
2.6	.989	.989	.989	.990	.990	.990	.990	.990	.991	.991
2.7	.991	.991	.991	.992	.992	.992	.992	.992	.992	.992
2.8	.993	.993	.993	.993	.993	.993	.993	.994	.994	.994
2.9	.994	.994	.994	.994	.994	.995	.995	.995	.995	.995

Tabular value is r. For example, if $\mathbf{z}_r = 1.72$, then $r = 0.938$.

TABLE A.13a. Critical Values for Duncan's New Multiple Range Test $\alpha = 0.05$

$\nu \backslash r$	2	3	4	5	6	7	8	9	10
1	17.97	17.97	17.97	17.97	17.97	17.97	17.97	17.97	17.97
2	6.085	6.085	6.085	6.085	6.085	6.085	6.085	6.085	6.085
3	4.501	4.516	4.516	4.516	4.516	4.516	4.516	4.516	4.516
4	3.927	4.013	4.033	4.033	4.033	4.033	4.033	4.033	4.033
5	3.635	3.749	3.797	3.814	3.814	3.814	3.814	3.814	3.814
6	3.461	3.587	3.649	3.680	3.694	3.697	3.697	3.697	3.697
7	3.344	3.477	3.548	3.588	3.611	3.622	3.626	3.626	3.626
8	3.261	3.399	3.475	3.521	3.549	3.566	3.575	3.579	3.579
9	3.199	3.339	3.420	3.470	3.502	3.523	3.536	3.544	3.547
10	3.151	3.293	3.376	3.430	3.465	3.489	3.505	3.516	3.522
11	3.113	3.256	3.342	3.397	3.435	3.462	3.480	3.493	3.501
12	3.082	3.225	3.313	3.370	3.410	3.439	3.459	3.474	3.484
13	3.055	3.200	3.289	3.348	3.389	3.419	3.442	3.458	3.470
14	3.033	3.178	3.268	3.329	3.372	3.403	3.426	3.444	3.457
15	3.014	3.160	3.250	3.312	3.356	3.389	3.413	3.432	3.446
16	2.998	3.144	3.235	3.298	3.343	3.376	3.402	3.422	3.437
17	2.984	3.130	3.222	3.285	3.331	3.366	3.392	3.412	3.429
18	2.971	3.118	3.210	3.274	3.321	3.356	3.383	3.405	3.421
19	2.960	3.107	3.199	3.264	3.311	3.347	3.375	3.397	3.415
20	2.950	3.097	3.190	3.255	3.303	3.339	3.368	3.391	3.409
24	2.919	3.066	3.160	3.226	3.276	3.315	3.345	3.370	3.390
30	2.888	3.035	3.131	3.199	3.250	3.290	3.322	3.349	3.371
40	2.858	3.006	3.102	3.171	3.224	3.266	3.300	3.328	3.352
60	2.829	2.976	3.073	3.143	3.198	3.241	3.277	3.307	3.333
120	2.800	2.947	3.045	3.116	3.172	3.217	3.254	3.287	3.314
INF	2.772	2.918	3.017	3.089	3.146	3.193	3.232	3.265	3.294

$\nu \backslash r$	11	12	13	14	15	16	17	18	19
1	17.97	17.97	17.97	17.97	17.97	17.97	17.97	17.97	17.97
2	6.085	6.085	6.085	6.085	6.085	6.085	6.085	6.085	6.085
3	4.516	4.516	4.516	4.516	4.516	4.516	4.516	4.516	4.516
4	4.033	4.033	4.033	4.033	4.033	4.033	4.033	4.033	4.033
5	3.814	3.814	3.814	3.814	3.814	3.814	3.814	3.814	3.814
6	3.697	3.697	3.697	3.697	3.697	3.697	3.697	3.697	3.697
7	3.626	3.626	3.626	3.626	3.626	3.626	3.626	3.626	3.626
8	3.579	3.579	3.579	3.579	3.579	3.579	3.579	3.579	3.579
9	3.547	3.547	3.547	3.547	3.547	3.547	3.547	3.547	3.547
10	3.525	3.526	3.526	3.526	3.526	3.526	3.526	3.526	3.526
11	3.506	3.509	3.510	3.510	3.510	3.510	3.510	3.510	3.510
12	3.491	3.496	3.498	3.499	3.499	3.499	3.499	3.499	3.499
13	3.478	3.484	3.488	3.490	3.490	3.490	3.490	3.490	3.490
14	3.467	3.474	3.479	3.482	3.484	3.484	3.485	3.485	3.485
15	3.457	3.465	3.471	3.476	3.478	3.480	3.481	3.481	3.481
16	3.449	3.458	3.465	3.470	3.473	3.477	3.478	3.478	3.478
17	3.441	3.451	3.459	3.465	3.469	3.473	3.475	3.476	3.476
18	3.435	3.445	3.454	3.460	3.465	3.470	3.472	3.474	3.474
19	3.429	3.440	3.449	3.456	3.462	3.467	3.470	3.472	3.473
20	3.424	3.436	3.445	3.453	3.459	3.464	3.467	3.470	3.472
24	3.406	3.420	3.432	3.441	3.449	3.456	3.461	3.465	3.469
30	3.389	3.405	3.418	3.430	3.439	3.447	3.454	3.460	3.466
40	3.373	3.390	3.405	3.418	3.429	3.439	3.448	3.456	3.463
60	3.355	3.374	3.391	3.406	3.419	3.431	3.442	3.451	3.460
120	3.337	3.359	3.377	3.394	3.409	3.423	3.435	3.446	3.457
INF	3.320	3.343	3.363	3.382	3.399	3.414	3.428	3.442	3.454

TABLE A.13a.—*Continued*

$\nu \backslash r$	20	22	24	26	28	30	32	34	36
1	17.97	17.97	17.97	17.97	17.97	17.97	17.97	17.97	17.97
2	6.085	6.085	6.085	6.085	6.085	6.085	6.085	6.085	6.085
3	4.516	4.516	4.516	4.516	4.516	4.516	4.516	4.516	4.516
4	4.033	4.033	4.033	4.033	4.033	4.033	4.033	4.033	4.033
5	3.814	3.814	3.814	3.814	3.814	3.814	3.814	3.814	3.814
6	3.697	3.697	3.697	3.697	3.697	3.697	3.697	3.697	3.697
7	3.626	3.626	3.626	3.626	3.626	3.626	3.626	3.626	3.626
8	3.579	3.579	3.579	3.579	3.579	3.579	3.579	3.579	3.579
9	3.547	3.547	3.547	3.547	3.547	3.547	3.547	3.547	3.547
10	3.526	3.526	3.526	3.526	3.526	3.526	3.526	3.526	3.526
11	3.510	3.510	3.510	3.510	3.510	3.510	3.510	3.510	3.510
12	3.499	3.499	3.499	3.499	3.499	3.499	3.499	3.499	3.499
13	3.490	3.490	3.490	3.490	3.490	3.490	3.490	3.490	3.490
14	3.485	3.485	3.485	3.485	3.485	3.485	3.485	3.485	3.485
15	3.481	3.481	3.481	3.481	3.481	3.481	3.481	3.481	3.481
16	3.478	3.478	3.478	3.478	3.478	3.478	3.478	3.478	3.478
17	3.476	3.476	3.476	3.476	3.476	3.476	3.476	3.476	3.476
18	3.474	3.474	3.474	3.474	3.474	3.474	3.474	3.474	3.474
19	3.474	3.474	3.474	3.474	3.474	3.474	3.474	3.474	3.474
20	3.473	3.474	3.474	3.474	3.474	3.474	3.474	3.474	3.474
24	3.471	3.475	3.477	3.477	3.477	3.477	3.477	3.477	3.477
30	3.470	3.477	3.481	3.484	3.486	3.486	3.486	3.486	3.486
40	3.469	3.479	3.486	3.492	3.497	3.500	3.503	3.504	3.504
60	3.467	3.481	3.492	3.501	3.509	3.515	3.521	3.525	3.529
120	3.466	3.483	3.498	3.511	3.522	3.532	3.541	3.548	3.555
INF	3.466	3.486	3.505	3.522	3.536	3.550	3.562	3.574	3.584

$\nu \backslash r$	38	40	50	60	70	80	90	100
1	17.97	17.97	17.97	17.97	17.97	17.97	17.97	17.97
2	6.085	6.085	6.085	6.085	6.085	6.085	6.085	6.085
3	4.516	4.516	4.516	4.516	4.516	4.516	4.516	4.516
4	4.033	4.033	4.033	4.033	4.033	4.033	4.033	4.033
5	3.814	3.814	3.814	3.814	3.814	3.814	3.814	3.814
6	3.697	3.697	3.697	3.697	3.697	3.697	3.697	3.697
7	3.626	3.626	3.626	3.626	3.626	3.626	3.626	3.626
8	3.579	3.579	3.579	3.579	3.579	3.579	3.579	3.579
9	3.547	3.547	3.547	3.547	3.547	3.547	3.547	3.547
10	3.526	3.526	3.526	3.526	3.526	3.526	3.526	3.526
11	3.510	3.510	3.510	3.510	3.510	3.510	3.510	3.510
12	3.499	3.499	3.499	3.499	3.499	3.499	3.499	3.499
13	3.490	3.490	3.490	3.490	3.490	3.490	3.490	3.490
14	3.485	3.485	3.485	3.485	3.485	3.485	3.485	3.485
15	3.481	3.481	3.481	3.481	3.481	3.481	3.481	3.481
16	3.478	3.478	3.478	3.478	3.478	3.478	3.478	3.478
17	3.476	3.476	3.476	3.476	3.476	3.476	3.476	3.476
18	3.474	3.474	3.474	3.474	3.474	3.474	3.474	3.474
19	3.474	3.474	3.474	3.474	3.474	3.474	3.474	3.474
20	3.474	3.474	3.474	3.474	3.474	3.474	3.474	3.474
24	3.477	3.477	3.477	3.477	3.477	3.477	3.477	3.477
30	3.486	3.486	3.486	3.486	3.486	3.486	3.486	3.486
40	3.504	3.504	3.504	3.504	3.504	3.504	3.504	3.504
60	3.531	3.534	3.537	3.537	3.537	3.537	3.537	3.537
120	3.561	3.566	3.585	3.596	3.600	3.601	3.601	3.601
INF	3.594	3.603	3.640	3.668	3.690	3.708	3.722	3.735

NOTE: Tables A.13a and A.13b are reproduced, with the author's permission, from H. Leon Harter's *Order Statistics and Their Use in Testing and Estimation*, Vol. 1, U.S. Government Printing Office, Washington, D.C., 1970.

TABLE A.13b. Critical Values for Duncan's New Multiple Range Test $\alpha = 0.01$

$\nu \backslash r$	2	3	4	5	6	7	8	9	10
1	90.03	90.03	90.03	90.03	90.03	90.03	90.03	90.03	90.03
2	14.04	14.04	14.04	14.04	14.04	14.04	14.04	14.04	14.04
3	8.261	8.321	8.321	8.321	8.321	8.321	8.321	8.321	8.321
4	6.512	6.677	6.740	6.756	6.756	6.756	6.756	6.756	6.756
5	5.702	5.893	5.989	6.040	6.065	6.074	6.074	6.074	6.074
6	5.243	5.439	5.549	5.614	5.655	5.680	5.694	5.701	5.703
7	4.949	5.145	5.260	5.334	5.383	5.416	5.439	5.454	5.464
8	4.746	4.939	5.057	5.135	5.189	5.227	5.256	5.276	5.291
9	4.596	4.787	4.906	4.986	5.043	5.086	5.118	5.142	5.160
10	4.482	4.671	4.790	4.871	4.931	4.975	5.010	5.037	5.058
11	4.392	4.579	4.697	4.780	4.841	4.887	4.924	4.952	4.975
12	4.320	4.504	4.622	4.706	4.767	4.815	4.852	4.883	4.907
13	4.260	4.442	4.560	4.644	4.706	4.755	4.793	4.824	4.850
14	4.210	4.391	4.508	4.591	4.654	4.704	4.743	4.775	4.802
15	4.168	4.347	4.463	4.547	4.610	4.660	4.700	4.733	4.760
16	4.131	4.309	4.425	4.509	4.572	4.622	4.663	4.696	4.724
17	4.099	4.275	4.391	4.475	4.539	4.589	4.630	4.664	4.693
18	4.071	4.246	4.362	4.445	4.509	4.560	4.601	4.635	4.664
19	4.046	4.220	4.335	4.419	4.483	4.534	4.575	4.610	4.639
20	4.024	4.197	4.312	4.395	4.459	4.510	4.552	4.587	4.617
24	3.956	4.126	4.239	4.322	4.386	4.437	4.480	4.516	4.546
30	3.889	4.056	4.168	4.250	4.314	4.366	4.409	4.445	4.477
40	3.825	3.988	4.098	4.180	4.244	4.296	4.339	4.376	4.408
60	3.762	3.922	4.031	4.111	4.174	4.226	4.270	4.307	4.340
120	3.702	3.858	3.965	4.044	4.107	4.158	4.202	4.239	4.272
INF	3.643	3.796	3.900	3.978	4.040	4.091	4.135	4.172	4.205

$\nu \backslash r$	11	12	13	14	15	16	17	18	19
1	90.03	90.03	90.03	90.03	90.03	90.03	90.03	90.03	90.03
2	14.04	14.04	14.04	14.04	14.04	14.04	14.04	14.04	14.04
3	8.321	8.321	8.321	8.321	8.321	8.321	8.321	8.321	8.321
4	6.756	6.756	6.756	6.756	6.756	6.756	6.756	6.756	6.756
5	6.074	6.074	6.074	6.074	6.074	6.074	6.074	6.074	6.074
6	5.703	5.703	5.703	5.703	5.703	5.703	5.703	5.703	5.703
7	5.470	5.472	5.472	5.472	5.472	5.472	5.472	5.472	5.472
8	5.302	5.309	5.314	5.316	5.317	5.317	5.317	5.317	5.317
9	5.174	5.185	5.193	5.199	5.203	5.205	5.206	5.206	5.206
10	5.074	5.088	5.098	5.106	5.112	5.117	5.120	5.122	5.124
11	4.994	5.009	5.021	5.031	5.039	5.045	5.050	5.054	5.057
12	4.927	4.944	4.958	4.969	4.978	4.986	4.993	4.998	5.002
13	4.872	4.889	4.904	4.917	4.928	4.937	4.944	4.950	4.956
14	4.824	4.843	4.859	4.872	4.884	4.894	4.902	4.910	4.916
15	4.783	4.803	4.820	4.834	4.846	4.857	4.866	4.874	4.881
16	4.748	4.768	4.786	4.800	4.813	4.825	4.835	4.844	4.851
17	4.717	4.738	4.756	4.771	4.785	4.797	4.807	4.816	4.824
18	4.689	4.711	4.729	4.745	4.759	4.772	4.783	4.792	4.801
19	4.665	4.686	4.705	4.722	4.736	4.749	4.761	4.771	4.780
20	4.642	4.664	4.684	4.701	4.716	4.729	4.741	4.751	4.761
24	4.573	4.596	4.616	4.634	4.651	4.665	4.678	4.690	4.700
30	4.504	4.528	4.550	4.569	4.586	4.601	4.615	4.628	4.640
40	4.436	4.461	4.483	4.503	4.521	4.537	4.553	4.566	4.579
60	4.368	4.394	4.417	4.438	4.456	4.474	4.490	4.504	4.518
120	4.301	4.327	4.351	4.372	4.392	4.410	4.426	4.442	4.456
INF	4.235	4.261	4.285	4.307	4.327	4.345	4.363	4.379	4.394

TABLE A.13b.—*Continued*

ν\r	20	22	24	26	28	30	32	34	36
1	90.03	90.03	90.03	90.03	90.03	90.03	90.03	90.03	90.03
2	14.04	14.04	14.04	14.04	14.04	14.04	14.04	14.04	14.04
3	8.321	8.321	8.321	8.321	8.321	8.321	8.321	8.321	8.321
4	6.756	6.756	6.756	6.756	6.756	6.756	6.756	6.756	6.756
5	6.074	6.074	6.074	6.074	6.074	6.074	6.074	6.074	6.074
6	5.703	5.703	5.703	5.703	5.703	5.703	5.703	5.703	5.703
7	5.472	5.472	5.472	5.472	5.472	5.472	5.472	5.472	5.472
8	5.317	5.317	5.317	5.317	5.317	5.317	5.317	5.317	5.317
9	5.206	5.206	5.206	5.206	5.206	5.206	5.206	5.206	5.206
10	5.124	5.124	5.124	5.124	5.124	5.124	5.124	5.124	5.124
11	5.059	5.061	5.061	5.061	5.061	5.061	5.061	5.061	5.061
12	5.006	5.010	5.011	5.011	5.011	5.011	5.011	5.011	5.011
13	4.960	4.966	4.970	4.972	4.972	4.972	4.972	4.972	4.972
14	4.921	4.929	4.935	4.938	4.940	4.940	4.940	4.940	4.940
15	4.887	4.897	4.904	4.909	4.912	4.914	4.914	4.914	4.914
16	4.858	4.869	4.877	4.883	4.887	4.890	4.892	4.892	4.892
17	4.832	4.844	4.853	4.860	4.865	4.869	4.872	4.873	4.874
18	4.808	4.821	4.832	4.839	4.846	4.850	4.854	4.856	4.857
19	4.788	4.802	4.812	4.821	4.828	4.833	4.838	4.841	4.843
20	4.769	4.784	4.795	4.805	4.813	4.818	4.823	4.827	4.830
24	4.710	4.727	4.741	4.752	4.762	4.770	4.777	4.783	4.788
30	4.650	4.669	4.685	4.699	4.711	4.721	4.730	4.738	4.744
40	4.591	4.611	4.630	4.645	4.659	4.671	4.682	4.692	4.700
60	4.530	4.553	4.573	4.591	4.607	4.620	4.633	4.645	4.655
120	4.469	4.494	4.516	4.535	4.552	4.568	4.583	4.596	4.609
INF	4.408	4.434	4.457	4.478	4.497	4.514	4.530	4.545	4.559

ν\r	38	40	50	60	70	80	90	100
1	90.03	90.03	90.03	90.03	90.03	90.03	90.03	90.03
2	14.04	14.04	14.04	14.04	14.04	14.04	14.04	14.04
3	8.321	8.321	8.321	8.321	8.321	8.321	8.321	8.321
4	6.756	6.756	6.756	6.756	6.756	6.756	6.756	6.756
5	6.074	6.074	6.074	6.074	6.074	6.074	6.074	6.074
6	5.703	5.703	5.703	5.703	5.703	5.703	5.703	5.703
7	5.472	5.472	5.472	5.472	5.472	5.472	5.472	5.472
8	5.317	5.317	5.317	5.317	5.317	5.317	5.317	5.317
9	5.206	5.206	5.206	5.206	5.206	5.206	5.206	5.206
10	5.124	5.124	5.124	5.124	5.124	5.124	5.124	5.124
11	5.061	5.061	5.061	5.061	5.061	5.061	5.061	5.061
12	5.011	5.011	5.011	5.011	5.011	5.011	5.011	5.011
13	4.972	4.972	4.972	4.972	4.972	4.972	4.972	4.972
14	4.940	4.940	4.940	4.940	4.940	4.940	4.940	4.940
15	4.914	4.914	4.914	4.914	4.914	4.914	4.914	4.914
16	4.892	4.892	4.892	4.892	4.892	4.892	4.892	4.892
17	4.874	4.874	4.874	4.874	4.874	4.874	4.874	4.874
18	4.858	4.858	4.858	4.858	4.858	4.858	4.858	4.858
19	4.844	4.845	4.845	4.845	4.845	4.845	4.845	4.845
20	4.832	4.833	4.833	4.833	4.833	4.833	4.833	4.833
24	4.791	4.794	4.802	4.802	4.802	4.802	4.802	4.802
30	4.750	4.755	4.772	4.777	4.777	4.777	4.777	4.777
40	4.708	4.715	4.740	4.754	4.761	4.764	4.764	4.764
60	4.665	4.673	4.707	4.730	4.745	4.755	4.761	4.765
120	4.619	4.630	4.673	4.703	4.727	4.745	4.759	4.770
INF	4.572	4.584	4.635	4.675	4.707	4.734	4.756	4.776

TABLE A.14a. Critical Values for the Studentized Range $\alpha = 0.05$

$\nu \backslash r$	2	3	4	5	6	7	8	9	10
1	17.97	26.98	32.82	37.08	40.41	43.12	45.40	47.36	49.07
2	6.085	8.331	9.798	10.88	11.74	12.44	13.03	13.54	13.99
3	4.501	5.910	6.825	7.502	8.037	8.478	8.853	9.177	9.462
4	3.927	5.040	5.757	6.287	6.707	7.053	7.347	7.602	7.826
5	3.635	4.602	5.218	5.673	6.033	6.330	6.582	6.802	6.995
6	3.461	4.339	4.896	5.305	5.628	5.895	6.122	6.319	6.493
7	3.344	4.165	4.681	5.060	5.359	5.606	5.815	5.998	6.158
8	3.261	4.041	4.529	4.886	5.167	5.399	5.597	5.767	5.918
9	3.199	3.949	4.415	4.756	5.024	5.244	5.432	5.595	5.739
10	3.151	3.877	4.327	4.654	4.912	5.124	5.305	5.461	5.599
11	3.113	3.820	4.256	4.574	4.823	5.028	5.202	5.353	5.487
12	3.082	3.773	4.199	4.508	4.751	4.950	5.119	5.265	5.395
13	3.055	3.735	4.151	4.453	4.690	4.885	5.049	5.192	5.318
14	3.033	3.702	4.111	4.407	4.639	4.829	4.990	5.131	5.254
15	3.014	3.674	4.076	4.367	4.595	4.782	4.940	5.077	5.198
16	2.998	3.649	4.046	4.333	4.557	4.741	4.897	5.031	5.150
17	2.984	3.628	4.020	4.303	4.524	4.705	4.858	4.991	5.108
18	2.971	3.609	3.997	4.277	4.495	4.673	4.824	4.956	5.071
19	2.960	3.593	3.977	4.253	4.469	4.645	4.794	4.924	5.038
20	2.950	3.578	3.958	4.232	4.445	4.620	4.768	4.896	5.008
24	2.919	3.532	3.901	4.166	4.373	4.541	4.684	4.807	4.915
30	2.888	3.486	3.845	4.102	4.302	4.464	4.602	4.720	4.824
40	2.858	3.442	3.791	4.039	4.232	4.389	4.521	4.635	4.735
60	2.829	3.399	3.737	3.977	4.163	4.314	4.441	4.550	4.646
120	2.800	3.356	3.685	3.917	4.096	4.241	4.363	4.468	4.560
INF	2.772	3.314	3.633	3.858	4.030	4.170	4.286	4.387	4.474

$\nu \backslash r$	11	12	13	14	15	16	17	18	19
1	50.59	51.96	53.20	54.33	55.36	56.32	57.22	58.04	58.83
2	14.39	14.75	15.08	15.38	15.65	15.91	16.14	16.37	16.57
3	9.717	9.946	10.15	10.35	10.53	10.69	10.84	10.98	11.11
4	8.027	8.208	8.373	8.525	8.664	8.794	8.914	9.028	9.134
5	7.168	7.324	7.466	7.596	7.717	7.828	7.932	8.030	8.122
6	6.649	6.789	6.917	7.034	7.143	7.244	7.338	7.426	7.508
7	6.302	6.431	6.550	6.658	6.759	6.852	6.939	7.020	7.097
8	6.054	6.175	6.287	6.389	6.483	6.571	6.653	6.729	6.802
9	5.867	5.983	6.089	6.186	6.276	6.359	6.437	6.510	6.579
10	5.722	5.833	5.935	6.028	6.114	6.194	6.269	6.339	6.405
11	5.605	5.713	5.811	5.901	5.984	6.062	6.134	6.202	6.265
12	5.511	5.615	5.710	5.798	5.878	5.953	6.023	6.089	6.151
13	5.431	5.533	5.625	5.711	5.789	5.862	5.931	5.995	6.055
14	5.364	5.463	5.554	5.637	5.714	5.786	5.852	5.915	5.974
15	5.306	5.404	5.493	5.574	5.649	5.720	5.785	5.846	5.904
16	5.256	5.352	5.439	5.520	5.593	5.662	5.727	5.786	5.843
17	5.212	5.307	5.392	5.471	5.544	5.612	5.675	5.734	5.790
18	5.174	5.267	5.352	5.429	5.501	5.568	5.630	5.688	5.743
19	5.140	5.231	5.315	5.391	5.462	5.528	5.589	5.647	5.701
20	5.108	5.199	5.282	5.357	5.427	5.493	5.553	5.610	5.663
24	5.012	5.099	5.179	5.251	5.319	5.381	5.439	5.494	5.545
30	4.917	5.001	5.077	5.147	5.211	5.271	5.327	5.379	5.429
40	4.824	4.904	4.977	5.044	5.106	5.163	5.216	5.266	5.313
60	4.732	4.808	4.878	4.942	5.001	5.056	5.107	5.154	5.199
120	4.641	4.714	4.781	4.842	4.898	4.950	4.998	5.044	5.086
INF	4.552	4.622	4.685	4.743	4.796	4.845	4.891	4.934	4.974

TABLE A.14a.—*Continued*

ν\r	20	22	24	26	28	30	32	34	36
1	59.56	60.91	62.12	63.22	64.23	65.15	66.01	66.81	67.56
2	16.77	17.13	17.45	17.75	18.02	18.27	18.50	18.72	18.92
3	11.24	11.47	11.68	11.87	12.05	12.21	12.36	12.50	12.63
4	9.233	9.418	9.584	9.736	9.875	10.00	10.12	10.23	10.34
5	8.208	8.368	8.512	8.643	8.764	8.875	8.979	9.075	9.165
6	7.587	7.730	7.861	7.979	8.088	8.189	8.283	8.370	8.452
7	7.170	7.303	7.423	7.533	7.634	7.728	7.814	7.895	7.972
8	6.870	6.995	7.109	7.212	7.307	7.395	7.477	7.554	7.625
9	6.644	6.763	6.871	6.970	7.061	7.145	7.222	7.295	7.363
10	6.467	6.582	6.686	6.781	6.868	6.948	7.023	7.093	7.159
11	6.326	6.436	6.536	6.628	6.712	6.790	6.863	6.930	6.994
12	6.209	6.317	6.414	6.503	6.585	6.660	6.731	6.796	6.858
13	6.112	6.217	6.312	6.398	6.478	6.551	6.620	6.684	6.744
14	6.029	6.132	6.224	6.309	6.387	6.459	6.526	6.588	6.647
15	5.958	6.059	6.149	6.233	6.309	6.379	6.445	6.506	6.564
16	5.897	5.995	6.084	6.166	6.241	6.310	6.374	6.434	6.491
17	5.842	5.940	6.027	6.107	6.181	6.249	6.313	6.372	6.427
18	5.794	5.890	5.977	6.055	6.128	6.195	6.258	6.316	6.371
19	5.752	5.846	5.932	6.009	6.081	6.147	6.209	6.267	6.321
20	5.714	5.807	5.891	5.968	6.039	6.104	6.165	6.222	6.275
24	5.594	5.683	5.764	5.838	5.906	5.968	6.027	6.081	6.132
30	5.475	5.561	5.638	5.709	5.774	5.833	5.889	5.941	5.990
40	5.358	5.439	5.513	5.581	5.642	5.700	5.753	5.803	5.849
60	5.241	5.319	5.389	5.453	5.512	5.566	5.617	5.664	5.708
120	5.126	5.200	5.266	5.327	5.382	5.434	5.481	5.526	5.568
INF	5.012	5.081	5.144	5.201	5.253	5.301	5.346	5.388	5.427

ν\r	38	40	50	60	70	80	90	100
1	68.26	68.92	71.73	73.97	75.82	77.40	78.77	79.98
2	19.11	19.28	20.05	20.66	21.16	21.59	21.96	22.29
3	12.75	12.87	13.36	13.76	14.08	14.36	14.61	14.82
4	10.44	10.53	10.93	11.24	11.51	11.73	11.92	12.09
5	9.250	9.330	9.674	9.949	10.18	10.38	10.54	10.69
6	8.529	8.601	8.913	9.163	9.370	9.548	9.702	9.839
7	8.043	8.110	8.400	8.632	8.824	8.989	9.133	9.261
8	7.693	7.756	8.029	8.248	8.430	8.586	8.722	8.843
9	7.428	7.488	7.749	7.958	8.132	8.281	8.410	8.526
10	7.220	7.279	7.529	7.730	7.897	8.041	8.166	8.276
11	7.053	7.110	7.352	7.546	7.708	7.847	7.968	8.075
12	6.916	6.970	7.205	7.394	7.552	7.687	7.804	7.909
13	6.800	7.854	7.083	7.267	7.421	7.552	7.667	7.769
14	6.702	6.754	6.979	7.159	7.309	7.438	7.550	7.650
15	6.618	6.669	6.888	7.065	7.212	7.339	7.449	7.546
16	6.544	6.594	6.810	6.984	7.128	7.252	7.360	7.457
17	6.479	6.529	6.741	6.912	7.054	7.176	7.283	7.377
18	6.422	6.471	6.680	6.848	6.989	7.109	7.213	7.307
19	6.371	6.419	6.626	6.792	6.930	7.048	7.152	7.244
20	6.325	6.373	6.576	6.740	6.877	6.994	7.097	7.187
24	6.181	6.226	6.421	6.579	6.710	6.822	6.920	7.008
30	6.037	6.080	6.267	6.417	6.543	6.650	6.744	6.827
40	5.893	5.934	6.112	6.255	6.375	6.477	6.566	6.645
60	5.750	5.789	5.958	6.093	6.206	6.303	6.387	6.462
120	5.607	5.644	5.802	5.929	6.035	6.126	6.205	6.275
INF	5.463	5.498	5.646	5.764	5.863	5.947	6.020	6.085

NOTE: Tables A.14a and A.14b are reproduced, with the author's permission, from H. Leon Harter's *Order Statistics and Their Use in Testing and Estimation*, Vol. 1, U.S. Government Printing Office, Washington, D.C., 1970.

TABLE A.14b. Critical Values for the Studentized Range $\alpha = 0.01$

$\nu \backslash r$	2	3	4	5	6	7	8	9	10
1	90.03	135.0	164.3	185.6	202.2	215.8	227.2	237.0	245.6
2	14.04	19.02	22.29	24.72	26.63	28.20	29.53	30.68	31.69
3	8.261	10.62	12.17	13.33	14.24	15.00	15.64	16.20	16.69
4	6.512	8.120	9.173	9.958	10.58	11.10	11.55	11.93	12.27
5	5.702	6.976	7.804	8.421	8.913	9.321	9.669	9.972	10.24
6	5.243	6.331	7.033	7.556	7.973	8.318	8.613	8.869	9.097
7	4.949	5.919	6.543	7.005	7.373	7.679	7.939	8.166	8.368
8	4.746	5.635	6.204	6.625	6.960	7.237	7.474	7.681	7.863
9	4.596	5.428	5.957	6.348	6.658	6.915	7.134	7.325	7.495
10	4.482	5.270	5.769	6.136	6.428	6.669	6.875	7.055	7.213
11	4.392	5.146	5.621	5.970	6.247	6.476	6.672	6.842	6.992
12	4.320	5.046	5.502	5.836	6.101	6.321	6.507	6.670	6.814
13	4.260	4.964	5.404	5.727	5.981	6.192	6.372	6.528	6.667
14	4.210	4.895	5.322	5.634	5.881	6.085	6.258	6.409	6.543
15	4.168	4.836	5.252	5.556	5.796	5.994	6.162	6.309	6.439
16	4.131	4.786	5.192	5.489	5.722	5.915	6.079	6.222	6.349
17	4.099	4.742	5.140	5.430	5.659	5.847	6.007	6.147	6.270
18	4.071	4.703	5.094	5.379	5.603	5.788	5.944	6.081	6.201
19	4.046	4.670	5.054	5.334	5.554	5.735	5.889	6.022	6.141
20	4.024	4.639	5.018	5.294	5.510	5.688	5.839	5.970	6.087
24	3.956	4.546	4.907	5.168	5.374	5.542	5.685	5.809	5.919
30	3.889	4.455	4.799	5.048	5.242	5.401	5.536	5.653	5.756
40	3.825	4.367	4.696	4.931	5.114	5.265	5.392	5.502	5.599
60	3.762	4.282	4.595	4.818	4.991	5.133	5.253	5.356	5.447
120	3.702	4.200	4.497	4.709	4.872	5.005	5.118	5.214	5.299
INF	3.643	4.120	4.403	4.603	4.757	4.882	4.987	5.078	5.157

$\nu \backslash r$	11	12	13	14	15	16	17	18	19
1	253.2	260.0	266.2	271.8	277.0	281.8	286.3	290.4	294.3
2	32.59	33.40	34.13	34.81	35.43	36.00	36.53	37.03	37.50
3	17.13	17.53	17.89	18.22	18.52	18.81	19.07	19.32	19.55
4	12.57	12.84	13.09	13.32	13.53	13.73	13.91	14.08	14.24
5	10.48	10.70	10.89	11.08	11.24	11.40	11.55	11.68	11.81
6	9.301	9.485	9.653	9.808	9.951	10.08	10.21	10.32	10.43
7	8.548	8.711	8.860	8.997	9.124	9.242	9.353	9.456	9.554
8	8.027	8.176	8.312	8.436	8.552	8.659	8.760	8.854	8.943
9	7.647	7.784	7.910	8.025	8.132	8.232	8.325	8.412	8.495
10	7.356	7.485	7.603	7.712	7.812	7.906	7.993	8.076	8.153
11	7.128	7.250	7.362	7.465	7.560	7.649	7.732	7.809	7.883
12	6.943	7.060	7.167	7.265	7.356	7.441	7.520	7.594	7.665
13	6.791	6.903	7.006	7.101	7.188	7.269	7.345	7.417	7.485
14	6.664	6.772	6.871	6.962	7.047	7.126	7.199	7.268	7.333
15	6.555	6.660	6.757	6.845	6.927	7.003	7.074	7.142	7.204
16	6.462	6.564	6.658	6.744	6.823	6.898	6.967	7.032	7.093
17	6.381	6.480	6.572	6.656	6.734	6.806	6.873	6.937	6.997
18	6.310	6.407	6.497	6.579	6.655	6.725	6.792	6.854	6.912
19	6.247	6.342	6.430	6.510	6.585	6.654	6.719	6.780	6.837
20	6.191	6.285	6.371	6.450	6.523	6.591	6.654	6.714	6.771
24	6.017	6.106	6.186	6.261	6.330	6.394	6.453	6.510	6.563
30	5.849	5.932	6.008	6.078	6.143	6.203	6.259	6.311	6.361
40	5.686	5.764	5.835	5.900	5.961	6.017	6.069	6.119	6.165
60	5.528	5.601	5.667	5.728	5.785	5.837	5.886	5.931	5.974
120	5.375	5.443	5.505	5.562	5.614	5.662	5.708	5.750	5.790
INF	5.227	5.290	5.348	5.400	5.448	5.493	5.535	5.574	5.611

TABLE A.14b.—*Continued*

$\nu \backslash r$	20	22	24	26	28	30	32	34	36
1	298.0	304.7	310.8	316.3	321.3	326.0	330.3	334.3	338.0
2	37.95	38.76	39.49	40.15	40.76	41.32	41.84	42.33	42.78
3	19.77	20.17	20.53	20.86	21.16	21.44	21.70	21.95	22.17
4	14.40	14.68	14.93	15.16	15.37	15.57	15.75	15.92	16.08
5	11.93	12.16	12.36	12.54	12.71	12.87	13.02	13.15	13.28
6	10.54	10.73	10.91	11.06	11.21	11.34	11.47	11.58	11.69
7	9.646	9.815	9.970	10.11	10.24	10.36	10.47	10.58	10.67
8	9.027	9.182	9.322	9.450	9.569	9.678	9.779	9.874	9.964
9	8.573	8.717	8.847	8.966	9.075	9.177	9.271	9.360	9.443
10	8.226	8.361	8.483	8.595	8.698	8.794	8.883	8.966	9.044
11	7.952	8.080	8.196	8.303	8.400	8.491	8.575	8.654	8.728
12	7.731	7.853	7.964	8.066	8.159	8.246	8.327	8.402	8.473
13	7.548	7.665	7.772	7.870	7.960	8.043	8.121	8.193	8.262
14	7.395	7.508	7.611	7.705	7.792	7.873	7.948	8.018	8.084
15	7.264	7.374	7.474	7.566	7.650	7.728	7.800	7.869	7.932
16	7.152	7.258	7.356	7.445	7.527	7.602	7.673	7.739	7.802
17	7.053	7.158	7.253	7.340	7.420	7.493	7.563	7.627	7.687
18	6.968	7.070	7.163	7.247	7.325	7.398	7.465	7.528	7.587
19	6.891	6.992	7.082	7.166	7.242	7.313	7.379	7.440	7.498
20	6.823	6.922	7.011	7.092	7.168	7.237	7.302	7.362	7.419
24	6.612	6.705	6.789	6.865	6.936	7.001	7.062	7.119	7.173
30	6.407	6.494	6.572	6.644	6.710	6.772	6.828	6.881	6.932
40	6.209	6.289	6.362	6.429	6.490	6.547	6.600	6.650	6.697
60	6.015	6.090	6.158	6.220	6.277	6.330	6.378	6.424	6.467
120	5.827	5.897	5.959	6.016	6.069	6.117	6.162	6.204	6.244
INF	5.645	5.709	5.766	5.818	5.866	5.911	5.952	5.990	6.026

$\nu \backslash r$	38	40	50	60	70	80	90	100
1	341.5	344.8	358.9	370.1	379.4	387.3	394.1	400.1
2	43.21	43.61	45.33	46.70	47.83	48.80	49.64	50.38
3	22.39	22.59	23.45	24.13	24.71	25.19	25.62	25.99
4	16.23	16.37	16.98	17.46	17.86	18.20	18.50	18.77
5	13.40	13.52	14.00	14.39	14.72	14.99	15.23	15.45
6	11.80	11.90	12.31	12.65	12.92	13.16	13.37	13.55
7	10.77	10.85	11.23	11.52	11.77	11.99	12.17	12.34
8	10.05	10.13	10.47	10.75	10.97	11.17	11.34	11.49
9	9.521	9.594	9.912	10.17	10.38	10.57	10.73	10.87
10	9.117	9.187	9.486	9.726	9.927	10.10	10.25	10.39
11	8.798	8.864	9.148	9.377	9.568	9.732	9.875	10.00
12	8.539	8.603	8.875	9.094	9.277	9.434	9.571	9.693
13	8.326	8.387	8.648	8.859	9.035	9.187	9.318	9.436
14	8.146	8.204	8.457	8.661	8.832	8.978	9.106	9.219
15	7.992	8.049	8.295	8.492	8.658	8.800	8.924	9.035
16	7.860	7.916	8.154	8.347	8.507	8.646	8.767	8.874
17	7.745	7.799	8.031	8.219	8.377	8.511	8.630	8.735
18	7.643	7.696	7.924	8.107	8.261	8.393	8.508	8.611
19	7.553	7.605	7.828	8.008	8.159	8.288	8.401	8.502
20	7.473	7.523	7.742	7.919	8.067	8.194	8.305	8.404
24	7.223	7.270	7.476	7.642	7.780	7.900	8.004	8.097
30	6.978	7.023	7.215	7.370	7.500	7.611	7.709	7.796
40	6.740	6.782	6.960	7.104	7.225	7.328	7.419	7.500
60	6.507	6.546	6.710	6.843	6.954	7.050	7.133	7.207
120	6.281	6.316	6.467	6.588	6.689	6.776	6.852	6.919
INF	6.060	6.092	6.228	6.338	6.429	6.507	6.575	6.636

TABLE A.15. Critical Values of the Ratio F_{max}

$\alpha = 0.05$

$\nu \backslash a$	2	3	4	5	6	7	8	9	10	11	12
2	39.0	87.5	142	202	266	333	403	475	550	626	704
3	15.4	27.8	39.2	50.7	62.0	72.9	83.5	93.9	104	114	124
4	9.60	15.5	20.6	25.2	29.5	33.6	37.5	41.1	44.6	48.0	51.4
5	7.15	10.8	13.7	16.3	18.7	20.8	22.9	24.7	26.5	28.2	29.9
6	5.82	8.38	10.4	12.1	13.7	15.0	16.3	17.5	18.6	19.7	20.7
7	4.99	6.94	8.44	9.70	10.8	11.8	12.7	13.5	14.3	15.1	15.8
8	4.43	6.00	7.18	8.12	9.03	9.78	10.5	11.1	11.7	12.2	12.7
9	4.03	5.34	6.31	7.11	7.80	8.41	8.95	9.45	9.91	10.3	10.7
10	3.72	4.85	5.67	6.34	6.92	7.42	7.87	8.28	8.66	9.01	9.34
12	3.28	4.16	4.79	5.30	5.72	6.09	6.42	6.72	7.00	7.25	7.48
15	2.86	3.54	4.01	4.37	4.68	4.95	5.19	5.40	5.59	5.77	5.93
20	2.46	2.95	3.29	3.54	3.76	3.94	4.10	4.24	4.37	4.49	4.59
30	2.07	2.40	2.61	2.78	2.91	3.02	3.12	3.21	3.29	3.36	3.39
60	1.67	1.85	1.96	2.04	2.11	2.17	2.22	2.26	2.30	2.33	2.36
INF	1.00	1.00	1.00	1.00	1.00	1.00	1.00	1.00	1.00	1.00	1.00

$\alpha = 0.01$

$\nu \backslash a$	2	3	4	5	6	7	8	9	10	11	12
2	199	448	729	1036	1362	1705	2063	2432	2813	3204	3605
3	47.5	85	120	151	184	21(6)	24(9)	28(1)	31(0)	33(7)	36(1)
4	23.2	37	49	59	69	79	89	97	106	113	120
5	14.9	22	28	33	38	42	46	50	54	57	60
6	11.1	15.5	19.1	22	25	27	30	32	34	36	37
7	8.89	12.1	14.5	16.5	18.4	20	22	23	24	26	27
8	7.50	9.9	11.7	13.2	14.5	15.8	16.9	17.9	18.9	19.8	21
9	6.54	8.5	9.9	11.1	12.1	13.1	13.9	14.7	15.3	16.0	16.6
10	5.85	7.4	8.6	9.6	10.4	11.1	11.8	12.4	12.9	13.4	13.9
12	4.91	6.1	6.9	7.6	8.2	8.7	9.1	9.5	9.9	10.2	10.6
15	4.07	4.9	5.5	6.0	6.4	6.7	7.1	7.3	7.5	7.8	8.0
20	3.32	3.8	4.3	4.6	4.9	5.1	5.3	5.5	5.6	5.8	5.9
30	2.63	3.0	3.3	3.4	3.6	3.7	3.8	3.9	4.0	4.1	4.2
60	1.96	2.2	2.3	2.4	2.4	2.5	2.5	2.6	2.6	2.7	2.7
INF	1.00	1.0	1.0	1.0	1.0	1.0	1.0	1.0	1.0	1.0	1.0

Reproduced with permission of the Biometrika Trust, from *Biometrika Tables for Statisticians*, Vol. 1, 3rd edition, 1966, edited by E.S. Pearson and H.O. Hartley.

TABLE A.16. Logs Base Ten

	.00	.01	.02	.03	.04	.05	.06	.07	.08	.09
1.0	.0000	.0043	.0086	.0128	.0170	.0212	.0253	.0294	.0334	.0374
1.1	.0414	.0453	.0492	.0531	.0569	.0607	.0645	.0682	.0719	.0755
1.2	.0792	.0828	.0864	.0899	.0934	.0969	.1004	.1038	.1072	.1106
1.3	.1139	.1173	.1206	.1239	.1271	.1303	.1335	.1367	.1399	.1430
1.4	.1461	.1492	.1523	.1553	.1584	.1614	.1644	.1673	.1703	.1732
1.5	.1761	.1790	.1818	.1847	.1875	.1903	.1931	.1959	.1987	.2014
1.6	.2041	.2068	.2095	.2122	.2148	.2175	.2201	.2227	.2253	.2279
1.7	.2304	.2330	.2355	.2380	.2405	.2430	.2455	.2480	.2504	.2529
1.8	.2553	.2577	.2601	.2625	.2648	.2672	.2695	.2718	.2742	.2765
1.9	.2788	.2810	.2833	.2856	.2878	.2900	.2923	.2945	.2967	.2989
2.0	.3010	.3032	.3054	.3075	.3096	.3118	.3139	.3160	.3181	.3201
2.1	.3222	.3243	.3263	.3284	.3304	.3324	.3345	.3365	.3385	.3404
2.2	.3424	.3444	.3464	.3483	.3502	.3522	.3541	.3560	.3579	.3598
2.3	.3617	.3636	.3655	.3674	.3692	.3711	.3729	.3747	.3766	.3784
2.4	.3802	.3820	.3838	.3856	.3874	.3892	.3909	.3927	.3945	.3962
2.5	.3979	.3997	.4014	.4031	.4048	.4065	.4082	.4099	.4116	.4133
2.6	.4150	.4166	.4183	.4200	.4216	.4232	.4249	.4265	.4281	.4298
2.7	.4314	.4330	.4346	.4362	.4378	.4393	.4409	.4425	.4440	.4456
2.8	.4472	.4487	.4502	.4518	.4533	.4548	.4564	.4579	.4594	.4609
2.9	.4624	.4639	.4654	.4669	.4683	.4698	.4713	.4728	.4742	.4757
3.0	.4771	.4786	.4800	.4814	.4829	.4843	.4857	.4871	.4886	.4900
3.1	.4914	.4928	.4942	.4955	.4969	.4983	.4997	.5011	.5024	.5038
3.2	.5051	.5065	.5079	.5092	.5105	.5119	.5132	.5145	.5159	.5172
3.3	.5185	.5198	.5211	.5224	.5237	.5250	.5263	.5276	.5289	.5302
3.4	.5315	.5328	.5340	.5353	.5366	.5378	.5391	.5403	.5416	.5428
3.5	.5441	.5453	.5465	.5478	.5490	.5502	.5514	.5527	.5539	.5551
3.6	.5563	.5575	.5587	.5599	.5611	.5623	.5635	.5647	.5658	.5670
3.7	.5682	.5694	.5705	.5717	.5729	.5740	.5752	.5763	.5775	.5786
3.8	.5798	.5809	.5821	.5832	.5843	.5855	.5866	.5877	.5888	.5899
3.9	.5911	.5922	.5933	.5944	.5955	.5966	.5977	.5988	.5999	.6010
4.0	.6021	.6031	.6042	.6053	.6064	.6075	.6085	.6096	.6107	.6117
4.1	.6128	.6138	.6149	.6159	.6170	.6180	.6191	.6201	.6212	.6222
4.2	.6232	.6243	.6253	.6263	.6274	.6284	.6294	.6304	.6314	.6325
4.3	.6335	.6345	.6355	.6365	.6375	.6385	.6395	.6405	.6415	.6425
4.4	.6435	.6444	.6454	.6464	.6474	.6484	.6493	.6503	.6513	.6522
4.5	.6532	.6542	.6551	.6561	.6571	.6580	.6590	.6599	.6609	.6618
4.6	.6628	.6637	.6646	.6656	.6665	.6675	.6684	.6693	.6702	.6712
4.7	.6721	.6730	.6739	.6749	.6758	.6767	.6776	.6785	.6794	.6803
4.8	.6812	.6821	.6830	.6839	.6848	.6857	.6866	.6875	.6884	.6893
4.9	.6902	.6911	.6920	.6928	.6937	.6946	.6955	.6964	.6972	.6981
5.0	.6990	.6998	.7007	.7016	.7024	.7033	.7042	.7050	.7059	.7067
5.1	.7076	.7084	.7093	.7101	.7110	.7118	.7126	.7135	.7143	.7152
5.2	.7160	.7168	.7177	.7185	.7193	.7202	.7210	.7218	.7226	.7235
5.3	.7243	.7251	.7259	.7267	.7275	.7284	.7292	.7300	.7308	.7316
5.4	.7324	.7332	.7340	.7348	.7356	.7364	.7372	.7380	.7388	.7396

TABLE A.16.—*Continued*

	.00	.01	.02	.03	.04	.05	.06	.07	.08	.09
5.5	.7404	.7412	.7419	.7427	.7435	.7443	.7451	.7459	.7466	.7474
5.6	.7482	.7490	.7497	.7505	.7513	.7520	.7528	.7536	.7543	.7551
5.7	.7559	.7566	.7574	.7582	.7589	.7597	.7604	.7612	.7619	.7627
5.8	.7634	.7642	.7649	.7657	.7664	.7672	.7679	.7686	.7694	.7701
5.9	.7709	.7716	.7723	.7731	.7738	.7745	.7752	.7760	.7767	.7774
6.0	.7782	.7789	.7796	.7803	.7810	.7818	.7825	.7832	.7839	.7846
6.1	.7853	.7860	.7868	.7875	.7882	.7889	.7896	.7903	.7910	.7917
6.2	.7924	.7931	.7938	.7945	.7952	.7959	.7966	.7973	.7980	.7987
6.3	.7993	.8000	.8007	.8014	.8021	.8028	.8035	.8041	.8048	.8055
6.4	.8062	.8069	.8075	.8082	.8089	.8096	.8102	.8109	.8116	.8122
6.5	.8129	.8136	.8142	.8149	.8156	.8162	.8169	.8176	.8182	.8189
6.6	.8195	.8202	.8209	.8215	.8222	.8228	.8235	.8241	.8248	.8254
6.7	.8261	.8267	.8274	.8280	.8287	.8293	.8299	.8306	.8312	.8319
6.8	.8325	.8331	.8338	.8344	.8351	.8357	.8363	.8370	.8376	.8382
6.9	.8388	.8395	.8401	.8407	.8414	.8420	.8426	.8432	.8439	.8445
7.0	.8451	.8457	.8463	.8470	.8476	.8482	.8488	.8494	.8500	.8506
7.1	.8513	.8519	.8525	.8531	.8537	.8543	.8549	.8555	.8561	.8567
7.2	.8573	.8579	.8585	.8591	.8597	.8603	.8609	.8615	.8621	.8627
7.3	.8633	.8639	.8645	.8651	.8657	.8663	.8669	.8675	.8681	.8686
7.4	.8692	.8698	.8704	.8710	.8716	.8722	.8727	.8733	.8739	.8745
7.5	.8751	.8756	.8762	.8768	.8774	.8779	.8785	.8791	.8797	.8802
7.6	.8808	.8814	.8820	.8825	.8831	.8837	.8842	.8848	.8854	.8859
7.7	.8865	.8871	.8876	.8882	.8887	.8893	.8899	.8904	.8910	.8915
7.8	.8921	.8927	.8932	.8938	.8943	.8949	.8954	.8960	.8965	.8971
7.9	.8976	.8982	.8987	.8993	.8998	.9004	.9009	.9015	.9020	.9025
8.0	.9031	.9036	.9042	.9047	.9053	.9058	.9063	.9069	.9074	.9079
8.1	.9085	.9090	.9096	.9101	.9106	.9112	.9117	.9122	.9128	.9133
8.2	.9138	.9143	.9149	.9154	.9159	.9165	.9170	.9175	.9180	.9186
8.3	.9191	.9196	.9201	.9206	.9212	.9217	.9222	.9227	.9232	.9238
8.4	.9243	.9248	.9253	.9258	.9263	.9269	.9274	.9279	.9284	.9289
8.5	.9294	.9299	.9304	.9309	.9315	.9320	.9325	.9330	.9335	.9340
8.6	.9345	.9350	.9355	.9360	.9365	.9370	.9375	.9380	.9385	.9390
8.7	.9395	.9400	.9405	.9410	.9415	.9420	.9425	.9430	.9435	.9440
8.8	.9445	.9450	.9455	.9460	.9465	.9469	.9474	.9479	.9484	.9489
8.9	.9494	.9499	.9504	.9509	.9513	.9518	.9523	.9528	.9533	.9538
9.0	.9542	.9547	.9552	.9557	.9562	.9566	.9571	.9576	.9581	.9586
9.1	.9590	.9595	.9600	.9605	.9609	.9614	.9619	.9624	.9628	.9633
9.2	.9638	.9643	.9647	.9652	.9657	.9661	.9666	.9671	.9675	.9680
9.3	.9685	.9689	.9694	.9699	.9703	.9708	.9713	.9717	.9722	.9727
9.4	.9731	.9736	.9741	.9745	.9750	.9754	.9759	.9763	.9768	.9773
9.5	.9777	.9782	.9786	.9791	.9795	.9800	.9805	.9809	.9814	.9818
9.6	.9823	.9827	.9832	.9836	.9841	.9845	.9850	.9854	.9859	.9863
9.7	.9868	.9872	.9877	.9881	.9886	.9890	.9894	.9899	.9903	.9908
9.8	.9912	.9917	.9921	.9926	.9930	.9934	.9939	.9943	.9948	.9952
9.9	.9956	.9961	.9965	.9969	.9974	.9978	.9983	.9987	.9991	.9996

TABLE A.17. Angular Transformation arc sin $\sqrt{\%}$

%	.0	.1	.2	.3	.4	.5	.6	.7	.8	.9
0	0.00	1.81	2.56	3.14	3.63	4.05	4.44	4.80	5.13	5.44
1	5.74	6.02	6.29	6.55	6.80	7.03	7.27	7.49	7.71	7.92
2	8.13	8.33	8.53	8.72	8.91	9.10	9.28	9.46	9.63	9.80
3	9.97	10.14	10.30	10.47	10.63	10.78	10.94	11.09	11.24	11.39
4	11.54	11.68	11.83	11.97	12.11	12.25	12.38	12.52	12.66	12.79
5	12.92	13.05	13.18	13.31	13.44	13.56	13.69	13.81	13.94	14.06
6	14.18	14.30	14.42	14.54	14.65	14.77	14.89	15.00	15.12	15.23
7	15.34	15.45	15.56	15.68	15.79	15.89	16.00	16.11	16.22	16.32
8	16.43	16.54	16.64	16.74	16.85	16.95	17.05	17.15	17.26	17.36
9	17.46	17.56	17.66	17.76	17.85	17.95	18.05	18.15	18.24	18.34
10	18.43	18.53	18.63	18.72	18.81	18.91	19.00	19.09	19.19	19.28
11	19.37	19.46	19.55	19.64	19.73	19.82	19.91	20.00	20.09	20.18
12	20.27	20.36	20.44	20.53	20.62	20.70	20.79	20.88	20.96	21.05
13	21.13	21.22	21.30	21.39	21.47	21.56	21.64	21.72	21.81	21.89
14	21.97	22.06	22.14	22.22	22.30	22.38	22.46	22.54	22.63	22.71
15	22.79	22.87	22.95	23.03	23.11	23.18	23.26	23.34	23.42	23.50
16	23.58	23.66	23.73	23.81	23.89	23.97	24.04	24.12	24.20	24.27
17	24.35	24.43	24.50	24.58	24.65	24.73	24.80	24.88	24.95	25.03
18	25.10	25.18	25.25	25.33	25.40	25.47	25.55	25.62	25.70	25.77
19	25.84	25.91	25.99	26.06	26.13	26.21	26.28	26.35	26.42	26.49
20	26.57	26.64	26.71	26.78	26.85	26.92	26.99	27.06	27.13	27.20
21	27.27	27.34	27.42	27.49	27.56	27.62	27.69	27.76	27.83	27.90
22	27.97	28.04	28.11	28.18	28.25	28.32	28.39	28.45	28.52	28.59
23	28.66	28.73	28.79	28.86	28.93	29.00	29.06	29.13	29.20	29.27
24	29.33	29.40	29.47	29.53	29.60	29.67	29.73	29.80	29.87	29.93
25	30.00	30.07	30.13	30.20	30.26	30.33	30.40	30.46	30.53	30.59
26	30.66	30.72	30.79	30.85	30.92	30.98	31.05	31.11	31.18	31.24
27	31.31	31.37	31.44	31.50	31.56	31.63	31.69	31.76	31.82	31.88
28	31.95	32.01	32.08	32.14	32.20	32.27	32.33	32.39	32.46	32.52
29	32.58	32.65	32.71	32.77	32.83	32.90	32.96	33.02	33.09	33.15
30	33.21	33.27	33.34	33.40	33.46	33.52	33.58	33.65	33.71	33.77
31	33.83	33.90	33.96	34.02	34.08	34.14	34.20	34.27	34.33	34.39
32	34.45	34.51	34.57	34.63	34.70	34.76	34.82	34.88	34.94	35.00
33	35.06	35.12	35.18	35.24	35.30	35.37	35.43	35.49	35.55	35.61
34	35.67	35.73	35.79	35.85	35.91	35.97	36.03	36.09	36.15	36.21
35	36.27	36.33	36.39	36.45	36.51	36.57	36.63	36.69	36.75	36.81
36	36.87	36.93	36.99	37.05	37.11	37.17	37.23	37.29	37.35	37.41
37	37.46	37.52	37.58	37.64	37.70	37.76	37.82	37.88	37.94	38.00
38	38.06	38.12	38.17	38.23	38.29	38.35	38.41	38.47	38.53	38.59
39	38.65	38.70	38.76	38.82	38.88	38.94	39.00	39.06	39.11	39.17
40	39.23	39.29	39.35	39.41	39.47	39.52	39.58	39.64	39.70	39.76
41	39.82	39.87	39.93	39.99	40.05	40.11	40.16	40.22	40.28	40.34
42	40.40	40.45	40.51	40.57	40.63	40.69	40.74	40.80	40.86	40.92
43	40.98	41.03	41.09	41.15	41.21	41.27	41.32	41.38	41.44	41.50
44	41.55	41.61	41.67	41.73	41.78	41.84	41.90	41.96	42.02	42.07
45	42.13	42.19	42.25	42.30	42.36	42.42	42.48	42.53	42.59	42.65
46	42.71	42.76	42.82	42.88	42.94	42.99	43.05	43.11	43.17	43.22
47	43.28	43.34	43.39	43.45	43.51	43.57	43.62	43.68	43.74	43.80
48	43.85	43.91	43.97	44.03	44.08	44.14	44.20	44.26	44.31	44.37
49	44.43	44.48	44.54	44.60	44.66	44.71	44.77	44.83	44.89	44.94

TABLE A.17.—*Continued*

%	.0	.1	.2	.3	.4	.5	.6	.7	.8	.9
50	45.00	45.06	45.11	45.17	45.23	45.29	45.34	45.40	45.46	45.52
51	45.57	45.63	45.69	45.74	45.80	45.86	45.92	45.97	46.03	46.09
52	46.15	46.20	46.26	46.32	46.38	46.43	46.49	46.55	46.61	46.66
53	46.72	46.78	46.83	46.89	46.95	47.01	47.06	47.12	47.18	47.24
54	47.29	47.35	47.41	47.47	47.52	47.58	47.64	47.70	47.75	47.81
55	47.87	47.93	47.98	48.04	48.10	48.16	48.22	48.27	48.33	48.39
56	48.45	48.50	48.56	48.62	48.68	48.73	48.79	48.85	48.91	48.97
57	49.02	49.08	49.14	49.20	49.26	49.31	49.37	49.43	49.49	49.55
58	49.60	49.66	49.72	49.78	49.84	49.89	49.95	50.01	50.07	50.13
59	50.18	50.24	50.30	50.36	50.42	50.48	50.53	50.59	50.65	50.71
60	50.77	50.83	50.89	50.94	51.00	51.06	51.12	51.18	51.24	51.30
61	51.35	51.41	51.47	51.53	51.59	51.65	51.71	51.77	51.83	51.88
62	51.94	52.00	52.06	52.12	52.18	52.24	52.30	52.36	52.42	52.48
63	52.54	52.59	52.65	52.71	52.77	52.83	52.89	52.95	53.01	53.07
64	53.13	53.19	53.25	53.31	53.37	53.43	53.49	53.55	53.61	53.67
65	53.73	53.79	53.85	53.91	53.97	54.03	54.09	54.15	54.21	54.27
66	54.33	54.39	54.45	54.51	54.57	54.63	54.70	54.76	54.82	54.88
67	54.94	55.00	55.06	55.12	55.18	55.24	55.30	55.37	55.43	55.49
68	55.55	55.61	55.67	55.73	55.80	55.86	55.92	55.98	56.04	56.10
69	56.17	56.23	56.29	56.35	56.42	56.48	56.54	56.60	56.66	56.73
70	56.79	56.85	56.91	56.98	57.04	57.10	57.17	57.23	57.29	57.35
71	57.42	57.48	57.54	57.61	57.67	57.73	57.80	57.86	57.92	57.99
72	58.05	58.12	58.18	58.24	58.31	58.37	58.44	58.50	58.56	58.63
73	58.69	58.76	58.82	58.89	58.95	59.02	59.08	59.15	59.21	59.28
74	59.34	59.41	59.47	59.54	59.60	59.67	59.74	59.80	59.87	59.93
75	60.00	60.07	60.13	60.20	60.27	60.33	60.40	60.47	60.53	60.60
76	60.67	60.73	60.80	60.87	60.94	61.00	61.07	61.14	61.21	61.27
77	61.34	61.41	61.48	61.55	61.61	61.68	61.75	61.82	61.89	61.96
78	62.03	62.10	62.17	62.24	62.31	62.38	62.44	62.51	62.58	62.65
79	62.73	62.80	62.87	62.94	63.01	63.08	63.15	63.22	63.29	63.36
80	63.43	63.51	63.58	63.65	63.72	63.79	63.87	63.94	64.01	64.09
81	64.16	64.23	64.30	64.38	64.45	64.52	64.60	64.67	64.75	64.82
82	64.90	64.97	65.05	65.12	65.20	65.27	65.35	65.42	65.50	65.57
83	65.65	65.73	65.80	65.88	65.96	66.03	66.11	66.19	66.27	66.34
84	66.42	66.50	66.58	66.66	66.74	66.82	66.89	66.97	67.05	67.13
85	67.21	67.29	67.37	67.46	67.54	67.62	67.70	67.78	67.86	67.94
86	68.03	68.11	68.19	68.28	68.36	68.44	68.53	68.61	68.70	68.78
87	68.87	68.95	69.04	69.12	69.21	69.30	69.38	69.47	69.56	69.64
88	69.73	69.82	69.91	70.00	70.09	70.18	70.27	70.36	70.45	70.54
89	70.63	70.72	70.81	70.91	71.00	71.09	71.19	71.28	71.37	71.47
90	71.57	71.66	71.76	71.85	71.95	72.05	72.15	72.24	72.34	72.44
91	72.54	72.64	72.74	72.84	72.95	73.05	73.15	73.26	73.36	73.46
92	73.57	73.68	73.78	73.89	74.00	74.11	74.21	74.32	74.44	74.55
93	74.66	74.77	74.88	75.00	75.11	75.23	75.35	75.46	75.58	75.70
94	75.82	75.94	76.06	76.19	76.31	76.44	76.56	76.69	76.82	76.95
95	77.08	77.21	77.34	77.48	77.62	77.75	77.89	78.03	78.17	78.32
96	78.46	78.61	78.76	78.91	79.06	79.22	79.37	79.53	79.70	79.86
97	80.03	80.20	80.37	80.54	80.72	80.90	81.09	81.28	81.47	81.67
98	81.87	82.08	82.29	82.51	82.73	82.97	83.20	83.45	83.71	83.98
99	84.26	84.56	84.87	85.20	85.56	85.95	86.37	86.86	87.44	88.19

TABLE A.18. Orthogonal Polynomials

$a = 3$	
-1	+1
0	-2
+1	+1
2	6

$a = 4$		
-3	+1	-1
-1	-1	+3
+1	-1	-3
+3	+1	+1
20	4	20

$a = 5$			
-2	+2	-1	+1
-1	-1	+2	-4
0	-2	0	+6
+1	-1	-2	-4
+2	+2	+1	+1
10	14	10	70

$a = 6$				
-5	+5	-5	+1	- 1
-3	-1	+7	-3	+ 5
-1	-4	+4	+2	-10
+1	-4	-4	+2	+10
+3	-1	-7	-3	- 5
+5	+5	+5	+1	+ 1
70	84	180	28	252

$a = 7$				
0	-4	0	+6	0
+1	-3	-1	+1	+5
+2	0	-1	-7	-4
+3	+5	+1	+3	+1
28	84	6	154	84

$a = 8$				
+1	-5	-3	+ 9	+15
+3	-3	-7	- 3	+17
+5	+1	-5	-13	-23
+7	+7	+7	+ 7	+ 7
168	168	264	616	2184

$a = 9$				
0	-20	0	+18	0
+1	-17	- 9	+ 9	+ 9
+2	- 8	-13	-11	+ 4
+3	+ 7	- 7	-21	-11
+4	+28	+14	+14	+ 4
60	2772	990	2002	468

$a = 10$				
+1	-4	-12	+18	+ 6
+3	-3	-31	+ 3	+11
+5	-1	-35	-17	+ 1
+7	+2	-14	-22	-14
+9	+6	+42	+18	+ 6
330	132	8580	2860	780

$a = 11$				
0	-10	0	+6	0
+1	- 9	-14	+4	+4
+2	- 6	-23	-1	+4
+3	- 1	-22	-6	-1
+4	+ 6	- 6	-6	-6
+5	+15	+30	+6	+3
110	858	4290	286	156

$a = 12$				
+ 1	-35	- 7	+28	+20
+ 3	-29	-19	+12	+44
+ 5	-17	-25	-13	+29
+ 7	+ 1	-21	-33	-21
+ 9	+25	- 3	-27	-57
+11	+55	+33	+33	+33
572	12012	5148	8008	15912

Answers to Most Odd-Numbered Exercises and All Review Exercises

CHAPTER 1

Exercises

1.1.1. 0.020, 0.019, 0.009, 0.034, 0.047, *E*.

1.2.3. a. Theoretical results are insufficient; he wants to prevent cases of paralytic polio.
b. The vaccine should be used.

1.2.5. a. H_0: $\pi = 0.5$.
b. H_a: $\pi > 0.5$.
c. 8, 9, or 10 correct.
d. 0, 1, . . . , 7 correct.
e. i, iii, v.

1.2.7. a. H_0: $\pi = 0.25$. H_a: $\pi > 0.25$.
b. Reject H_0 if 9, 10, . . . , 20 pick the new brand; accept otherwise.
c. iv.

1.3.1. a. Low-income people may not have phones, and the wealthy may have unlisted numbers.

b. Monday-morning customers may include many affluent owners of businesses open on weekends, and these people are more likely to be Republicans than the general population.

c. i. Anglers may exaggerate the length.
   ii. Many of the dead fish will be large old ones.

d. The largest school is not necessarily representative.

e. The students may be more talented than average in tone differentiation.

1.3.3. His conjecture was based on a survey with no control of other variables.

Review Exercises

False: 1.1, 1.5, 1.6, 1.8, 1.9, 1.10, 1.11, 1.12, 1.13, 1.15, 1.20, 1.22, 1.23

CHAPTER 2

Exercises

2.1.1. a. 2, 8, 1.
b. 3, 2.
c. 18, 43, 6, 3, 39.
d. 8, 14, 20, 9.

2.1.7. A random sample can be approximated by choosing every hundredth name on the list after a random start in the first hundred names.

2.2.1. a. Continuous numerical.
b. Nominal.
c. Nominal.
d. Nominal.
e. Continuous numerical.
f. Discrete numerical.
g. Nominal.

2.2.3. a. Female, male.
c. Less than 3, 3, more than 3.
e. Blue-eyed, not blue-eyed.

2.3.1. a. 1/6.
b. 4/6, 3/6, 1/6, 2/6, 3/6.

2.3.3. a. 1.
b. 1/4.
c. 1/4.
d. 3/4.
e. 3/4.

2.4.1. A: 1, 1/2;
B: 7, 2;
C: 1.6250, 0.2969;
D: 57/32, 1775/1024.

2.4.3. a. $p(0) = 0.94$, $p(5) = 0.03$, $p(10) = 0.02$, $p(25) = 0.01$.
b. 0.60.
c. No.
d. 8.64.
e. 0.97.

2.4.5. a. 2.5.
b. $(a + b)/2$.

Review Exercises

False:			
	2.3	2.8	2.20
	2.4	2.11	2.24
	2.6	2.12	2.25
	2.7	2.16	

CHAPTER 3

Exercises

3.1.1. a. 1/5.
b. 2/5.
c. 3/5.
d. 1.
e. 0.
f. 0.
g. 4/5.
h. 3/5.

3.1.3. a. 90/1024.
b. 918/1024.
c. 376/1024.

3.1.5. a. 25/7776.
b. 1/1296.

3.1.7. a. 1.
b. 3.
c. 1.
d. 10.
e. 5.
f. 4.

3.1.9. a. 0.11.
b. 6.6×10^{-5}.
c. 3.7×10^{-7}.

3.1.11. $p(0) = 0.01$, $p(1) = 0.08$, $p(2) = 0.26$, $p(3) = 0.41$, $p(4) = 0.24$.

3.1.13. 1.6, 0.96.

3.1.15. 32.

3.1.17. a. 1/12.
b. 1/6.
c. 1/66.

3.1.19. a. 1/2.
b. 1/32.

3.1.21 a. There are nine choices for the first station, eight for the second, and so on, and the total number of possibilities is the product.
b. 362,880.
c. 90,720.

3.1.23. a. 252.
b. 1.
c. 1/252.
d. The examiner might inadvertently indicate the pictures of the dead subjects.

3.2.1. a. 0.000.
b. 0.000.
c. 0.904.
d. 0.238.

e. 1.000.
f. 0.548.

3.2.3. a. H_0: $\pi = 0.30$.
b. H_a: $\pi \neq 0.30$.
c. 0.053.
d. Accept H_0; the game may be operating as desired. He must assume the players are random.

3.2.5. a. 0.417.
b. Increase the sample size.

3.2.7. a. Twenty or fewer miles per gallon, more than 20 miles per gallon.
b. H_0: $\pi = 0.70$. (π is the proportion of Type B cars that average more than 20 miles per gallon.)
c. H_a: $\pi > 0.70$.
d. Type II. Use a large sample size.

3.2.9. a. 0, 1, 2, 11, 12, . . . , 20.
b. 10, 11, . . . , 20.
c. 0.176.
d. iv.

3.2.11. a. Accept.
b. Reject.

3.2.15. a. H_0: $\pi = 0.20$. H_a: $\pi > 0.20$.
b. 9, 10, . . . , 25.
c. No. Eight is not in the region of rejection. The condition does not occur more frequently in children born with cleft palates.

3.2.17. a. H_0: $\pi_M = 0.50$.
b. H_0: $\pi_C = 0.20$.
c. 10.
d. i. H_0: $\pi_C = 0.20$, H_a: $\pi_C < 0.20$.
ii. 0, 1, 2.
iii. H_0 is rejected. Color blindness is less prevalent in females.
e. i. H_0: $\pi_C = 0.20$, H_a: $\pi_C > 0.20$.
ii. 8, 9, . . . , 25
iii. H_0 is not rejected. The proportion of color-blind males may be 0.20.

3.2.19. a. H_0: $\pi = 0.20$.
b. H_a: $\pi > 0.20$.
c. 8, 9, . . . , 20.

3.3.1. a. 0.70.
b. $0.639 \le \pi \le 0.756$.
c. Yes.

3.3.3. a. 0.75.
b. $0.653 \le \pi \le 0.831$.

3.3.5. a. 0.14.

3.3.7. a. 0.52.
b. $0.456 \le \pi \le 0.583$.

3.3.9. a. 0.81.
b. $0.722 \le \pi \le 0.880$.

3.3.11. (1) 0.229, 0.591.
(2) 250, 90.
(3) 0.263, 0.382.
(4) 0.816, 0.897.
(5) 0.164, 0.511.
(6) 100, 17.
(7) 0.074, 0.157.
(8) 29, 0.90.
(9) 500, 0.99.
(10) 8, 0.25, 0.55.

Review Exercises

False: 3.2, 3.4, 3.5, 3.6, 3.7, 3.9, 3.12, 3.14, 3.16, 3.17, 3.18, 3.20, 3.24, 3.25, 3.26, 3.27, 3.28, 3.29, 3.30, 3.33, 3.34

CHAPTER 4

Exercises

4.1.1. 0.21.

4.1.3. 0.1336.

4.1.5. a. 0.0820.
b. 0.205.
c. 0.918.
d. 1, 2, 3, 4.

4.1.7. a. For $\lambda = 0.25$, $p(0) = 0.7788$, $p(1) = 0.1947$, $p(2) = 0.0243$, $p(3)$

$= 0.0020$, $p(4) = 0.0001$, $p(5) = 0.0000, \ldots$. For $\lambda = 0.50$, 1.00, and 10.00 use Table A.7.

4.2.1. H_0: $\lambda = 12$ per 3 milliseconds, H_a: $\lambda > 12$ per 3 milliseconds. Reject H_0 if $y = 19, 20, 21, \ldots$. Accept H_0. There is no evidence that the level is higher than 4 per millisecond.

4.2.3. H_0: $\lambda = 10$, H_a: $\lambda < 10$; reject H_0. There is evidence of a reduction.

4.2.5. a. H_0: $\lambda = 1$ per 100 cells.
b. H_a: $\lambda > 1$ per 100 cells.
c. 0.0190.
d. 0.7787. It seems necessary because the probability of four or more basophils is 0.2213.

4.3.1. $1.4 \leq \lambda \leq 6.0$.

4.3.3. $0.0175 \leq \lambda \leq 0.0450$.

4.4.1. 0.0047.

4.4.3. H_0: $\lambda = 4$, H_a: $\lambda < 4$; reject H_0 if $y = 0$ or 1. H_0 is accepted. No evidence of a reduction in the proportion of defective sets.

4.4.5. $\lambda \leq 9.0$.

Review Exercises

False:		
	4.1	4.6
	4.3	4.8
	4.4	4.9
	4.5	4.10

CHAPTER 5

Exercises

5.1.1. a. 18.475.
b. 2.156.
c. 95.023.
d. 0.05.
e. 0.975.
f. 18.307.
g. 42.796.
h. 5.

5.1.3. $\chi^2 = 9.336$, no evidence that the table is not random.

5.1.5. $\chi^2 = 1.240$, 75% of the plants may be red-flowering. We assume the nongerminating seeds would have produced the same proportion of plants with red flowers.

5.1.7. a. 2.
b. 90, 72, 18.
c. $\chi^2 = 5.56$, there is evidence that the claim is wrong.

5.1.9. a. H_0: $\pi_Y = 0.50$, $\pi_J = 0.30$, $\pi_M = 0.20$.
H_a: At least one inequality.
b. Type II.
c. $\chi^2 = 8.177$, reject H_0.
d. Fishing should be closed.
e. iii.

5.1.11. a. 0.44, 0.11, 0.36, 0.09.
b. 3.
c. 44, 11, 36, 9.
d. 11.345.
e. H_0: $\pi = 0.20$.
f. H_0: $\pi = 0.45$.
g. 6.635.

5.2.1. $\chi^2 = 8.290$, $b(y; 4, 0.40)$ may be the correct distribution.

5.2.3. $\chi^2 = 0.0222$, this may be from a binomial distribution.

5.2.5. $\hat{\lambda} = 0.5246$, $\chi^2 = 2.52$, this seems to be from a Poisson distribution.

5.3.1. a. H_0: Stomach ulcers are independent of the first initial of a person's last name.
H_a: Stomach ulcers are related to the first initial of a person's last name.
b. $\chi^2 = 6.25$; reject H_0.
c. There is evidence of a relationship between stomach ulcers and the first initial of a person's last name. (This should not be interpreted to mean that the initial *causes* the ulcer; it is also possible that a Type I error has occurred.)

5.3.3. a. H_0: Parents' approval or disapproval of young people "living together" is independent of the age of their youngest child.
b. $\chi^2 = 48.611$; reject H_0.

5.3.5. a. H_0: Success in the training program is independent of performance on the screening test.

H_a: Success in the training program is related to performance on the screening test.

b. 6.635.

c. $\chi^2 = 70.96$; reject H_0, the screening test is effective in predicting success.

d. The sample must be random and from the group of people who will be in the program.

5.3.7. a. H_0: $\pi_i = \pi_B = \pi_C$. (π_i is the proportion of dead black flies for each insecticide.)

b. 9.210.

c. $\chi^2 = 1.49$.

d. Greater than 0.05.

e. Do not reject H_0; the insecticides are equally effective.

Review Exercises

False:			
	5.1	5.10	5.18
	5.3	5.11	5.19
	5.6	5.13	5.23
	5.7	5.14	5.24
	5.8	5.16	5.25
	5.9		

CHAPTER 6

Exercises

6.1.1. 70.

6.1.3. a. 2.0.

c. 2.0.

d. $\hat{\mu} = \bar{y} = 2.0$.

6.2.1. a. 6.

6.2.3. a. 1.68.

b. 1.68.

6.2.5. b. 3, 9, 3.

c. 1.5, 1.5, 2.8.

6.2.7. $\mu = 0.238$, $\sigma = 0.740$; $\mu \pm 2\sigma$ is -1.242 to 1.718, which contains 0.941 of the data; $\mu \pm 3\sigma$ is -1.982 to 2.458, which contains 0.972 of the data.

6.3.1. b. 22/3, 38/9.

6.3.5. c. 65.
d. 65.
e. 3.33.
f. 1.67.

Review Exercises

False: 6.1, 6.2, 6.5, 6.7, 6.10, 6.11, 6.13, 6.15, 6.16, 6.17

CHAPTER 7

Exercises

7.1.1. a. 0.818.
b. 0.499.
c. 0.382.
d. 0.010.
e. 0.500.
f. 0.943.
g. 0.124.
h. 0.445.
i. 0.318.
j. 0.046.
k. 0.002.

7.1.3. a. 0.933.
b. 67.2 to 132.8.
c. 120.8.
d. 95.0.

7.1.5. a. 0.001.
b. 0.159.

7.1.7. 70 in.

7.2.1. a. 1.64.
b. −1.64.
c. 2.33.
d. −2.33.

e. 2.56.
f. −2.56.

7.2.3. a. H_0: $\mu = 19.3$.
b. H_a: $\mu < 19.3$.
c. $z \leq -1.64$ or $y \leq 18.808$.
d. 0.359.

7.3.1. a. 0.106.
b. $(0.106)^5$.
c. 0.003.

7.4.1. 0.023.

7.4.3. a. 0.50, 0.371.
b. 0.50, 0.159.
c. i. H_0: $\mu = 90$, H_a: $\mu < 90$.
ii. $\bar{y} \leq 83.44$.
iii. 0.739.

7.4.5. a. 84.
c. $z = 1.7$. There is evidence that it will be profitable.

7.4.7. a. 3.06.
b. 1.94 to 5.53.
c. $\chi^2 = 27.73$; do not reject H_0. The variance may be 3.2.
d. 5.27 to 6.53.

7.5.1. a. i. 0.369.
ii. 0.302.
iii. 0.378.
b. i. 0.147.
ii. 0.174.
c. The continuity correction is more important for small samples.

7.5.3. a. 25%
b. H_0: $\pi = 0.25$, H_a: $\pi \neq 0.25$.
c. $z = 2.47$; reject H_0. The disorder appears to be genetic.

7.5.5. a. 0.64.
b. 0.0023.
c. 0.55 to 0.73.
d. There is evidence that people can distinguish because $\pi = 0.50$ is not in the confidence interval.

7.5.7. a. 0.808.
b. 0.191.

Review Exercises

False:			
	7.1	7.12	7.21
	7.2	7.13	7.22
	7.3	7.16	7.23
	7.4	7.17	7.28
	7.5	7.18	7.29
	7.6	7.20	7.30
	7.7		

CHAPTER 8

Exercises

8.1.1. a. 2.764.
b. −2.764.
c. 2.365.
d. −2.365.
e. 2.807.
f. −2.807.

8.1.3. a. $8000.
b. 10,240,000.
c. 2.00.
d. Between 0.025 and 0.05.

8.2.1. a. 1000.
b. 100.
c. 786.9 to 1213.1.

8.2.3. a. 1.704.
b. 34.7 to 41.7.

8.2.5. a. 3.7 to 4.7.
b. $\mu \leq 4.7$.

8.2.7. a. H_0: $\mu_d = 0$, H_a: $\mu_d > 0$.
b. $t = 2.00$; H_0 is rejected. There is evidence of improvement on the second test.

8.2.9. a. The design removes extraneous variability introduced by soil conditions, climate, and farming methods.
b. 3.0.
d. $t = 3.242$, reject H_0.
e. There is evidence that the seed company's claim is correct.
f. 2.24 to 3.76. H_0 is rejected because 2.0 is not in this interval.

8.2.11. a. 40.5.

b. H_0: $\mu_d = 0$, H_a: $\mu_d \neq 0$.

c. $t = 1.4$; do not reject H_0. There is no evidence of a difference in weight gain.

d. -1.9 to 7.9. Since this interval contains 0 the null hypothesis is accepted.

8.3.1. a. 105.

b. H_0: $\mu_U = \mu_R$, H_a: $\mu_U > \mu_R$.

c. $t = 2.30$; reject H_0. Urban pollution is higher.

d. $\mu_U - \mu_R \leq 24.7$.

8.3.3. -3.439 to -0.561.

8.3.5. $t = 1.80$; reject H_0. There is evidence that those who finish on time score higher. However, since this was obtained from a survey without control for other factors, it should be applied cautiously.

8.3.7. a. -2.18 to -0.02.

b. Since the interval does not contain zero, there is evidence of inequality. However, the evidence is weak because 0 is close to the upper limit -0.02.

8.4.1. a. 6.538.

b. 4.886.

c. 2328.

d. 0.430.

e. 0.132.

8.4.3. $F = 4.00$; reject the hypothesis of equal variances. Use the t' test for means. $t' = -2.50$ with $\nu = 14$; reject H_0. There is evidence of a difference in the mean resin content.

8.4.5. a. $F = 0.444$; do not reject H_0. There is no evidence of a difference in variability.

b. 0.111 to 2.449.

Review Exercises

False:			
	8.1	8.16	8.30
	8.5	8.18	8.31
	8.8	8.19	8.32
	8.9	8.21	8.34
	8.12	8.22	8.36
	8.13	8.26	8.37
	8.14	8.27	8.40
	8.15		

CHAPTER 9

Exercises

9.1.1. c.

9.1.3. Days per pound.

9.1.5. a. 180.
b. 18.
c. $\hat{y} = -208 + 18x$.
d. 80.

9.1.7. a. i. Positive.
ii. Yes.
iii. 19.2.
iv. Minutes per staff hour, per patient.
v. $\hat{y} = -1 + 19.2x$.
vi. 18.2, 95.0.
b. i. Negative.
ii. Not intuitive prior to the survey.
iii. $\hat{y} = 19.15 - 1.43x$.
iv. $x = 10$.

9.1.9. a. 68.
b. 68.5.
c. 0.50.

9.1.11. $\hat{y} = 53.69 + 0.6187x$.

9.2.1. c. i. H_0: $\beta = 0$, H_a: $\beta > 0$.
ii. 1.895.
iii. $t = 6.0$; reject H_0. There is evidence that increase in study is linearly related to higher grades.

9.2.3. a. Fish per hour.
b. Fish.
c. Fish.

9.2.5. a. i. H_0: $\beta = 0$. There is no linear relationship between time spent on patient care and patient load.
ii. Time would seem to increase as number of patients increases.
iii. 2.132.
iv. $t = 16.0$; reject H_0. There is a linear relationship between time spent on patient care and patient load.
b. i. H_0: $\beta = 0$.

ii. It is not clear prior to the survey whether the relationship is positive or negative.

iii. $t = -5.39$; reject H_0. There is a linear relationship; time for reports decreases as patient load increases.

9.2.7. a. 0.14.

b. 0.14.

c. i. H_0: $\beta = 0$.

ii. Radioactivity disappears over time.

iii. -2.353.

iv. $t = -1.754$; do not reject H_0. There is no evidence of a linear relationship.

9.3.1. a. 40.

b. 9.

c. 72 ± 3.5.

9.33. a. H_0: $\beta = 0$.

b. H_a: $\beta \neq 0$, since it is not obvious whether a larger number of fillings in the previous two years indicates that there will be little left to do or a very fast decay rate.

c. ± 2.306.

9.3.5. a. 3.0.

b. 1.44.

c. H_0: $\beta = 0$, H_a: $\beta > 0$, $t = 1.00$; do not reject H_0. There is no evidence of a linear relationship.

d. 20.06.

e. 20.06 ± 0.78.

f. 19.7.

g. 19.7 ± 0.83.

h. Parts d through g are invalid because there is no linear relationship.

9.3.7. a. 16.65 to 1.

b. -2.17 to -0.69.

9.3.9. a. 1.17 ± 0.53.

b. 1.17 ± 0.87.

c. It is in both intervals. However, there is no evidence of a linear trend (Exercise 9.2.7) so the line should not be used for prediction.

d. i. 1.77.

ii. 1.77 ± 1.32.

iii. Since there is no evidence of a linear trend, this estimate is invalid.

9.4.1. a. $-1, +1$.
b. 10, 1.
c. $-0.9, +0.4$.
d. Significant; nonsignificant.

9.4.3. $t = 4.0$; reject H_0. Length explains a significant portion of the variability in weight.

9.4.5. a. 2.
b. 1.
c. i. 2.
ii. 1.

9.4.7. 0.694 to 0.992.

Review Exercises

False:			
	9.2	9.16	9.35
	9.3	9.17	9.36
	9.4	9.18	9.37
	9.5	9.19	9.39
	9.7	9.23	9.40
	9.8	9.26	9.41
	9.11	9.28	9.43
	9.12	9.30	9.44
	9.13	9.31	9.45
	9.14		

CHAPTER 10

Exercises

10.1.1. 72/5, 30/5, 42/5.

10.1.3. a. $F = 11.61$; reject H_0. There is a difference in the mean heights of the two groups.
b. $t = -3.41$; reject H_0.
c. $(t_{0.025,6})^2 = F_{0.05,1,6}$.

10.2.1. $F = 1.55$, H_0 is not rejected. There is no significant difference among the diets.

10.2.3. H_0: $\mu_A = \mu_B = \mu_C$, $F = 5.14$; reject H_0. There is at least one difference among the mean lifetimes.

10.2.5. H_0: $\mu_A = \mu_B = \mu_C$; this does not appear to be true from the graph. 3.885. $F = 216.7$; reject H_0. The mean amount of vitamin C differs for at least two of the methods.

10.2.7. $F = 23.56$; reject H_0. There is evidence of different mean weights at different locations.

10.2.9.
a. $a = 7$, $n = 5$, total degrees of freedom $= 7(5) - 1 = 34$.
b. Trial.
c. Normality, independence, equal variances.
d.

df	SS	MS
6	330	
28	644	23

e. H_0: $\mu_1 = \mu_2 = \cdots = \mu_7$, H_a: At least one inequality.
f. $F_{0.05,6,28} = 2.445$. $F = 2.39$; do not reject H_0. There is no significant difference among the insecticides.

10.2.11.
a. δ_h, ξ_{hi}.
b. $\sum_h \delta_h = 0$, ξ_{hi} is NID$(0, \sigma^2)$.
c. $h = 1, 2, 3 = a$. $i = 1, 2, \ldots, 5 = n$.
d. $F = 4.00$; reject H_0. There is a significant difference among the mean weight-bearing capacities.

10.3.1.
a.

df	SS	MS
4	2392	
		180

c. Yes, F is significant.
d. $\bar{y}_C$ $\bar{y}_A$ $\bar{y}_D$ $\bar{y}_B$ $\bar{y}_E$ (underline spans $\bar{y}_A$ through $\bar{y}_E$)

10.3.3.
a. $F = 2.88$; accept H_0.
b. No, F is not significant.
c. No significant differences.

10.3.5.
a. $\bar{y}_{I}$ $\bar{y}_{II}$ $\bar{y}_{III}$ (underlines: $\bar{y}_{I}$–$\bar{y}_{II}$; $\bar{y}_{II}$–$\bar{y}_{III}$)

b. 3.682. $F = 6.25$, which exceeds the critical value.
c. H_0: $\mu_3 = (\mu_1 + \mu_2)/2$, H_a: $\mu_3 \neq (\mu_1 + \mu_2)/2$. Critical value 10.4, $\bar{y}_3 - (\bar{y}_1 + \bar{y}_2)/2 = 15$, so the yield with III is significantly different from the average of I and II.

10.4.1. There is a significant difference between the home type and the industrial type, $F = 9.68$.

10.4.3. a. H_0: $\mu_1 = \mu_2 = \cdots = \mu_6$, H_a: At least one inequality.
b. $F = 7.12$; reject H_0.
c. The placebo is significantly different from the analgesics.
e. 14%.
f. Pain relief is obtained more quickly with aspirin in any form than with the placebo.

10.5.1. a. 6.48 to 10.82.
b. -7.40 to 4.60.
c. -17.65 to -6.35.
d. 3.65 to 13.35.

10.5.3. a. $F = 4.0$; do not reject H_0.
b. 4.97 to 19.03.
c. 37.66 to 42.34.
d. -10.06 to -1.94.
e. 4.61 to 15.39.

Review Exercises

False:			
	10.2	10.10	10.16
	10.5	10.11	10.17
	10.6	10.13	10.19
	10.8	10.14	10.20
	10.9		

CHAPTER 11

Exercises

11.1.1. a. Fixed.
b. Random.
c. Random.
d. Fixed.
e. Fixed.

11.1.3. a. $F = 0.69$; accept H_0: $\sigma^2{}_A = 0$. There is evidence of significant variability among families.
b. $r_I = 0$.
c. Obesity is not explained by family membership.

11.1.5. a. REM.
b. H_0: $\sigma^2{}_A = 0$.

c. $F = 19$; reject H_0.

d. 0.90.

e. Ten percent of the variability is due to the lab technique, and this may not be reliable enough for medical decisions.

11.2.1. a. $F_{max} = 19.75$; reject H_0. There is at least one inequality among the variances.

b. σ^2_{NY} σ^2_{SK} σ^2_{LN} σ^2_{CD} σ^2_{DA} σ^2_{RN}

11.2.3. $F_{max} = 7.4$; do not reject H_0.

11.3.1. a. $F_{max} = 11.8$; reject H_0.

c. Yes. $F_{max} = 8.2$; accept H_0.

d. $F = 3.65$; do not reject H_0.

e. There is no evidence of difference among the means of the square root of time.

Review Exercises

False: 11.2 11.9 11.14
11.4 11.12 11.15
11.6

CHAPTER 12

Exercises

12.1.1. a. i. The cock effect. Random.
ii. The hen effect. Random.

b. $F = 0.05$; do not reject H_0. There is no evidence of significant variability due to males.

12.1.3. *C A E B D*

B or *D* should be purchased.

12.2.1. b. Among hybrids $F = 38.97$; reject H_0.
Among locations $F = 5.82$; reject H_0.

c. Yes.

d. Yes.

e. RC-3 DBC FR-11 BCM

Any hybrid except RC-3 should be used.

12.2.3.
- b. Fixed.
- c. Random.
- d. H_0: $\alpha_1 = \alpha_2 = \alpha_3 = \alpha_4 = \alpha_5$.
- e. Among models $F = 3.60$; reject H_0.
  Among cities $F = 2.59$; do not reject H_0.
- f. Yes.
- g. Since Type I error is not serious, use Fisher's least significant difference.
- h. $D \quad B \quad C \quad A \quad E$

  C, A, and E get the best mileage.
- i. No.
- j. 17%.

12.2.5.
- a. 4, 5.
- b. 1.2.
- c. $\bar{y}_1 \quad \bar{y}_2 \quad \bar{y}_3 \quad \bar{y}_4$

12.3.1.
- b. For covers $F = 0.94$; do not reject H_0.
  For newsstands $F = 2.92$; do not reject H_0.
  For weeks $F = 1.29$; do not reject H_0.
- c. The mean sales among covers do not differ.
- d. Without this design, 125 repetitions of the experiment would be necessary.

12.3.3.
- c. For weeks $F = 0.22$; do not reject H_0.
  For days $F = 0.32$; do not reject H_0.
  For operations $F = 0.35$; do not reject H_0.
- e. Weeks are random, days are fixed, and operations are fixed.
- f. None of the effects analyzed contribute significantly to differences in the number of unsafe incidents.

12.3.5. SS_e would have zero degrees of freedom, so MS_e does not exist.

12.4.1.
- a. Fixed.
- b. Fixed.
- c. For diets $F = 12.6$, for jogging $F = 69.1$, for interaction $F = 1.6$.
- e. Yes.
- f. Yes.
- g. No.

h. Use Fisher's least significant difference to locate the best diet and the best amount of jogging. Either a high protein or a high carbohydrate diet should be combined with two miles of jogging.

12.4.3. a.

Source	df	E(MS)	F
Plant species		$\sigma^2 + 5\sigma_{AB}^2 + 25\sigma_A^2$	5.125
Hillsides	4	$\sigma^2 + 5\sigma_{AB}^2 + 30\sigma_B^2$	5.200
P × H		$\sigma^2 + 5\sigma_{AB}^2$	6.667
Error	120	σ^2	

b. $6.667 > F_{0.05,20,120} = 1.662$, so there is a significant interaction.

c. $\hat{\sigma}_A^2 = 13.12$, $\hat{\sigma}_B^2 = 11.12$, so species contributes more to the total variability.

12.5.1. b. All effects fixed.

c. $F = 11.49$; reject H_0.

d. $SS_a = 1{,}302.1$ $SS_b = 351{,}939.6$ $SS_c = 112{,}266.7$
$SS_{ab} = 2{,}572.9$ $SS_{ac} = 2{,}002.7$ $SS_{bc} = 15{,}366.7$
$SS_{abc} = 7{,}927.3$, $SS_e = 44{,}800.0$

e.
$$E(MS_a) = \sigma^2 + bcn \sum \alpha_i^2/(a-1)$$
$$E(MS_b) = \sigma^2 + acn \sum \beta_j^2/(b-1)$$
$$E(MS_c) = \sigma^2 + abn \sum \gamma_k^2/(c-1)$$
$$E(MS_{ab}) = \sigma^2 + nc \sum\sum \alpha\beta_{ij}^2/(a-1)(b-1)$$
$$E(MS_{ac}) = \sigma^2 + nb \sum\sum \alpha\gamma_{ik}^2/(a-1)(c-1)$$
$$E(MS_{bc}) = \sigma^2 + na \sum\sum \beta\gamma_{jk}^2/(b-1)(c-1)$$
$$E(MS_{abc}) = \sigma^2 + n \sum\sum\sum \alpha\beta\gamma_{ijk}^2/(a-1)(b-1)(c-1)$$
$$E(MS_e) = \sigma^2$$

f. Only the nitrogen levels and phosphorus levels are related to significant differences. There are no interactions.

12.5.3. a. Seed treatment (A), fixed.
Male (B), random.
Female (C), random.

b. F for Treatments 4.68; reject H_0.
F for Crosses 17.74; reject H_0.
F for $T \times C$ 13.00; reject H_0.

c. $SS_m = 26.09$, $SS_f = 13.93$, $SS_{mf} = 45.11$.

d. $SS_{tm} = 1.14$, $SS_{tf} = 29.34$, $SS_{tmf} = 31.93$.

e.

Source	df	F
Treatment (A)	1	no exact test
Male (B)	3	$MS_b/MS_{bc} = 2.27$

e.

Source	df	F
Treatment (A)	1	no exact test
Female (C)	3	$MS_c/MS_{bc} = 1.21$
$A \times B$	9	$MS_{ab}/MS_{abc} = 0.08$
$A \times C$	3	$MS_{ac}/MS_{abc} = 2.07$
$B \times C$	3	$MS_{bc}/MS_e = 15.66$*
$A \times B \times C$	9	$MS_{abc}/MS_e = 11.09$*
Error	32	

f. 36%

g. Because of the significant interactions which reverse the effects of scarification, the treatment has different effects on different crosses; scarification cannot be recommended in general.

12.6.1.

Source	df	F
Whole Units		
Wash temperature	1	80.34*
Brands	3	31.14*
Whole unit remainder	3	
Subunits		
Dry temperature	2	117.22*
Wash temp. × Dry temp.	2	17.51*
Subunit remainder	12	

Review Exercises

False:	12.1	12.9	12.21
	12.3	12.12	12.25
	12.5	12.13	12.26
	12.7	12.16	12.29
	12.8	12.18	

CHAPTER 13

Exercises

13.1.1. b. (3,7), (4,8), (5,7).

c. $\bar{x}_{..} = 4$

d. $\hat{y}_{1j} = 1 + 2x_{1j}$, $\hat{y}_{2j} = 2x_{2j}$, $\hat{y}_{3j} = -3 + 2x_{3j}$.

e. (3,9), (4,8), (5,5).

f. Increase. Order is changed.

13.1.3. (1) e (5) c
(2) g (6) f
(3) h (7) d
(4) not indicated

13.2.1. c. $F = 4.93$; reject H_0. The adjusted alloy averages are significantly different.

13.3.1. a. 4.
b. $F = 33.78$; reject H_0. The slope is not zero.

13.3.3. Yes.

13.4.1. b. 0.80.
e. adj $\bar{y}_{1.} = 22.4$, adj $\bar{y}_{2.} = 18.0$, adj $\bar{y}_{3.} = 22.6$.
f. $21.52 \leq \mu_1 \leq 23.28$, $17.26 \leq \mu_2 \leq 18.74$, $21.72 \leq \mu_3 \leq 23.48$.
g. 18.0 $\underline{22.4 \quad 22.6}$

Review Exercises

False: 13.1 13.7 13.11
13.3 13.10 13.12
13.5

CHAPTER 14

Exercises

14.1.1. a. $\begin{bmatrix} 13 & 27 \\ 24 & 25 \end{bmatrix}$

b. $\begin{bmatrix} -30 & 10 \\ 54 & -18 \end{bmatrix}$

c. $\begin{bmatrix} 16 \\ 24 \end{bmatrix}$

d. $\begin{bmatrix} -1 \\ 50 \\ 7 \end{bmatrix}$

14.1.3. $\begin{bmatrix} 5 & -2 \\ -2 & 1 \end{bmatrix}$

14.2.1. a. $\left[\begin{array}{cc|c|cc} 10 & 20 & 4 & 1 & 0 \\ 20 & 40 & 2 & 0 & 1 \end{array}\right]$ $\left[\begin{array}{cc|c|cc} 1 & 0 & 12.4 & 4.1 & -2.0 \\ 0 & 1 & -6.0 & -2.0 & 1.0 \end{array}\right]$
b. $F = 36.15$; reject H_0.

14.2.3. a. 0.7524.
b. $F = 22.79$; R^2 is significant.
c. Reject.
d. $\hat{y} = 5852.06 - 2.563x_1 + 1.224x_2$.
e. Decreased by -2.563.

14.3.1. a. $r_{12} = -0.70$, $r_{y1} = -0.50$, $r_{y2} = 0.55$.
b. $t = 1.079$; do not reject H_0. There is no relationship between graying and age when tooth condition is held constant.

14.5.1. a. $\hat{y} = -71.6 + 48.5 \log x$. $H_0\colon \beta = 0$ is rejected with $t = 4.156$. There is a linear relationship.

14.5.3. b. i. -0.342.
ii. 0.998.
iii. 0.996.
c. i. She expects increased cooking time to reduce the number of salmonella colonies.
ii. $t = -15.42$; reject H_0.
d. i. 4.500.
ii. 2.852 to 7.099.
iii. Since $ae^{bx} = 0$ is impossible, solve $ae^{bx} = 1$. More than 19.4 minutes are required for an expected survival of zero.

14.6.1. a. 1.9401, -0.1125; both terms contribute significantly.
b. Yes, $F = 6.2$.
c. 43.02.

14.6.3. a. $F = 5.78$; reject H_0. There is a significant difference among fertilizers.
b. The linear and quadratic trends are significant.
c. From the group totals it seems to be included.
d. $R^2 = 0.683$ for the quadratic model.
$R^2 = 0.684$ for the cubic model.

Review Exercises

False:			
	14.1	14.10	14.16
	14.2	14.12	14.17
	14.3	14.13	14.19
	14.6	14.15	14.20
	14.7		

Index

FLEISS • Statistical Methods for Rates and Proportions, *Second Edition*

FRANKEN, KÖNIG, ARNDT, and SCHMIDT • Queues and Point Processes

GALAMBOS • The Asymptotic Theory of Extreme Order Statistics

GIBBONS, OLKIN, and SOBEL • Selecting and Ordering Populations: A New Statistical Methodology

GNANADESIKAN • Methods for Statistical Data Analysis of Multivariate Observations

GOLDBERGER • Econometric Theory

GOLDSTEIN and DILLON • Discrete Discriminant Analysis

GROSS and CLARK • Survival Distributions: Reliability Applications in the Biomedical Sciences

GROSS and HARRIS • Fundamentals of Queueing Theory

GUPTA and PANCHAPAKESAN • Multiple Decision Procedures: Theory and Methodology of Selecting and Ranking Populations

GUTTMAN, WILKS, and HUNTER • Introductory Engineering Statistics, *Third Edition*

HAHN and SHAPIRO • Statistical Models in Engineering

HALD • Statistical Tables and Formulas

HALD • Statistical Theory with Engineering Applications

HAND • Discrimination and Classification

HARTIGAN • Clustering Algorithms

HILDEBRAND, LAING, and ROSENTHAL • Prediction Analysis of Cross Classifications

HOAGLIN, MOSTELLER, and TUKEY • Understanding Robust and Exploratory Data Analysis

HOEL • Elementary Statistics, *Fourth Edition*

HOEL and JESSEN • Basic Statistics for Business and Economics, *Third Edition*

HOLLANDER and WOLFE • Nonparametric Statistical Methods

IMAN and CONOVER • Modern Business Statistics

JAGERS • Branching Processes with Biological Applications

JESSEN • Statistical Survey Techniques

JOHNSON and KOTZ • Distributions in Statistics
Discrete Distributions
Continuous Univariate Distributions—1
Continuous Univariate Distributions—2
Continuous Multivariate Distributions

JOHNSON and KOTZ • Urn Models and Their Application: An Approach to Modern Discrete Probability Theory

JOHNSON and LEONE • Statistics and Experimental Design in Engineering and the Physical Sciences, Volumes I and II, *Second Edition*

JUDGE, HILL, GRIFFITHS, LÜTKEPOHL and LEE • Introduction to the Theory and Practice of Econometrics

JUDGE, GRIFFITHS, HILL and LEE • The Theory and Practice of Econometrics

KALBFLEISCH and PRENTICE • The Statistical Analysis of Failure Time Data

KEENEY and RAIFFA • Decisions with Multiple Objectives

LAWLESS • Statistical Models and Methods for Lifetime Data

LEAMER • Specification Searches: Ad Hoc Inference with Nonexperimental Data

McNEIL • Interactive Data Analysis

MANN, SCHAFER and SINGPURWALLA • Methods for Statistical Analysis of Reliability and Life Data

MARTZ and WALLER • Bayesian Reliability Analysis

MEYER • Data Analysis for Scientists and Engineers

MIKÉ and STANLEY • Statistics in Medical Research: Methods and Issues with Applications in Cancer Research

MILLER • Survival Analysis

MILLER, EFRON, BROWN, and MOSES • Biostatistics Casebook

MONTGOMERY and PECK • Introduction to Linear Regression Analysis